Electron Microscopy in Human Medicine

Volume 3

Infectious Agents

Electron Microscopy in Human Medicine

Volume 3
Infectious Agents

Edited by
Jan Vincents Johannessen

Departments of Pathology,
The Norwegian Radium Hospital and Norsk Hydro's Institute for Cancer Research,
Oslo, Norway
and Columbia University College of Physicians and Surgeons
New York

McGRAW-HILL International Book Company

New York · St Louis · San Francisco · Auckland · Bogotá
Guatemala · Hamburg · Johannesburg · Lisbon · London · Madrid
Mexico · Montreal · New Delhi · Panama · Paris · San Juan
São Paulo · Singapore · Sydney · Tokyo · Toronto

This book has been set in Times New Roman 327 and Univers

British Library Cataloguing in Publication Data

Electron Microscopy in Human Medicine.
Vol. 3: Infectious Agents
1. Diagnosis, Electron microscopic
I. Johannessen, Jan Vincents
616.07'58 RB43.5 79–42633

ISBN 0–07–032503–0

Electron Microscopy in Human Medicine
Volume 3

12345 WJM 83210

Printed and bound in Great Britain

Contents of Volume Three

Contributors ix
Preface to series xi
Preface xiii
Acknowledgements xv

Part One
Viruses

1 Viruses 3
  1.1 Introduction 3
  1.2 Techniques of negative staining 6
    1.2.1 General methods 7
    1.2.2 "On the grid" method 10
    1.2.3 Immune electron microscopy 11
  1.3 Types of specimen examined by negative staining 12
    1.3.1 Tissue culture specimens 13
    1.3.2 Egg-grown material 14
    1.3.3 Urine 15
    1.3.4 Serum 15
    1.3.5 Vesicle fluid and scabs 16
    1.3.6 Solid tissue 17
    1.3.7 Feces 18
  1.4 Morphology of the major virus groups 19
    1.4.1 Helical symmetry 19
    1.4.2 Cubic viruses 25
    1.4.3 Nonsymmetrical RNA viruses 38
    1.4.4 Nonsymmetrical DNA viruses 39
  1.5 Results obtained from negative staining 39
References 41

Part Two
Bacteriophages

2 Bacteriophages 49
  2.1 Introduction 49
  2.2 The virion 50
    2.2.1 Viral morphogenesis 53
  2.3 Heteroduplex phage DNA molecules 55
  2.4 Viral entry 57
References 62

Part Three
Mycoplasmas

3 Mycoplasmas 69
- 3.1 Introduction 69
- 3.2 Electron microscopy of mycoplasmas 70
- 3.3 Cellular morphology and reproduction 71
- 3.4 Morphology and structure of the mycoplasma colony 72
- 3.5 Ultrastructure of the mycoplasma cell 73
  - 3.5.1 Basic structural components 73
  - 3.5.2 Specialized structures 78
- 3.6 Visualization of mycoplasmas in tissues and cell cultures 81

References 83

Part Four
Bacteria

4 Bacteria 91
- 4.1 Introduction 91
  - 4.1.1 General techniques 91
  - 4.1.2 Comparison of procaryotic and eucaryotic ultrastructure 101
- 4.2 Gram-positive procaryotes 105
  - 4.2.1 Clostridia and bacilli 105
  - 4.2.2 *Listeria* 112
  - 4.2.3 Actinomycetes 112
  - 4.2.4 *Nocardia* 120
  - 4.2.5 Mycobacteria 122
  - 4.2.6 Bifidobacteria 124
  - 4.2.7 Corynebacteria 128
  - 4.2.8 *Rothia* 132
  - 4.2.9 Lactobacilli 135
  - 4.2.10 Staphylococci 135
  - 4.2.11 Micrococci: *Sarcina, Gaffkya, Micrococcus* 140
  - 4.2.12 Streptococci 143
- 4.3 Gram-negative procaryotes 154
  - 4.3.1 *Brucella* 154
  - 4.3.2 *Haemophilus–Bordetella* 154
  - 4.3.3 *Bacteroides* 162
  - 4.3.4 *Neisseria* 164
  - 4.3.5 *Branhamella* 173
  - 4.3.6 *Moraxella* 173
  - 4.3.7 *Veillonella* 180
  - 4.3.8 Fusobacteria 180
  - 4.3.9 *Capnocytophaga* 184
  - 4.3.10 *Escherichia* and related genera 186
  - 4.3.11 *Selenomonas–Alysiella–Simonsiella* 200
  - 4.3.12 *Vibrio*, *Spirilla*, and related genera 200
  - 4.3.13 *Pseudomonas* 214

4.3.14 Spirochetes 218
4.3.15 Mycoplasmas 226
4.3.16 *Spiroplasma* 232
4.3.17 *Rickettsia*, *Bartonella*, *Chlamydia*, and related genera 232
4.3.18 Legionnaires bacterium–*Legionella* 252
References 266

Part Five
Protozoa

5 Protozoa 281
5.1 Introduction 281
5.2 Classification 282
5.3 Class Rhizopodea 283
5.4 Genus *Entamoeba* 283
5.4.1 External appearance and life cycle 284
5.4.2 Ultrastructure 284
5.5 Genus *Naegleria* 288
5.5.1 Primary amoebic meningoencephalitis 288
5.5.2 Other pathological roles of *Naegleria* 289
5.5.3 Cysts 289
5.5.4 Flagellate form 290
5.6 Class Zoomastigophorea 290
5.7 Genera *Trypanosoma* and *Leishmania* 292
5.7.1 Terminology 292
5.7.2 External appearance 296
5.7.3 Ultrastructure 296
5.7.4 Uptake and digestion of nutrients 307
5.7.5 The rosette forms 308
5.8 Class Telospora 308
5.9 Suborder Eimeriina 308
5.9.1 General fine-structural details 310
5.9.2 Infective stages 312
5.9.3 Sexual reproductive cells 314
5.10 Suborder Haemosporina 315
5.11 Genus *Plasmodium* 316
5.11.1 Erythrocytic stages 316
5.11.2 Sexual forms 319
5.11.3 Mosquito stages 320
References 321

Part Six
Fungi

6 Fungi 335
6.1 Introduction 335
6.2 The ultrastructure of fungi cultured *in vitro* 335
6.2.1 Keratinophilic dermatophytes 336

6.2.2 Causative agents of deep mycoses 336
6.2.3 Opportunistic pathogenic fungi 338
6.3 Ultrastructural features of host–fungus relationships in mycoses 342
6.3.1 Evidence of fungal effects on the host 342
6.3.2 Evidence of host effects on fungi 347
6.3.3 Evidence for pathogenetic concepts in mycoses 358
6.4 Action of drugs on fine structure of fungi 362
6.4.1 Griseofulvin 362
6.4.2 Polyene antimycotics 362
6.4.3 Imidazole antimycotics 364
6.4.4 5-Fluorocytosine 367
6.4.5 Other compounds 370
References 371
Index 391

The Contributors

June D. Almeida, one of the world's leading virologists, is at present a Principal Scientist with the Wellcome Research Laboratories at Beckenham, Kent. She previously worked with co-author Professor A. P. Waterson both at St. Thomas' Hospital and Hammersmith Hospital, London. Dr. Almeida has a M.Phil. degree in microbiology and a D.Sc. for published work in the field of virology. Since the introduction of negative staining in 1958, Dr. Almeida has used the electron microscope to study virus structure and the interaction of viruses and antibody.

Eyvind A. Freundt is a leader in the research on animal mycoplasmas. An M.D., Ph.D. from the University of Copenhagen, Dr. Freundt became Professor of Medical Microbiology (Virology) at the University of Aarhus in 1964 and has also been Dean of this university's Medical School. He is head of FAO/WHO Collaborating Centre for Animal Mycoplasmas. He is a member of numerous scientific societies, member of several international committees on viruses and mycoplasmas, and has won several awards, not only in medicine but also in archeology.

Stanley C. Holt, B.Sc. from New York University, M.Sc. from Michigan State University, and Ph.D. from the University of California, is Professor of Microbiology at the University of Massachusetts. He has held positions as visiting Professor in the U.S. and abroad and has been Chairman of the Ultrastructure and Morphology Division of The American Society of Microbiology.

As one of the world's leading microbiologists, Dr. Holt serves on the editorial boards of several journals and is Editor of the *Journal of Bacteriology*. The majority of his numerous papers and book chapters deal with differential and density gradient ultracentrifugation, thin layer chromatography, gas chromatography, electrophoresis, and electron microscopy of bacteria.

Jean-Marie Jadin was born in M'Bandaka (Zaïre) in 1940. He obtained his M.D. degree in Louvain and his D.Sc. in Paris. He is specialized in tropical diseases and in electron microscopy. Formerly he was assistant at the University of Louvain (1968–72), head of the Group of Electron Microscopy (1972–75) at the Nuclear Center of Mol, Belgium. Presently he is Professor of Parasitology at the Faculty of Sciences of Namur and Investigator at the International Institute of Cellular and Molecular Pathology, Brussels. His special interest in trypanosomiasis (pathology, chemotherapy, morphology, and biochemistry of the parasites) is reflected in his numerous papers.

Jean-Baptiste Jadin was born in Namur in 1906. He obtained his M.D. degree at Louvain and is Agrégé de 'Enseignement Supérieur, Louvain. For more than twenty years he was Director of State Research Laboratories in the Belgian Congo, now Zaïre. He has been Professor of Protozoology and Head of the Department of Medical Protozoology at the Institute of Tropical Diseases, Antwerp. Professor Jadin is one of the leading protozoologists in the world. He is Membre Honoraire de l'Académie Royale des Sciences d'Outre-Mer (Belgique) and Membre Correspondant de l'Académie National de Médecine (Paris) and local secretary of the Royal Society of Tropical Medicine and Hygiene. He is interested in the biology of Kinetoplastidae, free-living amoebae and of entamoebae.

Elinor Meynell obtained her B.A., M.B. and M.D. degrees from Trinity College, Dublin University and her Dip.Bact. from the University of London. She has had research fellowships followed by senior lectureships at the London School of Hygiene and Tropical Medicine. Dr. Meynell has been a Senior Lecturer at the Wright-Fleming Institute of Microbiology, member of the staff of the Medical Research Council Microbial Genetics Research Unit, and has also been a member of the staff of the Lister Institute of Preventive Medicine, and a Senior Research Fellow in Microbiology, University of Kent.

Her main interests in the field of microbiology have been bacteriophage and bacterial plasmids and she has published several standard reference papers and book chapters on these topics.

Johannes Müller is Dr.rer.nat. and Lecturer in Medical Microbiology and Hygiene at the Klinikum der Albert-Ludwigs-Universität in Freiburg where he holds the position of Akademischer Oberrat in the Department of Parasitology and Mycology. His special interests are medical mycology, in particular electron microscopy and immune-electron microscopy of mycotic lesions on which subjects he has published numerous standard reference papers.

Ole Vinther has a degree in Chemistry from the Institute of Chemistry, University of Aarhus, Denmark and has pursued research in radiation chemistry. Since 1972 he has been Research Fellow of the Danish Medical Research Council, working at the Institute of Medical Microbiology, University of Aarhus. His special fields of research are now the biochemistry, ultrastructure, and ultrahistochemistry of mycoplasmas.

Anthony P. Waterson is a senior statesman in the field of virology. He has an M.D. degree from Cambridge and is a Fellow of the Royal College of Physicians, London, and also a Fellow of the Royal College of Pathologists. He was previously Professor of Medical Microbiology in the University of London at St. Thomas' Hospital Medical School. In 1967 he became the first Professor of Virology in the United Kingdom when he was appointed to this chair in the University of London at the Royal Postgraduate Medical School, at the same time becoming Consultant Virologist to Hammersmith Hospital. His general interest is the development of clinical virology in the context of hospitals and his special interests are the detection and control of viral hepatitis, the assessment of virological risk to the fetus during pregnancy, and the role of viruses in cardiac disorders.

Preface to series

The electron microscope has made its way from the research laboratories into almost all fields of human medicine. In some disciplines, such as nephrology and virology, it has already become an established and indispensable tool. In others, such as oncology, it is rapidly becoming one.

The rapid expansion of electron microscopy in human medicine represents a challenge to most medical institutions. Their electron microscopy laboratories are often run by people without training in human pathology while most pathologists lack ultrastructural experience.

The present series is the first comprehensive attempt to bridge this gap by letting leading experts present the current state of the art in one all-embracing endeavour. This kind of information has previously been scattered as numerous papers in medical and nonmedical journals or books dealing with limited fields only.

Electron Microscopy in Human Medicine should provide a solid foundation for those who are in the process of building up experience in ultrastructural pathology, and also broaden the horizon of those with experience in one narrow area of human electron microscopy. The series, furthermore, should present clinicians with a dynamic ultrastructural view of the diseases they deal with and help them to decide when to save material for electron microscopical investigation.

Without the enthusiastic and idealistic support of all the authors of this series and the excellent cooperation provided by the publishers and my hospital, the editing of this venture would have been at best troublesome and at worst impossible.

Oslo, 1979.

Jan Vincents Johannessen

Preface

The identification of infectious agents has always represented a challenge to morphologists studying human disease.

The resolving power of the light microscope limited the hunt to protozoa, fungi, and bacteria. The only exceptions were those viral disorders where the etiological agents were indirectly revealed through the appearance of inclusion bodies. The viral particles themselves, however, escaped detection. The electron microscope changed this situation. Not only did it resolve the single viral particles, but, when negative staining techniques were used, their fine stubstructures. It also disclosed previously unknown details of the larger infectious agents.

Electron microscopy is an important tool in basic research on infectious disorders. It is also an important diagnostic method, particularly when one is dealing with viruses that are difficult or impossible to culture.

This third volume of the series *Electron Microscopy in Human Medicine* contains the most comprehensive presentation of the ultrastructure of infectious agents ever published. The famous British virologists June Almeida and Anthony Waterson; the first lady of bacteriophage research Elinor Meynell; the Danish mycoplasma authorities Eyvind Freundt and Ole Vinther; the world-renowned American bacteriologist Stanley Holt; the leading German mycologist Johannes Müller, and the well-known Jean-Marie and Jean-Baptiste Jadin from Belgium have written this authoritative volume. The book will be of interest alike to diagnostic virologists, microbiologists, mycologists, protozoologists, and pathologists, and to all other working with infectious diseases in the clinic or in the research laboratory.

Oslo, 1980.

Jan Vincents Johannessen

Acknowledgements to Part One

The authors would like to thank the many clinicans and virologists who provided the specimens. They also thank Mrs. H. Powell and Mrs. L. King for their secretarial patience.

The micrograph for Figure 1.12 was kindly supplied by C. Smales, Animal Virus Research Institute, Pirbright, Surrey.

Acknowledgements to Part Two

The micrographs to Figure 2.1 were supplied by R. Newsam, University of Kent. Micrographs for Figure 2.2 were kindly provided by Dr. Manfred Bayer and Figure 2.2(a) has been reproduced from M. E. Bayer and A. F. Bocharov, "The capsid structure of bacteriophage lambda," *Virology*, **54**, 465, 1973, by kind permission of the editors of *Virology*. Figure 2.3 was reproduced with abbreviation from Y. Kikuchi and J. King, "Genetic control of bacteriophage T4 baseplate morphogenesis. III. Formation of the central plug and overall assembly pathway," *J. Molec. Biol.*, **99**, 695, 1975, by kind permission of Dr. J. King and the editors of the *Journal of Molecular Biology*. The micrographs to Figure 2.4 were kindly provided by Dr. J. King and Dr. Y. Kikuchi with permission from the editors of the *Journal of Molecular Biology*. Those to Figure 2.5 were kindly provided by Dr. R. Haselkorn and reproduced from S. S. Kells and R. Haselkorn, "Bacteriophage T4 short tail fibres are the product of gene 12," *J. molec. Biol.*, **83**, 473, 1974, by kind permission of Dr. Robley Williams and the editors of the *Journal of Molecular Biology*. The micrograph to Figure 2.6 was kindly supplied by Dr. S. Beckendorf with permission from the editors of the *Journal of Molecular Biology*. Figure 2.7 micrographs were reproduced from J-S. Kim and N. Davidson, "Electron microscope heteroduplex study of sequence relations of T2, T4 and T6 bacteriophage DNAs," *Virology*, **57**, 93, 1974, by kind permission of Dr. N. Davidson and the editors of *Virology*. The micrographs to Figure 2.8 were kindly provided by Dr. Manfred Bayer.

Acknowledgements to Part Three

The following figures have been reproduced with the kind permission of the authors and publishers: Figures 3.3 and 3.6 from F. T. Black, A. Birch-Andersen, and E. A. Freundt, "Morphology and ultrastructure of human T-mycoplasmas," *J. Bact.*, **111**, 254, 1972, published by the American Society for Microbiology, Washington D.C.; Figures 3.4 and 3.5 courtesy of Dr. F. T. Black, Institute of Medical Microbiology, University of Aarhus, Aarhus, Denmark; Figures 3.7(a) and (b) from T. W. Tillack, R. Carter, and S. Razin, "Native and reformed *Mycoplasma laidlawii* membranes compared by freeze-etching," *Biochim. Biophys. Acta*, **219**, 123, 1970, published by Elsevier Publishing Company, Amsterdam; Figures 3.8(a) and (b) from A. M. Collier, "Pathogenesis of *Mycoplasma pneumoniae* infection as studied in the human foetal trachea in organ culture," in *Pathogenic Mycoplasmas*, Ciba Foundation Symposium, 1972, p. 307, published by Elsevier · Excerpta Medica · North Holland, Amsterdam; Figures 3.9 and 3.10 from J. E. Peterson, A. W. Rodwell, and E. S. Rodwell, "Occurrence and ultrastructure of a variant (*rho*) form of *Mycoplasma*," *J. Bact.*, **115**, 411, 1973.

Acknowledgements to Part Four

The preparation of this work was supported in part by Public Health Service grants DE 03512, DE 04383 and DE 05123 from the National Institute of Dental Research.

The author is indebted to E. R. Leadbetter for his reading and criticism of the manuscript which could not have been completed without his kindness and thoughtfulness, and to Erika Musante for her continued excellence in the preparation of the photographic material. As usual Gloria Holt stood beside him during the long hours of preparation.

Acknowledgements to Part Six

The author gratefully acknowledges the kind advice given by colleagues S. Batsford and Annemarie Polak-Wyss. The following figures have been reproduced with the kind permission of the authors and publishers: Figures 6.1(a) and (b) from J. Müller, J. Tokunaga, and H. Takamiya, "Electron-microscopic analysis of *Candida albicans* cell wall antigens by the direct immuno-ferritin test," *6th International Congress ISHAM, Tokyo, 1975*; Figure 6.2 from S. De Nollin, F. Thone, and M. Borgers, "Enzyme cytochemistry of *Candida albicans*," *J. Histochem. Cytochem.*, **23**, 758, 1975; Figure 6.3 from D. Poulain and J. Biguet, "Etude en microscopie électronique de l'action kératinolytique de *Trichophyton mentagrophytes* dans le poil de cobaye infecté expérimentalement," *Sabouraudia*, **12**, 1, 1974; Figures 6.4 and 6.16 from J. Müller, C. Douchet, and S. Sakuma, "Etudes au microscope électronique des reins de souris infectées par *Candida albicans*," in E. Drouhet (ed.), *Proceedings 5th International Congress ISHAM, Paris, 1971*; Figure 6.5 from D. Pugh and R. A. Cawson, "The cytochemical localization of phospholipase in *Candida albicans* infecting the chick chorioallantoic membrane," *Sabouraudia*, **15**, 29, 1977; Figures 6.6, 6.7, 6.8, and 6.10 from J. Tokunaga, M. Tokunaga, J. Müller, H. Takamiya, and A. Vogt, "Ultrastructural host–fungus relationships: fungi pathogenic for warm-blooded hosts," *12th International Congress of Microbiology*, Munich, 1978; Figure 6.9 from T. Arai, Y. Mikami, and K. Yokoyama, "Phagocytosis of *Candida albicans* by rabbit alveolar macrophages and guinea pig neutrophils," *Sabouraudia*, **15**, 171, 1977; Figure 6.11 from D. Poulain, J. Biguet, and A. Vernes, "Aspects ultrastructuraux des relations hôte-parasite au niveau des téguments parasités par *Trichophyton mentagrophytes*," *Ann. Microbiol. Institut. Pasteur (Paris)*, **127B**, 473, 1976; Figure 6.12 from E. S. Kuttin, J. Müller, H. Takamiya, H. Melchinger-Müller, and R. Jaeger, "Electronmicroscopic studies of *Candida albicans* infected kidneys in non-immunized and immunized hosts," *14. Tagg. Deutschsprach. Mykolog. Ges. Tübingen, 1978* and *Mykosen* (in press); Figure 6.13 from H. Melchinger-Müller, J. Müller, H. Takamiya, and B. Nold, "*Candida* asteroid bodies in klinischen Untersuchungsmaterialien: Immunelektronenmikroskopische Untersuchungen an Vaginalmaterial von Kolpitis-Patieninnen," *Mykosen*, **23**, 161, 1980; Figure 6.14 from J. Müller, H. Takamiya, and R. Jaeger, "Elektronen-mikroskopische Darstellung von Immunreaktionen an *Candida*-Zellen; Asteroid Bodies bei *Candida albicans* im Urin von Nephritis-Patienten," *Sabouraudia*, **15**, 87, 1977; Figure 6.15 from H. Drexler, H. Melchinger, and J. Müller, "Asteroid bodies an Fadenpilzen in Organen einer Leukämie-Patientin," *Schwerpunkt-medizin*, **2**, 42, 1979; Figure 6.17 from W.-H. Voigt, "Zur Wirkung von Antimykotika auf die Ultrastruktur von Pilzzellen, in J. Hartung and D. Lubach (eds.), *Mykosen. Systematik, Klinik, Therapie*, Thieme, Stuttgart, 1975; Figure 6.18 from W.-H. Voigt and M. Plempel, "Elektronenmikroscopische Untersuchungen an human-pathogenen Pilzen. 1. Mitteilung: Ultrastrukturelle Veränderungen von *Candida albicans*-Zellen durch Clotrimazol im Tierexperiment," *Arzneim.-Forsch. (Drug. Res.)*, **24**, 508, 1974; Figures 6.19(a)–(c) from S. De Nollin and M. Borgers, "An ultrastructural and cytochemical study of *Candida albicans* after *in vitro* treatment with imidazoles," *Mykosen*, **19**, 317, 1976; Figures 6.20(a) and (b) from H.-J. Preusser, and H. Rostek, "Freeze-fracture studies of the plasmalemma of *Candida albicans* after treatment with econazole-nitrate", *Sabouraudia*, **17**, 389–398, 1980. Figure 6.21 from T. Arai, Y. Mikami, K. Yokoyamy, T. Kowata, and K. Masuda, "Morphological changes in yeasts as a result of the action of 5-fluorocytosine," *Antimicrob. Agents Chemother.*, **12**, 255, 1971.

Part One

Viruses

1. Viruses

June D. Almeida and Anthony P. Waterson

1.1. Introduction

When the essential nature of viruses came to be understood, they were found to be of interest not merely as causes of disease but also because they were the simplest form of life possible. It seemed natural to suppose that an organism so limited in genetic material as a virus would have to deal with morphological construction as simply and efficiently as possible. Before 1958 two techniques had been employed to investigate virus morphology. These were examination of the shadowcast appearance of the particles in the electron microscope, [83] which revealed little more than the overall shape of the viruses, and X-ray diffraction,[37] for those viruses that could be obtained in sufficient quantity and purity. Taken together, these techniques revealed that the viruses available were certainly limited in types of morphological construction, and that the majority fell into one or two major groups,[54] namely, those that were rod-shaped in the electron microscope and displayed helical symmetry by X-ray diffraction, and those that were more or less spherical in outline in the microscope and revealed the presence of cubic symmetry by X-ray diffraction. Working on this evidence, Crick and Watson[27] published a prophetic paper in 1956. They concluded that the symmetry displayed by viruses was present because the limited amount of nucleic acid possessed by them would be insufficient to code for large or elaborate molecules of protective protein and that, instead, the virus covering would be made of small, simple subunits which had to be arranged in a simple and efficient manner.[28] It was predicted that the two types of symmetry (helical and cubic) found in association with purified virus would turn out to be the expression of this arrangement. The confirmation of these predictions became possible in 1958 when Brenner and Horne[22] applied the technique of negative staining at the electron microscope level to the study of viral morphology.

Early negative stained micrographs showed that the protective outer shell of the adenovirus was built up of separate subunits arranged to form an icosahedron.[39] This fitted well with the previously known fact that cubic symmetry was associated with "spherical" viruses because the icosahedron is one of a group of regular solids that are described as being cubic. Basically, this term refers to the fact that the octahedron, the dodecahedron, and the icosahedron possess diagonals corresponding to those of the cube. Further electron microscopic studies then revealed that other "spherical" viruses had icosahedral form and symmetry, and a complex pattern of subdivision emerged.[24] The

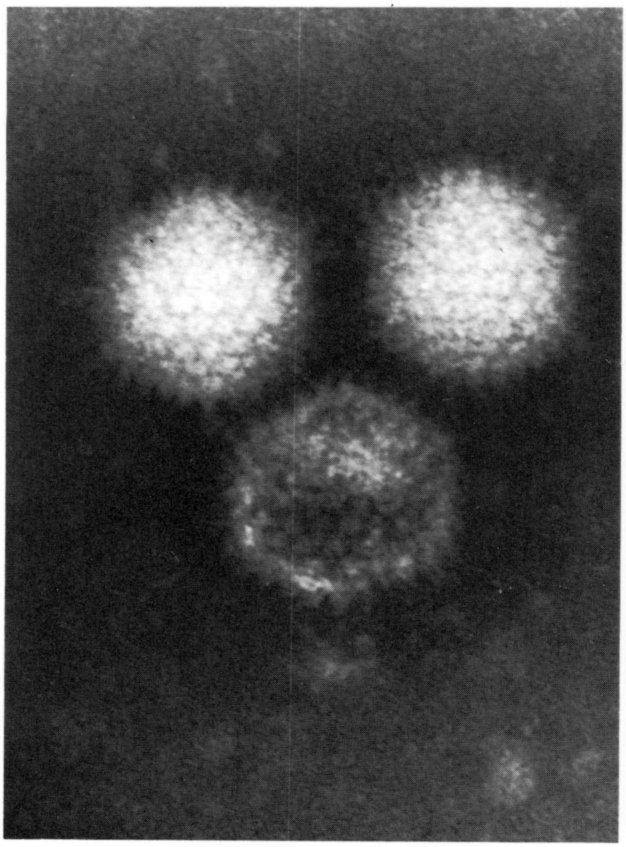

Figure 1.1. Adenovirus particles.
Three particles showing the subunit construction present in cubic viruses. On the two upper particles it is possible to see the triangular facets and overall icosahedral form of the adenovirus. (× 500,000)

icosahedral shell of the virus could be shown to consist of small identically shaped subunits that differed only with respect to their location on the icosahedron. Those at the apices were surrounded by five other subunits while those on the faces or edges of the icosahedron were surrounded by six. The terminology that emerged described the overall icosahedral shell as the capsid, and the individual subunits as capsomeres. Capsomeres were then further subdivided into pentons and hexons depending on the number of others surrounding them.

The actual arrangement of pentons and hexons to form an icosahedron can vary considerably. There are always twelve pentons, as these form the apices of the icosahedron, but the hexons can be arranged either in a face-centered or an edge-centered pattern for the simpler types, or with skew symmetry for the more complex capsids. All of these symmetry arrangements give rise to a series based on increasing numbers of hexons. For the simplest face-centered group, the number of capsomeres forming the capsid can be derived from the formula $10\ (n-1)^2 + 2$, where n is the number of subunits present between two five-fold axes or apices. Herpesvirus and adenovirus belong to this group with capsomere numbers of 162 and 252, respectively (Figure 1.1).

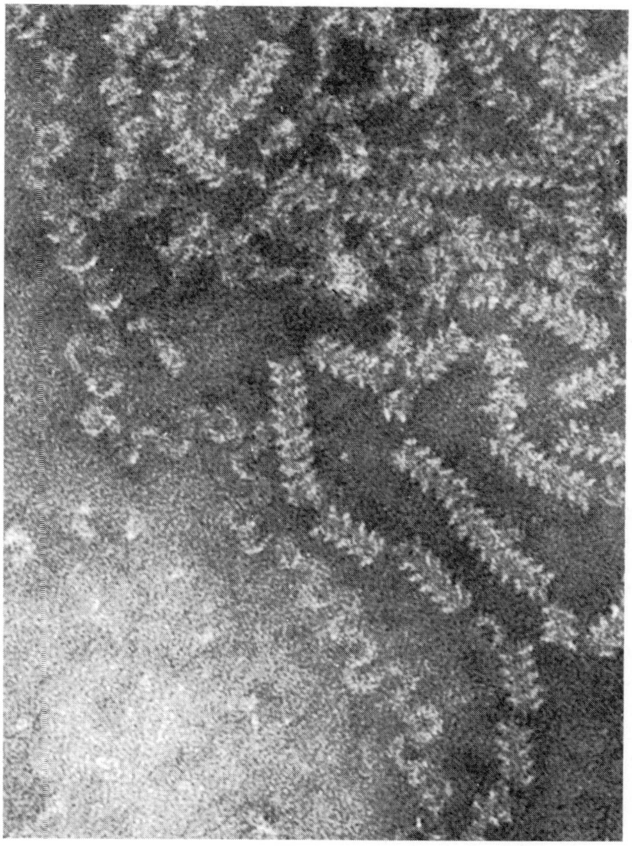

Figure 1.2. RNP of a parainfluenza virus.
Many RNA viruses have their nucleic acid in the form of an inner ribonucleoprotein (RNP) helix. This micrograph is unusual as it is possible to see that the commonly visualized herringbone pattern is derived from a single, helically arranged strand. (× 300,000)

Negative staining of rod-shaped helical viruses showed that they were indeed an arrangement of ribonucleic acid and identical protein subunits.[40] The subunits were joined together identically to form a helix[12, 41] and the nucleic acid lay in a protected position in an internal groove (Figure 1.2). Hence, the prediction of Crick and Watson was completely borne out for the helical viruses, as they consisted of identical subunits identically arranged in helical symmetry. For the cubic viruses it is almost true, as it would appear to be mainly a configurational difference which distinguishes pentons from hexons. However, returning to helical viruses, it was found that both plant and animal viruses used the same nucleic acid-protein component construction, but the animal viruses invariably added a cell-derived but virally altered lipoprotein membrane. This feature will be discussed in greater detail in Chapter 4, section 4.1. Terminology for the helical viruses[24] describes the internal nucleic acid and protein helix as the ribonucleoprotein (RNP) and the cell-derived outer membrane as the envelope.

It should be quite clear that the geometric virus morphology revealed by negative staining is that of the protein subunits which protect the nucleic acid from the environment. The nucleic acid of the virus is not demonstrated by negative staining.

These findings supplied by the negative staining technique confirmed in a practical fashion that viruses were morphologically limited.[9] It now appeared that this limitation could be put to good effect, as it was found that viruses that looked alike also possessed other features in common. Morphologically identical viruses always contained the same type of nucleic acid, shared a similar morphogenesis within the cell, and could be serologically related.

Up to this point, the results and findings of negative staining had been more or less academic, but with the realization that morphology could yield information on the nucleic acid type of a newly discovered virus, and even where to look for serological relationships, interest was aroused among clinical virologists. Additionally, in 1962, when it was found that viruses need not be purified from their cell substrate for successful negative staining to be carried out,[4] the technique became of even greater interest to those looking for viruses in animal or human disease. Nowadays it is accepted that the electron microscope technique of negative staining is one of the most important tools in the hands of the clinical virologist, both from the viewpoint of characterization of an unknown virus and for the positive and rapid identification of a virus in a situation calling for a differential diagnosis.[59]

Before looking at the morphological features of the major virus groups, the underlying mechanism of negative staining should be considered. The electron microscope is a transmission instrument, the electron beam passing through the specimen and being projected onto a phosphorescent screen. If untreated virus particles are placed on the specimen grid they have so little electron scattering ability that the final micrographe shows only the merest trace of particle outline. There are two possible approaches to making the virus particle more suitable for electron microscope study. One is to increase the electron density of the particles by attaching heavy metal atoms to them, and this is exactly what happens when thin sections are stained with lead hydroxide. The other approach is to leave the particles with their original, low electron density but to surround them with a material of high electron density. The particles then show up as light areas in a dark background, from which the term *negative* staining arises. The two important aspects of a negative stain are that it should have high electron scattering ability and that it should dry down to a film which appears amorphous at the resolving power of the electron microscope. Compounds of two heavy metals, tungsten and uranium, have been found to possess these qualities, and the negative stains in use at the moment almost invariably contain one or the other.[52]

1.2. Techniques of negative staining

The original reagent used for negative staining was 2 per cent phosphotungstic acid adjusted to neutrality with N potassium hydroxide. Now, almost twenty years later, although many variants have been described, a simple solution of phosphotungstic acid (PTA) still remains the solution of choice for the majority of negative staining procedures.[52] It has been found that if the stain is adjusted to pH 6 with the hydroxide, contrast is somewhat improved. However, for certain acid-sensitive viruses such as foot and mouth disease, it is necessary to use the stain at pH 8. We have also found it an advantage to use the PTA at the higher concentration of 4 percent as it can be diluted to give the correct density on the grid. The advantage of this higher concentration is that a low titer virus suspension does not need to be diluted to any great extent in staining because a relatively small volume of stain can be added.

One of the greatest advantages of the negative staining method is that it can be applied to the study of viruses that are still associated with material from the original cell in which they have been grown.[4] Obviously, the ratio of virus to cell material is an important one, but generally viruses are sufficiently distinctive to be recognized even when the background is a "noisy" one. However, although gross cell debris can be accepted in a negatively stained preparation, two other components of the virus-cell system must be carefully eliminated before good microscopy can be achieved, namely, the inorganic salts and the low molecular weight proteins present in all biological specimens. The inorganic salts have to be removed because they crystallize when the grid is dried and may completely obscure the specimen. Likewise, a high background content of low molecular weight protein can overlay a specimen and completely cover even larger virus particles. Satisfactory methods of specimen preparation should deal with both of these unwanted components and, luckily, a technique for ridding a specimen of one usually deals with the other also. The most frequently employed technique accomplishes their removal by centrifuging the specimen at a speed capable of depositing at least some of the virus while leaving low molecular proteins in the supernatant. This supernatant is then discarded and replaced by distilled water, a step which also removes the inorganic salts. An alternative method of specimen preparation is to place the specimen directly on the microscope grid and to wash away as much background material as possible. The first method has the widest application for specimen preparation and can be described as the general method of negative staining. The second technique is more limited and is of particular use in selected situations. It can be described as "on the grid" staining.

1.2.1. General methods

The starting material may be fluid, such as tissue culture supernatant, tissue culture cells, or it may be solid tissue. Both cells and solid tissue will have to be disrupted to make a suspension and this will be dealt with in detail when individual types of specimen are considered. Once a suspension has been formed all the various types of specimen can be processed by the same method. A suitable volume of specimen, which can be as little as 0·5 ml, but which is ideally in the 2–5 ml range, is clarified by a 10 min centrifugation in a clinical centrifuge. The supernatant from this stage is then centrifuged for one hour at a g-force between 10,000 and 15,000. The fluid is decanted and the tube maintained in an inverted position so that the last drop of fluid is not allowed to run back onto the pellet. Still holding the tube upside down it is placed in a beaker

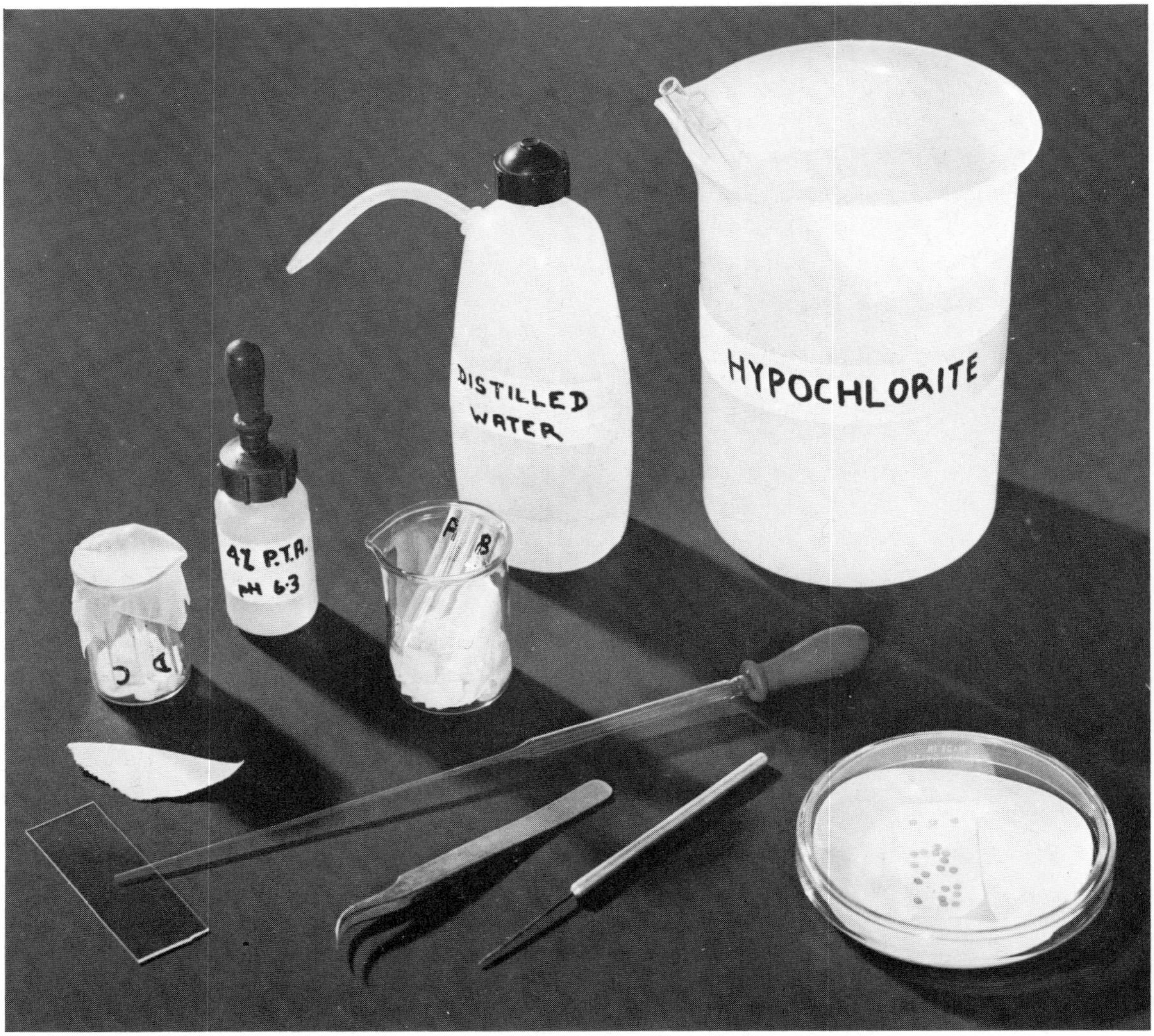

Figure 1.3. Equipment for negative staining.
All the necessary equipment is shown. Specimens still awaiting examination are in the beaker to the left. The tubes are inverted to give maximum drainage and the moisture content is maintained by sealing the beaker with parafilm.

with absorbent material such as a tissue at the bottom. This step ensures that fluid still remaining in the tube will continue to drain and not be returned to the pellet. If the tissue in the bottom of the beaker is kept moist and the whole beaker covered with foil or parafilm it is possible to leave the pellets in the inverted tubes for several hours at 4 °C (Figure 1.3). In this pelleted form virus appears to remain unchanged until it is possible to prepare the specimen for examination in the microscope.

Immediately before the specimen is to be examined, the tube is turned the right way up again and a small amount of distilled water added. The exact volume to be used depends on the individual specimen. A large, bulky pellet could take up to 1·0 ml while a tube that visually appears empty will need as little as 50 μl. The aim in adding the distilled water is to produce a suspension

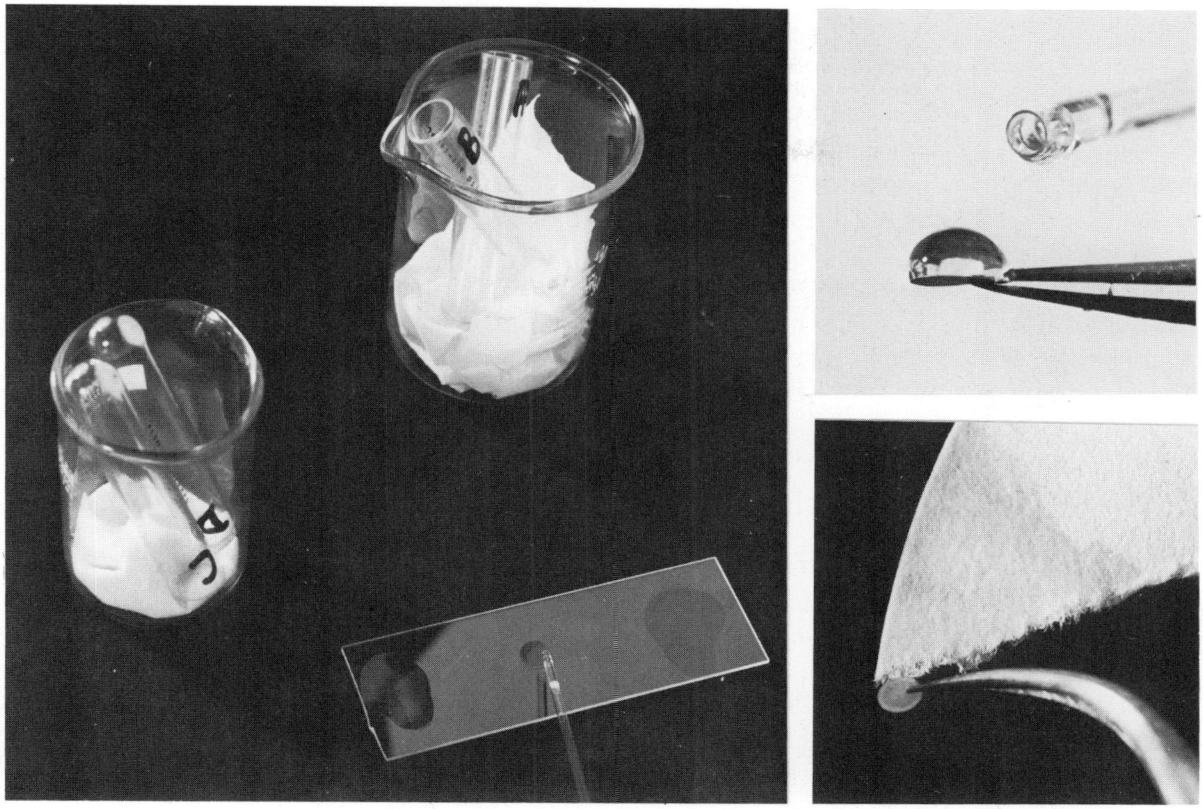

(b)

(c)

Figure 1.4. General method of negative staining.
(a) Pellets are resuspended in distilled water, and specimen and stain are mixed on a slide.
(b) A drop of the mixture is placed on a 400 mesh carbon–formvar coated grid.
(c) Excess fluid is withdrawn with the torn edge of a filter paper.

that exhibits only the slightest turbidity and frequently appears clear. The beginner will tend to add too little water to the pellet and the resulting specimen will contain too much material and hence be too thick for good microscopy. One drop of this aqueous suspension is then placed on a glass microscope slide, which makes a suitable base for this step and is completely disposable (Figures 1.3, 1.4). For all routine specimens, a standard stain of 4 percent phosphotungstic acid adjusted to pH 6 with normal potassium hydroxide is recommended. Specimen and stain are then mixed, the proportion of the two fluids generally being close to one part stain to two of specimen. However, this proportion will vary somewhat with individual specimens. For example, a purified preparation of a small virus will require relatively less stain than a crude clinical specimen containing a large particle such as a poxvirus. After the specimen and stain have been mixed, a drop is placed on a microscope grid using a Pasteur pipette. After a few seconds' interval to allow absorption to take place, excess fluid is withdrawn by means of a torn edge of filter paper (Figure 1.4). In our hands, the most satisfactory grids have been copper 400 mesh, coated with formvar and then stabilized with carbon. Immediately the grid is dry it can be examined in the electron microscope.

1.2.2. "On the grid" method

This technique is of particular use in dealing with specimens that are available in only small volumes or are contained in high molarity substrates. Also, if it is thought that a virus is present in high titer this method can be used for a very quick examination of clinical specimens. It is not good for specimens with a lot of background material.

A drop of the specimen is placed directly onto the microscope grid. Low molecular weight background material and/or high molarity buffers are removed by washing the grid gently with several drops of distilled water. The final drop of this is removed with filter paper and before the grid can dry a drop of 2 percent phosphotungstic acid adjusted to pH 6 is added. This drop is left for a few seconds and then withdrawn and the specimen allowed to dry. The method is of particular use in dealing with gradient fractions which are still in the original gradient substrate, e.g., cesium or sucrose. Such specimens are difficult for electron microscopy as even trace amounts of gradient material will spoil the preparation. However, the direct or the grid method can yield highly satisfactory preparations as long as it is remembered that the washing stage has to be more thorough than with ordinary biological material.

In reaching a decision as to which technique should be employed, it is obviously of importance to know the limit of sensitivity of negative staining. Early in the history of this technique it was felt that 10^9 particles were required before a virus could be visualized. Through the years it has become accepted that much smaller amounts of virus can yield positive results. Exactly how many particles are necessary to bring a specimen into range of the electron microscope is difficult to assess, as it would be necessary in each case to know accurately the physical to infectious particle ratio. In addition, the morphological features of the virus concerned play a part in determining the lower limit of sensitivity of the technique. For example, it is an easy matter to find one poxvirus on a grid square, but not so easy to identify a single enterovirus particle. However, it would appear that a lower limit for visualization would be in the range of 10^5–10^6 physical particles/ml of suspension. The fact that noninfectious particles are just as good as active virus for purposes of visualization acts in favor of the microscope. This in turn means that the infectivity of a positive specimen can be as low as 10^3 particles/ml, or even 10^2 for viruses with a high physical to infectious particle ratio. Since it is rarely possible to predict the particle-infectivity ratio of any particular specimen, it is worthwhile examining any specimen that has a titer of greater than 10^3 infectious units/ml. However, in cases where the infectivity level is unknown, and this will be in the majority of cases, investment of time in a negative staining examination is so small that it is a worthwhile undertaking anyway.

On many occasions the sensitivity of either of these methods for producing negatively stained specimens can be extended by employing the refinement of immune electron microscopy.

1.2.3. Immune electron microscopy

As will be seen, certain virus groups do not have distinctive morphology when viewed by negative staining. Rubella virus[19] is a good example of this (Figure 1.28). In addition, small viruses can be lost among background material in specimens such as fecal suspensions. Immune electron microscopy[6] tackles both of these problems by the addition of specific antibody to the specimen. This step pulls the viruses together into complexes thus making identification considerably easier or in some cases (e.g., hepatitis A virus)[48, 78] possible (Figure 1.17). The addition of specific antibody achieves in one maneuvre both identification and concentration. The technique of immune electron microscopy can be applied to all the various types of specimens that are encountered. As for general negative staining, solid tissue and cell material will have to be processed to form a suspension. The starting material is clarified by low-speed centrifugation, and an aliquot of the supernatant placed in a suitable container. Most frequently this container will be the centrifuge tube used to obtain the final pellet. The volume of material will be influenced by several factors, such as availability of virus specimen and suspected virus titer. The lower the titer, the greater the volume needed. In practice the volume is usually in the range of 0·5 to 5·0 ml. A suitable quantity of antiserum is then added to the viral suspension. A convenient volume appears to be 0·2 ml but it is important that the antiserum be used at an appropriate dilution level. The estimate of this dilution factor is the most important step in carrying out the technique. Too high titer antiserum leads to the condition of antibody excess, and instead of complex formation the particles will be surrounded by a heavy halo of antibody molecules (Figure 1.5). Too little antibody is unable to give rise to complex formation, and the specimen is no different from that of the virus alone. In order to establish the titer of antiserum to be used it is necessary to carry out experiments with a known virus preparation. This standard preparation is reacted with various dilutions of the antiserum. We have found 1:5, 1:20, and 1:50 a good starting range. After mixing, the two reagents are allowed to react for an hour at room temperature and then are centrifuged for an hour at 12,000 g. The resultant pellet is used for negative staining in the usual manner. Over the range described it should be possible to see changes in the size of aggregate formation. If 1:50 provides the best aggregates it would be worthwhile examining even higher dilutions. Once the optimal dilution on this control specimen is known it will usually be found that this will be suitable for most test specimens, as the equivalence range is a relatively wide one and virus concentration frequently remains within a fairly narrow range for one virus type.

In carrying out immune electron microscopy attention should be paid to a number of practical points. The antiserum should have a reasonably high titer as this allows a greater dilution factor with consequently less background protein. The antiserum should also be clean: serum with a high lipid content does not give good electron microscope results. An important but dirty antiserum can sometimes be rescued by centrifuging it at 40,000 g for an hour and taking the material that lies between the pellet at the bottom of the tube and the lipid layer at the top. The time that the viral antigen and antiserum is left to react

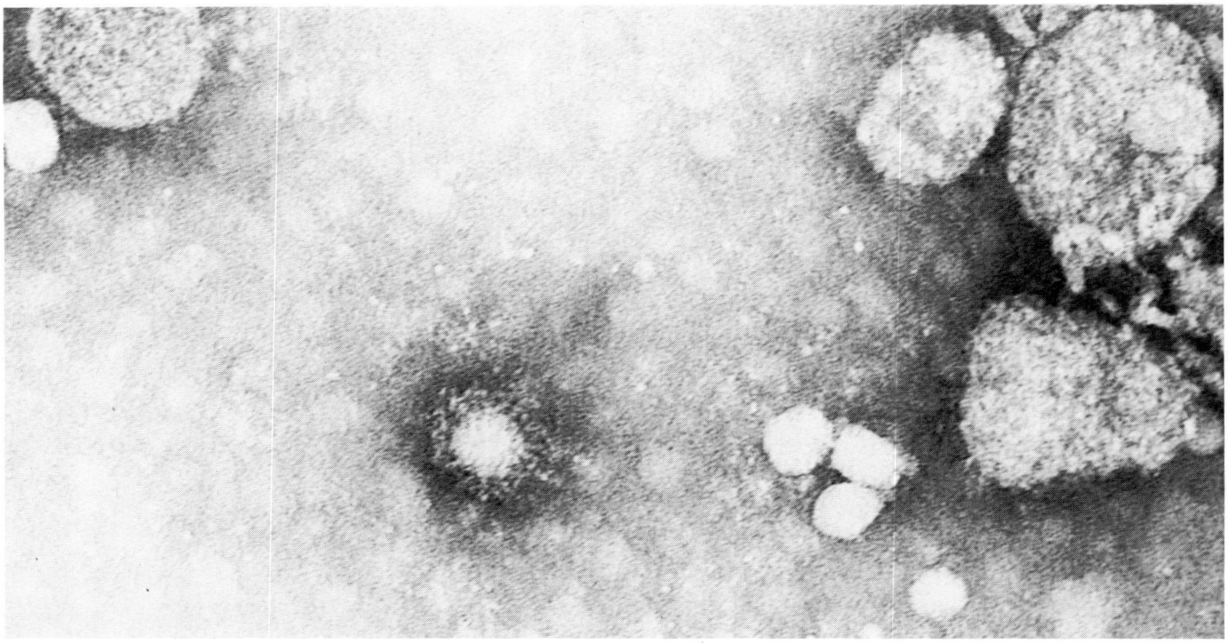

Figure 1.5. Antibody excess in immune electron microscopy.
A fecal specimen containing hepatitis A virus to which too much antiserum has been added. There is only a single virus particle in the field and it is surrounded by a heavy layer of antibody molecules. In these conditions aggregation cannot occur and virus particles are obscured. (Compare with Figure 1.17.) (× 300,000)

can be varied. Best aggregation is said to occur if the reagents are left overnight at 4 °C. Unfortunately, the need for a speedy answer sometimes dictates that this cannot be done, and either an hour at room temperature or even 30 minutes at 37 °C can be employed to give results that are only a little below optimal. As for simple negative staining, it is all-important that as little background material as possible is carried onto the grid. Antibody cannot be visualized when it is overlaid with low molecular weight material.

1.3. Types of specimen examined by negative staining

The range of specimens that can be sent for electron microscopic examination is wide, and they will be reviewed in the order of frequency most often encountered in a laboratory dealing with clinical situations. The specimens can be divided into two broad groups: clinical specimens sent directly for electron microscopy and specimens which have already been propagated by an *in vitro* technique. Probably the most frequent request is for help in assessing the type of virus that either is or is thought to be growing in cell culture or in eggs. The exact steps needed to deal with such specimens are as follows.

1.3.1. Tissue culture specimens

Since it will not be known whether the majority of virus remains within the cells or is present in the supernatant both must be prepared for microscopy. Tissue culture supernatant is best treated by the first method of preparation. Approximately 5 ml is centrifuged for one hour at 12,000–15,000 g, the supernatant discarded and the tube inverted and held in a beaker. As previously described, the pellet is resuspended in distilled water and this solution used for negative staining. It is sometimes difficult to decide whether the supernatant requires a clarifying spin or not before this main centrifugation. The best guideline is supplied by the supernatant itself. If it is a clear solution then clarification is not necessary. Slightly opalescent fluid will also be acceptable without clarification. Only if the supernatant is grossly contaminated by cell debris should it be necessary to clarify initially. Care must be taken that the clarification is minimal since viral material is frequently associated with the very cell fragments that have been shed into the medium. As a general rule, it is better to accept a rather dirtier specimen than risk losing the virus in well-meaning but misguided attempts at purification. Examination of tissue culture supernatants has been successful for almost all viruses that can be grown in culture. The main exceptions are those very cell-associated viruses, such as cytomegalovirus, that are best examined from the cells themselves.[26]

Examination of the cells is carried out as follows: discard most of the supernatant from the cell cultures leaving 2–5 ml depending on the type of culture being harvested. Scrape the cells off the glass into the remaining supernatant using a rubber or glass scraper. Since the cells will eventually be disrupted there is no need to worry about cell preservation. Spin the cells down in a bench centrifuge and discard the supernatant (or use it for a supernatant preparation). The cell pellet is then resuspended in a volume of distilled water that will yield a faintly turbid solution. At that level the conditions will be sufficiently hypotonic to disrupt the cells and a solution of cell fragments will be obtained. This suspension is now ready for negative staining. Figure 1.6 shows cytomegalovirus particles visualized in this way.

In order to save time, if a simple diagnosis is needed, it is possible to remove some tissue culture supernatant and start a centrifuge run of one hour. During this hour the cells are prepared as described and examined in the electron microscope. If the virus is visualized in association with the cell debris the supernatant need not be examined but if the cell preparation proved negative the supernatant preparation would then be ready for further examination.

The degree of cytopathic effect (CPE) present in the tissue culture preparation can also suggest whether cells or supernatant would be the best starting material. In a very degenerate culture there is a greater likelihood that virus will have been released into the medium while an early CPE is more likely to have virus still contained within the cells.

The direct examination of tissue culture material is of advantage in several respects. An unknown virus isolated in cell culture can be examined at the time the medium change is made and, if the morphology can be established, further

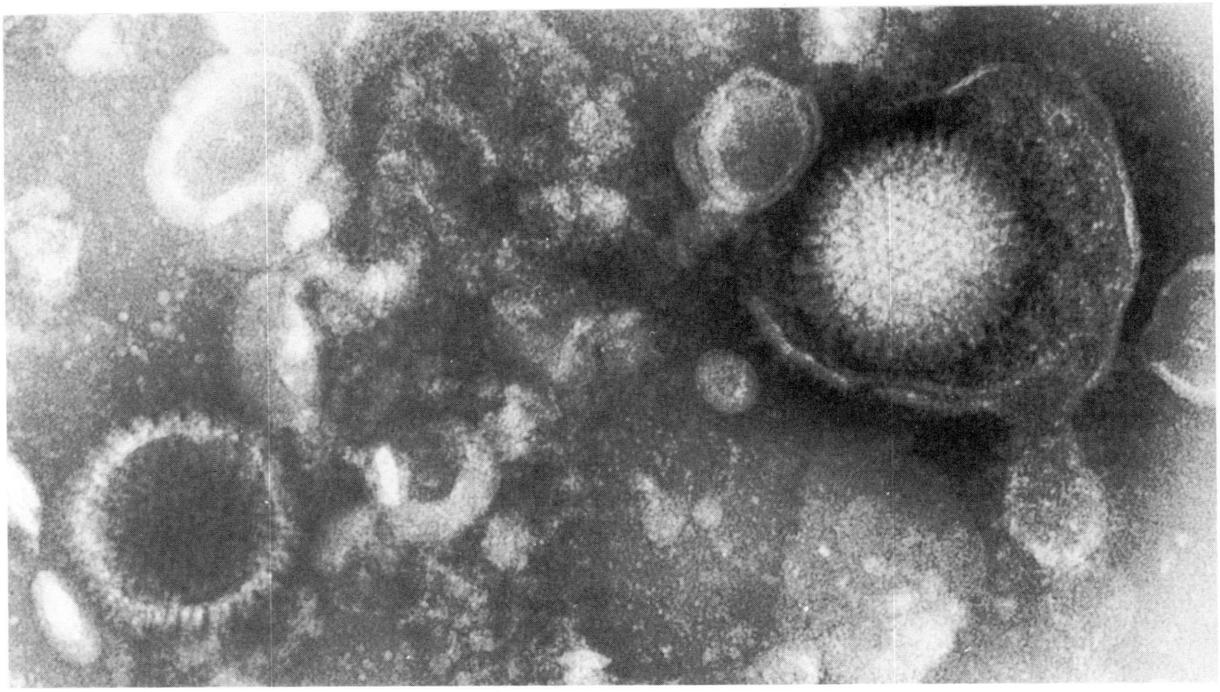

Figure 1.6. Cytomegalovirus from tissue culture specimen.
The majority of viruses have sufficiently distinctive form to be recognized even when they are contained within the original cell substrate. In this illustration a capsid and complete virion of cytomegalovirus can be seen clearly although they are surrounded by fragments of cell debris. (× 300,000)

characterization can be considerably simplified since the virus can be assigned immediately to a relatively small group.

On occasion it might be suspected that a cell culture is carrying a contaminating virus. This can be checked by negative staining and while a negative result cannot be relied on, as the amount of contaminating virus may be too small, a positive result will reveal the type of contaminant within hours, if not minutes. In our hands direct electron microscopy has revealed on various occasions contaminating adenovirus, SV40, and reovirus in cell culture.

1.3.2. Egg-grown material

Allantoic fluid

This is handled in the same manner as tissue culture supernatant. A large range of viruses, including influenza, parainfluenza, and coronaviruses[18] can be identified quickly and simply after centrifugation of the allantoic fluid.

Chorioallantoic membrane (CAM)

Initially, this material must be washed in an isotonic solution such as phosphate buffered saline (PBS) to rid it of as many red cells as possible. The membrane is then lightly homogenized with distilled water in a homogenizer such as the TenBroeck type. Homogenization should not be too thorough as the virus is usually contained in the superficial cells of the membrane which will be disrupted early in the course of homogenization. Further homogenization will only disrupt the deeper fibrous tissue, adding to background debris without contributing further virus. The homogenate can then be examined either directly by using it to carry out negative staining, or a cycle of centrifugation can be added and the final pellet used for microscopy. Examination of the CAM is particularly relevant in dealing with suspected poxvirus infections. If poxviruses are present it is usually a simple matter to identify them because of their size and characteristic shape.

1.3.3. Urine

Urine is the starting material of choice for the isolation of cytomegaloviruses which infect the kidney and are excreted in quantity in the urine. If sufficient is supplied, it is possible to centrifuge approximately 10 ml at the recommended speed of 15,000 g for an hour and use the pellet for negative staining. It would appear that in this situation the biological recovery of virus is more sensitive than electron microscopy, but it is still possible to identify the typical herpesvirus morphology in at least a percentage of suspected cases.

1.3.4. Serum

One of the most important conditions associated with a virus that has not yielded to *in vitro* culture methods is human viral hepatitis type B. [10, 78] Individuals in the early stages of hepatitis B, and other individuals who either have a chronic prolongation of the disease, or who are silent carriers, have in their serum a diagnostic antigen known as hepatitis B antigen (HBAg). This antigen can be detected by a range of serological tests of which the most sensitive are passive hemagglutination[25] and radioimmunoassay.[33] In addition, the HBAg has a most distinctive appearance by negative staining. On occasion it is necessary to make an absolute diagnosis on a serum specimen and, while serological tests are quick and reliable, direct visualization of the HBAg can be additionally assuring. In order to examine a serum specimen for HBAg the following procedure is adopted.

A small aliquot of serum, usually between 0·5 and 1 ml, is mixed with an equal volume of PBS. This is centrifuged for 1 hour at 15,000 g and the supernatant discarded. The pellet is resuspended in the original volume of PBS and centrifuged as before. The additional step of washing used in preparing serum is carried out in order to rid the specimen of the large quantity of low molecular weight serum proteins that would obscure any HBAg present. Even a serum giving only a low grade positive by serological tests should contain enough antigen for recognition. HBAg is found in three morphological forms, pleomorphic spheres and tubules of approximately 22 nm diameter and more uniform double-shelled particles of 42 nm diameter (Figure 1.25). These 42 nm forms are known as Dane particles[29] and at present are considered to be the most likely candidate for the actual infective particle of the hepatitis B virus. A further degree of sensitivity can be given to the EM test for HBAg by the use of immune electron microscopy. To carry this out, the starting aliquot of serum is mixed with 0·1 ml of a suitable dilution of antiserum. (The antiserum we use is at a dilution of 1:20.) Depending on the time available, the HBAg-antibody mixture is left for an hour at room temperature or half an hour at 37°C and then processed as described above.

A particular use of the microscope in this field is the examination of sera that might have the HBAg in the form of naturally occurring immune complexes.[7] When these do occur serological techniques are at a disadvantage as clumped antigen may not register in the test. The electron microscope, on the other hand, has an advantage since the complexes can be readily visualized.

1.3.5. Vesicle fluid and scabs

The most important single recurrent diagnostic problem that can be resolved by electron microscopy is the differentation between varicella (chickenpox) and atypical variola (smallpox[38, 65, 75]). Varicella virus is a member of the herpesvirus group (Figure 1.24) and has the cubic capsid and surrounding membrane typical of this group.[62, 81] Variola, on the other hand, is a poxvirus with quite a different appearance (Figure 1.29). Only a single particle of either type is needed to reach a diagnosis. Ideally the specimen from which diagnosis should be made is vesicle fluid, although occasionally it is necessary to employ homogenized scab as starting material. In this section only vesicle fluid will be considered. If the rash is at an early stage, it is a simple matter to find two or three unbroken vesicles and harvest the fluid. This can be carried out by several different methods but we have found the following technique simple and efficient. Approximately 2 cm length of distilled water is drawn into the shaft of an ordinary sterile Pasteur pipette with a teat. The tip is then used to rupture the vesicle and the water allowed to wash over the lesion, after which it is drawn back into the pipette. Material from one vesicle is sufficient for diagnostic purposes, but if other techniques are to be carried out it is worthwhile harvesting the material from several lesions at the same time. It must also be remembered that in breaking the vesicles to obtain fluid pathogenic virus will have been released onto the skin, and fluid may continue to leak. Broken vesicles should be swabbed with a detergent solution, dried with cotton wool, and covered with a plaster. The small amount of fluid obtained from the vesicles can be placed in any small screw-capped container if it is to be transported, or, if the electron microscopy is to be done within the hospital, a conical centrifuge tube of the capped variety makes an excellent container.

In preparing the specimen for microscopy the possibility that it might contain variola virus must be taken into account and the work carried out in a suitable safety hood with a chlorine-containing disinfectant (i.e., sodium hypochlorite) available for all the equipment used (Figure 1.3). A small aliquot of the vesicle fluid is further diluted with distilled water until the specimen is almost clear and the usual quantity of standard phosphotungstic acid added. This mixture is applied to a grid and excess fluid withdrawn. The grid is now ready for examination, and the time lapse between the vesicle fluid arriving in the EM laboratory and the grid being placed in the microscope need be no more than ten minutes. If unbroken vesicles have been available there is usually little problem in finding virus. Both herpesviruses and poxviruses are large and distinctive and, if present, are usually in high titer. When virus is found it is important to take micrographs, since considerable weight is likely to be placed on the EM diagnosis and a rapidly processed photograph supplies firm evidence of the virus implicated. In the case where unbroken vesicles are no longer available it is necessary to use scab. These are homogenized by any means that will not give rise to aerosol formation. A carefully used glass-Teflon homogenizer of the TenBroeck type will serve the purpose. A limited amount of distilled water should be used to carry out the homogenization and the resulting suspension can then be used directly for negative staining.

1.3.6. Solid tissue

Skin lesions are probably the most frequently encountered solid tissue specimens that are sent for electron microscopic diagnosis. Human common skin warts,[42] genital warts,[63] molluscum contagiosum,[43] and orf[60, 61] all require an answer as to whether the lesion in question has a viral cause, and since none of these viruses can be grown *in vitro* electron microscopy is the only positive means of identification. Luckily, in each case, the causative agent has a distinctive morphology that allows a definite diagnosis to be made. Since the epidermis in all these warty conditions has increased keratin, it is necessary to grind the tissue well so that virus is released. Except for very small lesions, which are best handled in a TenBroeck-type homogenizer, a porcelain mortar and pestle is efficient, and sterile silver sand can be added for better disruption.

The tissue is ground using distilled water as a diluent until a reasonable homogenate is obtained. This is placed in an ordinary bench centrifuge tube and clarified for approximately ten minutes. The supernatant is then ready for examination by negative staining. The amount of distilled water should be kept to a minimum, i.e., no more than 2 ml. If this has not been possible then it may be necessary to concentrate the virus by an additional higher speed centrifugation and to use the pellet from this procedure for microscopy. The skin lesions being considered usually have large amounts of virus that can be easily recognized. However, if a common wart is an old lesion then it can happen that wart architecture might be preserved but virus no longer be present. If the patient has warts of differing duration, a small, early lesion should be chosen for examination.

Although epidermal lesions may be the most frequently encountered of the solid tissue specimens presented for electron microscopy, a wide range of other organs will occasionally have to be processed. Brain biopsy in suspected SSPE,[1] or suspected acute herpes encephalitis,[56] are examples. Since this tissue is easily homogenized, a TenBroeck homogenizer is a simple and efficient means of obtaining a suspension. As before, the tissue is homogenized using distilled water and, if biological recovery is also to be undertaken, it should be done with sterile precautions and an aliquot removed for culture. Since brain is

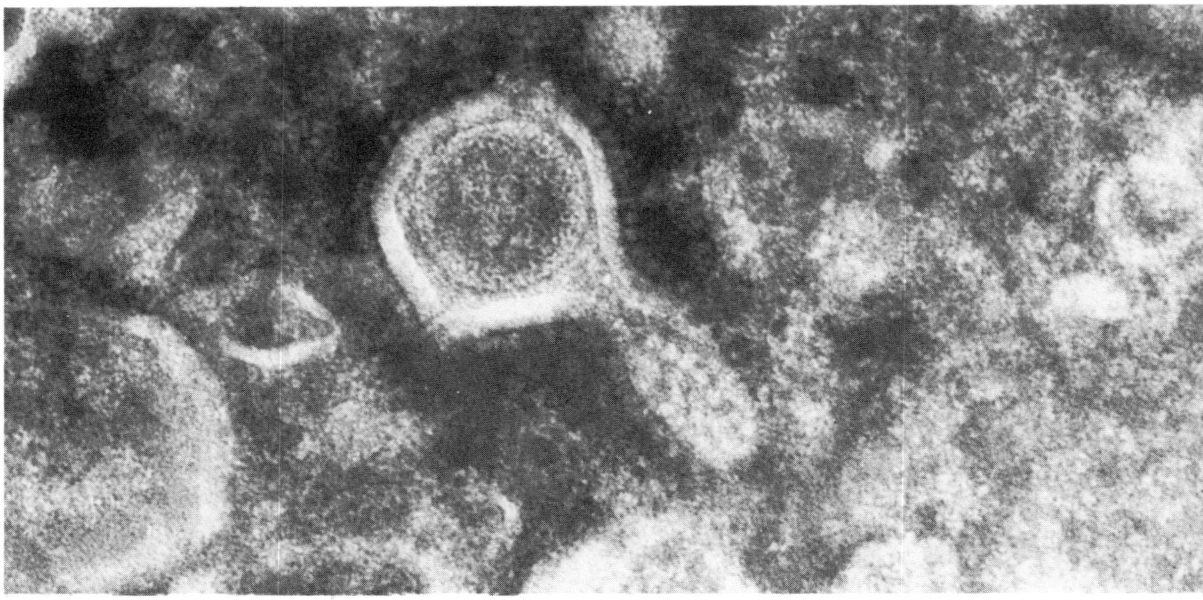

Figure 1.7. Bittner mouse mammary tumor virus from solid tissue specimen. Virus present in solid tissue can frequently be visualized after simple processing of the specimen. Here a virion can be seen within a background of tissue debris. (× 300,000)

highly cellular it is almost essential to undertake a cycle of low-speed centrifugation followed by high-speed centrifugation and to use the resulting pellet for microscopy. Whereas skin lesions yield a final preparation with a low background of cellular debris, brain tissue has a poor signal to noise ratio and even when considerable amounts of virus are present it is difficult to recognize them against the background. In spite of this it has proved possible to identify herpes particles, the measles-like virus of SSPE, and the papovavirus associated with progressive multifocal leukoencephalopathy (PML) in brain tissue.[44]

Turning from man to animals, using the same technique it has been possible to recognize Gumboro disease virus in the bursa of Fabricius in fowls,[5] infectious canine hepatitis virus in dog liver, mouse hepatitis virus in mouse liver, Sendai (parainfluenza I) virus from mouse lung, and Bittner mouse mammary tumor virus[11] from the solid growth (Figure 1.7).

1.3.7. Feces

The most recently recognized acceptable specimen for electron microscopy is the stool. At the time of writing, it might be described as the most fashionable of all materials for the purpose. The realization that direct examination of a simple fecal homogenate was a possible means of virus identification was made in 1973 by Fernelius *et al.*[33] Since that time a number of different human and animal viruses have been identified either by direct electron microscopy or by immune electron microscropy,[17, 34, 35, 49, 50, 84] and the technique is now well established. To prepare feces, it is first necessary to make a 10 percent suspension, either in distilled water or phosphate buffered saline. If the stool is liquid this is a simple if not altogether pleasant step, but if the stool is solid a suspension must be made by homogenization, preferably in a glass homogenizer of the TenBroeck type. The fecal suspension is then clarified by centrifugation in a bench centrifuge for 15 minutes at approximately 3,000 g. The supernatant is removed and if immune electron microscopy is to be employed antiserum is added at this point. After allowing these components to react the rest of the procedure is carried out as usual with centrifugation for an hour at 12,000 g. The tube is decanted and placed in the inverted position on absorbent material until negative staining can be carried out. Prepared in this way, considerable amounts of bacterial debris, particularly flagella, will be present but the background should be sufficiently clear to allow identification of virus (Figures 1.17, 1.19).

1.4. Morphology of the major virus groups

The original grouping of helical and cubic symmetry will be used to classify viruses, whilst nucleic-acid type, and presence or absence of an additional lipoprotein membrane, will be used to subdivide within these classes. In addition, it is now known that some virus groups do not seem to adhere to the cubic-helical system, and these will be discussed separately at the end.

The terminology used for the various virus groups is that of a recent classification produced by the International Committee on Taxonomy of Viruses.[32] After the generic name, a listing of the more commonly encountered human and veterinary viruses is given.

1.4.1. Helical symmetry

In addition to the nucleoprotein complex, all animal viruses with helical symmetry have a cell-derived but virally coded lipoprotein membrane. Only in plant and bacterial helical viruses does the simple arrangement of nucleic acid and protein subunits exist as complete virus. Furthermore, no animal virus displaying helical symmetry has yet been found to contain deoxyribonucleic acid. All such viruses have ribonucleic acid as their genetic material. Yet another generalization occurs with this virus group. Since all possess a lipoprotein outer envelope, they are all sensitive to organic solvents, a feature usually expressed as ether-sensitivity.

Family Paramyxoviridae

Genera

Paramyxovirus — Parainfluenza viruses, mumps virus, Newcastle disease virus.[55] Membranous sacs within which is a readily visualized internal helical component (Figure 1.8). Intact virus that retains a third dimension, i.e., not flattened on the grid, is approximately 100 nm in diameter. The majority of the particles are flattened and may measure up to 500 nm. The outer envelope, which bears fine projections (approximately 10 nm long) on the surface, is an unstable structure, rupturing readily

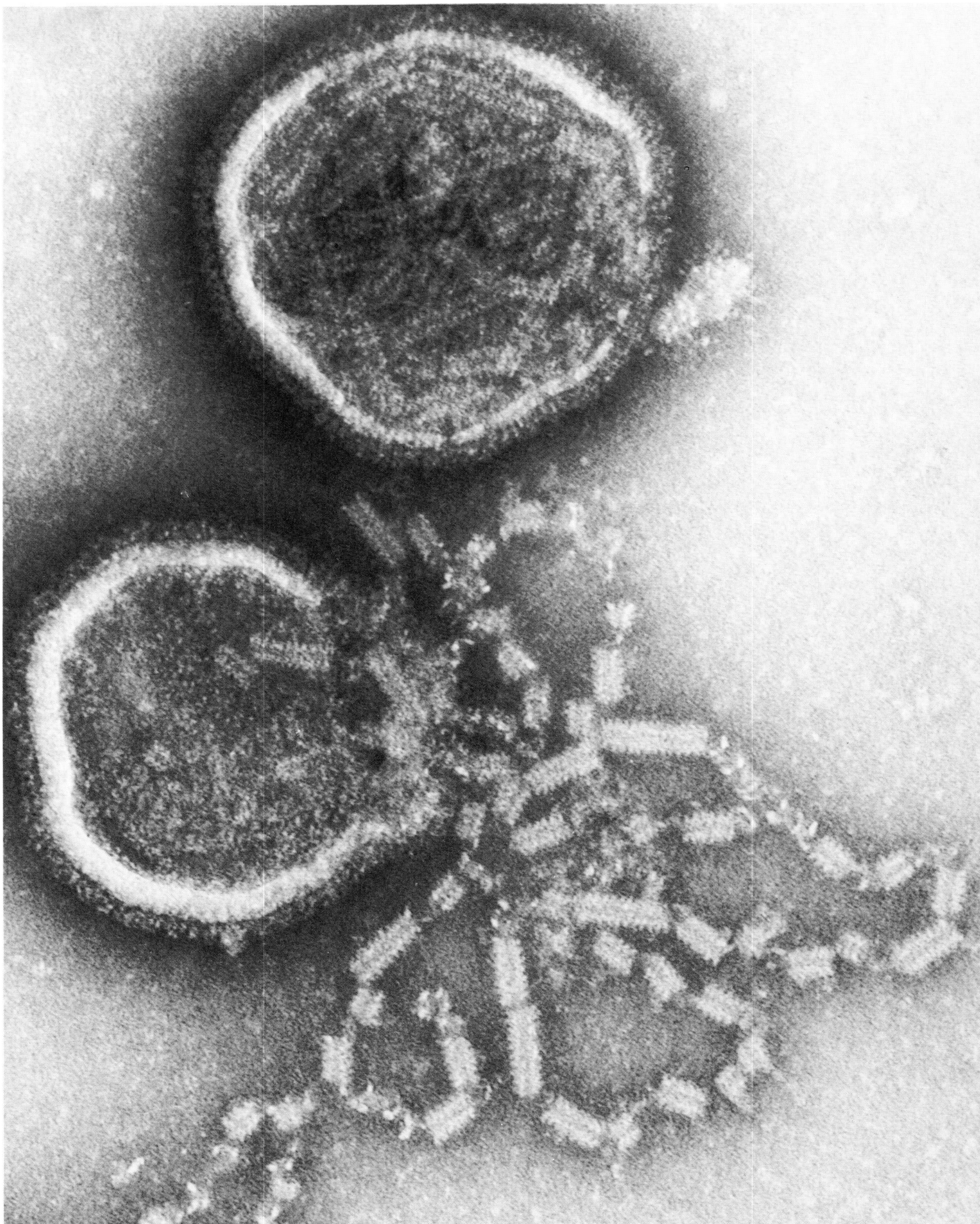

to reveal the internal ribonucleoprotein (RNP). This RNP appears as a linear structure with a regular herringbone pattern along it. Studies using transmission models have shown that this pattern is the one projected by a single helix.[12] Usually areas can be found showing this helix unwinding (Figure 1.2), and fragments of the helix can be found which show the cross-section of the structure. The diameter of the helix is 17–18 nm, and variable lengths up to 1,000 nm, and sometimes even longer, can be found.

Morbillivirus — Measles,[76, 77] rinderpest, and distemper viruses.
The structure is identical with that of the paramyxoviruses but it should be pointed out that, while the two genera are morphologically identical, they differ in certain other properties. The paramyxoviruses are true myxoviruses. They agglutinate red cells and also elute from them by means of neuraminidase. Of the morbillivirus group, only measles has been shown to cause hemagglutination, and none of the group possesses neuraminidase.

Pneumovirus — Respiratory syncytial virus.[85]
Membranous sacs with fragile, internal helical component. Size range as for paramyxovirus, but the outer envelope carries a more prominent projection 15 nm in length. Unlike the genus paramyxovirus the internal helical component is difficult to visualize and can be found only in short lengths and displaying a poorly defined substructure although a herringbone pattern can still be seen. The diameter of the helix is 12–15 nm.

Family Orthomyxoviridae

Genus

Influenza virus — Influenza types A and B.[42, 46]
Membrane-bounded particles of approximately 80 nm diameter but with considerable variation in size (Figure 1.9). Also filaments of approximately 70 nm diameter reaching lengths of several hundred nm. The surface of the virus envelope bears distinctive projections or spikes 10 nm in length. Degradation studies have shown that there are two morphologically different types of projection on the surface of the influenza virus: a rod-shaped subunit which is identified as the hemagglutinin, and a mushroom-shaped structure that represents the neuraminidase. On the intact virus these differences

Figure 1.8. Parainfluenza virus.
Two particles are visible. The upper particle still has a continuous limiting membrane although it has been penetrated by stain to reveal the internal helical RNP. The lower particle is disrupted and the helical component is spread on the grid. The diameter of such particles varies considerably but the majority will be in the 100–200 nm range. The RNP has a diameter of 18 nm. (× 300,000)

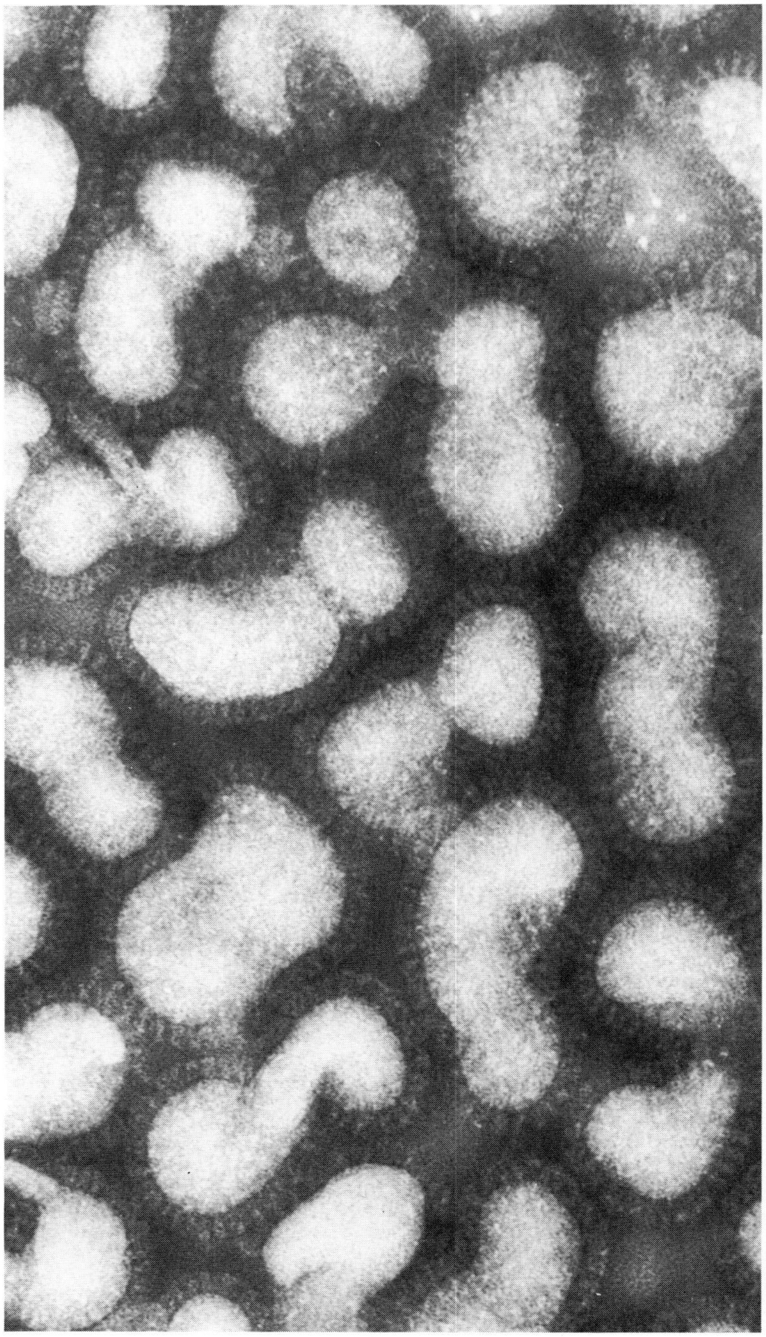

Figure 1.9. Influenza virus.
A typical field of virus particles. The virions are pleomorphic with distinctive 10 nm projections on the surface. No internal component is visible. (× 300,000)

cannot be resolved and the rod-shaped projection is the only one visualized. Unlike the paramyxoviruses, which are predominantly circular in outline, the most commonly occurring influenza particles are curved on one side and indented on the

(a) 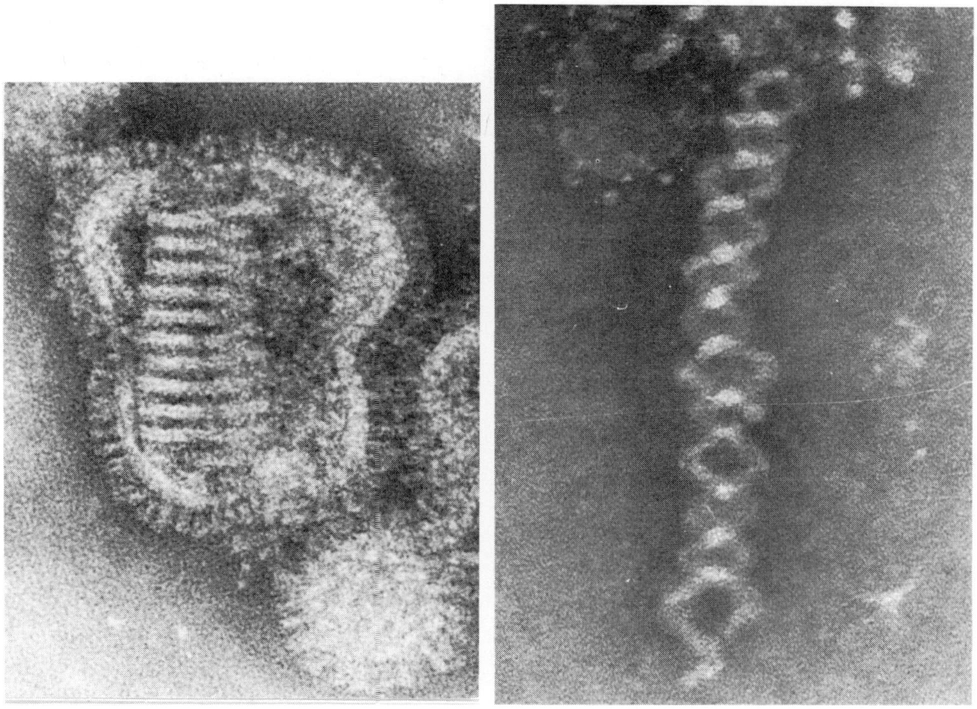 (b)

Figure 1.10. Internal component of influenza virus particle.
(a) One of the very occasional influenza particles showing an internal RNP component. This RNP is in the form of a strand 6 nm in diameter forming a helix approximately 50 nm diameter.
(b) Release of the internal component by detergent shows that it is a double helix. (× 300,000)

other. This pattern, taken with another frequently observed orientation, which shows the particles as having a central depression, suggests that the influenza virus is shaped in the form of a shallow bowl with very thick walls. The curved appearance would then be the particle in profile, while the indented pattern would be the 'bowl' lying flat.

Unlike the genus paramyxovirus, the envelope of the orthomyxoviridae is a stable one and the internal component is only rarely visualized.[8] The occasional particle penetrated by stain reveals an internal strand of 6 nm diameter tightly wound in a helical conformation. Unfortunately, if the particles are degraded by chemical means the internal component is seen as short linear fragments of 9 nm diameter which do not correspond to the component seen inside spontaneously disrupted virus. However, recent findings[2] have shown that if the influenza virus is degraded very gently it is possible to visualize an internal component arranged in the form of a double helix, the strand of the helix having a diameter of 6 nm (Figure 1.10(a) and (b)).

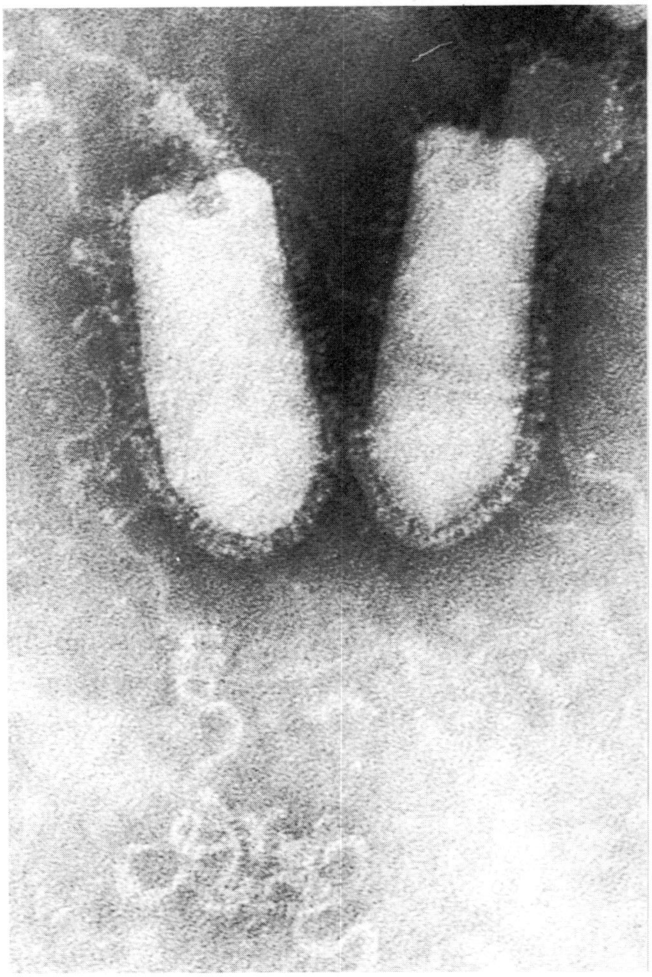

Figure 1.11. Vesicular stomatitis virus (VSV).
Two particles showing the bullet shape and surface projections. Some released helical RNP can be seen in the background of the micrograph. (× 300,000)

Family Rhabdoviridae

Genera

Vesiculovirus — Vesicular stomatitis virus.[45]
Bullet-shaped particles 60–80 × 130–220 nm having one rounded and one blunt end (Figure 1.11). The outer surface of the particles bears projections of 8–10 nm length. Removal of this lipoprotein envelope reveals a striated internal component and within this is the RNP packed in a helical fashion.

Lyssavirus — Rabies.[30]
Similar to vesicular stomatitis virus but having less distinctive surface projections. (Figure 1.12)

Unclassified helical virus

Marburg virus — This virus is sometimes erroneously grouped with the rhab-

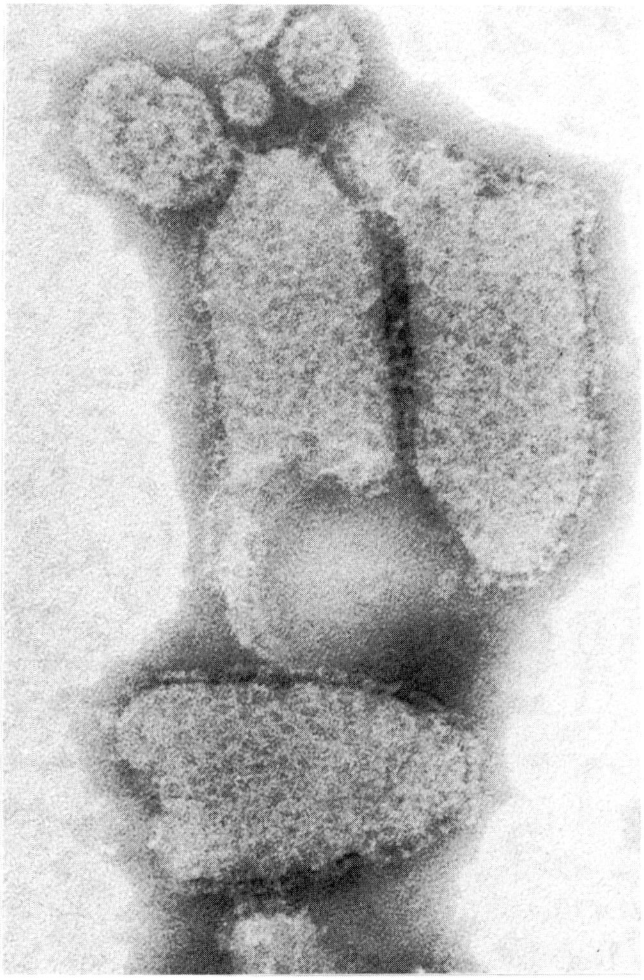

Figure 1.12. Rabies virus (C. Smales).
Three particles. The overall morphology is similar to VSV but the surface projections are less prominent. (× 300,000)

doviridae. However, although the virus has certain structural features in common with the rhabdoviruses, the overall form of the mature virion is not a linear bullet-shaped particle but a curved structure (Figure 1.13), sometimes appearing as a closed ring. Because of the presence of these doughnut-shaped forms, the name torovirus has been suggested.[16] Although not serologically identical the more recently encountered Ebola virus belongs in this group.[66]

1.4.2. Cubic viruses

Unlike viruses with helical symmetry which can contain only RNA, cubic viruses can be immediately divided into two groups, depending on the type of

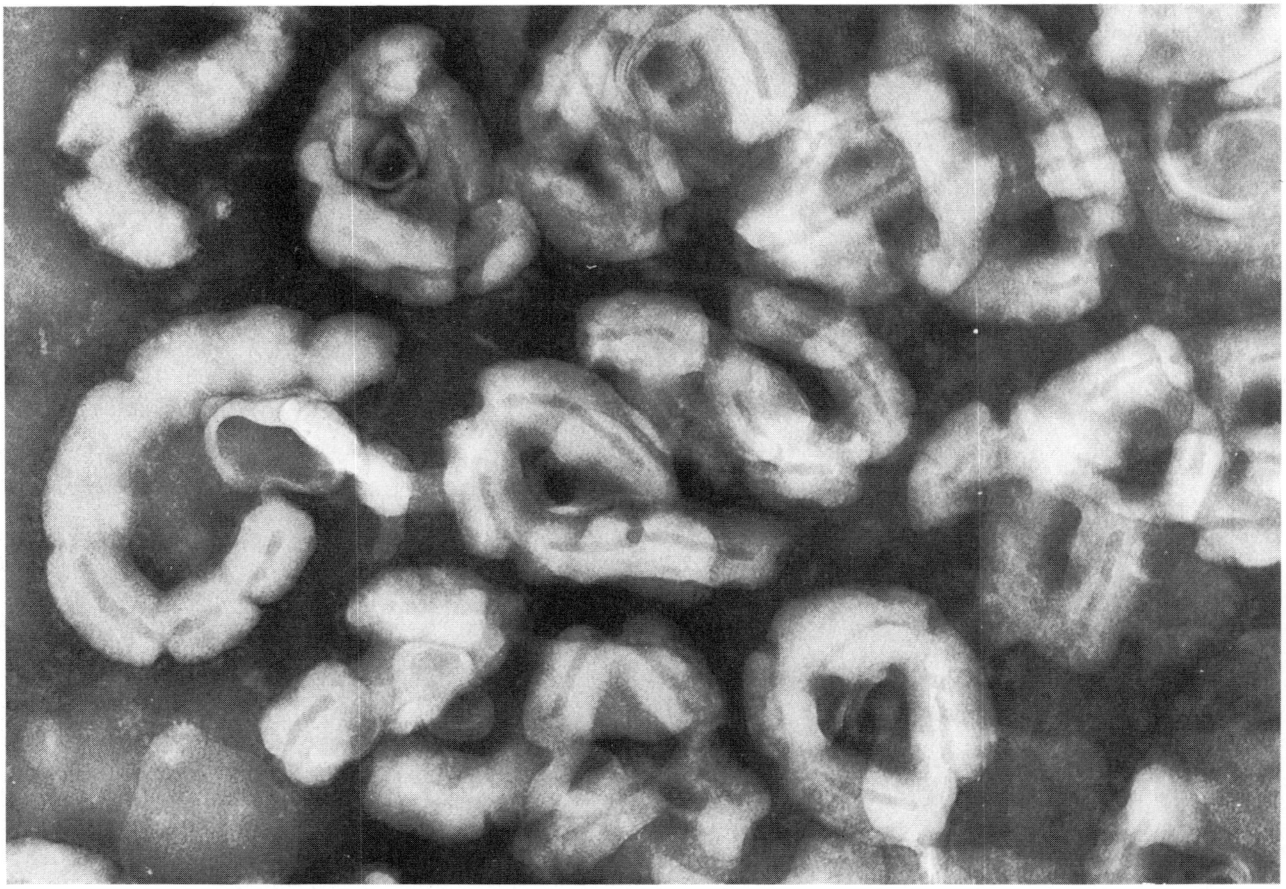

Figure 1.13. Marburg virus.
A group of the pleomorphic forms associated with the Marburg virus. The structures are linear with underlying helical symmetry but the mature particles appear curved, or comma shaped, and occasionally as closed rings. The basic filament is 100 nm in diameter with surface projections of 10 nm, and the length of the particles varies from 300 nm up to μm range. (× 120,000)

nucleic acid. The first to be considered are the RNA-containing cubic viruses, all of which are simple arrangements of nucleic acid and protein without the addition of a lipoprotein component. Hence all are ether resistant.

Cubic RNA viruses

Family Picornaviridae

Unfortunately, while the negative stains in current use allow visualization of this group, they have failed to reveal substructure in detail. Even the fact that these viruses have cubic symmetry derives from X-ray diffraction rather than electron microscopy.

Genera

Enterovirus — Poliovirus, ECHOvirus, Coxsackie virus.
Spherical particles of 27 nm diameter. The surface of the

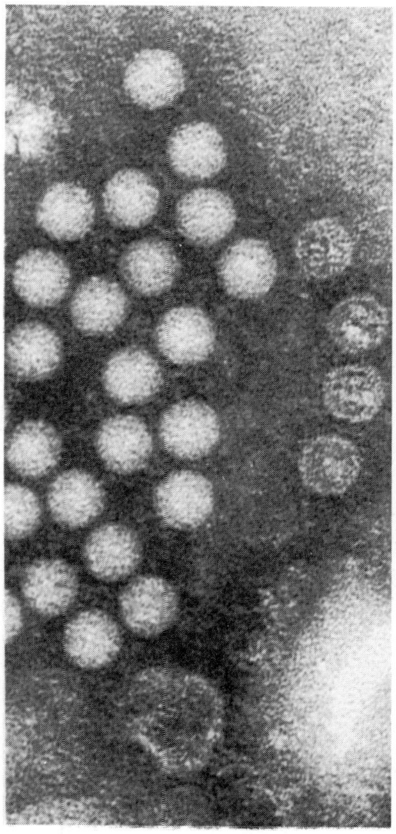

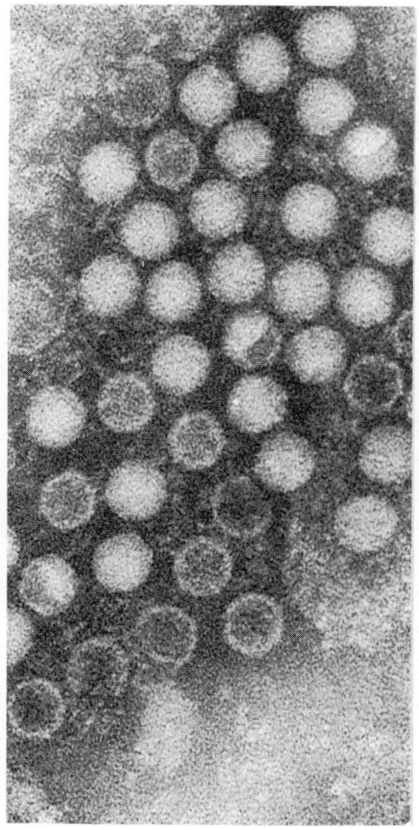

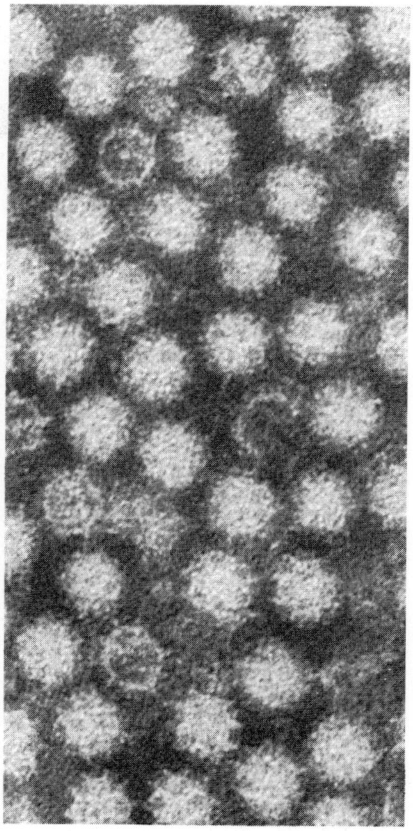

Figure 1.14. Poliomyelitis particles.
A group of particles aggregated by the addition of a small amount of specific antibody. The capsids of 27 nm diameter appear featureless and show either the white appearance of full particles on the left, or empty rim-stained forms on the right. Full and empty particles of polio virus differ antigenically and hence are segregated. (× 300,000)

Figure 1.15. Human rhinovirus.
A group of particles produced by immune electron microscopy (IEM). As for the enterovirus in Figure 1.14, full and empty forms occur and the capsids are approximately 27 nm diameter. Being pH labile this virus must be stained with phosphotungstic acid adjusted to pH 8. (× 300,000)

Figure 1.16. Feline respiratory virus.
Although included with the picornaviridae, caliciviruses show distinctive substructure by the negative staining technique. The capsids are 35 nm in diameter and appear to be built up from 32 negative capsomeres. (× 300,000)

particles reveals a fine, particulate appearance that suggests regular substructure, but further examination fails to elucidate an identifiable pattern (Figure 1.14). The particles show one of two types of morphology depending on whether or not they contain nucleic acid. Intact virions appear as round white discs as the stain is unable to penetrate to the interior. Damaged virus that has lost its nucleic acid allows the stain to penetrate and only the protein capsid is delineated. This means that the particles appear as rings in a transmission

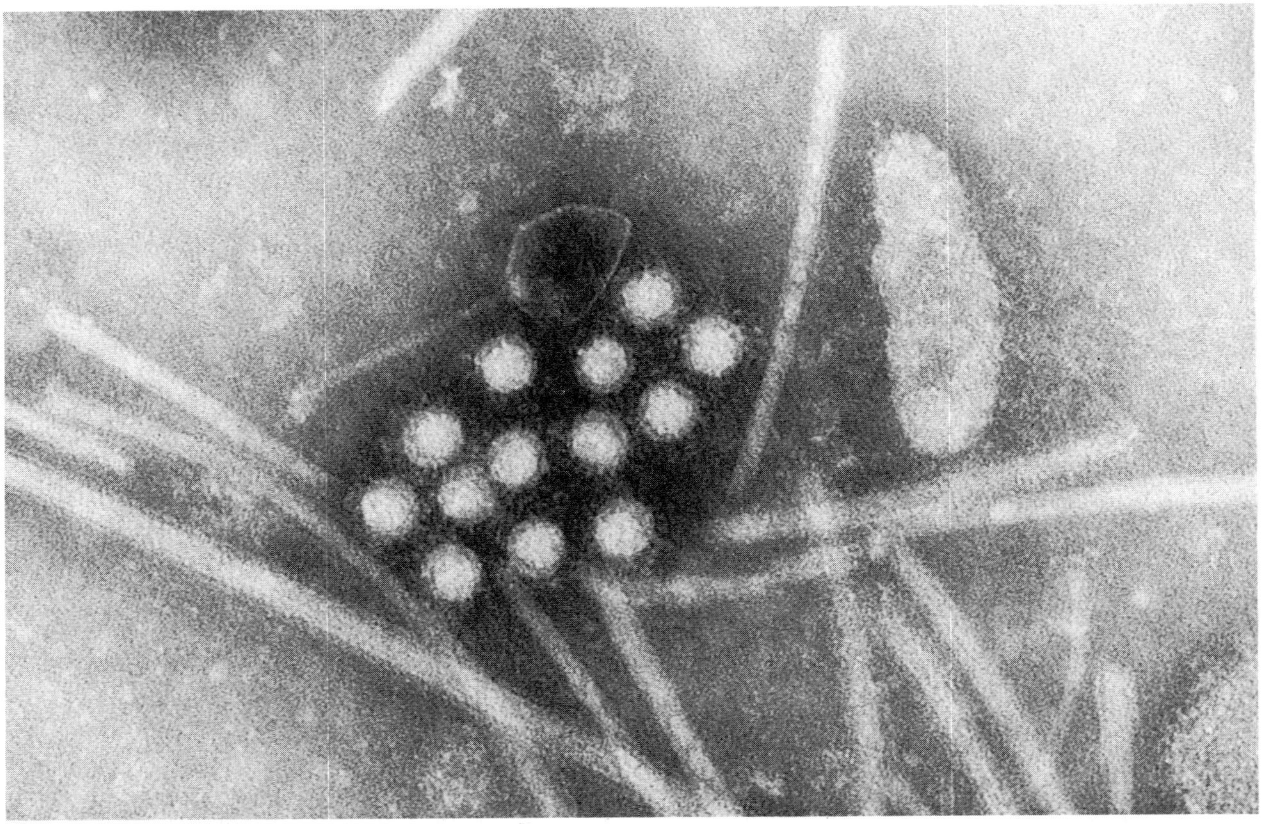

Figure 1.17. Hepatitis A virus.
A group of particles complexed by the addition of antibody (compare with Figure 1.5). The particles are similar to the enteroviruses although the capsids are slightly bigger, being in the size range 28–30 nm. In addition a small rim around the particles can sometimes be seen. (× 300,000)

image. This phenomenon is frequently described as 'rim' staining, and in a preparation of enterovirus the 'rim-stained' particles are usually the easiest to recognize.

Rhinovirus

More than 70 strains of human rhinovirus, animal rhinovirus. Morphologically the rhinoviruses exhibit the same characteristics as the enteroviruses (Figure 1.15), having a diameter of 27 nm and displaying both full and empty partices.[47] However, unlike the enteroviruses, the rhinoviruses are acid-labile and this is one of the few occasions when it is necessary to use phosphotungstic acid that has been adjusted to pH 8 rather than pH 6. It might be remarked that rhinoviruses are slightly, and perhaps subjectively, more delicate in appearance than enteroviruses.

Podoravirus

Foot and mouth virus.
Another pH-sensitive virus that must be stained using phosphotungstic acid at pH 8. Like the previous members of this family it has featureless, spherical particles displaying both

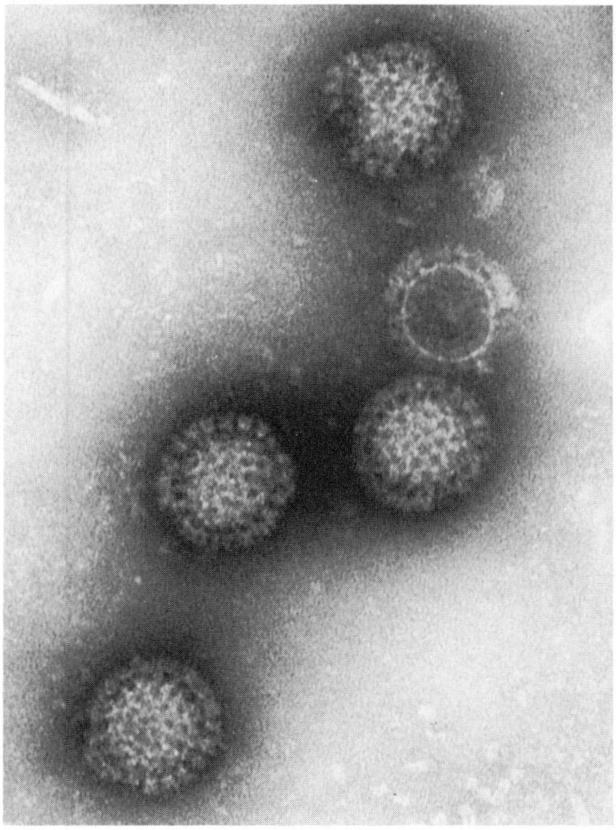

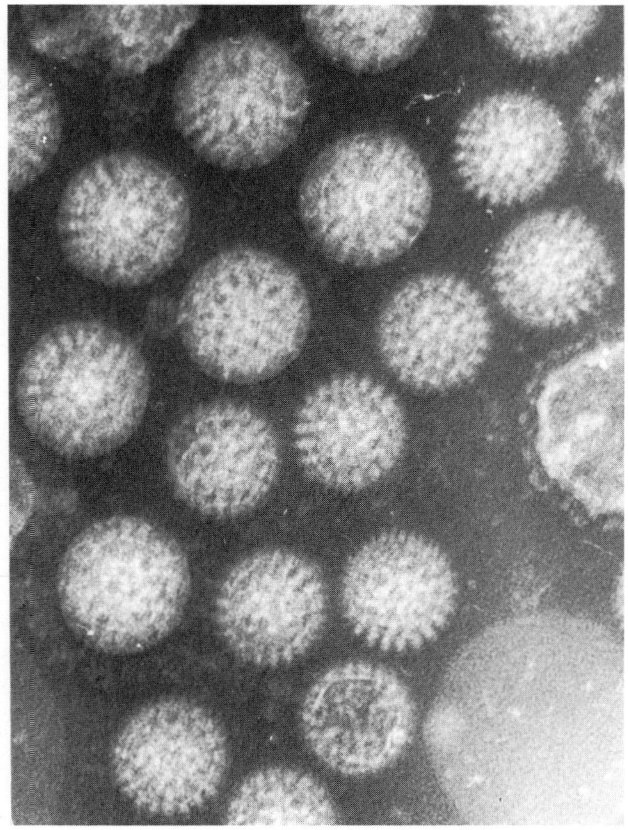

Figure 1.18. Reovirus.
Double capsid construction showing inner and outer capsids. The overall capsid measures 70–80 nm in diameter. Particular attention should be paid to the outer capsomeres which can be individually resolved. It is this feature that distinguishes the reoviruses from the rotaviruses. (× 300,00)

Figure 1.19. Rotavirus.
Like the reovirus this virus has a double capsid construction with an inner capsid of 55 nm and an outer of 65–70 nm diameter. The outer surface of the rotavirus is smooth, and individual subunits cannot be resolved. Specimen from calf feces. (× 300,000)

full and empty forms. However, foot and mouth disease virus is somewhat smaller than the groups previously described, the particles having an average diameter of 24 nm.[70]

Calicivirus

Vesicular exanthema of swine, infectious respiratory disease of cats.[15]

Although belonging to the family picornaviridae, caliciviruses are unlike the other genera so far described in that negative staining does reveal the arrangement of subunits forming the capsid. The capsid is 30–35 nm diameter, and the capsomeres are unusual in that they appear as depressions rather than the projections that occur with other viruses (Figure 1.16). This means that the arrangement of subunits appears as a pattern of dark areas which can be interpreted as forming a capsid of

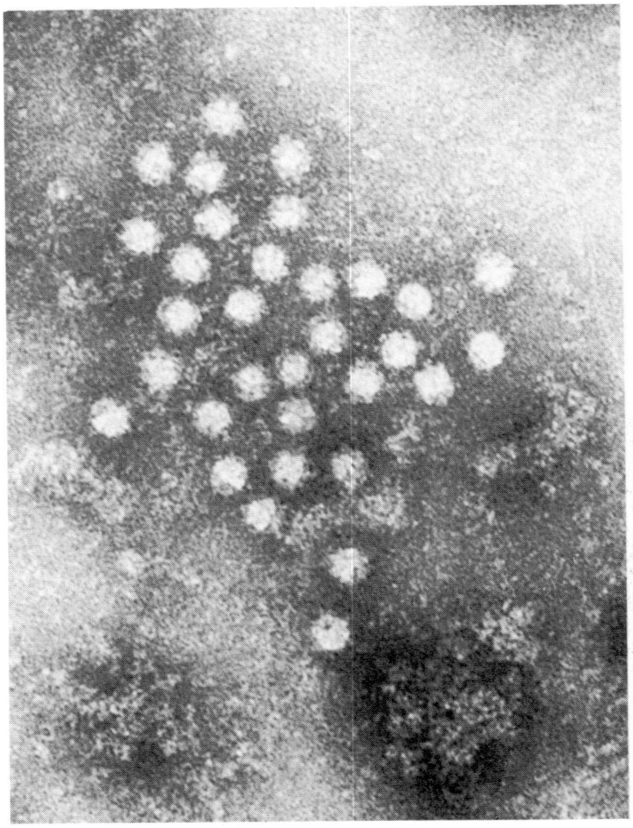

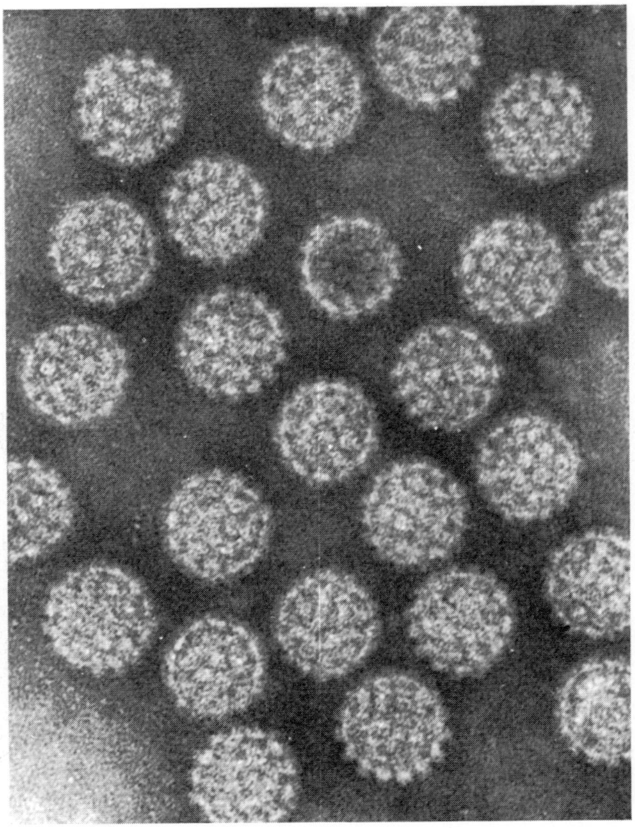

Figure 1.20. Parvovirus.
Satellite virus associated with the major virus of infectious bursal disease. Parvovirus (22 nm in diameter) is the smallest of the viruses illustrated in this chapter. Well-resolved particles can be seen to be built up of 12 subunits. It is a frequent contaminant of tissue cultures but difficult to identify morphologically without the use of IEM. (× 300,000)

Figure 1.21. Common wart virus.
A simple extract of either human or animal warts reveals the presence of the distinctive virus illustrated here. The majority of virions have a diameter of 55 nm. (× 300,000)

32 capsomeres. The name "negameres" has been suggested for these visually negative capsomeres.[15]

Unnamed genus

It is possible that yet another genus will be included in the picornaviridae since recent findings have shown that the virus associated with human hepatitis type A is cubic in form and is an RNA virus.[3, 31] The name "icterovirus" might be suggested for this genus. At present the only source of the virus is infectious feces and in order to visualize the virus it is necessary to carry out immune electron microscopy. This reveals that the virus has a diameter of 28–30 nm and a slightly fuzzy edge (Figure 1.17).

Family **Reoviridae**

Genera

Reovirus — Reovirus types 1–3.[69]
Distinctive, double-shelled capsids of overall diameter 70–80 nm (Figure 1.18). By negative staining the reovirus particles have been described as resembling lace doyleys as they display an open arrangement of capsomeres. The complete particle displays radiating subunits and the innner capsid can be distinguished within the outer capsid. Examination of a reovirus specimen invariably reveals the presence of incomplete particles consisting only of the inner capsid. The diameter of these inner capsids is 55 nm. The nucleic acid of the reovirus consists of 10 segments of double-stranded RNA.

Orbivirus — Blue tongue virus.[21, 58]
Orbiviruses are similar to reoviruses in some respects, including the possession of segmented double-stranded RNA. The members differ from reoviruses in size, ranging from 65–80 nm, and having an inner capsid of about 55 nm, revealed, for example, after centrifugation in casium chloride. Orbiviruses bear some morphological resemblance both to reoviruses and rotaviruses, but are distinguishable from both these groups.

Rotavirus[36, 51] — Gastroenteritis virus of numerous species including man, cattle, pigs, and mice (EDIM virus).
Double capsid construction similar to reovirus with an inner component of 55 nm and a complete virion of 60–66 nm (Figure 1.19). Rotavirus differs from reovirus inasmuch as the capsomeres of the outer capsid appear to form a continuous rim. This has led to the inner and outer capsids being termed rough and smooth, respectively.

The next group of viruses to be considered are those with cubic symmetry and DNA as their nucleic acid. Within this group a further criterion for classification is supplied by the presence or absence of an outer lipoprotein envelope.

Cubic DNA viruses without a lipoprotein component

Family **Parvoviridae**

Genus

Parvovirus — Human parvovirus, minute virus of mice, adeno-satellite virus,[72] small particle associated with Gumboro disease virus.
Small particles of buoyant density 1·43 and diameter 22 nm. Structural studies have shown that the particles are built up of 12 capsomeres. As with other cubic viruses, full and empty forms are found (Figure 1.20).

Family **Papovaviridae**

Genera

Papillomavirus — Human common wart virus,[82] papilloma viruses of various species.

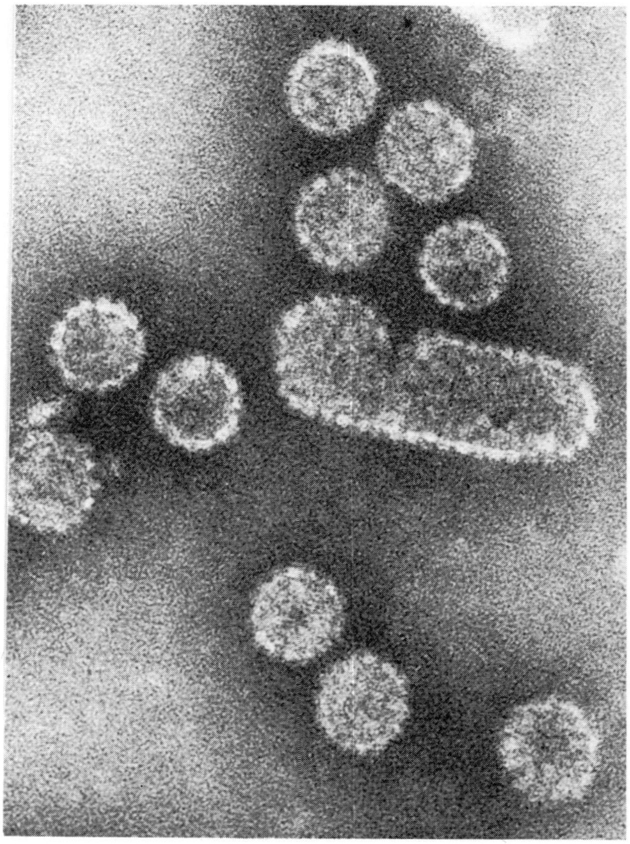

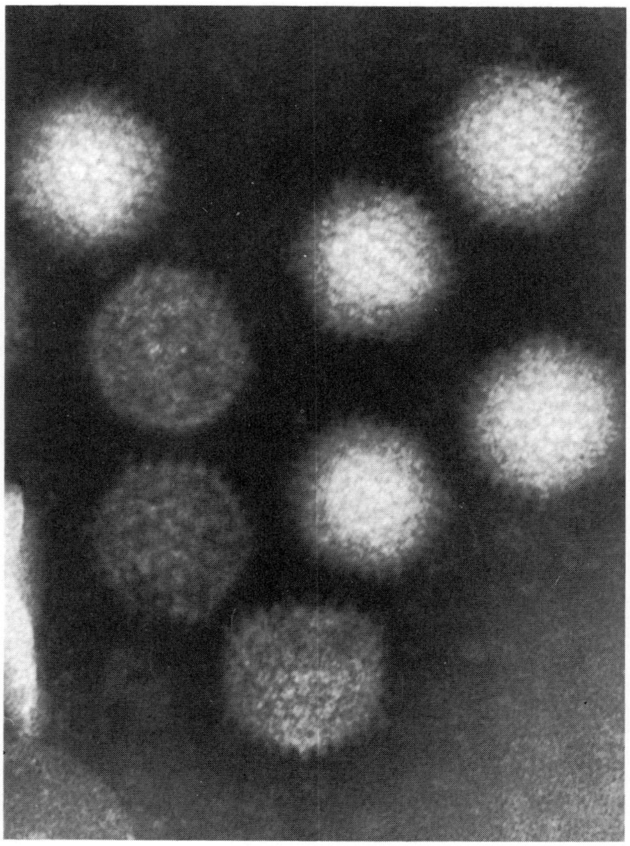

Figure 1.22. Polyomavirus.
This virus is similar to wart virus but 10 nm smaller, being 45 nm in diameter. The particles are also less robust in appearance than wart virus. Another morphological feature of the papovaviridae is shown by the tubule which is one of the many aberrant forms that can occur. (× 300,000)

Figure 1.23. Adenovirus particles.
A group of particles, showing both full and empty forms. Many display the triangular facets of the basic icosahedron and the subunits are clearly visible. The capsid is formed from 252 capsomeres. (× 300,000)

Spherical particles of 55 nm diameter with a distinctive subunit construction (Figure 1.21). Fine structure studies suggest that the wart virus capsid is built up from 72 capsomeres, but 42 and 92 capsomeres have also been suggested for this group.[53]

Polyomavirus[13,64,80] Polyoma, SV40, viruses of progressive multifocal leucoencephalopathy (PML).

Spherical particles of 45 nm diameter, again with well-visualized subunits (Figure 1.22). The overall appearance of the particles is less robust than that of the wart virus and the 10 nm difference in size is easily recognized. Aberrant particles, such as tubular forms, occur frequently in this family.

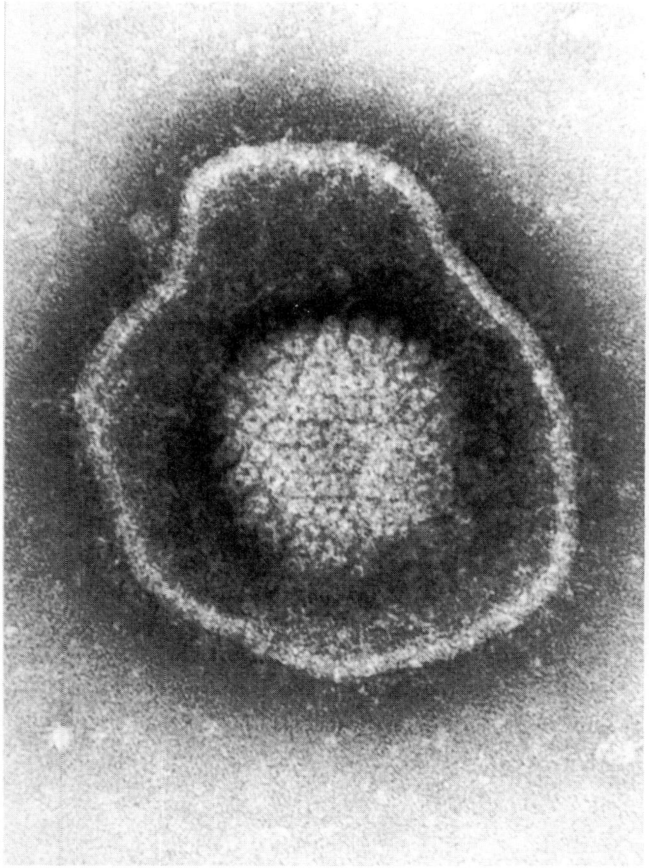

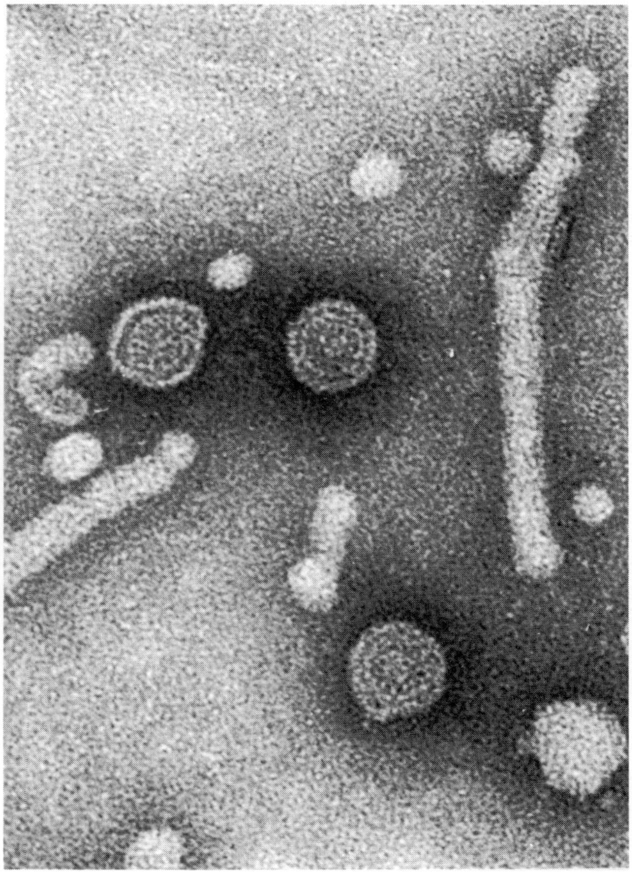

Figure 1.24. Herpesvirus.
In addition to the nucleic acid-containing protein capsid, these viruses have an additional cell-derived lipoprotein membrane. The internal capsid which is 100 nm in diameter is composed of 162 capsomeres and the lipoprotein membrane carries virally coded projections on the surface. (× 300,000)

Figure 1.25. Hepatitis B antigen.
A field showing the various morphological forms of hepatitis B antigen (HBAg). The three double shelled particles are the Dane particles now thought to be hepatitis B virus (HBV). The small spheres and the tubules consist only of the surface antigen (HBsAg) and do not contain nucleic acid. The Dane particles have HBsAg on the outer membrane and HBcAg (core) antigen on the inner capsid. (× 300,000)

Family Adenoviridae

Genera

Mastadenovirus — Adenoviruses of mammals.

Aviadenovirus — Adenoviruses of birds.

Both genera have outstanding hexagonal particles (Figure 1.23) of 80 nm diameter with distinctive and geometrically arranged capsomeres.[39, 74] Those particles that are orientated well as regards the electron beam allow the substructure arrangement of the adenovirus to be determined. Such particles can be seen to possess two types of structural subunits,

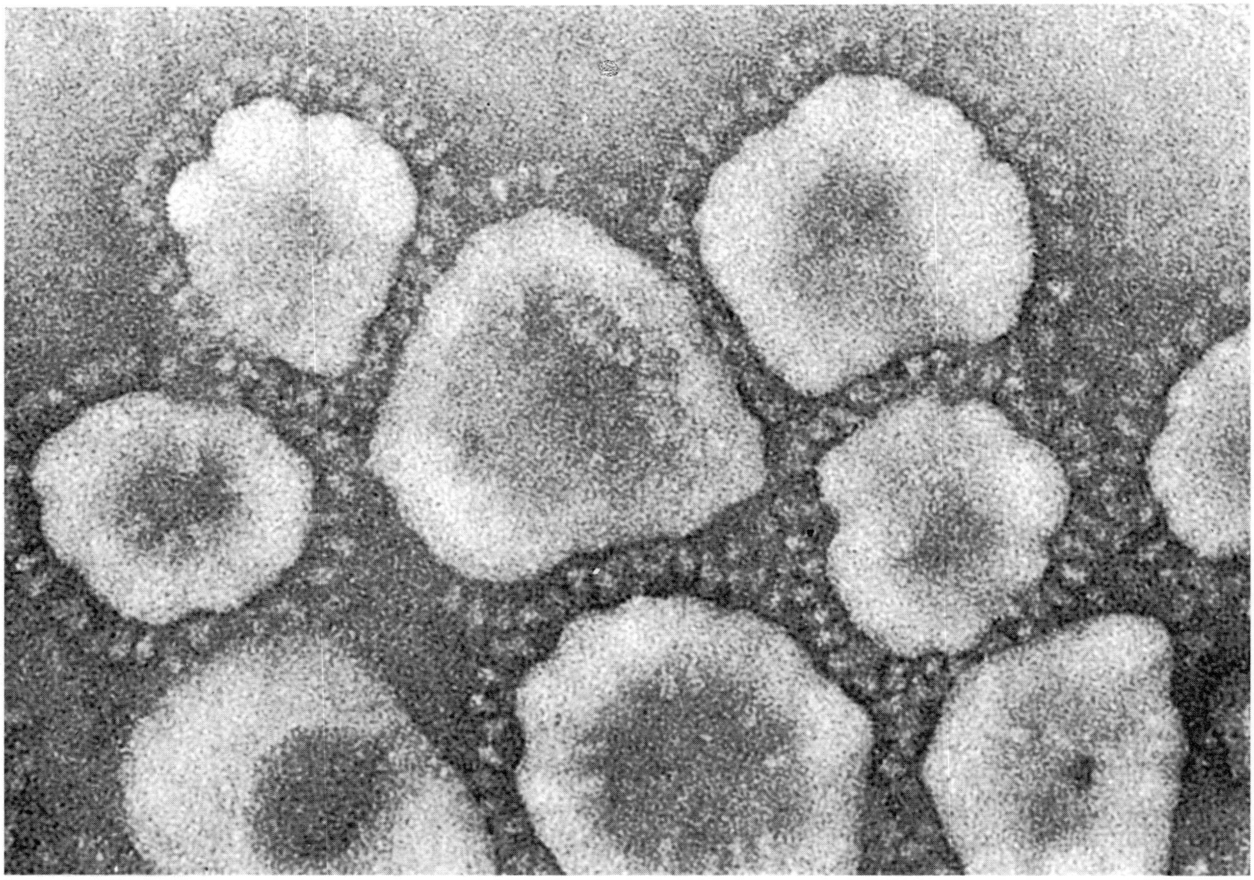

Figure 1.26. Avian infectious bronchitis.
Example of coronavirus showing pleomorphic sacs of approximately 100 nm diameter with distinctive 25 nm projections or peplomeres on the surface. (× 300,000)

those surrounded by five others, the pentons, and those surrounded by six others, the hexons. Pentons are located at the apices of the icosahedron, while hexons form the faces. From one penton to the next there are six subunits and application of the formula $10(n-1)^2 + 2$ gives the result that the adenovirus capsid is built up of 252 subunits.

Cubic DNA viruses with lipoprotein envelope

Family Herpetoviridae

Genus

Herpesvirus — Herpes simplex types I and II (herpesvirus hominis I and II), varicella-zoster virus, cytomegalovirus, Epstein-Barr (EB) virus, Marek's disease virus.
Hexagonal capsids of 100 nm diameter (Figure 1.24) that appear to flatten on the microscope grid and can be shown to have 162 capsomeres.[62, 81] Surrounding this nucleocapsid is a

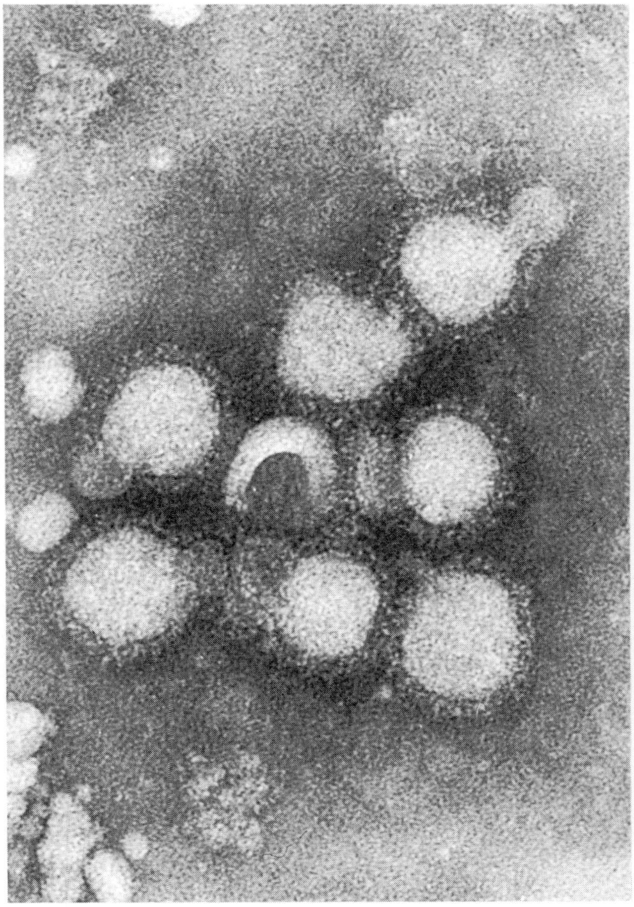

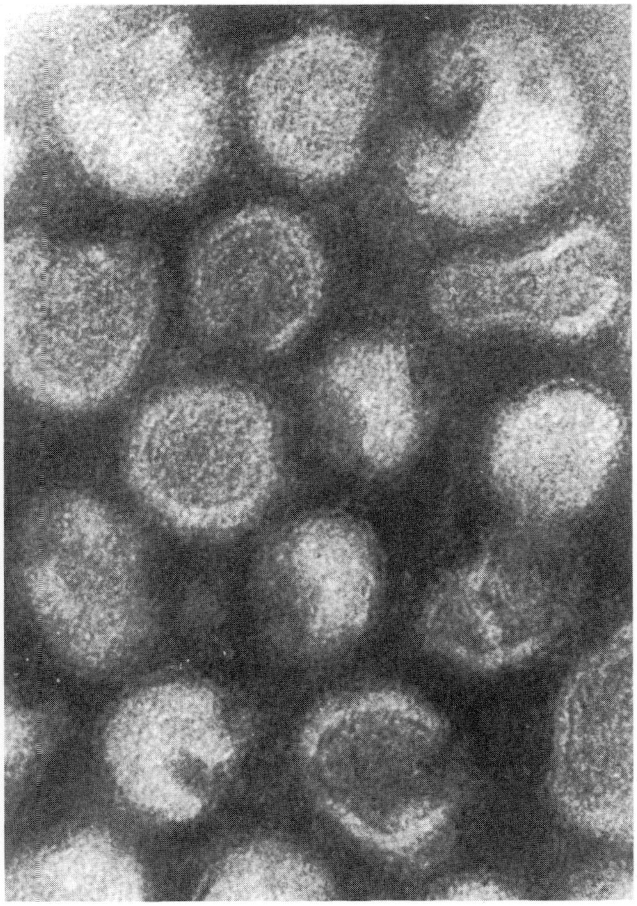

Figure 1.27. Sindbis virus.
Group A arboviruses (alphaviruses) are well illustrated by Sindbis virus. The particles consist of approximately spherical lipoprotein sacs with a halo of fine projections. (× 300,000)

Figure 1.28. Rubella virus.
Like Sindbis virus, rubella consists of approximately round lipoprotein sacs in the size range 60–80 nm. However, the togaviridae, to which rubella belongs, have only a small fringe of 5 nm on the surface of the particle and it is almost essential to emply immune electron microscopy in order to visualize the particles. (× 300,000)

cell-derived lipoprotein envelope that has fine projections of 12 nm on the surface. This membrane differs from the RNA helical viruses in that it is derived from the nuclear rather than the plasma membrane of the host cell. The most commonly encountered appearance of the herpes virion has been likened to a fried egg, with the capsid representing the yolk and the envelope spread out around it representing the white. In this form the diameter is variable and can be as large as 200 nm. However, it should be realized that the virus in this form is degraded and occasionally particles are seen in a more intact state. The envelope is then closely applied to the internal capsid and no structural detail is visible. The only morphological features of such particles are a markedly hexagonal or pentagonal outline and the fine projections on the surface. Particles of this type are a little over 100 nm in diameter.

Family Iridoviridae

Genus

Iridovirus — Tipila iridescent virus.
A large double-shelled icosahedral virus 250 nm in diameter having a proposed 812 capsomeres on the inner component.[68]

Another genus of this family will probably be needed for a number of animal viruses, the best known of which is African Swine fever,[14] which has a morphology similar to that of Tipula iridescent virus.

Unclassified DNA cubic virus with lipoprotein envelope

Hepatitis B virus (HBV) of man.[10]
Although not yet officially classified, hepatitis B virus would appear to belong within this virus group. The nucleic acid has been characterized as DNA and the internal core component of the Dane particle (HBcAg) is known to be synthesized in the nucleus. In addition, the Dane particle has the double-shelled construction associated with the herpesvirus group, although it is a much smaller virus, having an internal core of 27 nm diameter and an overall size of 42 nm (Figure 1.25). Unfortunately, although the Dane particle has the morphological characteristics of a virus it has not yet been shown to have biological activity. The other outstanding feature of hepatitis type B is the association of large amounts of non-nucleic acid-containing particles in association with the putative virus. These particles are formed from the same antigenic material as the outer covering of the Dane particle, and are known as hepatitis B surface antigen (HBsAg). Morphologically, this antigenic component is present in the form of pleomorphic spheres and rods having an average diameter of 20 nm and variable length (Figure 1.25).

Figure 1.29. Vaccinia virus.
Two particles are illustrated, the mature form of the virus being visible at bottom right. The dimensions are roughly 200 × 250 nm and the particle has overall rectangular form with a random pattern of ridges on the surface. The particle at the upper left of the micrograph is the immature form of the virus. It is larger than the mature particle and the surface appears smooth. Such particles have been penetrated by stain and reveal an internal membrane. (× 220,000)

Figure 1.30. Orf particles.
Three particles are visible. This virus and the virus of milkers' nodules (both pox viruses) exhibit a distinct helical arrangement of the surface ridges. They are 300 nm in length and approximately 150 nm in diameter. (× 220,000)

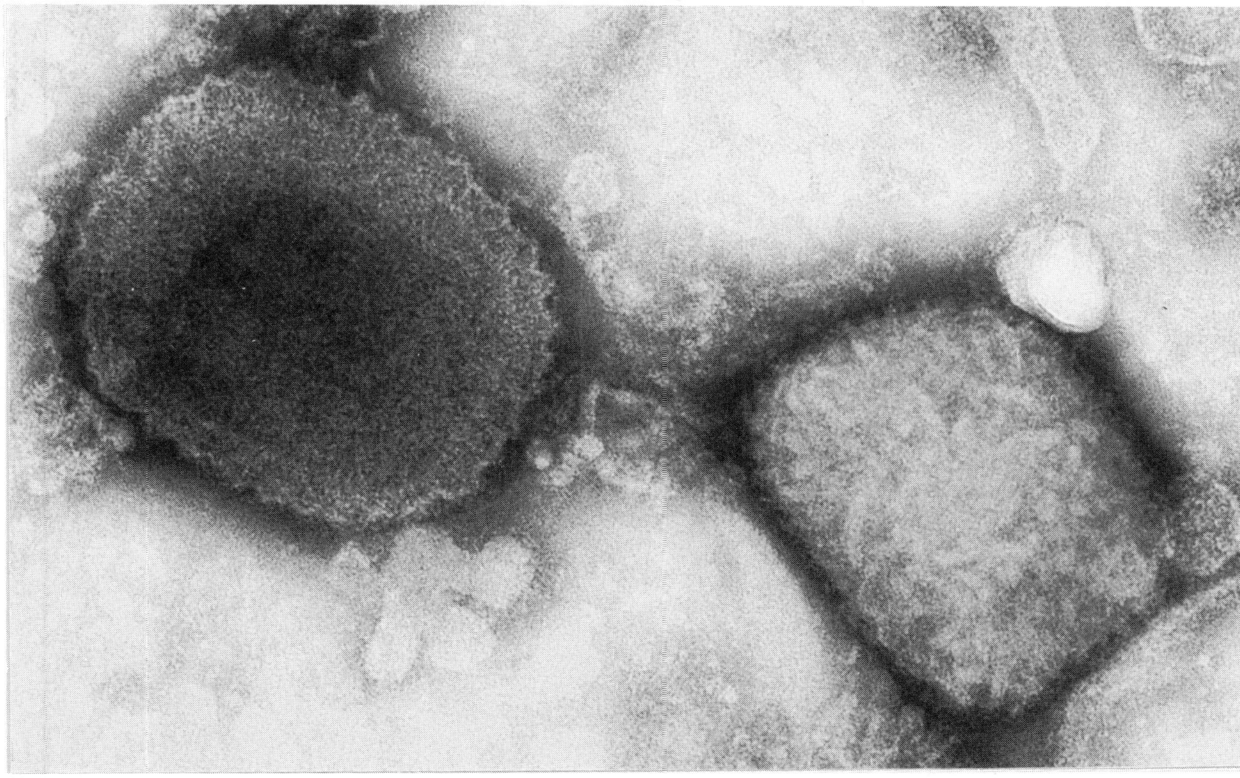

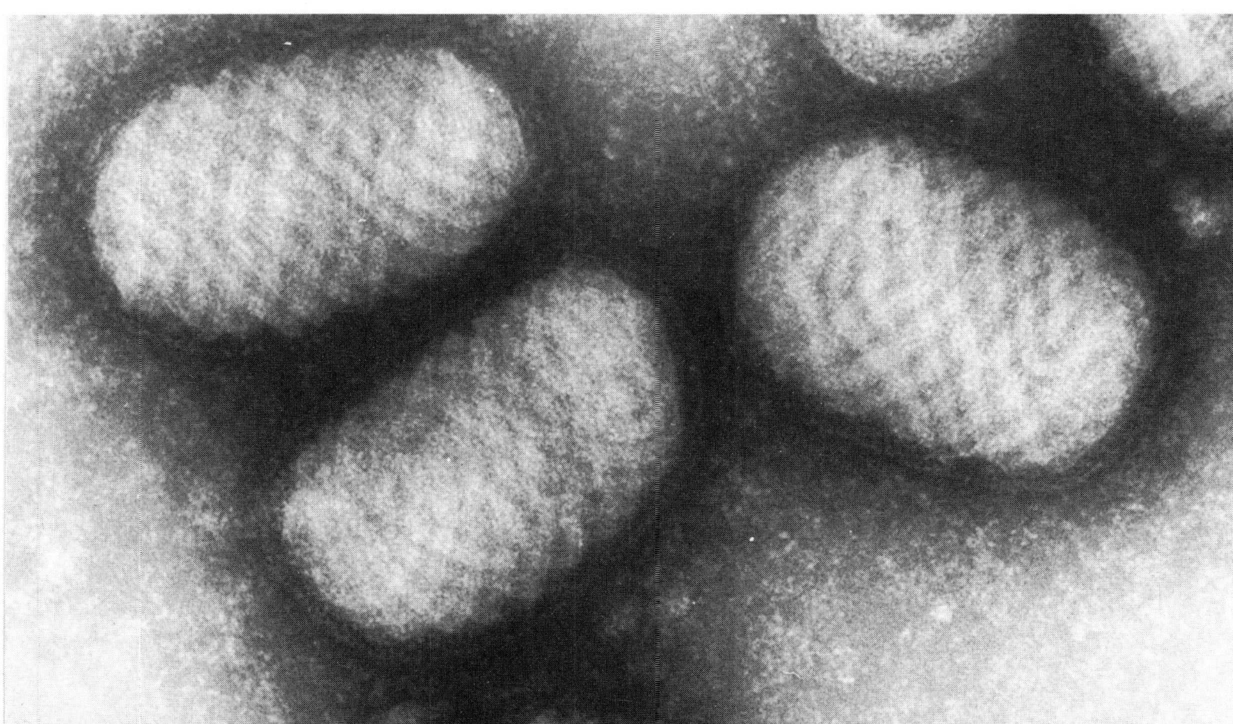

1.4.3. Nonsymmetrical RNA viruses

Family Coronaviridae

Genus

Coronavirus[57] — Several serotypes of coronavirus associated with human respiratory infection, avian infectious bronchitis, mouse hepatitis virus, transmissible gastroenteritis of pigs.

More or less spherical particles, 80–160 nm, frequently with a central depression giving an umbilicate appearance (Figure 1.26). The feature that has given this family its name is the halo of long projections that radiate from the surface. These projections are 25 nm in length and are leaf-shaped, being attached to the virus by a narrow, barely resolvable stem. Frequently the projections are missing, but in any preparation some of the particles at least will retain them and a specimen in which they cannot be seen is suspect. Recent studies have shown that there is a tongue-shaped internal component in the virion with no evidence of either cubic or helical symmetry.[20]

Family Togaviridae

This family contains arboviruses of groups A and B. Unfortunately, neither of these two groups is morphologically homogeneous and in some cases viruses are united only by the property of human infection combined with multiplication in an arthropod host. Therefore it has been decided to describe only those viruses that have been well-characterized and it must be accepted that others within the same group may have different morphological features.

Genera

Alphavirus — Sindbis virus.[67]

More or less spherical particles in the size range 60–70 nm with a halo of fine projections surrounding the membrane. The internal component is not well-characterized although a helix has been suggested (Figure 1.27).

Flavivirus — Yellow fever, Louping ill.

Rubivirus — Rubella.

Although flaviviruses are group B arboviruses and there is no real evidence to suggest that rubella virus is an arbovirus, virions of these two genera look alike and can be discussed together.

Structurally this is probably the least satisfactory of all the virus groups. The particles appear as nondescript membranous sacs in the size range 50–70 nm. Well-preserved particles display a narrow fringe of surface projections extending a mere 5 nm from the particle surface (Figure 1.28). This group possesses so few distinctive morphological features that it is almost mandatory that immune electron microscopy be used to identify the virus.[19]

1.4.4. Nonsymmetrical DNA viruses

Family Poxviridae

Genera

Orthopoxvirus — Variola (smallpox); Vaccinia, various poxviruses of birds and mammals.

Rectangular particles of 200 × 250 nm.[60, 79] Two different types of particle morphology can be found in most preparations and these are associated with mature and immature virions (Figure 1.29). The immature particles are slightly larger than the mature and are penetrated by stain so that they have a grey appearance. The overall outline of these particles is smooth and the main membrane is surrounded by an outer one that is frilled and more delicate. Mature particles appear light against the phosphotungstic background and are not penetrated by the stain. The surface of these particles appears wrinkled, with the heights of the ridges forming strands that cover the particles in a random pattern.

Parapoxvirus — Orf virus, and the virus of milkers' nodules.[60, 61]

Morphology is similar to that of the orthopoxviruses except that particles are longer, having dimensions of 200 × 300 nm, and the ridges described for orthopoxviruses here take on a distinctive helical pattern (Figure 1.30).

1.5. Results obtained from negative staining

The technique of negative staining has basically changed the attitude of virologists to the electron microscope. Before the introduction of the method in 1958 the electron miscroscope was a powerful tool for elucidating virological problems of academic importance, but since the thin sectioning techniques employed were both lengthy and fastidious they could not assist in the daily problems of clinical virology. A technique that could be carried out in minutes rather than days, and that could be used with crude material, was an advantage of major significance.[73] Negative staining was of particular value in dealing with viruses that could not be grown in culture. Shortly after the introduction of the method, the morphological characteristics of human common wart virus, molluscum contagiosum, and orf were established. Now, almost twenty years later, these viruses still cannot be grown *in vitro*, and the electron microscope remains the simplest means of identifying the causative virus of these epidermal lesions.

Similarly, two conditions of the central nervous system, progressive multifocal leukoencephalopathy (PML)[44, 86] and subacute sclerosing panencephalitis (SSPE),[71] were both recognized as viral conditions by means of electron microscopy, although in each case the initial clue came from the technique of thin sectioning, with later confirmation from negative staining. Biological studies of these two viruses, following EM recognition, were more successful and both can now be grown in the laboratory in cell systems.

Returning to viruses that cannot be grown *in vitro*, much of the information that has been obtained about the hepatitis B antigen came from electron microscopy of HBAg-positive sera, and it was the method of immune electron microscopy that identified the causative virus of human viral hepatitis A.

In the rotavirus field, the first association of human gastroenteritis with this virus group came from electron microscope studies. Subsequent examination of fecal specimens by direct electron microscopy has shown that rotaviruses are associated with gastroenteritis in a wide range of species.

The technique of negative staining is now almost twenty years old. Unlike many new approaches that start with a great amount of enthusiasm and then sink slowly into obscurity, this method has gained popularity as the years proceed. Proof of this can be gleaned from the fact that the majority of new viruses found in the last decade have been given names based on their morphological appearance, for example, coronavirus and rotavirus. The extension of immune electron microscopy has widened the scope of the method even further. Drawbacks of the method are that it requires capital outlay for an expensive item of equipment, the electron microscope, and the need for skilled personnel in running it. Obviously, not every virus laboratory can bear this expense, but at present there are few virologists who cannot gain access to a microscope or find someone willing to examine a specimen.

References to Part One

1. Agnarsdottir, G., "Subacute sclerosing panencephalitis," *Recent Adv. clin. Virol.*, **1**, 21–50, 1976.
2. Almeida, J. D. and Brand, C. M., "A morphological study of the internal component of influenza virus," *J. gen. Virol.*, **27**, 313–318, 1975.
3. ——, Deinhardt, F., and Zuckerman, A. J., "Virus-like particles in hepatitis A-positive faecal extracts," *Lancet*, **ii**, 1083–84, 1974.
4. —— and Howatson, A. F., "A negative staining method for cell-associated virus," *J. Cell Biol.*, **16**, 616–620, 1963.
5. —— and Morris, R., "Antigenically-related viruses associated with infectious bursal disease," *J. gen. Virol.*, **20**, 369–375, 1973.
6. —— and Waterson, A. P., "The morphology of virus-antibody interaction," *Adv. Virus. Res.*, **15**, 307–338, 1969.
7. —— and ——, "Immune complexes in hepatitis," *Lancet*, **ii**, 983–986, 1969.
8. —— and ——, "Two morphological aspects of influenza virus," in R. D. Barry and B. W. Mahy (eds.), *The Biology of Large RNA Viruses*, Academic Press, London, 1970, pp. 27–51.
9. —— and ——, "Some implications of a morphologically orientated classification of viruses," *Arch. ges. Virusforsch.*, **32**, 66–72, 1970.
10. —— and ——, "Hepatitis B antigen – an incomplete history," *Am. J. med. Sci.*, **270**, 105–114, 1975.
11. ——, ——, and Drewe, J. A., "A morphological comparison of Bittner and influenza viruses," *J. Hyg., Camb.*, **65**, 467–474, 1967.
12. ——, ——, and Fletcher, E. W. L., "Simulation of viral sub-structure by radiography of models," *Nature, Lond.*, **206**, 1125–1127, 1965.
13. ——, ——, and —— "Interpretation of structural detail in viruses of the polyoma group," *Prog. exp. Tumor Res.*, **8**, 95–124, 1966.
14. ——, ——, and Plowright, W., "The morphological characteristics of African swine fever virus and its resemblance to Tipula iridescent virus," *Arch. ges. Virusforsch.*, **20**, 392–395, 1967.
15. ——, ——, Prydie, J., and Fletcher, E. W. L., "The structure of a feline picornavirus and its relevance to cubic viruses in general," *Arch. ges. Virusforsch.*, **25**, 105–114, 1968.
16. ——, ——, and Simpson, D. I. H., "Morphology and morphogenesis of the Marburg agent," in G. A. Martini and R. Siegert (eds.), *Marburg Virus Disease*, Springer, Berlin, 1971, pp. 84–97.
17. —— and Zuckerman, A. J., "The importance of being rotavirus," *Nature, Lond.*, **260**, 393, 1976.
18. Berry, D. M. and Almeida, J. D., "The morphological and biological

effects of various antisera on avian infectious bronchitis virus," *J. gen. Virol.*, **3**, 97–102, 1968.

19. Best, J. M., Banatvala, J. E., Almeida, J. D., and Waterson, A. P., "Morphological characteristics of rubella virus," *Lancet*, **ii**, 237–239, 1967.
20. Bingham, R. and Almeida, J. D., "Morphological studies on the internal component of avian infectious bronchitis virus," *J. gen. Virol.*, **36**, 495–502, 1977.
21. Borden, E. C., Shope, R. E., and Murphy, F. A., "Physicochemical and morphological relationships of some arthropod-borne viruses to bluetongue virus – a new taxonomic group. Physicochemical and serological studies," *J. gen. Virol.*, **13**, 261–271, 1971.
22. Brenner, S. and Horne, R. W., "A negative staining method for high resolution electron microscropy of viruses," *Biochim. biophys. Acta*, **34**, 103–110, 1959.
23. Cameron, C. H. and Dane, D. S., "Radioimmunoassay and saturation analysis," *Br. med. Bull.*, **30**, 90–92, 1974.
24. Caspar, D. L. D., Dulbecco, R., Klug, A., Lwoff, A., Stoker, M. G. P., Tournier, P., and Wildy, P., "Proposals," *Cold Spring Harb. Symp. quant. Biol.*, **27**, 49–50, 1962.
25. Cayzer, I., Dane, D. S., Cameron, C. H., and Denning, J. V., "A rapid haemagglutination test for hepatitis B antigen," *Lancet*, **i**, 947–949, 1974.
26. Craighead, J. E., Kanich, R. E., and Almeida, J. D., "Nonviral microbodies with viral antigenicity produced in cytomegalovirus-infected cells," *J. Virol.*, **10**, 766–775, 1972.
27. Crick, F. H. C. and Watson, J. D., "Structure of small viruses," *Nature, Lond.*, **177**, 473–475, 1956.
28. —— and ——, in G. E. W. Wolstenholme and E. C. P. Miller (eds.), *The Nature of Viruses*, Churchill, London, 1957, pp. 5–13.
29. Dane, D. S., Cameron, C. H., and Briggs, M., "Virus-like particles in serum of patients with Australia antigen-associated hepatitis," *Lancet*, **i**, 695–698, 1970.
30. Davies, M. C., Englert, M. E., Sharpless, G. R., and Cabbasso, V. D., "The electron microscopy of rabies virus in cultures of chicken embryo tissues," *Virology*, **21**, 642–651, 1963.
31. Feinstone, S. M., Kapikian, A. Z., and Purcell, R. H., "Hepatitis A: detection by immune electron microscopy of a virus-like antigen associated with acute illness," *Science*, **182**, 1026–1028, 1973.
32. Fenner, F., "The classification and nomenclature of viruses," *J. gen. Virol.*, **31**, 463–470, 1976.
33. Fernelius, A. L., Ritchie, A. E., Classick, L. G., Norman, J. O., and Mebus, C. A., "Cell culture adaptation and propagation of a reovirus-like agent of calf diarrhoea from a field outbreak in Nebraska," *Arch. ges. Virusforsch.*, **37**, 114–130, 1972.
34. Flewett, T. H., Bryden, A. S., and Davies, H., "Virus particles in gastroenteritis," *Lancet*, **ii**, 1497, 1973.
35. ——, ——, and ——, "Diagnostic electron microscopy of faeces. I.

The viral flora of the faeces as seen by electron microscopy," *J. clin. Path.*, **27**, 603–608, 1974.

36. ———, ———, ———, Woode, G. N., Bridger, J. C., and Derrick, J. M., "Relation between viruses from acute gastroenteritis of children and newborn calves," *Lancet*, **ii**, 61–63, 1974.
37. Glauert, A. M. and Phillips, R., in D. Kay (ed.), *Techniques for Electron Microscopy*, Blackwell, Oxford, 1961, pp. 194–228.
38. Gordon, C. W., Donnelly, J. D., Fothergill, R., Ker, F. L., Millar, E. L. M., Flewett, T. H., Bedson, M. S., and Cruickshank, J. G., "Variola minor: a preliminary report from the Birmingham hospital region," *Lancet*, **i**, 1311–1313, 1966.
39. Horne, R. W., Brenner, S., Waterson, A. P., and Wildy, P., "The icosahedral form of an adenovirus," *J. molec. Biol.*, **1**, 84–86, 1959.
40. ———, Russell, G., and Trim, A. R., "High resolution electron microscopy of beet yellows virus filaments," *J. molec. Biol.*, **1**, 234–236, 1959.
41. ———, and Waterson, A. P., "A helical structure in mumps, Newcastle disease and Sendai viruses," *J. molec. Biol.*, **2**, 75–77. 1960.
42. ———, ———, Wildy, P., and Farnham, A. E., "The structure and composition of the myxoviruses. I. Electron microscope studies of myxovirus particles by negative staining techniques," *Virology*, **11**, 79–98, 1960.
43. Howatson, A. F., "Architecture and development of tumor viruses and their relation to viruses in general," *Fed. Proc.*, **21**, 947–953, 1962.
44. ———, Nagai, M., and Zu Rhein, G. M., "Polyomalike virions in human demyelinating brain disease," *Can. med. Ass. J.*, **93**, 379–386, 1965.
45. ——— and Whitmore, G. F., "The development and structure of vesicular stomatitis virus," *Virology*, **16**, 466–478, 1962.
46. Hoyle, L., Horne, R. W., and Waterson, A. P., "The structure and composition of the myxoviruses. II. Components released from the influenza virus particle by ether," *Virology*, **13**, 448–459, 1961.
47. Kapikian, A. Z., Almeida, J. D., and Stott, E. J., "Immune electron microscopy of rhinoviruses," *J. Virol.*, **10**, 142–146, 1972.
48. ———, Dienstag, J. L., and Purcell, R. H. "Immune electron microscopy as a method for the detection, identification and characterisation of agents not cultivable in an *in vitro* system," *Manual of Clinical Immunology*, American Society for Microbiology, 1976, pp. 467–480.
49. ———, Kim, H. W., Wyatt, R. G., Rodriguez, W. J., Ross, S., Cline, W. L., Parrott, R. H., and Chanock, R. M., "Reovirus-like agent in stools: association with infantile diarrhoea and development of serologic tests," *Science*, **185**, 1049–1053, 1974.
50. ——— *et al.*, "Recent advances in the aetiology of viral gastroenteritis," in *Acute Diarrhoea in Childhood*, Ciba Foundation Symposium, 1976, pp. 273–309,
51. ——— *et al.*, "Human reovirus-like agent as the major pathogen associated with 'winter' gastroenteritis in hospitalized infants and young children," *New Engl. J. Med.*, **294**, 965–972, 1976.
52. Kaschmeyer, R. H. and Meyers, R. J., in M. A. Hayat (ed.), *Principles and Techniques of Electron Microscopy*, vol. 2, *Biological Applications*, Van

Nostrand Rheinhold, London, 1972, chap. 3.

53. Klug, A., "Structure of viruses of the papilloma-polyoma type," *J. molec. Biol.*, **II**, 424–431, 1965.
54. ——— and Caspar, D. L. D., "The structure of small viruses," *Adv. Virus Res.*, **7**, 225–235, 1960.
55. Lancaster, J. E. and Alexander, D. J., *Newcastle Disease Virus and Spread*, monograph no. 11, Canada Department of Agriculture, Ottowa, 1975.
56. Longson, M. and Bailey, A. S., "Herpes encephalitis," *Recent Adv. clin. Virol.*, **1**, 1–20, 1976.
57. McIntosh, K., "Coronaviruses: a comparative review," *Curr. Top. Microbiol. Immun.*, **63**, 86–129, 1974.
58. Murphy, F. A., Borden, E. C., Shope, R. E., and Harrison, A., "Physicochemical and morphological relationships of some arthropod-borne viruses to blue-tongue virus – a new taxonomic group. Electron microscope studies," *J. gen. Virol.*, **13**, 273–288, 1971.
59. Nagington, J., "Electron microscopy in differential diagnosis of poxvirus infections," *Br. med. J.*, **2**, 1499–1500, 1964.
60. ——— and Horne, R. W., "Morphological studies of orf and vaccinia viruses," *Virology*, **16**, 248–260, 1962.
61. ———, Newton, A. A., and Horne, R. W., "The structure of orf virus," *Virology*, **23**, 461–472, 1964.
62. O'Callaghan, D. J. and Randall, C. C., "Molecular anatomy of herpesvirus: recent studies," *Prog. med. Virol.*, **22**, 152–210, 1976.
63. Oriel, J. D. and Almeida, J. D., "Demonstration of virus particles in human genital warts," *Br. J. vener. Dis.*, **46**, 37–42, 1970.
64. Padgett, B. L. and Walker, D. L., "New human papovaviruses," *Prog. med. Virol.*, **22**, 1–35, 1976.
65. Peters, D., Nielsen, G., and Bayer, M. E., "Variola: die Zuverlässigkeit der elektronenmikroskopischen Schnell-diagnostik," *Dt. med. Wschr.*, **87**, 2240–2246, 1962.
66. Simpson, D. I. H. and Zuckerman, A. J., "Marburg and Ebola: viruses in search of a relation," *Nature, Lond.*, **266**, 217–218, 1977.
67. Simpson, R. W. and Hauser, R. E., "Basic structure of group A arbovirus strains Middelburg, Sindbis and Semliki Forest examined by negative staining," *Virology*, **34**, 358–361, 1968.
68. Smith, K. M. and Hills, E. J., "Further studies on the electron microscopy of the tipula iridescent virus," *J. Mol. Biol.*, **1**, 227–280, 1959.
69. Stanley, N. F., "Reovirus," *Br. med. Bull.*, **23**, 150–154, 1967.
70. Talbot, P. and Brown, F., "A model for foot-and-mouth disease virus," *J. gen. Virol.*, **15**, 163–170, 1972.
71. Tellez-Nagel, I. and Harter, D. H., "Subacute sclerosing leucoencephalitis: ultrastructure of intranuclear and intracytoplasmic inclusions," *Science*, **154**, 899–901, 1966.
72. Tinsley, T. W. and Longworth, J. F., "Parvoviruses," *J. gen. Virol.*, **20**, 71–75, 1973.
73. Tyrrell, D. A. J. and Almeida, J. D., "Direct electron microscopy of organ cultures for the detection and characterization of viruses," *Arch. ges.*

Virusforsch., **22**, 417–425, 1967.
74. Valentine, R. C. and Pereira, H. G., "Antigens and structure of the adenoviruses," *J. molec. Biol.*, **13**, 13–20, 1965.
75. Van Rooyen, C. E. and Scott, G. D., "Smallpox diagnosis with special reference to electron microscopy," *Can. J. publ. Hlth*, **39**, 467–477, 1948.
76. Waterson, A. P., "Measles virus," *Arch. ges. Virusforsch.*, **16**, 57–80, 1965.
77. ———, "Measles," *Proc. R. Soc. Med.*, **67**, 1109–1112, 1974.
78. ———, "Infectious particles in hepatitis," *A. Rev. Med.*, **27**, 23–35, 1976.
79. Westwood, J. C. N., Harris, W. J., Zwartouw, H. T., Titmuss, D. H. J., and Appleyard, G., "Studies on the structure of vaccinia virus," *Jnl. gen. Microbiol.*, **34**, 67–78, 1964.
80. Wildy, P., Stoker, M. G. P., Macpherson, I. A., and Horne, R. W., "The fine structure of polyoma virus," *Virology*, **11**, 444–457, 1960.
81. ———, Russell, W. C., and Horne, R. W., "The morphology of herpes-virus," *Virology*, **12**, 204–222, 1960.
82. Williams, M. G., Howatson, A. F., and Almeida, J. D., "Morphological characterization of the virus of the human common wart (verruca vulgaris)," *Nature, Lond.*, **189**, 895–897, 1961.
83. Williams, R. C. and Wyckoff, R. W. G., "Applications of metallic shadow-casting to microscopy," *J. appl. Physiol.*, 1723–33, 1946.
84. Woode, G. N. and Bridger, J. C., "Viral enteritis of calves," *Vet. Rec.*, **96**, 85–88, 1975.
85. Zakstelskaya, L. Y., Almeida, J. D., and Bradstreet, C. M. P., "The morphological characterisation of respiratory syncytial virus by a simple electron microscope technique," *Acta virol.*, **11**, 420–423, 1967.
86. Zu Rhein, G. M. and Chou, S-M., "Particles resembling papovaviruses in human cerebral demyelinating disease," *Science*, **148**, 1477–1479, 1965.

Part Two

Bacteriophages

2. Bacteriophages

Elinor Meynell

2.1. Introduction

The role of electron microscopy in phage research has been determined by the nature of phage itself. Like all viruses, phages depend for their continued existence on a host, here a bacterium; as genetic adjuncts of bacteria, they are distinguished from similar agents, like plasmids, by protein coats which envelope their genome and which are essential for maximal infectivity. Thus, there are two principal structures to which electron microscopy can be applied: to the phage particle or its protein coat, in whole or in part; and to the nucleic acid constituting the phage genome, either within the virion or in its replicating form within the cell. Electron microscopy has been used with the primary object of providing morphological descriptions and as a method of experimental analysis guided by genetics and biochemistry.

The problems of interpretation encountered in analysis of phage structure are those found with other viruses, and the same theoretical considerations and techniques apply (see Chapter 1).[14, 16] Similarly, direct electron microscopy of phage DNA, for example, to identify its mode of replication,[29] or the use of denaturation mapping to identify its structure,[28, 53, 54] can be applied to all DNAs. The same is true of heteroduplex analysis.[18] However, genetic analysis in phage has been developed to the point where it can present a view of viral morphogenesis as a process depending exclusively on structural components; the control of this process is therefore inherent in the process of assembly itself.[13] Furthermore, the physical map of the DNA molecule revealed by heteroduplex analysis can be correlated precisely with the genetic map.[39] Genetic analysis has been based largely on the isolation of conditional lethal phage mutants, either temperature-sensitive (ts) where the phage is normal within a narrow temperature range, outside which the mutation is expressed; or suppressor-sensitive (sus) where the mutation is expressed in some hosts but not in others which carry the appropriate suppressor gene.[20]

A property peculiar to phage, as distinct from other viruses, is its ability to penetrate the bacterial cell wall. Entry occurs in a complex manner specific for each type of phage, whose host range is primarily determined by complementary sites on phage and bacterium, which electron microscopy or biochemical analysis have identified successfully in many systems.

2.2. The virion

Phages, like other viruses, are classified initially by the morphology of the virion (summarized in Figure 2.1) and the type of nucleic acid. With the exception of the lipid-containing phages[23] and *Mycoplasma* phage,[11] no new classes of particle have been reported since the review by Bradley,[8] although there are numerous variations within most classes, in size, relative dimensions of head and tail, and presence of additional tail structures.[24] Since the capsid incorporates the mechanism by which infection occurs, a particular morphology is necessarily associated with the manner in which the nucleic acid penetrates the bacterial envelope. With tailed particles containing double-stranded DNA, the virion most frequently found in nature, the nucleic acid of the particle enters the bacterium through the tail, which forms the organ of attachment. The recognized exceptions simultaneously affect morphology and type of nucleic acid, with corresponding differences in the mechanism of infection. Thus, phage ØX174, which contains single-stranded DNA, adsorbs by means of one of twelve short spikes protruding from each of the twelve vertices of the icosahedral particles.[6] The spike protein acts as a pilot protein, accompanying the DNA to its intracellular site of replication and initiating its conversion to the double-stranded replicative form.[31, 32] Filamentous phages also contain single-stranded DNA and small isometric phages contain RNA. Many, if not all, members of both classes adsorb to receptors genetically determined by plasmids, which microscopy shows to be, in many cases, the filamentous appendages known as sex pili.[48]

Capsids, for which a detailed morphology has been identified by electron microscopy and by X-ray diffraction analysis, include filamentous DNA phages[44, 45] and isometric RNA phages,[50, 64] as well as tailed phages such as lambda[46] and T4.[16] With lambda and T4, there are two structures, the head and the tail, to consider as well as the head precursor, the prohead, formed before incorporation of the DNA. Morphological subunits are difficult to distinguish by negative staining in the phage head, which has a relatively smooth surface. Polyheads, which are long tubular assemblies of head subunits, provide a more extensive repeating pattern than normal heads for diffraction analysis,[10, 13] and

Figure 2.1. Morphological classes of phage particle (R. Newsam).
(a)–(d) Phages with contractile tails. (a) T-even phage with extended tail and tail fibres. (× 200,000) (b) T-even phage with contracted tail. (× 140,000) (c) Phage P1 small head variant.[61] (× 155,000) (d) Phage for *Erwinia* sp. (× 165,000)
(e)–(g) Phages with long noncontractile tails. (e) Phage T5 showing appearance of such tails characteristic of empty particles.[46] (× 200,000) (f) Flagellar phage with extended tail fiber beside a flagellum (see p. 58). (× 165,000) (g) The same with tail fiber wound round a flagellum. (× 165,000)
(h) P22-like phage showing baseplate with pins. (× 160,000)
(i) Phage T7, a short-tailed phage. (× 200,000)
(j),(k) Phage ØX174, showing spikes at vertices. ((j) × 95,000; (k) × 280,000)
(l),(m) Isometric phage MS2 attached to an F-like sex pilus. ((l) × 95,000, (m) × 240,000)
(n) A segment about one half the length of filamentous phage Ike. (× 125,000)

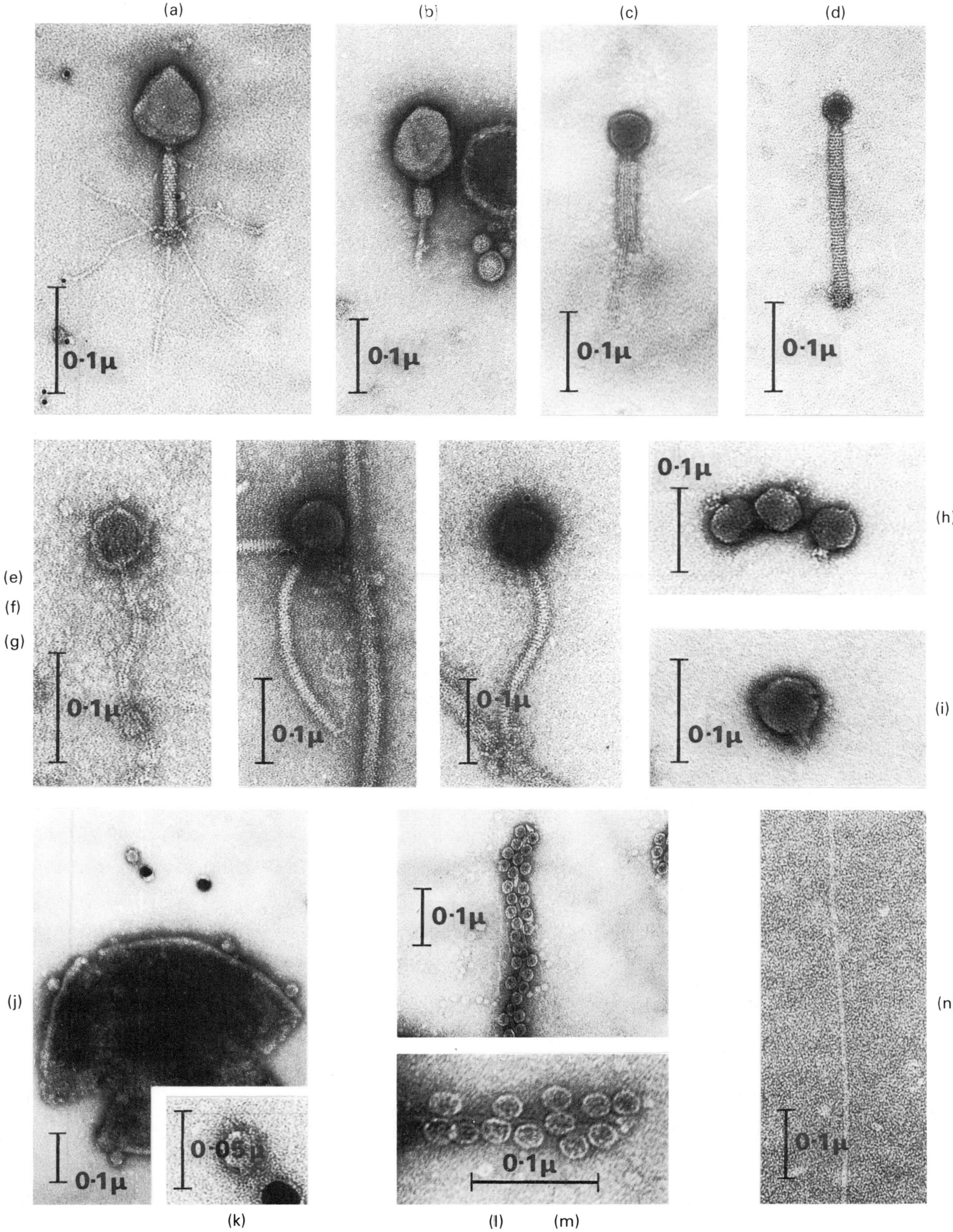
(a)
(b)
(c)
(d)
(e)
(f)
(g)
(h)
(i)
(j)
(k)
(l)
(m)
(n)
0·1μ
0·05μ

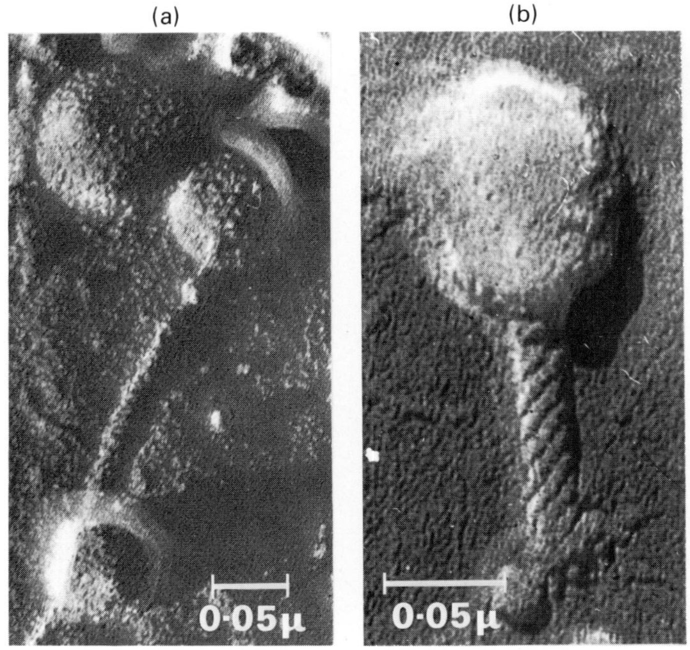

Figure 2.2. Phage particles by freeze-etching (M. Bayer).
(a) Heads of phages T2 and lambda. The individual capsomeres of the lambda head (the smaller of the two) appear as cylinders with a slight depression, in contrast to those of phage T2 which appear as rings.
(b) Tail of phage T4.

the type of structural rearrangement of capsid subunits occurring during the transition between prohead and mature head has been identified using the corresponding types of polyhead.[16, 26]

The problems associated with superposition of images at several levels as in negative staining do not apply to freeze-etching which shows directly the arrangement of capsomeres (Figure 2.2).[4, 5, 6, 10] Furthermore, the direction of turn in the helical phage tail is shown unambiguously (Figure 2.2). The image of the cylindrical shape of polyheads is preserved and comparison of the patterns on the convex surface seen by freeze-etching with the flattened surface seen by negative staining indicates that bonding between capsid subunits occurs nearest the interior, to leave an exterior consisting of a lattice formed by unbonded protrusions.[56]

Perhaps the least complex example of virus reproduction is represented by the filamentous single-stranded DNA phages, where no organized capsid intermediates accumulate but each individual chromosome of single-stranded DNA acquires its capsid as it is extruded from the cell[44] from protein subunits present in the inner bacterial membrane.[59] Electron microscopy shows two phage-related structures: intracellular DNA complexed with a protein which is not transmitted to the virion,[51] and recognizable virions being extruded from the bacterium. At the opposite extreme of complexity are the highly differentiated tailed phages, such as lambda and T4, and it is these phages, in whose reproduction many intermediates can be distinguished, that have provided the material for studies on the fundamental mechanism of viral morphogenesis.

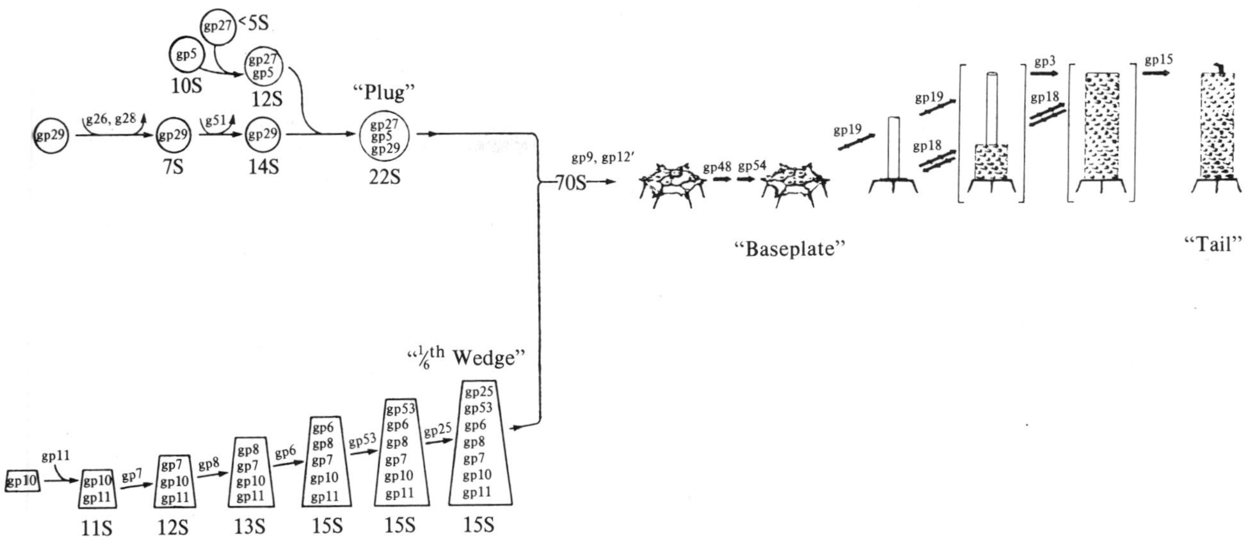

Figure 2.3. The pathway of phage T4 tail assembly (J. King).
Gene product (gp): protein identified as the product of the gene of that number. Intermediates that can be visualized in the electron microscope are shown as they appear. Other intermediates are shown schematically.

2.2.1. Viral morphogenesis

One of the most important questions in relation to assembly of virus particles is how far it is controlled specifically, or how far assembly of preformed protein molecules progresses spontaneously without further regulation. Formulated another way, the question is the level at which genetic control of assembly is exerted. Investigations with phage indicate that control operates at the level of assembly, since, by identifying the product of each gene known to affect morphology, the determining factor has been found to be simply the provision of the structural component. Assembly is sequential, not because the individual components are synthesized sequentially, but because of the necessity for each molecule to become part of the growing complex before it becomes capable of combining with another. Thus, in the construction of even the most complex morphological structure, such as the tail of phage T4, chosen here as the example, assembly is a process following the same basic rules as relatively simple structures like bacterial flagella.[2] The essential difference is that the final product may result from convergence of several pathways.

The analysis of sequential steps in assembly depends essentially on two properties of the phage which enable morphogenesis to be studied with naturally occurring precursors and not merely components of disaggregated virions. Phage precursors, which assemble *in vitro*, allow complementation to be detected between nonallelic phage mutants upon mixing their individual lysates[19] and structural intermediates, which accumulate when assembly is blocked by mutation, are identifiable morphologically. The sequential assembly of the tail of phage T4 is presented schematically in Figure 2.3.[38]

Briefly, the tail of phage T4 consists of a hollow tube, surrounded by a

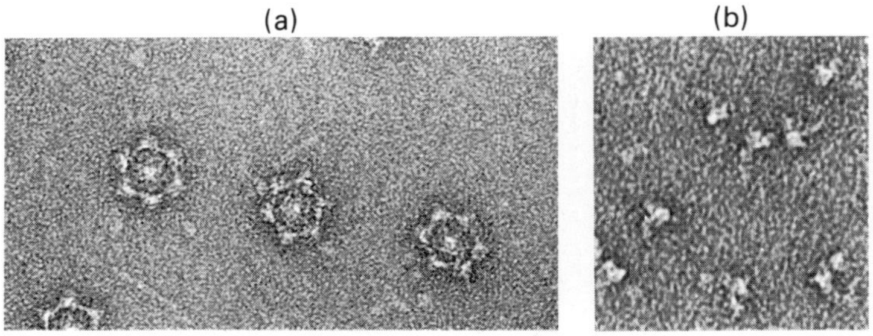

Figure 2.4. Tail baseplate structures.
(a) Baseplate of phage T4. (× 250,000)
(b) Comma-shaped baseplate subunit. (× 300,000)
Calibration bar: 50nm.

sheath connected at its proximal end to the phage head and at its distal end to a base plate which is hexagonal in axial view (Figure 2.4). The tube is composed of only one major protein, the product of a single gene. It assembles on the base plate, which must be completed beforehand, and its assembly is followed by assembly of the sheath (similarly composed of only one species of protein, the product of a single gene) which is stabilized after addition of the site for attachment of the head.[40] The base plate is the component of greatest complexity and at least 18 of the 21 genes involved in tail assembly are concerned with its formation.[36] The finished base plate has a spike at each apex and 6 evenly spaced comma-shaped subunits linking the rim to a 22S central plug. Each of the comma-shaped subunits exists before assembly as a 15S structure, identifiable by electron microscopy (Figure 2.4), formed by sequential assembly of 5 different proteins. A sixth protein converts them to a form in which they can polymerize. The genes specifying these 6 proteins are located in the first of 2 clusters of base plate genes. The second cluster is predominantly responsible for the central plug,[37] which is the product of 2 converging pathways and is required to stabilize the completed base plate. When mutations in "plug" genes are expressed, only occasional morphological structures are observed which not only lack a dense center but are abnormal in shape.[37]

The distal face of the base plate carries 6 long tail fibers responsible for specific attachment of phage to sensitive bacteria, as well as 6 tail pins and 6 short tail fibers (Figure 2.5),[35] whose independence of one another is proved by each being determined by a different gene.[34] The morphogenetic analysis of the long tail fibers, each bent at an angle about its midpoint and requiring the products of 6 genes,[13] illustrates the application of specific antibody to locate individual gene products in a structure where no morphological distinctions can otherwise be made. Specific antibody is produced by raising antiserum against normal whole phage and is absorbed either with mutant phage or with recombinant phage lacking the product of the gene under test. The fibers to be analyzed are then allowed to react with the residual specific antibody and examined by electron microscopy after negative staining. In this way, antigens located in either the distal or the proximal portions of the tail fibers are equated with the products of known genes.[63] Furthermore, the method could detect recombination within gene 37, which specifies a single molecule of a protein

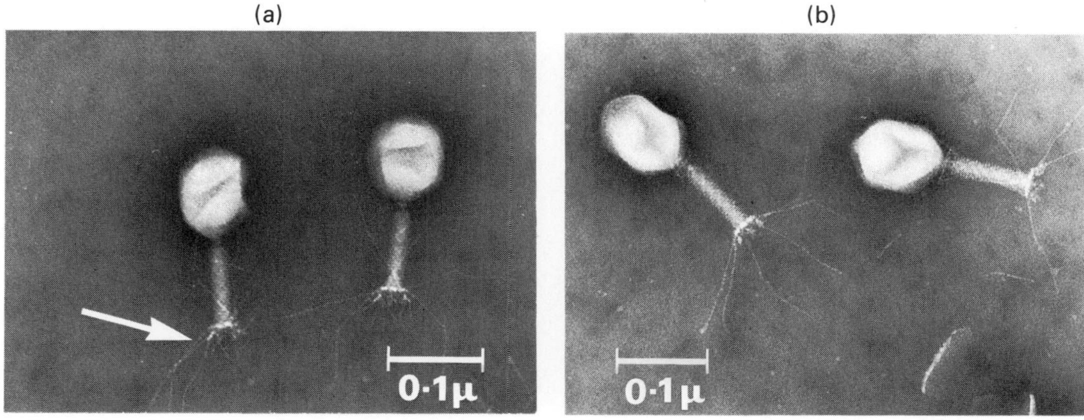

Figure 2.5. Tail fibers and tail pins (R. Haselkorn).
(a) Phage T4 with short tail fibers (indicated by arrow).
(b) The same without short tail fibers, but with long tail fibers and tail pins.

aligned along the fiber providing an array of different antigenic determinants. In recombinants between phage T4 and the related phage T2, the genetic contribution of each parental phage to the structure of this protein was reflected in the distribution of specific antibodies along the fiber (Figure 2.6). From the point of view of the evolutionary development of phages T2 and T4, it is interesting that their antigenic differences are almost exclusively confined to the distal segment of the protein, in agreement with their differences in host range which depends on whether or not the distal ends of the fibers have the appropriate specificity. Within a single gene, therefore, regions that have undergone major DNA base sequence changes during evolutionary divergence of T2 from T4 can be distinguished from regions that have been largely conserved.[7]

2.3. Heteroduplex phage DNA molecules

Heteroduplex analysis allows the presence and distribution of complementary base sequences to be recognized in double-stranded DNA molecules formed by hybridizing single strands obtained from different DNAs by denaturation.[18] The physical map thus obtained can be precisely correlated with the genetic map for T-even phages[39] (Figure 2.7) and phage lambda,[22, 58] amongst others.

Double-stranded and single-stranded DNA are distinguishable by electron microscopy and mark regions of homology and nonhomology, respectively, in the heteroduplexes. The lengths of both can be measured under appropriate conditions with reference to internal standards.[18] Nonhomologous regions take two basic forms. Deletion loops, which consist of only one strand of single-stranded DNA, occur where a difference between strands results either from a deletion in one strand or from an insertion of additional DNA in the other, such as results from a cryptic prophage[25] or from bacterial DNA in transducing phage.[22] Substitution loops occur where there are nonhomologous sequences in

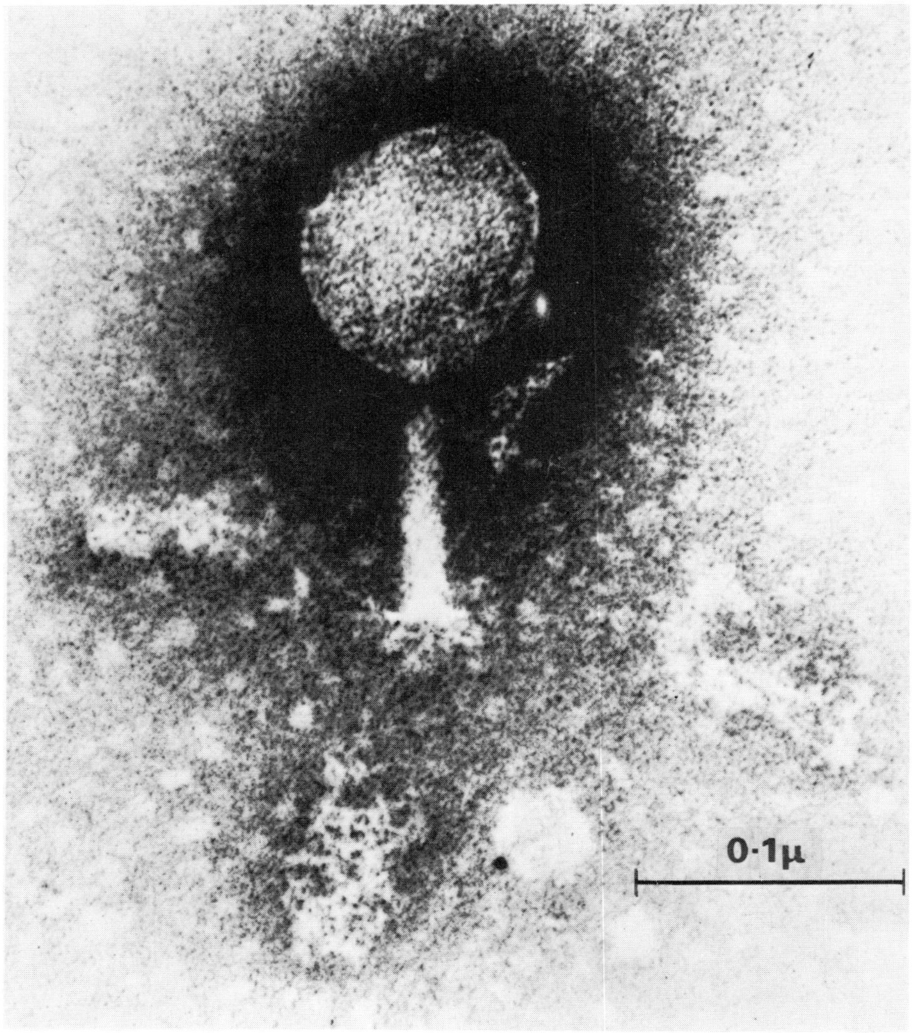

Figure 2.6. Location of specific gene products by antibody attachment (S. Beckendorf).
Phage T4 with homologous antibody attached to the distal halves of the tail fibers.

both strands; both strands then show single-stranded regions, not necessarily of equal length. Figure 2.7(a) includes examples of each of these basic forms, the positions of which have been genetically located, as shown in Figure 2.7(b). In addition, the entire molecule provides a visual indication of the way in which mature T-even phage DNA molecules are formed. As previously inferred from genetic and sedimentation data, the mature molecules are excised from a concatameric replicating form in lengths slightly exceeding the length of a single genome.[10, 60] The two original DNA molecules which provided the individual strands for the heteroduplex must each have been a circular permutation of the other. Therefore, the resultant heteroduplex molecule is circular with two single-strand branches, each corresponding to one end of each original strand.

Such branches are not always seen at their original site owing to secondary rearrangements within the molecule known as branch migration.[39]

A further example of the application of heteroduplex analysis to phage provides physical confirmation for the two broad types of evolutionary divergence inferred from genetic experiments. In one type, exemplified by T-even and lambdoid phages, where genetic recombination is relatively frequent, the predominant differences between individual phages are physically located in specific chromosomal regions, just as they are reflected in the activities of specific genes.[39, 58] In the alternative types of divergence, seen between phage T3 and T7-related phages, the differences comprise long stretches of chromosome having only a low degree of complementarity, as revealed by high sensitivity to the mildly denaturing conditions used in forming heteroduplexes.[17, 27]

2.4. Viral entry

One of the dominant features distinguishing phages from other categories of virus is the nature of their host and, hence, their special capacity for bringing about penetration of the bacterial envelope. There is little evidence, particularly with the larger tailed phages, that the host participates actively in the entry process other than by providing the appropriate receptor. Individual phages show considerable diversity in initial attachment but a possible indication of how final penetration may come about is provided by the appearance in ultrathin sections of *Escherichia coli* of numerous small (25–50 nm) sites where the inner bacterial membrane remains attached to the outer membrane after plasmolysis of the cells. It is at points on the surface corresponding to these sites that attached phage particles are localized (Figure 2.8(a),(b)).[3, 6] This is observed with particles of complex phages like T4, where adsorption has been examined electron microscopically in greatest detail,[1, 3, 57] as well as with small phages like ØX174 where, except for the demonstration that attachment occurs by a spike at one vertex of the capsid,[6] the sequence of events at infection has been followed mainly biochemically.[31, 32]

The morphological changes which occur during infection by T-even phages are well known,[1, 57] although the details of the mechanism at the molecular level are not understood.[62] These filaments, or whiskers, at the base of the phage head[63] control the extension[15] of the long tail fibers which make the initial contact with the bacterium. Contact is made through specific receptors in the cell wall, the fibers opposite one another attaching 180 nm apart, so that the phage particles are oriented with their tails pointing towards the bacterial surface. In the next step, the base plate is brought close to the bacterial wall, where the tail pins make contact with a second set of receptors, and the angle in the already attached long tail fibers becomes more acute. The base plate undergoes a conformational change from hexagonal to star-shaped, losing its central plug,[1] rupturing its attachment to the distal end of the tail tube, and rising up the tube together with the sheath which contracts to about half its original length. During this process the base plate maintains contact with the

(a)

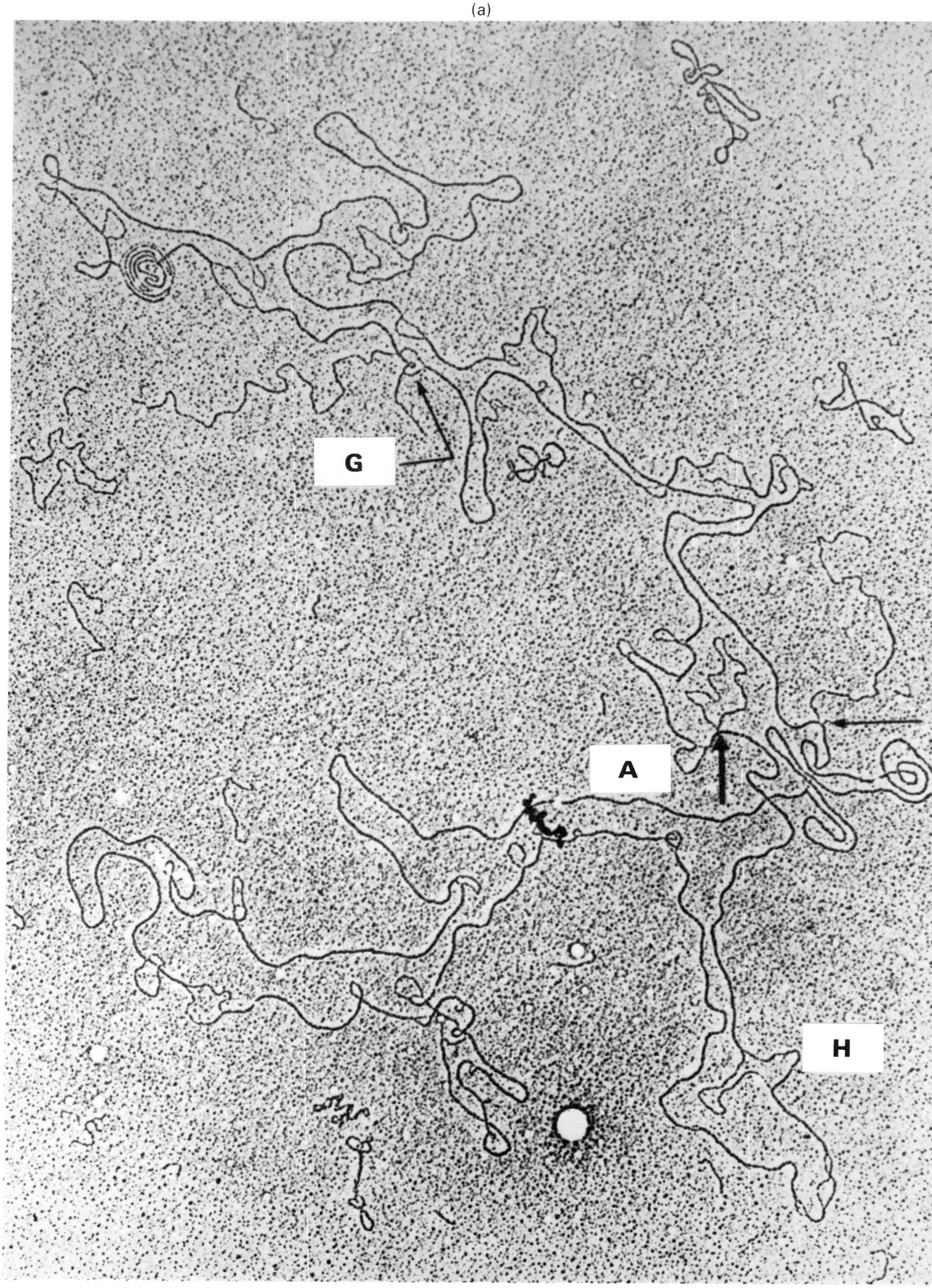

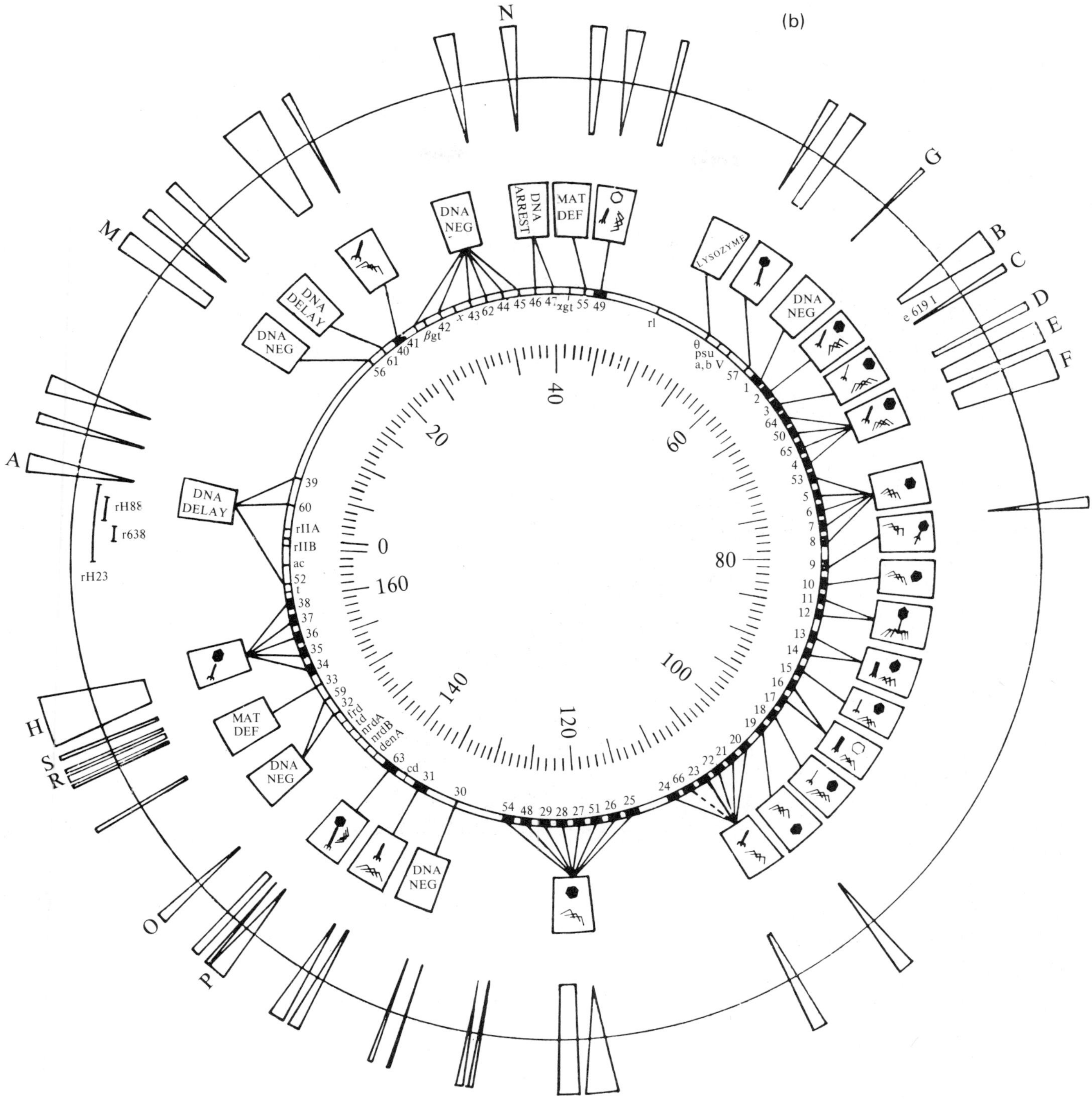

Figure 2.7. Heteroduplex between related T-even phages (N. Davidson).

(a) Heteroduplex molecule of wild type T2 and mutant T4. A, G: deletion loops, H: substitution loop.

(b) The outer circle represents the heteroduplex map and the inner circle the T-even genetic map. The innermost circle shows the physical scale in kilobase units. The T4 mutant has a large deletion in one region, indicated by the thick arrow in (a). This serves to orient the physical map with the genetic map in (b), where it is labeled rH23 at 163Kb. Similarly, the loops labeled A and H in (a) correspond to the triangle (deletion loop) and wedge (substition loop) labeled accordingly in (b). Two single-strand branches corresponding to the ends of the original strands are marked by thin arrows. (Both are forked because of branch migration and one has migrated to deletion loop G.)

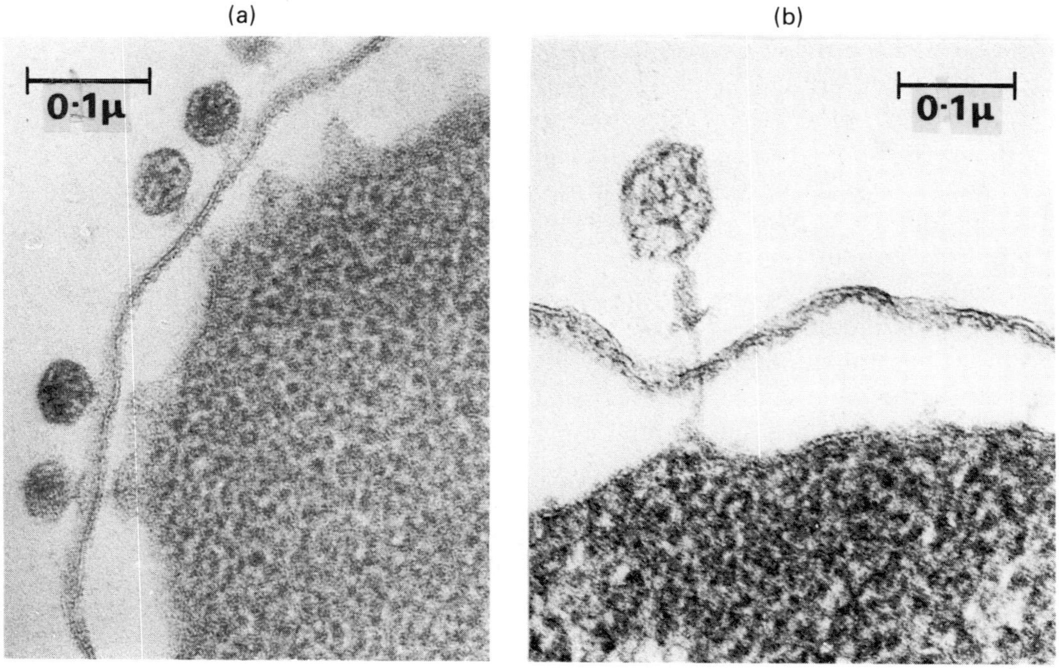

Figure 2.8. Adsorption to sites of membrane adhesion (M. Bayer).
(a) Ultrathin section of plasmolyzed *E. coli* showing particles of phage K29 at sites where the inner membrane has remained attached to the cell wall.
(b) A particle of phage T2 with contracted tail whose tail tube has penetrated the wall to two-thirds its thickness at a similar site.

cell wall by means of its short tail fibers,[34] which has the effect of causing the tube to penetrate the wall to a depth of 12 nm, as seen in ultrathin sections, releasing a strand of DNA into the cell.[57]

Electron microscopy can identify the site of the bacterial receptor in the case of phages attaching to flagella or pili. Flagellar phages, which are tailed phages (Figure 2.1), have a primary attachment site on the sides of the flagella[47] and subsequently migrate to a secondary site at the flagellar base, presumably by means of the tail fiber which is wound round the flagellar shaft.[55] Flagellar phages for *Caulobacter* are likely to behave similarly, although the particles show a different morphology.[33] The particular interest of flagellar phages in species such as *Caulobacter*, which undergo morphological differentiation, is that their presence maintains one cell type.

The pili acting as phage receptors in enterobacteria are sex pili determined by plasmids, and are seen to be located at sites of membrane adhesion.[3] These pili are also involved in gene transfer by conjugation.[48] In other bacterial species, pili act as phage receptors but no association with genetic transfer has yet been identified.[9] Isometric RNA phages attach along the pili (Figure 2.1). Microscopy shows filamentous DNA phages, in which one end is specialized for adsorption,[42] attached to the pilus tip.[12, 49] Additionally, with these phages, there is biochemical evidence that DNA and capsid protein enter as a unit to react with another type of receptor which is present in the inner membrane whether the bacteria contain the plasmid or not.[43] Uptake is linked to conversion of the single-stranded DNA of the virion to the double-stranded replicative form[43] and the capsid protein becomes incorporated in the bacterial membrane,[59] from where it is re-used in progeny virus.[59]

A large group of phages with a characteristic pattern of adsorption remains to be mentioned. These are phages in which the phage protein of the tail spike protein functions both as a structural component and an enzyme whose substrate is the primary bacterial receptor.[52] This may be one of the chemical groupings present in capsular,[21] surface lipopolysaccharide,[30] or Vi antigen,[41] depending on the phage. As shown for a capsule phage, contact with isolated receptor does not stimulate release of DNA; and one model, taking into account the possible significance of the sites of membrane adhesion seen in thin sections (Figure 2.8),[3] proposes that the phage particles are led to migrate by progressive enzyme activity along the polymer strands towards secondary receptors at these sites, where DNA release and penetration take place.[21]

If requirement for two-fold receptors is true of phages generally, it may exist as a means of avoiding accidental uncoating of nucleic acid in the special category of virus where this process occurs outside the cell.

References to Part Two

1. Amos, L. A. and Klug, A., "Three dimensional image reconstruction of the contractile tail of T4 bacteriophage," *J. molec. Biol.*, **99**, 51, 1975.
2. Asakura, S., Eguchi, G., and Iino, T., "Unidirectional growth of *Salmonella* flagella *in vitro*," *J. molec. Biol.*, **35**, 227, 1968.
3. Bayer, M. E., "Role of adhesion zones in bacterial cell surface function and biogenesis," in A. Tzagoloff (ed.), *Membrane Biogenesis*, Plenum, New York.
4. ——— and Bocharov, A. F., "The capsid structure of bacteriophage lambda," *Virology*, **54**, 465, 1973.
5. ——— and Remsen, C. C., "Bacteriophage T2 as seen with the freeze-etching technique," *Virology*, **40**, 703, 1970.
6. ——— and Starkey, T. W., "The adsorption of bacteriophage ØX174 and its interaction with *Escherichia coli*; a kinetic and morphological study," *Virology*, **49**, 236, 1972.
7. Beckendorf, S. K., "Structure of the distal half of the bacteriophage tail fibre," *J. molec. Biol.*, **73**, 37, 1973.
8. Bradley, D. E., "Ultrastructure of bacteriophages and bacteriocins," *Bact. Rev.*, **31**, 230, 1967.
9. ———, "A study of pili on *Pseudomonas aeruginosa*," *Genet. Res.*, **19**, 39, 1972.
10. Branton, D. and Klug, A., "Capsid geometry of bacteriophage T2: a freeze-etching study," *J. molec. Biol.*, **92**, 559, 1975.
11. Bruce, J., Gourlay, R. N., Hull, R., and Garwes, D. J., "Ultrastructure of Mycoplasmatales virus laidlawii I," *J. gen. Virol.*, **16**, 215, 1972.
12. Caro, L. G. and Schnös, M., "The attachment of the male-specific bacteriophage f1 to sensitive strains of *Escherichia coli*," *Proc. natn. Acad. Sci. U.S.A.*, **56**, 126, 1966.
13. Casjens, S. and King, J., "Virus assembly," *A. Rev. Biochem.*, **45**, 555, 1975.
14. Caspar, D. L. D. and Klug, A., "Physical principles in the construction of regular viruses," *Cold spring Harb. Symp. quant Biol.*, **27**, 1, 1962.
15. Conley, M. P. and Wood, W. B., "Bacteriophage T4 whiskers: a rudimentary environment-sensing device," *Proc. natn. Acad. Sci. U.S.A.*, **72**, 3701, 1975.
16. Crowther, R. A. and Klug, A., "Structural analysis of macromolecular assemblies by image reconstruction from electron micrographs," *A. Rev. Biochem.*, **45**, 161. 1975.
17. Davis, R. W. and Hyman, R. W., "A study in evolution: the DNA base sequence homology between coliphages T7 and T3," *J. molec. Biol.*, **62**, 287, 1971.

18. ———, Simon, M., and Davidson, N., "Electron microscope heteroduplex methods for mapping regions of base sequence homology in nucleic acids," S. P. Colowick and N. O. Kaplan (eds.), *Methods in Enzymology*, **21**, 413, 1971.
19. Edgar, R. S. and Wood, W. B., "Morphogenesis of bacteriophage T4 in extracts of mutant-infected cells," *Proc. natn. Acad. Sci. U.S.A.*, **55**, 498, 1966.
20. Epstein, R. H., Bolle, A., Steinberg, C. M., Kellenberger, E., Boy de la Tour, E., Chevalley, R., Edgar, R. S., Susman, M., Denhardt, G. H., and Lielausis, A., "Physiological studies of conditional lethal mutants of bacteriophage T4D," *Cold Spring Harb. Symp. quant. Biol.*, **28**, 375, 1963.
21. Fehmel, F., Feige, U., Niemann, H., and Stirm, S., "*Escherichia coli* capsule bacteriophages. VII. Bacteriophage 29 – host capsular polysaccharide interactions," *J. Virol.*, **16**, 591, 1975.
22. Fiandt, M., Hradecna, Z., Lozeron, H. A., and Szybalski, W., "Electron micrographic mapping of deletions, insertions, inversions, and homologies in the DNAs of coliphages lambda and Phi80," in A. D. Hershey (ed.), *The Bacteriophage Lambda*, Cold Spring Harbor Laboratory, New York, 1971.
23. Harrison, S. C., Caspar, D. L. D., Camerini-Otero, R. D., and Franklin, R. M., "Lipid and protein arrangement in bacteriophage PM2," *Nature New Biol.*, **229**, 197, 1971.
24. Hemphill, H. E. and Whiteley, H. R., "Bacteriophages of *Bacillus subtilis*," *Bact. Rev.*, **39**, 257, 1975.
25. Henderson, D. and Weil, J., "The nature and origin of a class of essential gene substitutions in bacteriophage λ," *Virology*, **67**, 124, 1975.
26. Howatson, A. F. and Kemp, C. L., "The structure of tubular head forms of bacteriophage λ : relation to the capsid structure of petit λ and normal λ heads," *Virology*, **67**, 80, 1975.
27. Hyman, R. W., Brunovskis, I., and Summers, W. C., "DNA base sequence homology between coliphages T7 and ØII and between T3 and ØII as determined by heteroduplex mapping in the electron microscope," *J. molec. Biol.*, **77**, 189, 1973.
28. Inman, R. B., "A denaturation map of the λ phage DNA molecule determined by electron microscopy," *J. molec. Biol.*, **18**, 464, 1966.
29. ——— and Schnös, M., "Structure of branch points in replicating DNA: presence of single-stranded connections in λ DNA branch points," *J. molec., Biol.*, **56**, 319, 1971.
30. Iwashita, S. and Kanegasaki, S., "Release of O antigen polysaccharide from *Salmonella newington* by phage ϵ^{34}," *Virology*, **68**, 27, 1975.
31. Jazwinski, S. M., Lindberg, A. A., and Kornberg, A., "The gene H spike protein of bacteriophages ØX174 and S13.1. Functions in phage-receptor recognition and in transfection," *Virology*, **66**, 283, 1975.
32. ———, Marco, R., and Kornberg, A., "The gene H spike protein of bacteriophages ØX174 and S13 II. Relation to synthesis of the parental replicative form," *Virology*, **66**, 294, 1975.
33. Jollick, J. D. and Wright, B. L., "A flagella specific bacteriophage for *Caulobacter*," *J. gen. Virol.*, **22**, 197, 1974.

34. Kells, S. S. and Haselkorn, R., "Bacteriophage T4 short tail fibres are the product of gene 12," *J. molec. Biol.*, **83**, 473, 1974.
35. ———, Ohtsuki, M., and Haselkorn, R., "The structure of bacteriophage T4 gene 12 protein," *J. molec. Biol.*, **99**, 349, 1975.
36. Kikuchi, Y. and King, J., "Genetic control of bacteriophage T4 baseplate morphogenesis. I. Sequential assembly of the major precursor, *in vivo* and *in vitro*," *J. molec., Biol.*, **99**, 645, 1975.
37. ——— and ———. "Genetic control of bacteriophage T4 baseplate morphogenesis. II. Mutants unable to form the central part of the baseplate," *J. molec. Biol.*, **99**, 673, 1975.
38. ——— and ———, "Genetic control of bacteriophage T4 baseplate morphogenesis. III. Formation of the central plug and overall assembly pathway," *J. molec. Biol.*, **99**, 695, 1975.
39. Kin, J-S. and Davidson, N., "Electron microscope heteroduplex study of sequence relations of T2, T4 and T6 bacteriophage DNAs," *Virology*, **57**, 93, 1974.
40. King, J., "Bacteriophage T4 tail assembly: four steps in core formation," *J. molec. Biol.*, **58**, 693, 1971.
41. Kwaitkowski, B., Beilharz, H., and Stirm, S., "Disruption of Vi bacteriophage III and localisation of its deacetylase activity," *J. gen. Virol.*, **29**, 267, 1975.
42. Marco, R., "The adsorption protein in mini M13 phages," *Virology*, **68**, 280, 1975.
43. ———, Jazwinski, S. M., and Kornberg, A., "Binding, eclipse, and penetration of the filamentous bacteriophage M13 in intact and disrupted cells," *Virology*, **62**, 209, 1974.
44. Marvin, D. A. and Hohn, B., "Filamentous bacterial viruses," *Bact. Rev.*, **33**, 172, 1969.
45. ———, Pigram, W. J., Wiseman, R. L., Wachtel, E. J., and Marvin, F. J., "Filamentous bacterial viruses. XII. Molecular architecture of the Class I (fd, Ifl, Ike) virion," *J. molec. Biol.*, **88**, 581, 1974.
46. Mazza, A. and Felluga, B., "Electron microscope observations on λ bacteriophage ultrastructure," *J. Ultrastruct. Res.*, **45**, 259, 1973.
47. Meynell, E. W., "A phage, $\O\chi$, which attacks motile bacteria," *J. gen. Microbiol.*, **25**, 253, 1961.
48. Meynell, G. G., *Bacterial Plasmids*, Macmillan, London, 1972.
49. ——— and Lawn, A. M., "Filamentous phages specific for the I sex factor," *Nature, Lond.*, **217**, 1184, 1968.
50. O'Callaghan, R., Bradley, R., and Paranchych, W., "Controlled alterations in the physical and biological properties of R17 bacteriophage induced by guanidine hydrochloride," *Virology*, **54**, 476, 1973.
51. Pratt, D., Laws, P., and Griffith, J., "Complex of bacteriophage M13 single-stranded DNA and gene 5 protein," *J. molec. Biol.*, **82**, 425, 1974.
52. Rieger, D., Freund-Molbert, E., and Stirm, S., "*Escherichia coli* capsule bacteriophages. III. Fragments of bacteriophage 29," *J. Virol.*, **15**, 964, 1975.
53. Saigo, K., "Denaturation mapping and chromosome structure in bacteriophage T5," *Virology*. **68**, 166, 1975.

54. —— and Uchida, H., "Connection of the right-hand terminus of DNA to the proximal end of the tail in bacteriophage lambda," *Virology*, **61**, 524, 1974.
55. Schade, S. Z., Adler, J., and Ris, H., "How bacteriophage χ attacks motile bacteria," *J. Virol.*, **1**, 599, 1967.
56. Scraba, D. G., Raska, I., Kellenberger, E., and Moor, H., "Electron microscopy of polyheads of bacteriophage T4 prepared by freeze-etching," *J. Ultrastruct. Res.*, **44**, 27, 1973.
57. Simon, L. D. and Anderson, T. F., "The infection of *Escherichia coli* by T2 and T4 bacteriophages as seen in the electron microscope. I Attachment and penetration. II Structure and function of the baseplate," *Virology*, **32**, 279 and 298, 1967.
58. Simon, M. N., Davis, R. W., and Davidson, N., "Heteroduplexes of DNA molecules of lamboid phages: physical mapping of their base sequence relationships by electron microscopy," in A. D. Hershey (ed.), *The Bacteriophage Lambda*, Cold Spring Harbor Laboratory, New York, 1971.
59. Smilowitz, H., "Bacteriophage f1 infection: fate of the parental major coat protein," *J. Virol.*, **13**, 94, 1974.
60. Streisinger, G., Emrich, J., and Stahl, M. M., "Chromosome structure in phage T4. III. Terminal redundancy and length determination," *Proc. natn. Acad. Sci. U.S.A.*, **57**, 292, 1967.
61. Walker, D. H. and Anderson, T. F., "Morphological variants of coliphage P1," *J. Virol.*, **5**, 765, 1970.
62. Yamamoto, M. and Uchida, H., "Organisation and function of the tail of bacteriophage T4. II. Structural control of the tail contraction," *J. molec. Biol.*, **92**, 207, 1975.
63. Yanagida, M. and Ahmed-Zadeh, C., "Determination of gene product positions in bacteriophage T4 by specific antibody association," *J. molec. Biol.*, **51**, 411, 1970.
64. Zinder, N. D. (ed.), *RNA-containing Bacteriophages*, Cold Spring Harbor Laboratory, New York, 1975.

Part Three

Mycoplasmas

3. Mycoplasmas

Eyvind A. Freundt and Ole Vinther

3.1. Introduction

The mycoplasmas are a group of prokaryotic microorganisms which in 1967 was assigned to a separate class, the Mollicutes.[21] This name, which means "soft skin," refers to an outstanding ultrastructural property of the proposed new class, the absence of a rigid true cell wall and the inability to synthesize biochemical cell wall precursors. In both respects mycoplasmas clearly differ from bacteria of the class Schizomycetes, but resemble the animal cell.

The mycoplasmas are characterized, moreover, by the typical "fried-egg" appearance of their minute colonies as visualized under the low-power light microscope (Figure 3.1), absolute resistance to penicillins, and susceptibility to broad-spectrum antibiotics.

The Mollicutes are divided into three families, the Mycoplasmataceae (genera *Mycoplasma* and *Ureaplasma*), Spiroplasmataceae (genus *Spiroplasma*), and Acholeplasmataceae (genus *Acholeplasma*). The two former are distinguished by requiring cholesterol for growth. At present a total of about fifty species are recognized within the Mollicutes, most of them belonging to the genus *Mycoplasma*. Less than half of these species are proven pathogens.

One species, *M. pneumoniae*, is of major importance in human medicine, inasmuch as this is the single most frequent cause of the well-known clinical syndrome "primary atypical pneumonia." Strong circumstantial evidence suggests that two species, *M. hominis*, and less frequently *M. fermentans*, may be responsible for a relatively high proportion of cases of salpingitis and related inflammatory gynecological conditions. *M. hominis* and *Ureaplasma urealyticum* (formerly known under the name of T-mycoplasma) have been incriminated also, although less convincingly, as possible etiological agents of nongonococcal urethritis, infertility, abortion, and prematurity. *M. salivarium, M. orale, M. buccale*, and *M. faucium* are more or less common members of the normal microbial flora of the oral cavity and throat of man.

The importance of mycoplasmas in veterinary medicine by far supersedes their role as human pathogens. The first representative of this group of microorganisms, isolated and described at the end of the last century, is the species now known as *M. mycoides*. One variety of this organism is the cause of contagious bovine pleuropneumonia, a disease of cattle which is still of considerable socioeconomic importance in many countries of Africa and in some areas of Asia. Infectious diseases of sheep and goats caused by the same or closely related organisms occur endemically, and are equally disastrous, in a great

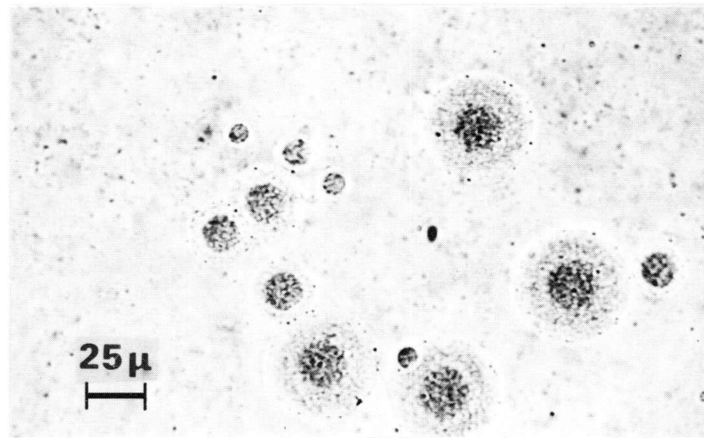

Figure 3.1. *Mycoplasma pneumoniae* colonies.
Six-day-old colonies on solid medium showing "fried-egg" appearance. (× 80)

many tropical and subtropical developing countries. Enzootic pneumonia of pigs and chronic respiratory disease of poultry are examples of mycoplasmal infections of domestic animals of widespread occurrence also in Europe and the United States.

A new era was heralded in mycoplasmology in the early seventies with the cultivation in artifical growth media of a mycoplasma, *Spiroplasma citri*, associated with and subsequently shown to cause "stubborn" disease of citrus plants. This discovery confirmed earlier electron microscopic observations of mycoplasma-like organisms in the phloem of a great variety of plants suffering from diseases hitherto believed to be caused by viruses.

The demonstration of similar organisms in insects and the more recent identification of the "suckling mouse cataract agent" as a spiroplasma[73] shows that mycoplasmas with characteristics of the genus *Spiroplasma* are not restricted to plants. This organism, which was originally isolated from a pooled extract of rabbit ticks, induces cataract, uveitis, and chronic encephalitis in newborn mice on intracerebral infection. It is serologically related to *S. citri*, although only distantly so.

For detailed information about the mycoplasmas, reference is made to more comprehensive surveys.[25, 29, 80]

3.2. Electron microscopy of mycoplasmas

Over the past 25 years the morphology and ultrastructure of the mycoplasmas have been subjected to intense studies by electron microscopy using a diversity of techniques. Centrifuged specimens of whole cells or cells grown on collodium films have been examined in the unfixed dried state,[23] by shadowcasting,[15, 16, 39] negative staining,[9, 11, 14, 15, 16, 31, 51, 56, 61] and scanning electron microscopy.[8, 11, 38, 40, 53]

Ultrathin sections of cultures grown in liquid or solid medium, fixed,

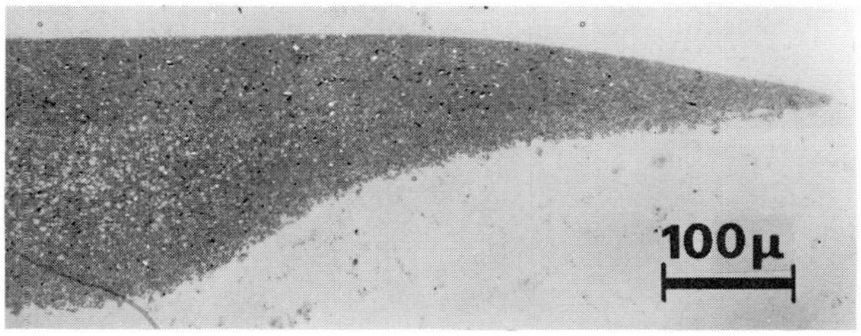

Figure 3.2. *Mycoplasma pneumoniae*, vertical section.
Five-day-old colony fixed and stained *in situ.* Vertical section showing flat superficial layer and hemispherical to flat lower part embedded in agar growth medium. (× 500)

stained, and embedded after centrifugation or *in situ*, have been studied by a great many authors.

The development of the freeze-etching technique for cleaving bilayered membranes provoked an explosion of studies of the fine structure of the mycoplasma membrane by this method.[6, 15, 27, 36, 43, 59, 64, 68, 70, 71, 74]

In the review that follows of electron microscope studies of mycoplasmas, observations made on the basis of nonhuman mycoplasmas are included to the extent that these can be supposed to be representative of the whole group. In addition, a few structures that are apparently unique for nonhuman mycoplasmas have been briefly described in order to illustrate the range of structural variability found in these basically very simple organisms.

More extensive surveys of the structure of the mycoplasmas may be found in recent reviews, papers and books.[24, 29, 43, 45, 80]

3.3. Cellular morphology and reproduction

The mycoplasma cells are highly pleomorphic, varying in shape from spherical to slightly ovoid, about 250–300 nm in diameter in the early exponential growth phase (Figures 3.4, 3.5, 3.6), to slender branched filaments ranging in length from a few to about 150 μm (Figure 3.3). Chains of coccoid bodies develop at a later growth phase, followed by fragmentation of the chains into individual cocci.[23, 24, 25] Helical morphology is typical of *S. citri* and related organisms.[15, 16, 73] Most mycoplasmas are nonmotile, although a few species, including *m. pneumoniae*, exhibit gliding motility. The spiroplasmas show undulating, rotatory and translational motility.

The mycoplasmas are the smallest organisms capable of replication on artificial growth media without the presence of other living cells. The replication appears to occur primarily by binary transversal fission of coccoid or filamentous cells, the development of chains of cocci probably depending on asynchrony between replication of the genome and cell division.

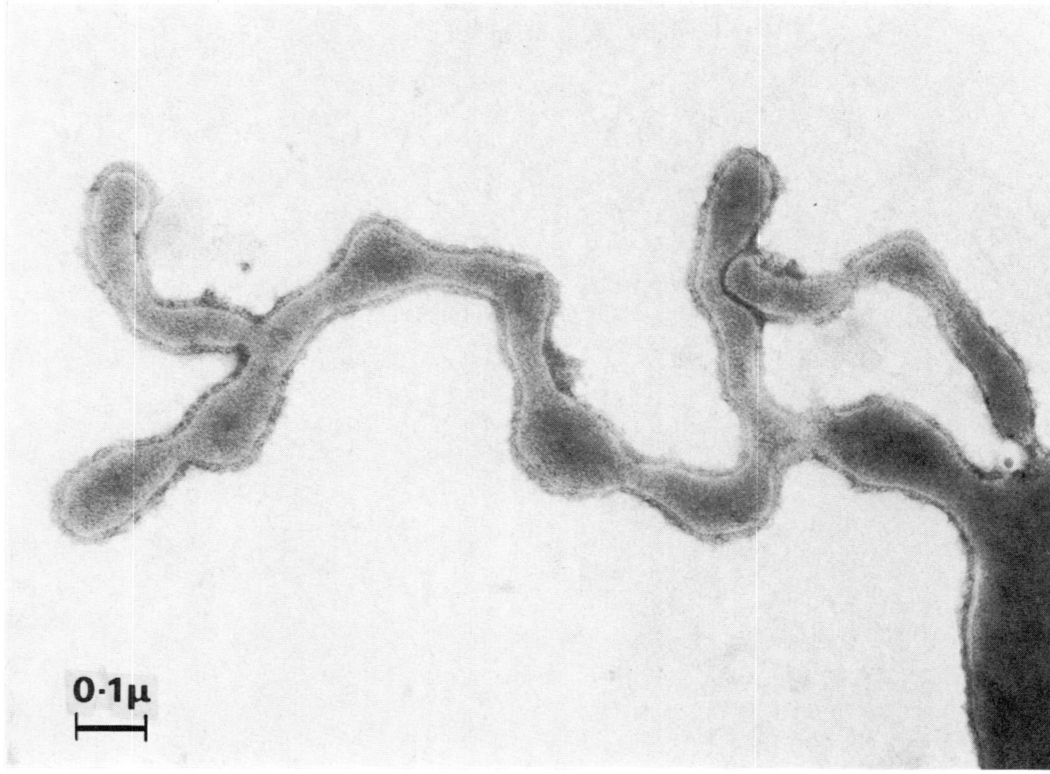

Figure 3.3. ***Ureaplasma urealyticum.***
Twelve-hour culture in liquid medium. Negatively stained specimen showing branching filamentous cell with well-defined hairy structure of surface layer. (× 90,000)

3.4. Morphology and structure of the mycoplasma colony

The "fried-egg" appearance, said to be typical of mycoplasma colonies grown on solid medium, is demonstrated in Figure 3.1. On vertical section the colony is seen to consist of a flat, centrally depressed or slightly convex superficial layer and a conical or hemispherical deeper part penetrating into the depth of the agar medium[19, 41, 54] (Figure 3.2). Based on the population density of the organisms and their organization, four distinct zones could be distinguished in colonies of *M. pneumoniae* and *M. salivarium*: (a) the surface growth exhibiting a close packing of the cells, (b) a central less populated area, (c) a band of dense growth, and (d) a peripheral, diffuse invasive front [41] (Figure 3.2). A similar zonal differentiation within the colony has been described in *M. orale.*[54] The size, shape, and general appearance of the individual cells contained in the colony vary from zone to zone and within each zone. In *M. orale*, filamentous cells were found exclusively in the surface layer, whereas small or large round cells occupied the lower part of the colonial area.[54] The surface growth of *M. salivarium* had a peculiar appearance due to the occurrence of elongated laminated structures, reminiscent of myelin structures, lying along the upper surface and also wrapped around individual organisms.[41] Since quite a high proportion of *M. salivarium* strains produce a surface film consisting, *inter alia*, of lipids it is conceivable that the said structures are lipid in nature.

The subcellular components of the cells in adequately fixed and stained agar colonies are usually well preserved and present themselves distinctly in thin sections.

Colonial growth in liquid medium has been described for *M. pneumoniae*, under the name of "spherules."[11] Scanning electron microscopy of spherules processed by the critical point method[5] to demonstrate three-dimensional morphology showed them to consist of entangled and occasionally branching short filaments.

3.5. Ultrastructure of the mycoplasma cell

3.5.1 Basic structural components

The basic cytological elements of the mycoplasma cell include the cell membrane, ribosomes, and DNA strands. Cytoplasmic membrane-bounded dense bodies or vacuoles of unknown significance are quite frequent.

Plasma membrane

The failure to demonstrate a true cell wall in thin sections of mycoplasma cells was first pointed out by van Iterson and Ruys[35] and Ruys and van Iterson,[65] and the absence of a cell wall has been amply confirmed in all subsequent electron microscopic studies of mycoplasmas.

Thus, the mycoplasma cell is bounded only by a single triple-layered membrane, about 75–100 Å thick (Figures 3.4, 3.5, 3.6).

The structure of the membrane, which in section appears as two electron-dense layers with a lucid intermediate layer, has been further studied by the freeze-etching technique as referred to in Section 3.2. Splitting of the membrane exposes the hydrophobic inner faces of the lipid bilayer, thereby revealing the presence of particles, usually about 75–150 Å in diameter (Figure 3.7 (a) and (b), page 75). These are more numerous on the convex than on the concave fracture face. Although the particles are largely unaffected by treatment with the broad spectrum proteolytic enzyme pronase, other observations strongly suggest that they do represent proteins. Thus their number was substantially decreased when *A. laidlawii* was incubated under conditions which abolished protein, but not lipid, synthesis.[71]

The existence of an additional surface structure outside the cell membrane was first suggested by Domermuth *et al.*[19] More significance can be attached, however, to the well-defined pattern of dense, minute spikes or hairlike structures, about 50–100 nm long, seen radiating from the cell surface in sectioned and/or negatively stained specimens of *M. pulmonis,*[34] *M. gallisepticum*,[14, 30] *U. urealyticum*[9, 76] (Figures 3.3, 3.4, 3.5, 3.6), and *M. hyopneumoniae.*[30] Staining of *M. meleagridis, M. mycoides* subsp. *mycoides*, and *M. dispar* with ruthenium red revealed a capsular substance appearing as a more or less dense amorphous layer extending 25–30 nm from the membrane.[27, 31] None of five other

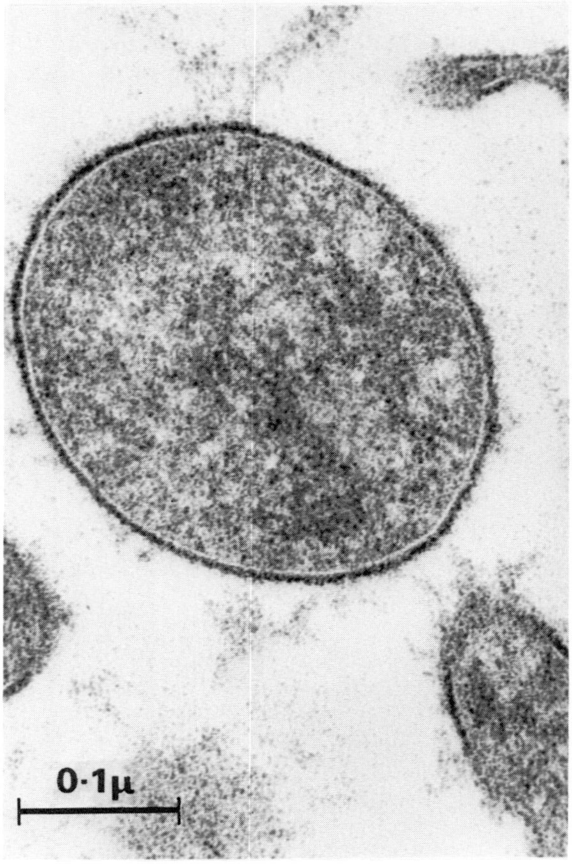

Figure 3.4. ***Ureaplasma urealyticum*: plasma membrane.**
Seven-day-old culture in solid medium. Thin section of round cell showing triple-layered limiting membrane, hairlike structure of surface layer, and densely packed ribosomes. Colony fixed *in situ* (glutaraldehyde, OsO_4), stained with uranyl acetate, and embedded in Vestopal-W. Sections poststained with magnesium uranyl acetate and lead citrate. (× 215,000)

mycoplasmas examined by this technique possessed any structure equivalent to the capsule of the said species. Cytochemical evidence of a capsule was likewise obtained by treating sections of *M. mycoides* subsp. *capri* with the concanavalin A-peroxidase-diaminobenzidine method.[66] Whereas ruthenium red does not react specifically with any group of chemical compounds, concanavalin A is known to bind to carbohydrates containing α-D-glucosyl or sterically closely related residues. The studies of Schiefer *et al.*[66] further indicate that the surface carbohydrate of *M. mycoides* may be bound to glycolipids.

Ribosomes

Closely packed in smaller young cells (Figures 3.4, 3.6) and in filaments,[3, 50] and more sparse in large nonviable cells, ribosomes measure approximately 170 Å in diameter and contain about 60 percent RNA and 40 percent protein.[45] Staining

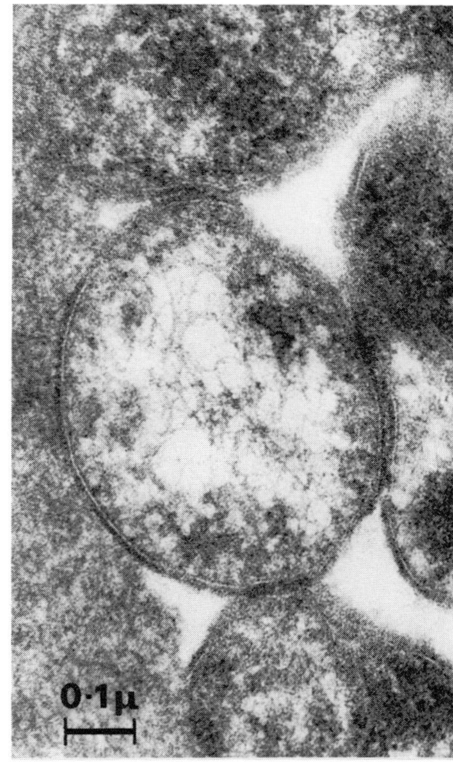

Figure 3.5. *Ureaplasma urealyticum*: nuclear area.
Fourteen-hour culture in liquid medium. Thin section showing network of DNA strands in central light area of round ce l. Fixation and staining essentially as for Figure 3.4. (× 93,000)

for total nucleic acid by the indium staining procedure, followed by extraction of RNA with perchloric acid, which does not affect DNA, confirmed the presence of RNA in the ribosome-like granules. In a few species (*M. gallisepticum*, *U. urealyticum*) a proportion of the ribosomes may be arranged in square-packed groups of four (Figure 3.6), often held together in regular geometric patterns in the shape of columns, cylinders, or helical structures containing up to one hundred ribosomes.[9, 19, 46, 47] Although these ribosomal arrangements have been regarded partly as preparatory artifacts,[42, 47] it has been suggested that they might be polyribosomes.[24]

Nuclear area

In unstained sections the nuclear area of mycoplasma cells presents itself as a more or less centrally located area, irregular in shape and containing a network of delicate strands (Figure 3.5). The thickness of these fibrils, which are more distinct in sections stained with uranyl acetate, varies from 25 to 40 Å. In specimens stained by the indium staining procedure the DNA remains located, after perchloric acid extraction, mainly in the central region, unbounded by

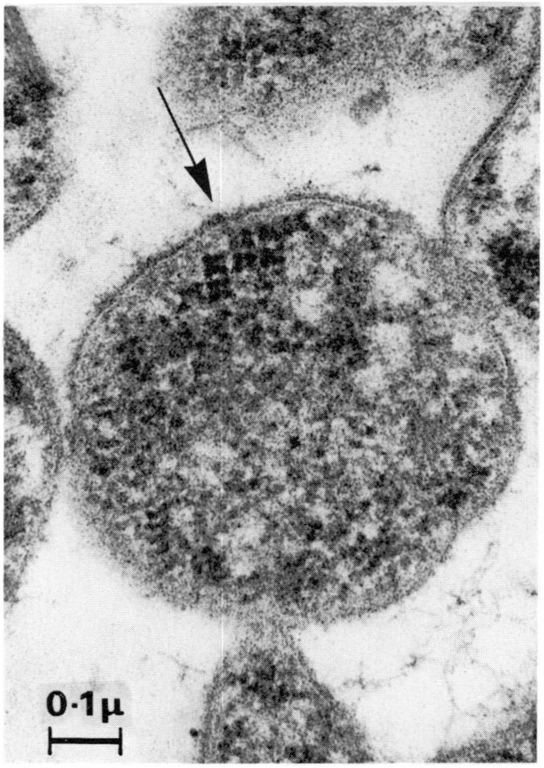

Figure 3.6. *Ureaplasma urealyticum*: ribosomes.
Twelve-hour culture in liquid medium. Thin section showing ribosomes partly arranged in square-packed groups of four (arrow). Fixation and staining essentially as for Figure 3.4. (× 91,000)

membranes and containing heavily stained granules and amorphous and fibrillar material.[47]

DNA has also been demonstrated by means of autoradiography of specimens grown in the presence of ^{3}H-thymidine in coccoid swellings of ρ-forms of *M. mycoides* (see below), but not in the filamentous or tubular structures connecting the coccoid bodies.[56]

Localization of enzymes

Cytochemical demonstration of the cellular sites of adenosine phosphatase and acid phosphatase in *M. gallisepticum*[52] showed a markedly restricted distribution in the cell, ATPase activity being localized exclusively along the inside of the membrane bounding the bleb and infrableb and that of acid phosphatase in the infrableb region. In experiments using tellurite as an electron acceptor, the infrableb region of *M. gallisepticum* was also shown to be the site of redox activity.[43] Thin sections of *M. meleagridis* cells grown in the presence of potassium tellurite exhibited a very significant increase in the contrast of the external track of the plasma membrane, but no cytoplasmic deposits of tellurite crystals as found by Maniloff[43] in *M. gallisepticum*. In a recent study combining cyto-

(a) (b)

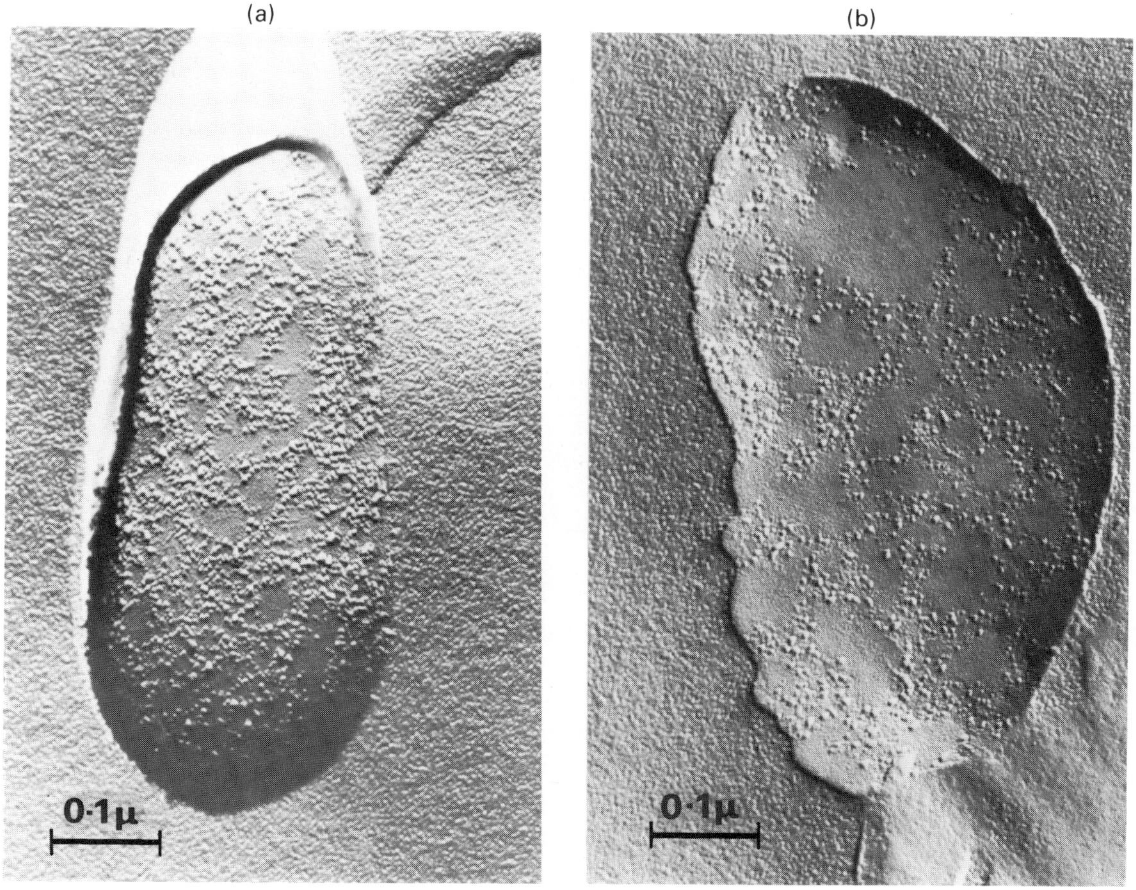

Figure 3.7. *Acholeplasma laidlawii.*
Platinum carbon replica of (a) convex and (b) concave fracture faces of the membrane of freeze-cleaved *A. laidlawii* cells suspended in 20 percent glycerol. Globular particles are seen on both fracture faces. (× 57,000)

chemistry with a biochemical method of localizing enzyme activity, the urease activity of *U. urealyticum* was found to be confined to the soluble fraction of the cytoplasma.[75] This is a confirmation of biochemical results reported by Masover *et al.*[48]

As suggested by Pollack,[57] a difference in the cellular localization of certain enzymes may well prove to be a useful distinguishing marker between the genera *Mycoplasma* and *Acholeplasma*. Thus, reduced nicotinamide adenine dinucleotide (NADH) oxidase activity was found by biochemical techniques to be associated with the membrane fraction of *Acholeplasma laidlawii* and with the soluble cytoplasmic fraction of *Mycoplasma* strains.[58] A recent study,[57] in which *A. granularum, A. modicum*, and *A. axanthum* were compared with a *Mycoplasma* strain with respect to the localization of NADH oxidase activity, confirmed and extended the original observation by Pollack *et al.*[58] Studies carried out in this laboratory suggest that the redox activity is confined to the soluble fraction of cells of *Mycoplasma* species and is membrane-associated in *Acholeplasma* species.[81]

3.5.2. Specialized structures

Terminal structures

A highly structured polar body protruding from one or both ends of *M. gallisepticum* cells, first described by Maniloff and his collaborators[43, 44, 45, 46, 47, 51] and later by others,[1, 13, 79] is known as the "bleb."

The bleb is hemispherical, about 800 × 1,250 Å. In sections, two dense areas that are intensely stained with uranyl salts, can be seen. The larger of these zones is apparently elliptical and located just beneath the plasma membrane, parallel to the convex surface of the bleb. It is connected with a lower circular flat plate by a series of fine, densely stained threads, approximately 25–50 Å thick. Between the bleb and the rest of the cell is the infrableb region which is about 2,000 Å in diameter. The bleb–infrableb structures are undoubtedly identical with the cell protrusions, containing aligned spherical or ovoid condensations that were demonstrated in 1964 in *M. gallisepticum* by Domermuth *et al.*[20]

Sections of *M. gallisepticum* cells stained for total nucleic acid or DNA did not take up the stain in the bleb–infrableb regions which thus appear to be free of nucleic acids.[45] On the other hand, the central core of the infrableb region and the periphery of the bleb have been found to be rich in basic proteins[43] and, as previously mentioned, these regions have been shown to be the sites of enzymatic activities. The bleb also seems to be the site of attachment to animal cells.[78]

A bleb-like structure has been described in another avian mycoplasma, *M. meleagridis*,[27] but since in this organism the supposed bleb only contains ribosomes (which do not occur in the *M. gallisepticum* bleb), it is doubful whether the cell protrusions of the two species are analogous structures.

A terminal organelle found in *M. pneumoniae* bears more resemblance to the bleb of *M. gallisepticum*. In broth cultures grown in bottles or Petri dishes young filamentous cells of *M. pneumoniae* attach to the glass or plastic surface, the terminal portion of the organisms being orientated perpendicular to the surface.[7] Sections of such organisms fixed and embedded *in situ* have revealed an electron-dense rod in the long axis of the tip, ending with a plate-like structure and surrounded by a mantel of electron-lucent cytoplasm limited by an outer membrane.[7, 8, 12] These observations first made by Biberfeld and Biberfeld[8] were confirmed and extended[17, 18] in electron microscopy studies of *M. pneumoniae* grown in organ cultures of human fetal trachea. Forty-eight hours after inoculation with a virulent strain of *M. pneumoniae* mycoplasma cells were seen among and at the base of the cilia of the respiratory epithelial cells. A terminal structure of essentially the same appearance as described earlier was found very frequently in the tapered terminal tip of the mycoplasma cells and always in that end of the organisms which appeared to be attached to the host cell membrane (Figure 3.8(a)). The contact between host cell and organism membranes was very intimate, and sometimes the epithelial membrane appeared indistinct at the point of contact, suggesting the possibility of local injury (Figure 3.8(b)).

Light microscopic examination of the tissue revealed cytoplasmic

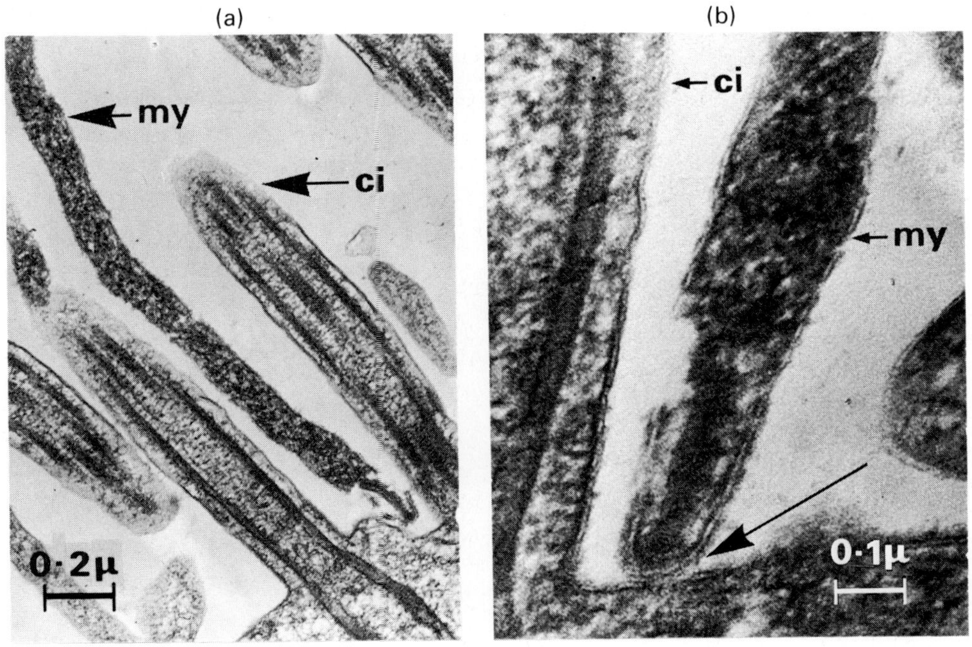

Figure 3.8. *Mycoplasma pneumoniae*: terminal structure.
(a) Human fetal trachea infected in organ culture with virulent *M. pneumoniae* for 48 hours. Filamentous *M. pneumoniae* organism with terminal structure in close association with cell membrane. Note dense central rod in terminal structure. ci: cilium; my: mycoplasma. (× 45,000)
(b) Arrow indicates site of attachment and area of possible injury and penetration of host cell membrane by terminal structure of *M. pneumoniae.* ci: cilium; my: mycoplasma. (× 160,000)

eosinophilia and vacuolization of the epithelial cell, nuclear swelling, and chromattin margination. These events were followed 1 to 2 days later by complete loss of cilia in many cells, disorganization of the entire epithelial layer, and loss of polarity of individual cells. Slowing of ciliary activity in live cultures was evident 3 days postinoculation, ending in complete immobilization on day 4.

The mechanism by which *M. pneumoniae* exerts its pathogenic effect is not known. The cytopathology has been ascribed in part to the toxic action of hydrogen peroxide generated in relatively large amounts by *M. pneumoniae.* Obviously, the close physical contact, possibly strengthened by a specific attachment, between mycoplasmas and host cells will allow peroxide to attack the cell membrane before being destroyed itself by tissue catalase. The fact, however, that no quantitative differences can be detected in peroxide production by virulent and avirulent strains of *M. pneumoniae* indicates that other mechanisms have to be considered as well.[17]

Rho (ρ)-forms

These are structural variants of certain mycoplasmas of bovine–caprine source.[56, 62, 63] The ρ-forms appear as relatively rigid filamentous structures, often with a number of discoid swellings (Figure 3.9) and containing an

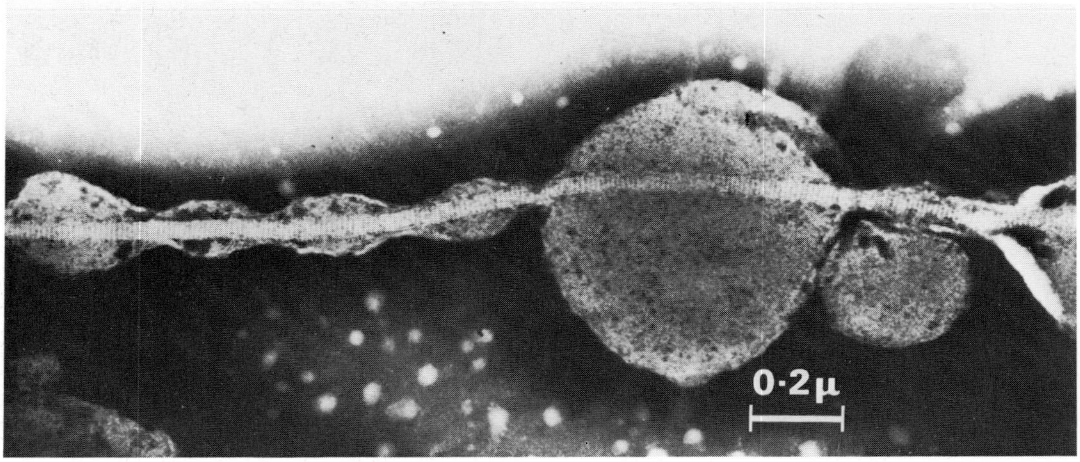

Figure 3.9. *Mycoplasma mycoides*: ρ-form.
M. mycoides subsp. *capri*, ρ-fiber in filamentous cell with discoid swellings. Fixed, negatively stained. (× 60,000)

intracytoplasmatic axial fiber (ρ-fibers are demonstrated most clearly in fixed, negatively stained cells (Figure 3.10) or in sectioned fixed organisms stained with lead and uranyl salts). Intact ρ-fibers can be released by lysis of the cells with the nonionic detergent Nonidet P.40.[63]

The diameter of the ρ-fiber is about 40–120 nm. It presents a pattern of transverse light and dark major bands, the dark band being divided by an intermediate more translucent band. In addition to the transverse banding pattern, continous longitudinal striations, probably due to parallel-aligned fibrils about 3 nm diameter, have been seen in some preparations. In most preparations the ρ-fiber terminates, in one or both ends, in a knob-like structure which presents itself as a translucent curved plate or circular structure located just beneath, and apparently firmly attached to the plasma membrane of the cell. As demonstrated in Figure 3.10, the plate is occasionally penetrated by a few fine fibrils that may be continuous with the longitudinal fibrils of the ρ-fiber.

The results of enzymatic treatment of isolated ρ-fibers suggest that they contain protein and are essentially devoid of nucleic acids.[56] Studies of the gel electrophoretic patterns revealed the presence in a ρ-form variant of two proteins which were not detected in non-ρ-form cells of the same strain.[62] In more recent studies of disassembled and reconstituted isolated fibers, only one of these proteins, protein B, was demonstrated in the reconstituted fibers.[63]

Rho-forms have been demonstrated with varying frequency in many *Mycoplasma* strains of bovine and caprine source. Their incidence is influenced by nutritional (unrestricted energy source, high tonicity) and very probably also by genetic factors.

Thus far, the possible functional significance of the ρ-fibers remains obscure. The theory has been advanced that it could simply be "a paracrystalline aggregation of a cell protein synthesized in excess of the requirement of the cell."[56]

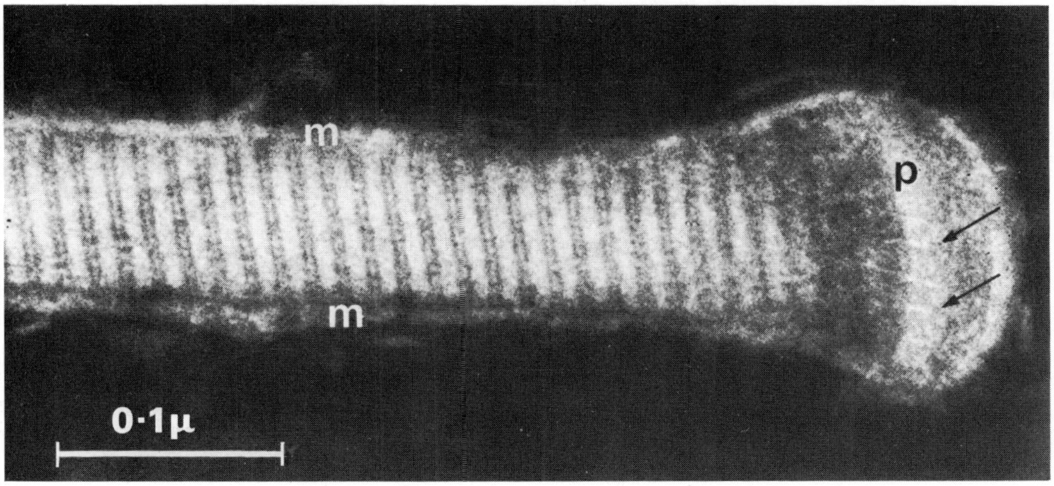

Figure 3.10. Rho-form cell.
Terminal portion of ρ-form cell showing transversal banding and, in some areas, suggestions of longitudinal striation. The ρ-fiber ends at the broad, electron-translucent plate (p) of the terminal structure. Several longitudinal fibrils (arrows) penetrate this plate. m: plasma membrane. (× 300,000)

However, more recent studies[62, 63] provide strong evidence in support of the view that the ρ-fibers are true organelles.

Phages

Virus-like particles (phages) have been demonstrated by electron microscopy in the cytoplasm of quite a few mycoplasmas, including the human *M. hominis*.[60] The significance of the phage infection regarding possible modification of the antigenic pattern and virulence of the mycoplasmas is not known.

3.6. Visualization of mycoplasmas in tissues and cell cultures

Considerable difficulties are involved in the direct electron microscopic demonstration in animal tissues of pleomorphic organisms which themselves lack a true cell wall. This problem was emphasized by Trung *et al.*[72] who discussed criteria by which mycoplasmas grown in tissue cultures may be distinguished from enveloped virus particles on one side, and from microvilli, etc., on the other side. The resemblance of mycoplasmas to membrane-bounded inclusions of the animal cell explains why it is so difficult to prove beyond doubt a possible intracellular occurrence of these organisms. Although the mycoplasmas are predominantly extracellular parasites, mycoplasma-like structures have occasionally been observed by electron microscopy either in cytoplasmic vacuoles of HeLa or other cells, which otherwise looked unimpaired, or free in the cytoplasm of necrotic or partly disintegrated cells.[4, 22, 32, 33, 67, 72] Phagocytic uptake of mycoplasmas can be achieved by neutrophils, eosinophils, monocytes, lymphocytes, and macrophages.[10, 37, 55, 78, 79] An increase in the number of lysosomes of alveolar macrophages in patients with *M. pneumoniae* chronic respiratory disease has been interpreted as an indication of intracellular digestion of mycoplasma cells.[77] Reports of chromosome changes in cell cultures infected with mycoplasmas lend further support to the assumption of a chance occurrence of intracellular parasitism of mycoplasmas. On the other hand, no intracellular mycoplasmas were found in tracheal organ cultures infected with *M. pneumoniae*[17, 18] or in HeLa cell cultures infected with another strain of this species.[11] Neither were mycoplasma-like structures found in an intracellular position in cultures of *Cercopithecus* kidney cells inoculated with *M. hominis.*[3] Finally, *in vivo* studies with *M. pulmonis* failed to reveal intracellular mycoplasmas in respiratory tract epithelial cells of mice, although large numbers of organisms were accumulating at the mucosal surface of the lower respiratory tract.

In any case, considerable caution must be exercised in interpreting mycoplasma-like structures seen in electron micrographs of animal tissues or tissue cultures.

The demonstration of mycoplasma-like structures in negatively stained preparations of the jejunal mucosa of man[69] or in calf feces[49] may also be regarded with some scepticism, although, admittedly, mycoplasmas have been isolated from bovine feces.[26]

Staining with ferritin-labeled antibody;[75b] for example, may facilitate the identification of mycoplasmas in tissues, but such identification attempts may be complicated by binding of the conjugates to the surface of respiratory tract epithelial cells whose membrane may have been antigenically modified by the interaction with mycoplasmas.[28]

It may be added, however, that the spiroplasmas now recognized as mycoplasmas, after they have been grown in artificial medium, were first suspected by the demonstration in the electron microscope of "mycoplasma-like organisms" in the phloem of plants.

References to Part Three

1. Allen, T. C., Stevens, J. O., Florance, E. R., and Hampton, R. O., "Ultrastructure of *Mycoplasma gallisepticum* isolate 1056," *J. Ultrastruct. Res.*, **33**, 318, 1970.
2. Anderson, D. R. and Barile, M. F., "Ultrastructure of *Mycoplasma orale* isolated from patients with leukemia," *J. natn. Cancer Inst.*, **36**, 161, 1966.
3. ———, Hopps, H. E., Barile, M. F., and Bernheim, B. C., "Comparison of the ultrastructure of several rickettsiae, ornithosis virus, and *Mycoplasma* in tissue culture," *J. Bact.*, **90**, 1387, 1965.
4. ——— and Manaker, R. A., "Electron microscopic studies of *Mycoplasma* (PPLO strain 880) in artificial medium and in tissue culture," *J. natn. Cancer Inst.*, **36**, 139, 1966.
5. Anderson, T. F., "Electron microscopy of microorganisms," in G. Oster and A. Pollister (eds.), *Physical Techniques in Biological Research*, vol. 3, Academic Press, New York, 1956, p. 177.
6. Bernstein-Ziv, R., "Cell division in *Mycoplasma gallisepticum*," *Can. J. Microbiol.*, **15**, 1125, 1969.
7. Biberfeld, G., "Discussion," in *Pathogenic Mycoplasmas, Ciba Foundation Symposium*, Elsevier-Excerpta Medica, Amsterdam, 1972, p. 322.
8. ——— and Biberfeld, P., "Ultrastructural features of *Mycoplasma pneumoniae*," *J. Bact.*, **102**, 855, 1970.
9. Black, F. T., Birch-Andersen, A., and Freundt, E. A., "Morphology and ultrastructure of human T-mycoplasmas," *J. Bact.*, **111**, 254, 1972.
10. Blom, J., Ernø, H., and Birch-Andersen, A., "Mycoplasmosis: experimental seminal vesiculitis, electron microscopy of infected tissue," *Acta path. microbiol. scand. B*, **81**, 176, 1973.
11. Boatman, E. S. and Kenny, G. E., "Morphology and ultrastructure of *Mycoplasma pneumoniae* spherules," *J. Bact.*, **106**, 115, 1971.
12. Bredt, W. and Bierther, F. W., "Licht- und elektronenmikroskopische Untersuchungen an *Mycoplasma pneumoniae*," *Zentbl. Parasitkde Abt. I, A*, **229**, 249, 1974.
13. Bruch, J. and Sethi, K. K., "Electron microscopy of ultrathin sections of *Mycoplasma gallisepticum* (strain TT) organisms grown on glass surface." *Pathologia Microbiol.*, **39**, 373, 1973.
14. Chu, H. P. and Horne, R. W., "Electron microscopy of *Mycoplasma gallisepticum* and *Mycoplasma mycoides* using the negative staining technique and their comparison with *Myxovirus*," *Ann. N. Y. Acad. Sci.*, **143**, 190, 1967.
15. Cole, R. M., Tully, J. G., Popkin, T. J., and Bové, J. M., "Ultrastructure of the agent of citrus 'stubborn' disease," *Ann. N. Y. Acad. Sci.*, **225**, 471, 1973.

16. ——, ——, ——, and ——, "Morphology, ultrastructure, and bacteriophage infection of the helical mycoplasma-like organism (*Spiroplasma citri* gen. nov. sp. nov.) cultured from 'stubborn' disease of citrus," *J. Bact.*, **115**, 367, 1973.
17. Collier, A. M., "Pathogenesis of *Mycoplasma pneumoniae* infection as studied in the human foetal trachea in organ culture," in *Pathogenic Mycoplasmas, Ciba Foundation Symposium*, Elsevier-Excerpta Medica, Amsterdam, 1972, p. 307.
18. —— and Clyde, W. A. C., "Relationships between *Mycoplasma pneumoniae* and human respiratory epithelium," *Infec. Immunity*, **3**, 694, 1971.
19. Domermuth, C. H., Nielsen, M. H., Freundt, E. A., and Birch-Andersen, A., "Ultrastructure of *Mycoplasma* species," *J. Bact.*, **88**, 727, 1964.
20. ——, ——, ——, and ——, "Gross morphology and ultrastructure of *Mycoplasma gallisepticum*," *J. Bact.* **88**, 1428, 1964.
21. Edward, D. G. ff. and Freundt, E. A., "Proposal for *Mollicutes* as name of the class established for the order *Mycoplasmatales*," *Int. J. syst. Bacteriol.*, **17**, 267, 1967.
22. Edwards, G. A. and Fogh, J., "Fine structure of pleuropneumonia-like organisms in pure culture and in infected tissue culture cells," *J. Bact.*, **79**, 267, 1960.
23. Freundt, E. A., "Morphological studies of the peripneumonia organism," *Acta path. microbiol. scand.*, **31**, 508, 1952.
24. ——, "Morphology and ultrastructure of the mycoplasmas," in Sharp, J. T. (ed.), *The Role of Mycoplasmas and L Forms of Bacteria in Disease*, Thomas, Springfield, Illinois, 1970.
25. ——, "The mycoplasmas," in R. E. Buchanan and N. E. Gibbons (eds.), *Bergey's Manual of Determinative Bacteriology*, 8 th ed., Williams and Wilkins, Baltimore, 1974, p. 929.
26. Gourlay, R. N. and Wyld, S. G., "Isolation of mycoplasmas from bovine fæces," *Lancet*, **2**, 231, 1975.
27. Green, F. and Hanson, R. P., "Ultrastructure and capsule of *Mycoplasma meleagridis*," *J. Bact.*, **116**, 1011, 1973.
28. Hara, K., Izumikawa, K., Kinoshita, I., Ota, M., Ikebe, A., Koika, M., and Hamada, M., "Experimental infection with *Mycoplasma pneumoniae* in the young hamster: location of ferritin-labelled antibody binding to infective tissue," *Tohuku J. exp. Med.*, **114**, 315, 1974.
29. Hayflick, L. (ed.), *The Mycoplasmatales and the L-Phase of Bacteria*, Appleton-Century-Crofts, New York, 1969.
30. Horne, R. W., "The ultrastructure of mycoplasma and mycoplasma-like organisms," *Micron*, **2**, 19, 1970.
31. Howard, C. J. and Gourlay, R. N., "An electron-microscopic examination of certain bovine mycoplasmas stained with ruthenium and the demonstration of a capsule on *Mycoplasma dispar*," *J. gen. Microbiol.*, **83**, 393, 1974.
32. Hummeler, K. and Armstrong, D., "Observations on mycoplasma strains in tissue cultures," *Ann. N. Y. Acad. Sci.*, **143**, 622, 1967.

33. ——, ——, and Tomassini, N., "Cytopathogenic mycoplasmas associated with two human tumors," *J. Bact.*, **90**, 511, 1965.
34. ——, Tomassini, N., and Hayflick, L., "Ultrastructure of a *Mycoplasma* (Negroni) isolated from human leukemia," *J. Bact.*, **90**, 517, 1965.
35. Iterson, W. van and Ruys, A. C., "The fine structure of the Mycoplasmataceae (microorganisms of the pleuropneumonia group = PPLO)," *J. Ultrastruct. Res.*, **3**, 282, 1960.
36. James, R. and Branton, D., "Lipid and temperature-dependent structural changes in *Acholeplasma laidlawii* cell membranes," *Biochim. biophys. Acta*, **323**, 378, 1973.
37. Jones, T. C. and Hirsch, J. G., "The interaction *in vitro* of *Mycoplasma pulmonis* with mouse peritoneal macrophages and L-cells," *J. exp. Med.*, **133**, 231, 1971.
38. Kammer, G. M., Pollack, J. D., and Klainer, A. S., "Scanning-beam electron microscopy of *Mycoplasma pneumoniae*," *J. Bact.*, **104**, 499, 1970.
39. Kim, K. S., Clyde, W. A., and Denny, F. W., "Physical properties of human *Mycoplasma* species," *J. Bact.*, **92**, 214, 1966.
40. Klainer, A. S. and Pollack, J. D., "Scanning electron microscopy techniques in the study of the surface structure of mycoplasmas," *Ann. N. Y. Acad. Sci.*, **225**, 236, 1973.
41. Knudson, D. L. and MacLeod, R., "*Mycoplasma pneumoniae* and *Mycoplasma salivarium*: electron microscopy of colony growth in agar," *J. Bact.*, **101**, 609, 1970.
42. Maniloff, J., "Subcellular organelles of mycoplasma," *Biophys. J.*, **9**, abstract A175, 1969.
43. ——, "Cytology of mycoplasmas," in *Pathogenic Mycoplasmas, Ciba Foundation Symposium*, Elsevier-Excerpta Medica, Amsterdam, 1972, p. 67.
44. —— and Morowitz, H. J., "Ultrastructure and life cycle of *Mycoplasma gallisepticum* A5969," *Ann. N. Y. Acad. Sci.*, **143**, 59, 1967.
45. —— and ——, "Cell biology of the mycoplasmas," *Bact. Rev.*, **36**, 263, 1972.
46. ——, ——, and Barrnett, R. J., "Studies of the ultrastructure and ribosomal arrangements of the pleuropneumonia-like organism A5969," *J. Cell Biol.*, **25**, 139, 1965.
47. ——, ——, and ——, "Ultrastructure and ribosomes of *M. gallisepticum*," *J. Bact.* , **90**, 193, 1965.
48. Masover, G. K., Sawyer, J. E., and Hayflick, L., "Urea-hydrolyzing activity of T-strain mycoplasma: *Ureaplasma urealyticum*," *J. Bact.*, **125**, 581, 1976.
49. McNulty, M. S., Curran, W. L., and McFerran, J. B., "Virus-like particles in calves fæces," *Lancet*, **2**, 78, 1975.
50. Metz, J. and Bredt, W., "Elektronenmikroskopische Untersuchungen an *Mycoplasma hominis* (Stamm W 463/691)," *Z. med. Mikrobiol. Immunol.*, **156**, 368, 1971.
51. Morowitz, H. J. and Maniloff, J., "Analysis of the life cycle of *Mycoplasma gallisepticum*," *J. Bact.*, **91**, 1638, 1966.

52. Munkres, M. and Wachtel, A., "Histochemical localization of phosphatases in *Mycoplasma gallisepticum*," *J. Bact.*, **93**, 1096, 1967.
53. Muse, K. E., Powell, D. A., and Collier, A. M., "*Mycoplasma pneumoniae* in hamster tracheal organ culture studied by scanning electron microscopy," *Infec. Immunity*, **13**, 229, 1976.
54. Nakamura, M. and Kawaguchi, M., "Ultrastructure of *Mycoplasma orale* serotype 1 in agar growth," *J. gen. Microbiol.*, **70**, 305, 1972.
55. Organick, A. B., Siegesmund, K. A., and Lutsky, I. I., "Pneumonia due to *Mycoplasma* in gnotobiotic mice," *J. Bact.*, **92**, 1164, 1966.
56. Peterson, J. E., Rodwell, A. W., and Rodwell, E. S., "Occurrence and ultrastructure of a variant (*rho*) form of *Mycoplasma*," *J. Bact.*, **115**, 411, 1973.
57. Pollack, J. D., "Localization of reduced nicotinamide adenine dinucleotide oxidase activity in *Acholeplasma* and *Mycoplasma* species," *Int. J. syst. Bacteriol.*, **25**, 108, 1975.
58. ———, Razin, S., and Cleverdon, R. C., "Localization of enzymes in *Mycoplasma*," *J. Bact.*, **90**, 617, 1965.
59. Razin, S., "The mycoplasma membrane," *Prog. Surf. Membrane Sci.*, **9**, 257, 1975.
60. Robertson, J., Gomersall, M., and Gill, P., "Virus-like particles in *Mycoplasma hominis*," *Can. J. Microbiol.*, **18**, 1971, 1972.
61. ———, ———, and ———, "*Mycoplasma hominis*: growth, reproduction, and isolation of small viable cells," *J. Bact.*, **124**, 1007, 1975.
62. Rodwell, A. W., Peterson, J. E., and Rodwell, E. S., "Nature of striated structures in mycoplasmas," *Ann. N. Y. Acad. Sci.*, **225**, 190, 1973.
63. ———, ———, and ———, "Striated fibers of the *rho* form of Mycoplasma: *in vitro* reassembly, composition, and structure," *J. Bact.*, **122**, 1216, 1975.
64. Rottem, S., Yashouv, Ne'eman, Z., and Razin, S., "Cholesterol in mycoplasma membranes. Composition, ultrastructure and biological properties of membranes from *Mycoplasma mycoides* var. *capri* cells adapted to grow with low cholesterol concentrations," *Biochim. biophys. Acta*, **323**, 495, 1973.
65. Ruys, A. C. and Iterson, W. van, "Some characteristics of pathogenic avian PPLO," *Antonie van Leeuwenhoek*, **27**, 129, 1961.
66. Schiefer, H.-G., Krauss, H., Brunner, H., and Gerhardt, U., "Ultrastructural visualization of surface carbohydrate structures on membranes by concanavalin A," *J. Bact.*, **124**, 1598, 1975.
67. Schwanz-Pfitzner, I. and Özel, M., "Elektronmikroskopischer Nachweis einer Mischinfektion in Zellkulturen: durch Egtved-Virus der Regenbogenforelle und Mykoplasmen (*Acholeplasma laidlawii*)," *Zentbl. Parasitkde Abt. I, A.*, **225**, 431, 1973.
68. Tillack, T. W., Carter, R., and Razin, S., "Native and reformed *Mycoplasma laidlawii* membranes compared by freeze-etching," *Biochim. biophys. Acta*, **219**, 123, 1970.
69. Tomkins, A. M., Wright, S. G., Bird, R. G., and James, W. P. T., "Virus-like particles in jejunal mucosa," *Lancet*, **2**, 36, 1975.
70. Tourtellotte, M. E., Branton, D., and Keith, A., "Membrane structure: spin

labeling and freeze etching of *Mycoplasma laidlawii*," *Proc. natn. Acad. Sci. U.S.A.*, **66**, 909, 1970.
71. —— and Zupnik, J. S., "Freeze-fractured *Acholeplasma laidlawii* membranes: nature of particles observed," *Science*, **179**, 84, 1973.
72. Trung, P. H., Boué, A., Lardemer, F., and Haguenau, F., "Ultrastructure du *Mycoplasme salivarium* et du *Mycoplasme orale* en culture de tissus," *Eur. J. Cancer*, **4**, 429, 1968.
73. Tully, J. G., Whitcomb, R. F., Williamson, D. L., and Clark, H. F., "Suckling mouse cataract agent is a helical wall-free prokaryote (spiroplasma) pathogenic for vertebrates," *Nature, Lond.*, **259**, 117, 1976.
74. Verkleij, A. J., Ververgaert, P. H. J., van Deenen, L. L. M., and Elbers, P. F., "Phase transitions of phospholipid bilayers and membranes of *Acholeplasma laidlawii* B visualized by freeze fracturing electron microscopy," *Biochim. biophys. Acta*, **288**, 326, 1972.
75. Vinther, O., "Localization of urease activity in *Ureaplasma urealyticum* cells," *Acta path. microbiol. scand. B*, **84**, 217, 1976.
76. Williams, M. H., "Electron microscopy of T-strains," *Ann. N. Y. Acad. sci.*, **143**, 397, 1967.
77. Woodruff, K. H., Schneider, E., Unger, L., and Coalson, J. J., "Ultrastructural changes in hamster tracheal ring cultures exposed to *Mycoplasma pneumoniae*," *Am. J. Path*, **72**, 91, 1973.
78. Zucker-Franklin, D., Davidson, M., and Thomas, L., "The interaction of mycoplasmas with mammalian cells. I. Hela cells, neutrophils, and eosinophils," *J. exp. Med.*, **124**, 521, 1966.
79. ——, ——, and ——, "The interaction of mycoplasmas with mammalian cells. II. Monocytes and lymphocytes," *J. exp. Med.*, **124**, 533, 1966.

Supplementary references

80. Barile, M. F., Razin, S., Tully, J. G., and Whitcomb, R. F. (eds.), *The Mycoplasmas*, Vol. 1–3, Academic Press, New York, 1979.
81. Vinther, O. and Freundt, E. A., "Ultrastructural localization of tellurite reduction in *Acholeplasma* species," *Acta path. microbiol. scand. B*, **85**, 184, 1977.
82. —— and ——, "Electron microscopical study of antibody binding to *Mycoplasma gallisepticum*: indirect immunoferritin labelling," *Acta path. microbiol. scand. B*, **87**, 37, 1979.

Part Four

Bacteria

4. Bacteria

Stanley C. Holt

4.1. Introduction

4.1.1. General techniques

The preparation of biological tissue for examination with the transmission electron microscope (TEM) involves a variety of techniques which include shadowcasting, surface replication, and freeze-fracture etching, positive and negative staining, and thin sectioning of chemically fixed material. These latter techniques have played significant roles in the elucidation of the ultrastructural anatomy of biological tissue. The advent of the scanning electron microscope (SEM) has provided a clearer three-dimensional representation of cellular organelles than was possible by other techniques (i.e., replication, freeze-etching).

The technique of shadowcasting first developed by Williams and Wyckoff[215] involves the coating of a specimen with a heavy or high density metal (i.e., uranium, gold, palladium–platinum) evaporated under high vacuum. This deposition of metal not only provides a clear picture of the cell in surface relief, but also, when deposited at a known inclined angle, enables an accurate measure of particle and organelle geometry (i.e., size and height; for example, Figures 4.1, 4.2(d), 4.6(b), 4.30(e), 4.34(g)). However, the procedure is limited in resolution both by the metal and the vacuum system employed. The metal not only must be capable of producing good contrast, but must do so when present as a thin film of low granularity. Attainment of the latter is a function of the vacuum system; high vacuums produce finer metal granularity than lower vacuums. The high melting metals and alloys such as iridium, platinum–carbon, and palladium–platinum provide excellent contrast and are of low granularity (see Table 1 in Margaritis *et al.*[135]). The vacuum system must be capable of high vacuum (10^{-5} Torr) to provide an unencumbered path of the evaporated metal to the specimen surface. Shielding the evaporation electrodes diminishes potential artifacts due to heating, especially shrinkage and surface alterations.

Surface detail of biological material can also be observed by the use of the surface replica technique; however, it has enjoyed most of its success in the metallurgical field in the examination of, for example, metals and crystals.[23] A solution for replicating (i.e., collodion) is layered onto the surface of an air or vacuum dried sample, and after drying is "stripped" from the surface. This stripping action provides an imprint of the material under study. The stripped

(a)

(b)

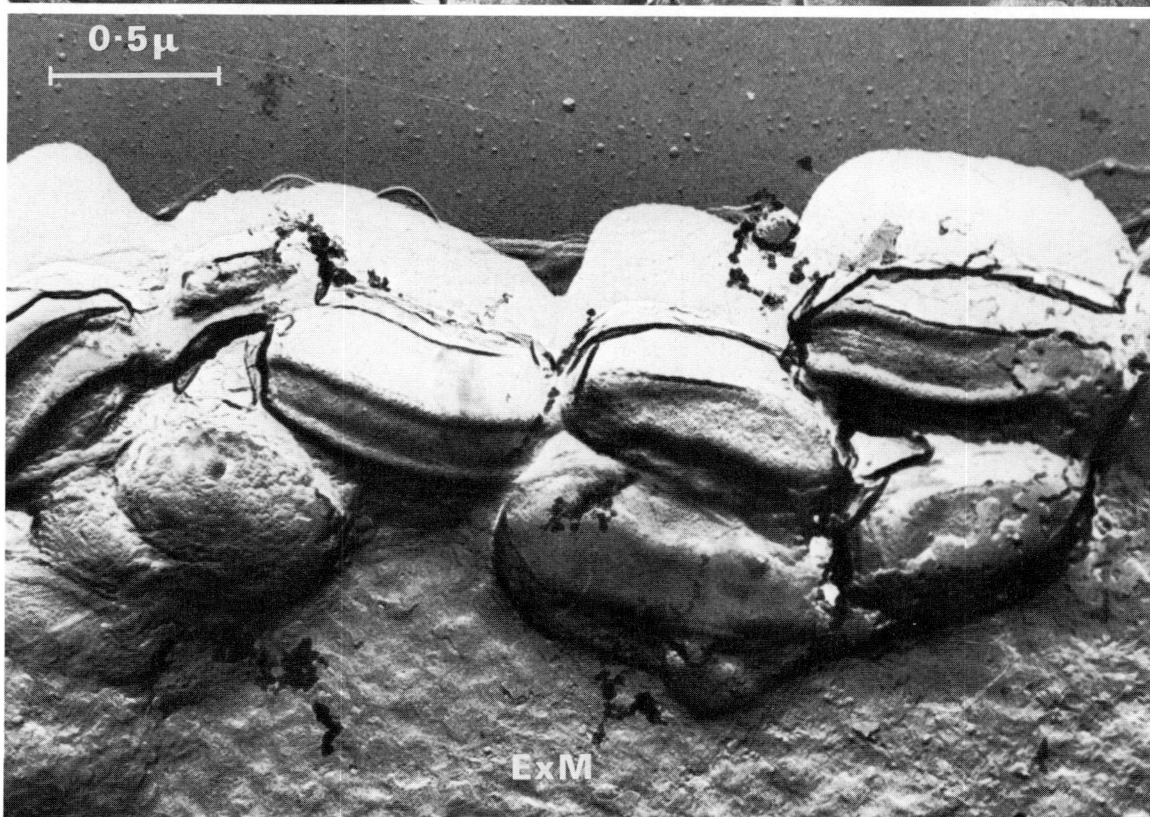

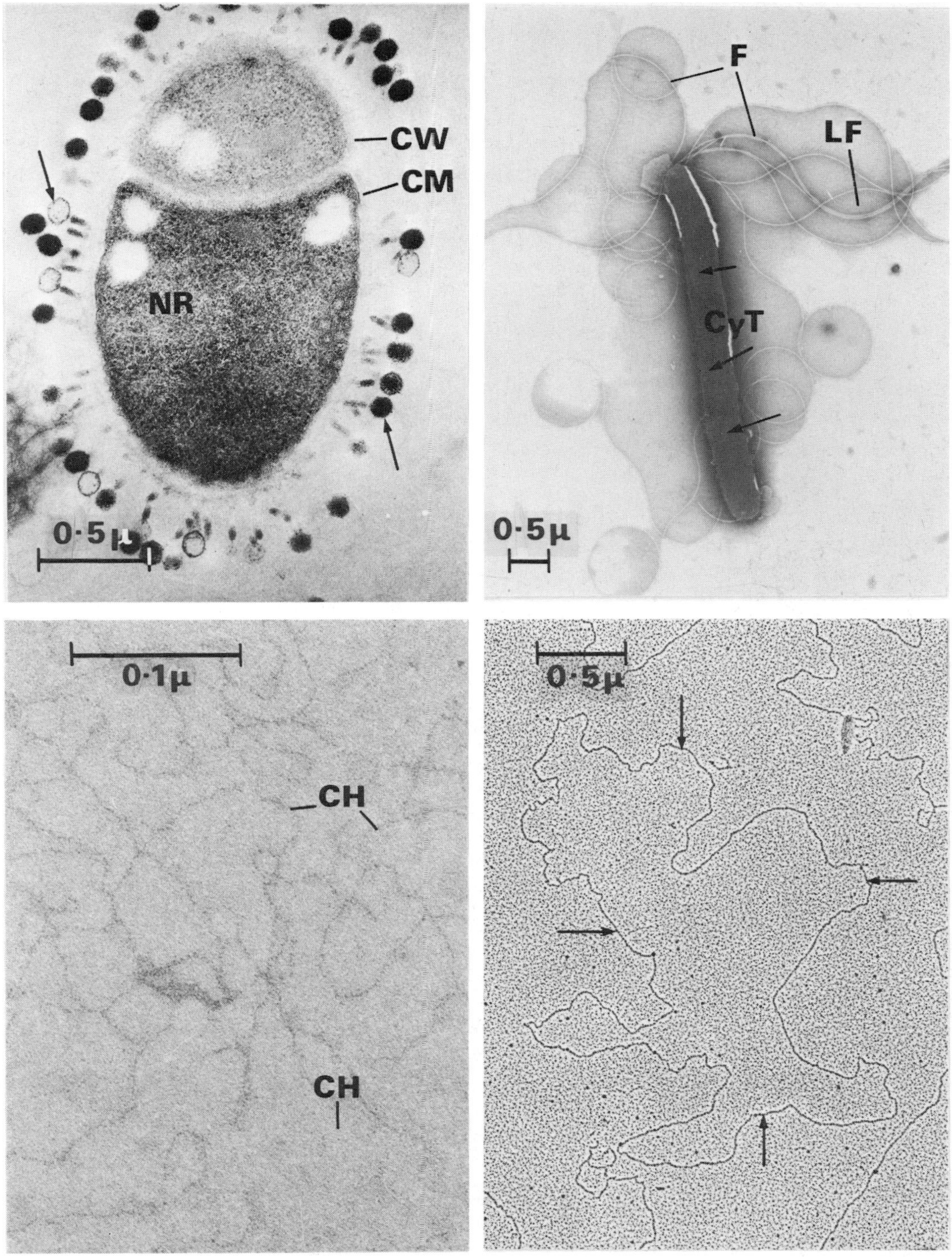

Figure 4.2. Positive and negative staining, and shadowcasting.

(a) Thin section of *Bacillus subtilis* infected with bacteriophage CP1 (arrows) after chemical fixation with glutaraldehyde-osmium and embedding in the epoxide Epon 812. CW: cell wall, CM: cell membrane, NR: nuclear region.

(b) Negatively stained (3 percent PTA, pH 6.8) *Protaminobacter ruber.* The cytoplasmic (CyT) region (arrow) stains electron dense (positive staining), while the flagella (F) are surrounded with stain in negative image (negative staining).

(c) Extracted chromatin (CH). Unfolded DNA from a chicken red blood cell positively stained with uranyl acetate appears as thin electron-dense threads.

(d) Rotary shadowcast (Pt–Pd, 3°) extracted lambda viral DNA. The complete circular chromosome (outlined by arrows) and its particulate fine detail is apparent.

Figure 4.1. Surface replication.

Freeze-fracture etched replicas of free spores of (a) *Bacillus fastidiosus*, and (b) human oral *Bacillus* isolate. The spores are covered with a granular fibrillar layer. The ridged coat is apparent through the exosporium. In (b) the spores are embedded in a thick matrix of extracellular material (ExM).

replica is then shadowcast by conventional shadowcasting techniques. By providing a faithful imprint of the object surface, the replica technique permits the discrimination of finer specimen detail than that obtained by direct shadowcasting. Several laboratories have adapted the replica technique to the examination of the surfaces of bacterial cells (Figures 4.1 and 4.17(c)).[22, 23, 99]

Two of the most useful techniques for the examination of biological samples by electron microscopy are positive[85] and negative staining.[162] In both procedures the sample to be examined is suspended in a diluent (distilled water, ammonium acetate, growth medium) and placed, in droplet form, on the surface of an electron microscope grid. Excess sample is removed, a drop of stain is mixed with the sample on the grid surface and allowed to react for a short period of time (30–60 sec). For positive staining, heavy metal ions (i.e., uranyl ions as uranyl acetate) are bound to the sample, most probably through charge interaction. The resulting "mordanted" sample is then examined in the electron microscope. An excellent example of the application of positive staining is the visualization of nucleic acid after interaction with uranyl ions (Figure 4.2(c)). The "naked" nucleic acid strands are so thin as to be essentially invisible when unstained, but "attachment" of uranyl ions permits their easy visualization. In contrast to positive staining, where stain-binding appears to result from charge interaction, in negative staining the sample on the electron microscope grid surface is surrounded by a material of high electron opacity (high atomic number). The stains which exhibit this characteristic electron opacity and high atomic number, and which have been used most often, and most successfully, include phosphotungstic acid, uranyl acetate and phosphomolybdic acid. In most instances (with the exception of uranyl acetate) negative stains are prepared as their potassium or ammonium salt. Since biological specimens usually consist of atoms of low atomic number, they do not, to any appreciable degree, scatter electrons produced in the electron microscope. The film of stain dries very rapidly, solidifying into a smooth film or glass. Not only does the stain surround the sample by surface tension, but it also penetrates crevices and macromolecular openings. The resulting image seen in the electron microscope is that of an almost electron-transparent biological specimen, the result of its low electron scattering capacity, surrounded by the electron opaque stain (e.g., Figures 4.2(b), 4.6(f), (g), 4.8(b), 4.13(a), (b)). Fine structural details as small as 0·015–0·02 μm in diameter are capable of resolution by this technique.

The most often used technique for the examination of biological material in the electron microscope is that of ultrathin sectioning of chemically fixed tissue.[77] The specimen to be examined is fixed either by submersion in the fixative or exposure to its vapors. Traditionally, osmium tetroxide or potassium permanganate have been used as fixatives.[113, 127] In recent years advantage has been taken of the rapid penetration and protein stabilizing properties provided by several aldehydes, notably formaldehyde–paraformaldehyde[109] and glutaraldehyde.[174] The use of these aldehyde fixatives requires a postfixation in osmium tetroxide in order to increase tissue contrast and to stabilize, as much as possible, lipid moieties which could be extracted by subsequent exposure to organic solvents (ethanol or acetone), prior to embedding in plastic resins, such as the Epons, Araldites, or Vestopals.[128] Increasingly popular, because of its ease of handling and excellent tissue penetration properties, has been Spurr's low viscosity embedding resin,[186] and the water-soluble glycol methacrylates for electron microscopic cytochemistry.[170] The sample is infiltrated with the embedding agent, cured and polymerized by heat or ultraviolet irradiation, and the embedded sample then cut into ultrathin sections, 0·4–0·6 μm in thickness.

Tissue contrast in sectioned material is enhanced with heavy metal stains, the most common of which are lead citrate[108, 166] and uranyl acetate. It is also possible to fix and stain specimens in order to localize specific enzymes and biological polymers. For example, acidic mucopolysaccharides have been localized with ruthenium red.[129, 130] The ruthenium red positive material has been found to be an integral constituent of some mammalian cell membranes, possibly involved in cell–cell, cell–surface interactions.[160] In some bacteria ruthenium red-positive polysaccharide polymers are found as the most external layer of the cell wall (e.g., Figures 4.9(b), 4.28(j), (k), 4.31(a), (b)). Thorium hydroxide has also been used successfully in the localization of acidic exopolysaccharides (Figure 3.24(n), (p)).

Esterases, phosphatases, and nucleosidases have also been localized in ultrathin cell sections by specific reaction product deposition (Figure 4.8(a), (d)).[39, 46] The technique of immuno-electron microscopy[181, 182] which employs antibody–ferritin or antibody–peroxidase conjugates is emerging as an especially powerful tool in the localization of specific cellular components (Figure 4.3(c), (d)). The technique is receiving increased attention in the determination of the distribution of bacteria within a heterogeneous population.[120]

Only in recent years has the technique of freeze-fracture etching emerged as a powerful tool for the examination of biological material (compare Figures 4.4(a),(b),(d),(e), 4.9(g), 4.29(d),(f))[21, 24, 99, 137, 143] The underlying significance of this technique is that the preparative procedures need not destroy tissue or cell viability, as is the case for chemical fixation. Thus, the number of artifacts attributed to specimen preparation are kept at a minimum. Samples are frozen on a gold-plated or copper grid in liquid freon which has been cooled to its freezing point with liquid N_2. The frozen specimen is fractured, or cut, with a sharp knife or razor blade either *in vacuo* or at atmospheric pressure. The fracturing process reveals surface domains which reflect the passage of the knife through theoretical fracture planes within the specimen (Figure 4.4(c)).[25] Since the ice level is at the plane of the fractured specimen, specimen contrast is not only poor, but many portions of the replica are "hidden" in the ice layer. A temperature differential between the fractured specimen and knife or heating block is maintained *in vacuo* for from several seconds to several minutes, resulting in the sublimation or vitrification of ice away from the specimen. This etching process permits the visualization of structural detail not apparent in unetched fractured samples. The final stage of the process involves replication with a heavy metal (e.g., platinum–carbon) which provides an exact replica and decoration of the fractured and etched surface (Figures 4.3(c),(d), 4.4(a),(d),(e), 4.24(b), 4.29(f), 4.43(g)–(i)), which is then viewed in the electron microscope.

With improvement in instrumentation, specifically with major increases in resolution, the scanning electron microscope has made a substantial and lasting

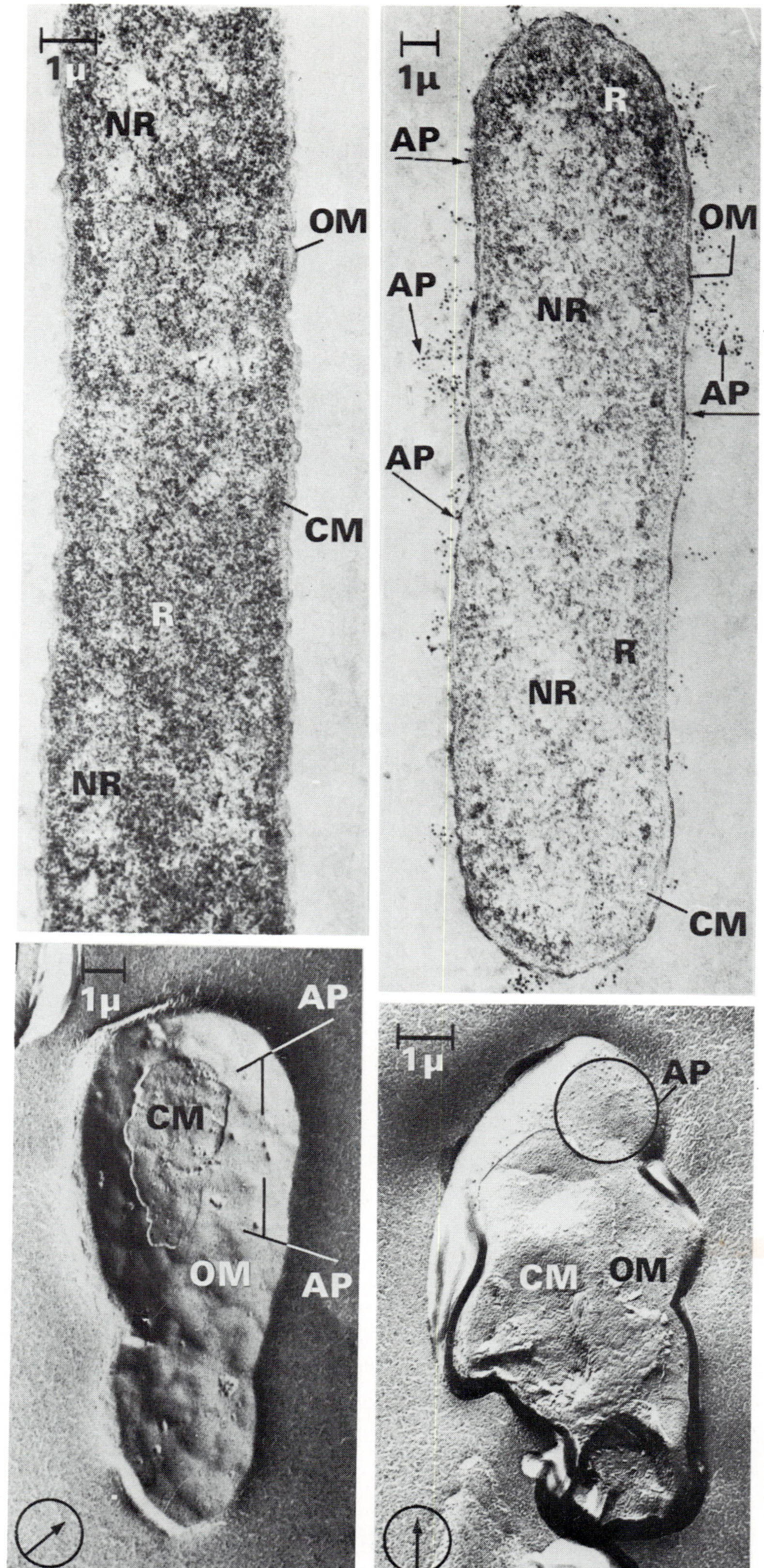

(a) (b) (c) (d)

Figure 4.3. Localization of enzymes.
Localization of alkaline phosphatase (APase) in (a), (b) thin sections of chemically fixed cells and (c), (d) freeze-fracture etched cells of *Escherichia coli*, strain F515. Control cell (a) shows no labeling with anti-APase–ferritin conjugate, while in (b) cells depressed for APase and reacted with anti-APase–ferritin conjugate show electron-dense ferritin particles (AP) at discrete regions along the surface of the outer cell membrane. By freeze-fracture-etching (c), (d), the anti-APase–ferritin conjugate particles (AP) are distributed over the outer cell membrane surface after 4·5 hour growth of F515 in phosphate-limiting conditions (c); 8·5 hour growth of F515 revealed the ferritin label to be more evenly distributed over the cell surface, and to a much higher density (d). NR: nuclear region, R: ribosome, CM: cell membrane, OM: outer cell membrane.

(a)

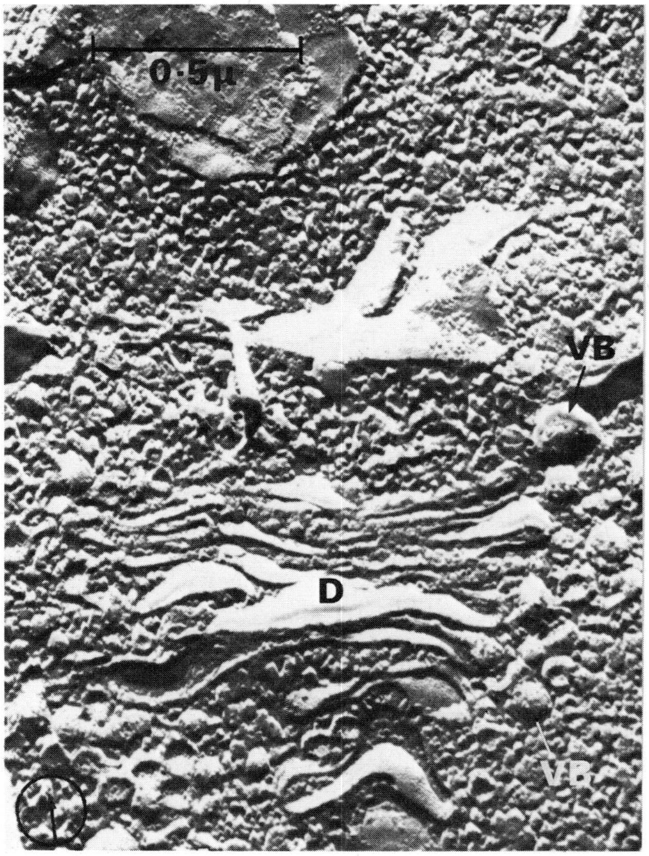

0·5μ
VB
D
VB

(b)

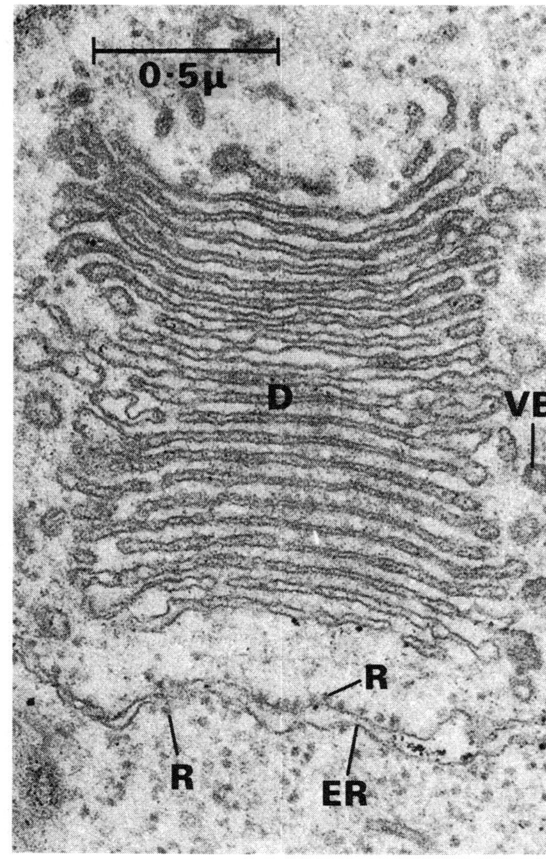

0·5μ
D
VB
R
R
ER

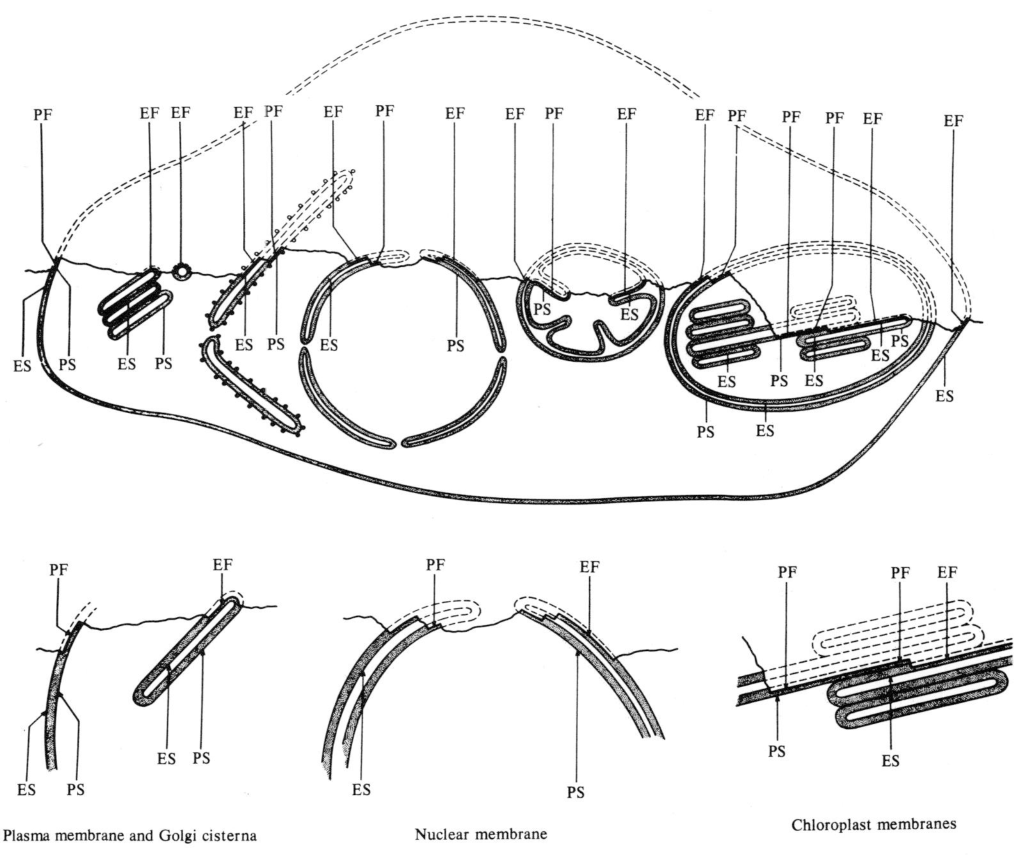

PF
EF
ES
PS
Plasma membrane and Golgi cisterna
Nuclear membrane
Chloroplast membranes

(c)

Figure 4.4. Freeze-fracture etching.
Comparison of (a) freeze-fracture-etched and (b) thin-section of chemically fixed dictyosome (Golgi complex) from *Trichomonas vaginalis.* In (a) the dictyosomes (D) is closely stacked, and contains vesicle bodies (VB) at the polar regions. In (b) the stacking of the dictyosome (D) is clearly defined, but its three-dimensional character is missing. R: ribosome, ER: endoplasmic reticulum. (c) Designations for frequently studied membrane fracture faces. The dark line through the cell and some of its organelles traces the course of hypothetical fracture prior to etching. PF and EF labels are used to distinguish the faces of the P and E halves of each membrane produced by fracture, while PS and ES indicate the true hydrophilic surfaces that could be exposed by etching.[25]

(d)

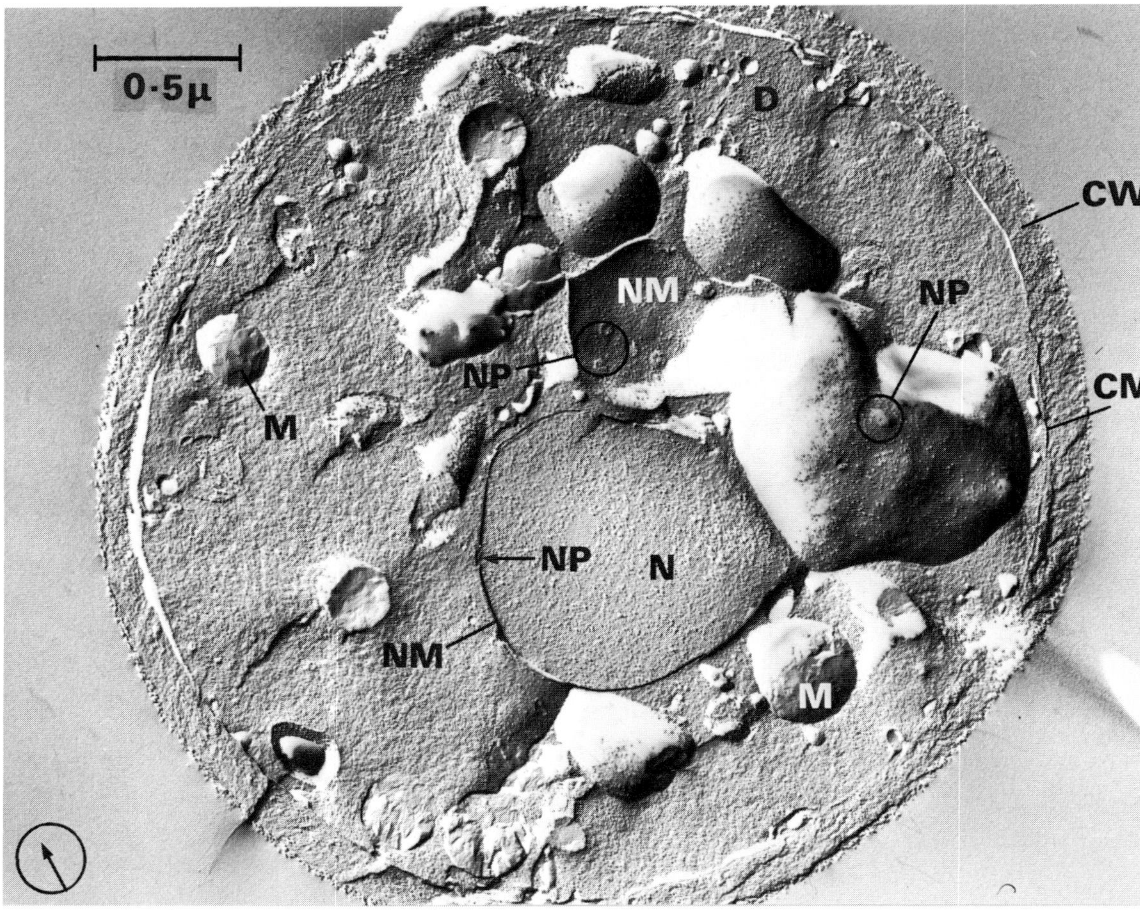

(e)

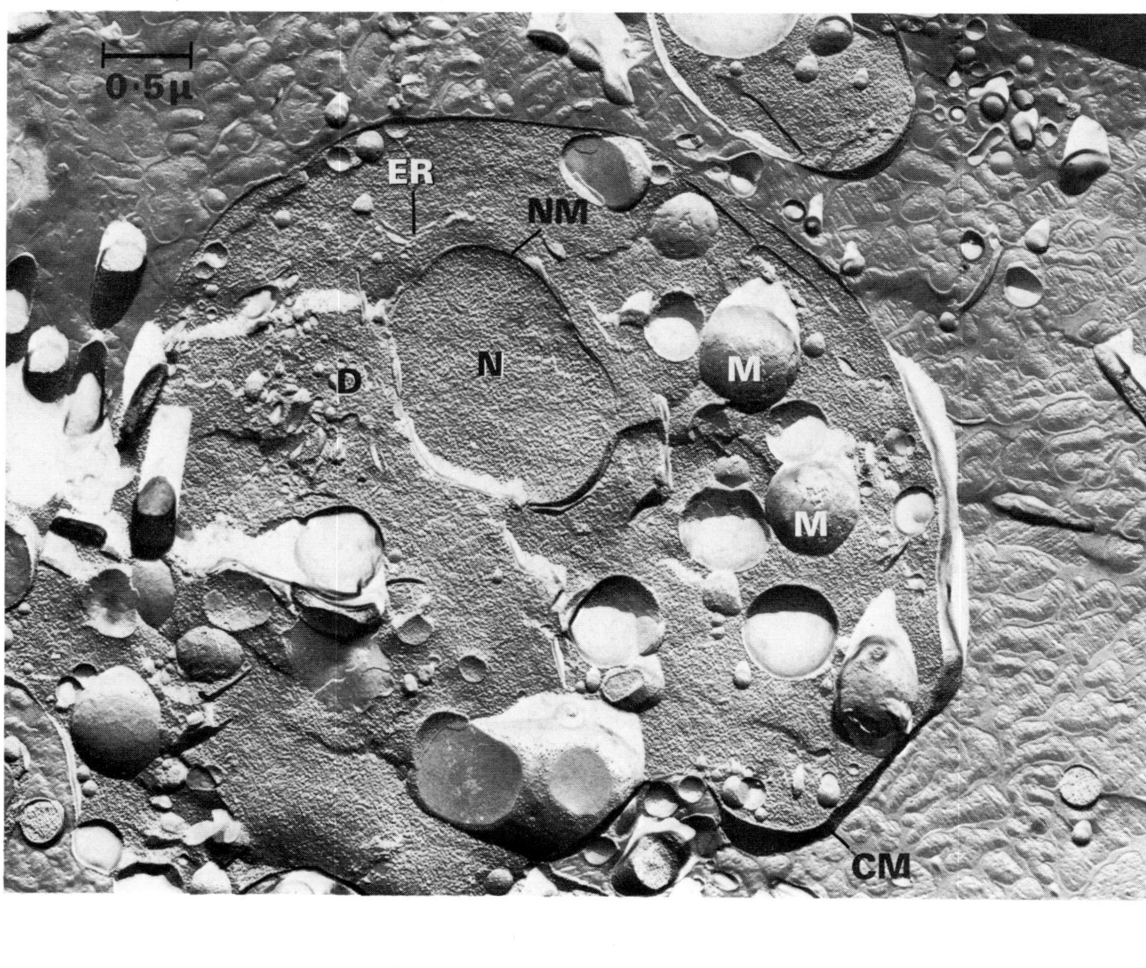

impact on the examination of both biological and metallurgical material.[53, 159] The preparative procedure for biological tissue for SEM, critical point drying (CPD), results in a sample with a minimum of collapse and deformation. Of considerable consequence is the fact that CPD tissue, especially bacteria, can be examined by both SEM and TEM (compare Figure 4.23(d),(m)). In practice, the normal drying of biological tissue with its numerous pits, crevices, and entrapped liquids (i.e., water) in air results in severe surface damage due to the extreme surface tension forces acting on the specimen surface. These surface tension forces are generated within the pits and crevices at liquid–gas interfaces. As the tissue dries the liquid–gas interface travels through the tissue surface, resulting in collapse. Elimination of the liquid–gas interface by raising the critical point of the dehydration liquid (acetone) above its critical temperature results in a transition from liquid to gas without an interface, and the elimination of any structural damage (Polaron Inc., personal communication).

4.1.2. Comparison of procaryotic and eucaryotic ultrastructure

Electron micrographs of thin sections of a metazoan (animal) cell (Figure 4.5(a)) reveal the presence of characteristic cellular organelles. Immediately apparent is the number of these membrane-limited organelles, especially the large membrane-limited nucleus containing chromatin material. Surrounding the nucleus is the rough endoplasmic reticulum as well as dictyosomes, or Golgi apparatus. Scattered throughout the cell cytoplasm are numerous mitochondria and inclusion granules. Table 4.1 (pp. 259–263) provides an outline of all the procaryotic and eucaryotic strains presented in this chapter with their sources of information.

Protozoa (Figure 4.5(b)) are similar in design and organization to the metazoan animal cell. There is, however, some structural differentiation and specialization which is almost unique to these organisms. This specialization is most clearly seen in the ciliary and flagellar apparatus.

The algae, as exemplified by *Porphyridium cruentum* (Figure 4.5(c)) contain a large, centrally disposed, membrane-limited pyrenoid surrounded by the large membrane-bound chloroplast. Numerous storage vacuoles and several mitochondria fill the electron-dense cytoplasm.

In marked contrast to the large number of membrane-bound organelles found in animal cells, protozoa, and algae, the procaryotic (bacterial) cell is reduced to a highly simplistic organization (e.g., Figures 4.3(a), 4.6(c), 4.8(a)). In some genera, notably *Mycoplasma* and *Spiroplasma*, the cell wall is absent, leaving the cell membrane as the most external and limiting layer (e.g., Figures

---

Figure 4.4. Freeze-fracture etching.
(d) *Cryptococcus neoformans* and (e) *Histomonas gallinae* after freeze-fracture etching showing the three-dimensional relief of the numerous organelles present in eucaryotic cells as well as the various membrane fracture faces (see Figure 4.4(c)) which occur during the preparative procedure. Various fracture planes have exposed the membrane (NM)-limited nucleus (N) at the level of chromatin and nuclear pores (NP), as well as revealing several membrane faces of various mitochondria (M). CW: cell wall, CM: cell membrane, D: dictyosome, ER: endoplasmic reticulum.

(a)

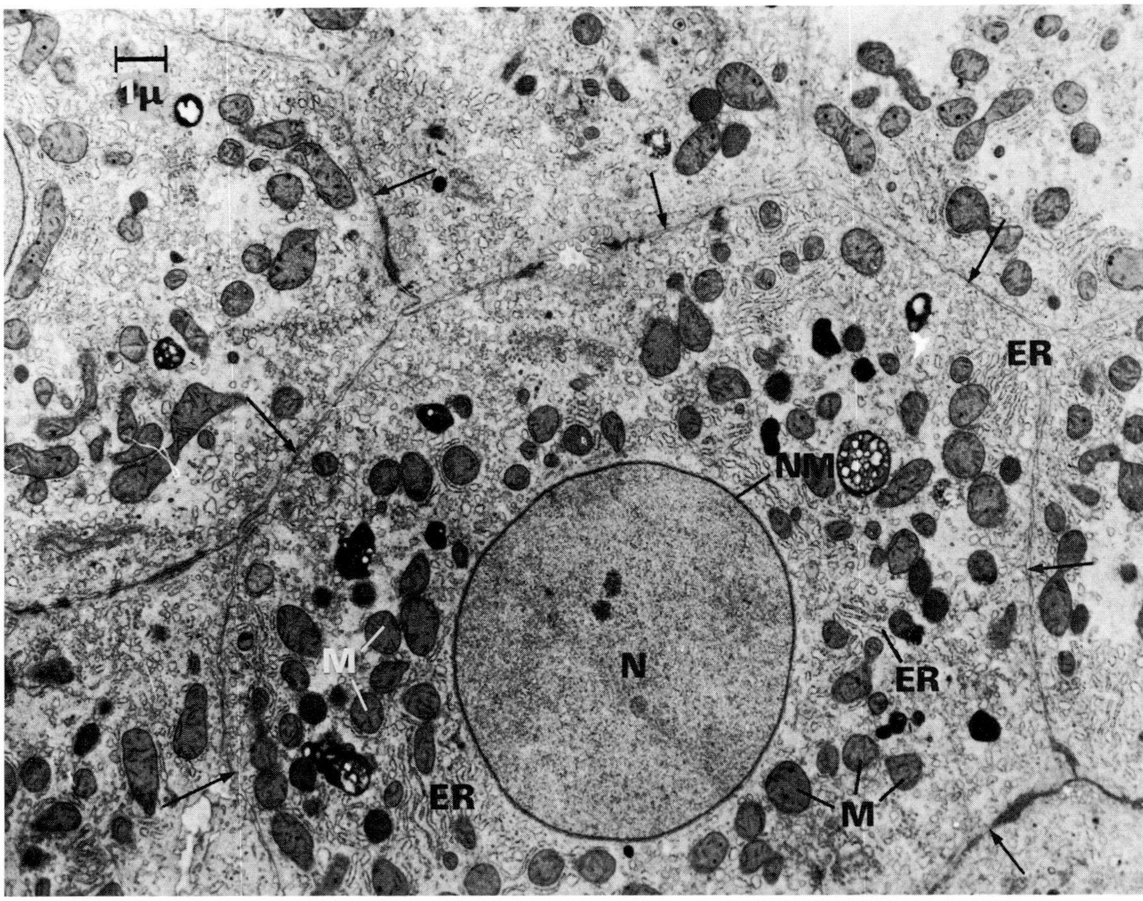

(b)

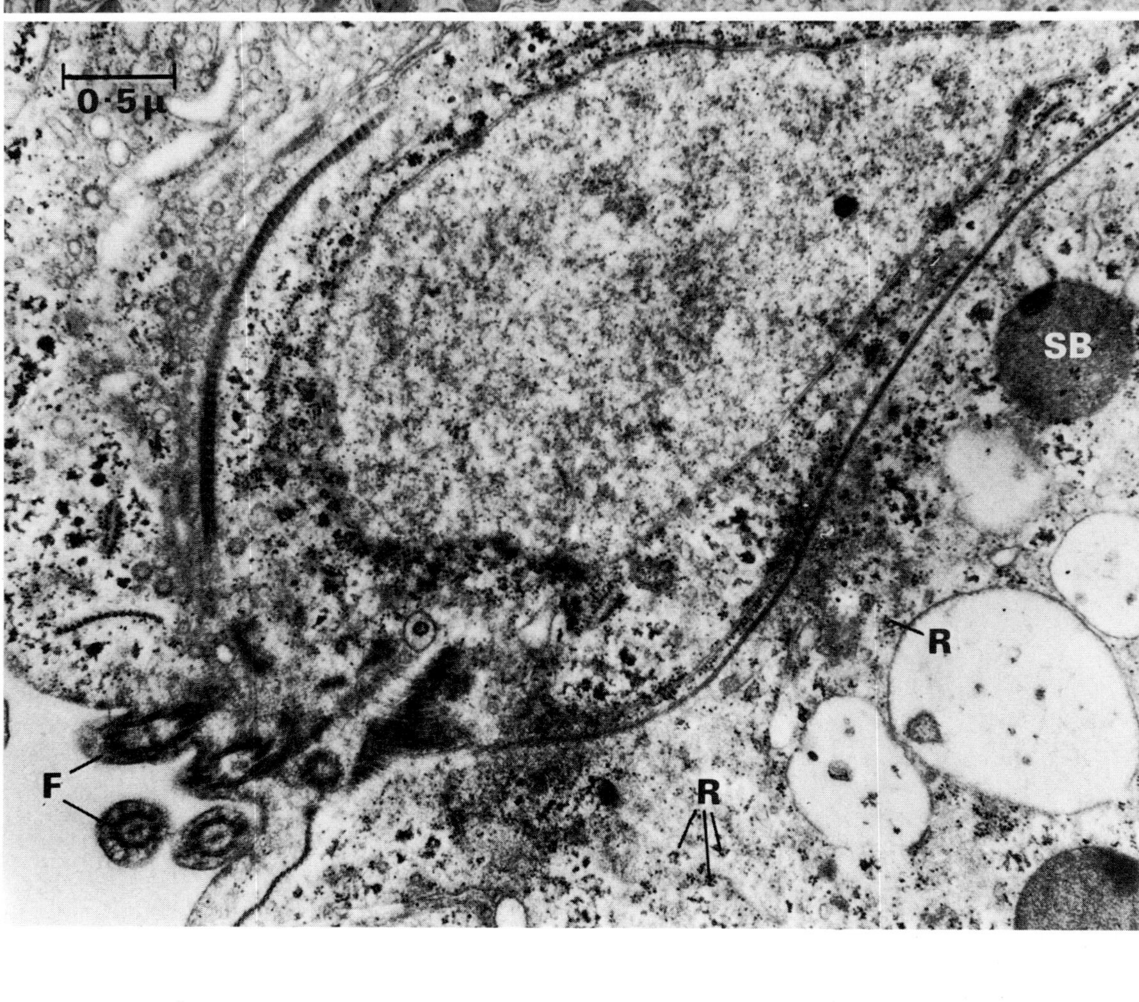

4.39(b),(u), 4.40(d)). The non-membrane-limited nucleus (commonly referred to as the nucleoid or nuclear region) usually occupies the central region of the cell. Surrounding and in some cases intermeshed with the nucleoid are small, electron-dense ribosomes. In some cases the outer layer of the cell wall has an associated fibrous or muclilagenous capsule. Procaryotic cellular specialization appears in the form of flagella (eg., Figures 4.2(b), 4.5(f),(g), 4.30(b),(c), 4.35(g),(h)), pili, or fimbriae (Figures 4.24(c),(k), 4.25(c), 4.30(d)), and in the elaboration of internal membranes such as mesosomes (e.g., Figures 4.8(a),(b), 4.18(d)), vesicles, or intracytoplasmic photosynthetic chromatophores. Heat resistant endospores (e.g., Figure 4.1, 4.6(h),(k)), of considerable importance to the food-processing industry (i.e., botulism) as well as being the causative agents of such diseases as tetanus and gas gangrene, are also formed in the interior of cells of certain bacteria. Depending upon the environmental conditions (C- or N-limitation, for example), procaryotic cells will store numerous granules of various sizes, dimensions, and consistency.[179]

The intent of this chapter is to provide a pictorial representation of the wide range of procaryotic cells frequently found associated with the human host. In some instances, particular genera of bacteria are presented which are known to be associated with human infection, while in other instances, procaryotes are described which are found as common (normal) inhabitants of the host. An attempt has been made to provide the reader with as much descriptive material as possible so that a clear morphological distinction can be made between the major groups of bacteria, especially between gram-positive and gram-negative cells. In addition, the presence of surface associated components (capsules, pili) have implications for the health scientist as a basis for the association(s) that may exist between a given bacterium, its host, and the production of a particular disease.[45]

The literature concerned with electron microscopic observations of human host-related bacteria is exhaustive, especially with regard to some of the more regularly studied genera (i.e., streptococci); however, for some of the less common, but medically important groups (i.e., *Bifidobacteria, Yersinia*), it is almost nonexistent. Wherever possible, electron micrographs of as wide an array of procaryotic cell types have been included in order to provide a basis for an understanding of the ultrastructural anatomy of procaryotes in general. A representative but *not* exhaustive bibliography is included which will direct the reader to specific areas of interest.

Figure 4.5. Comparison of mammalian, protozoan, and algal cell

(a) Thin section of a chemically fixed (glutaraldehyde-osmium) human liver cell from a patient with Whipple's disease. At least seven individual cells are separated from each other by unit cell membranes (arrows). The largest cell contains a large central nucleus (N), numerous mitochondria (M), endoplasmic reticulum (ER), and lysosomes. NM: nuclear membrane.

(b) Anterior portion of *Trichomonas gallinae* showing numerous cellular organelles and well-developed flagellar apparatus (F). R: ribosome, SB: storage body.

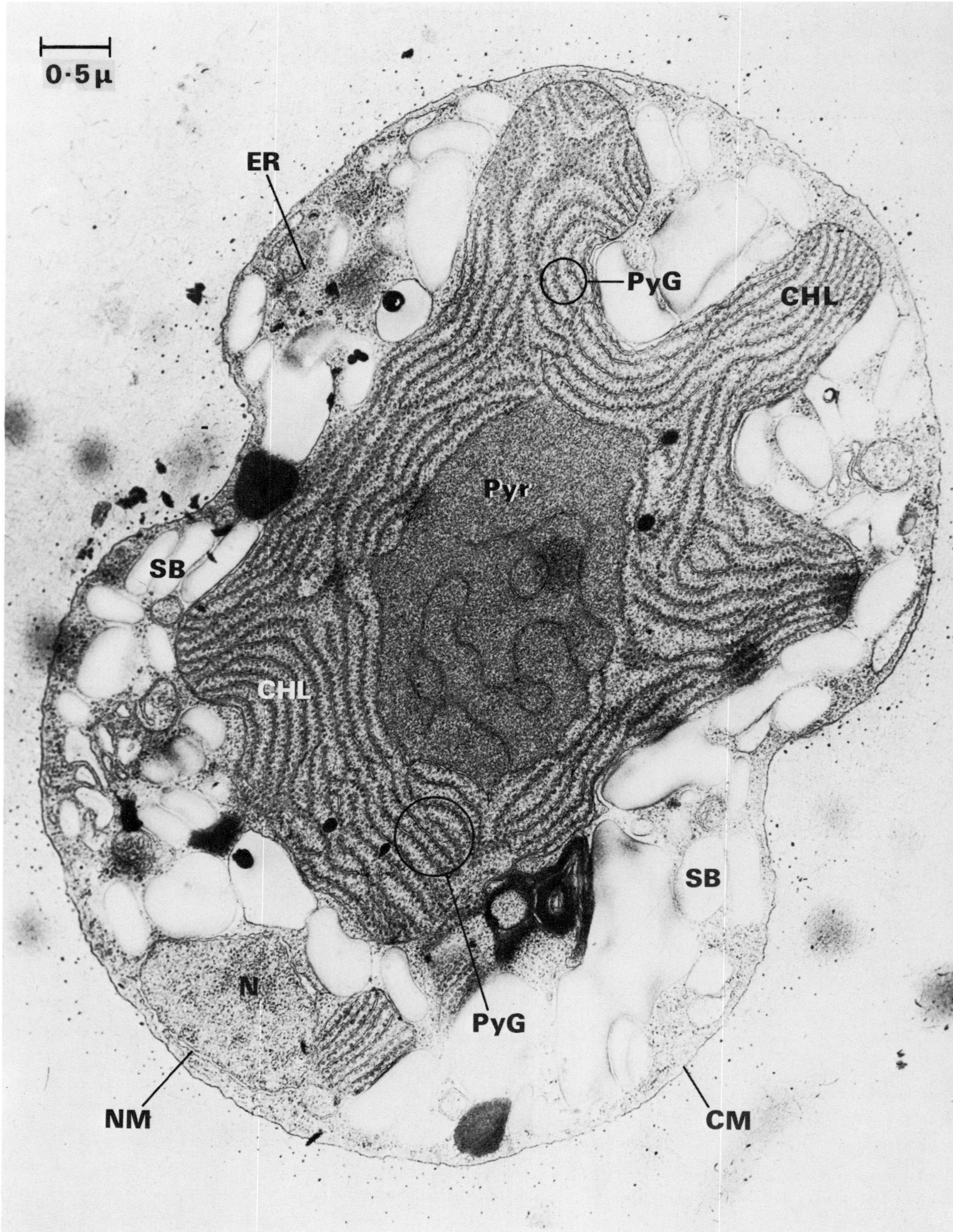
0·5μ
ER
PyG
CHL
Pyr
SB
CHL
SB
N
PyG
NM
CM

4.2. Gram-positive procaryotes

4.2.1. Clostridia and bacilli

Many of the anaerobic gram-positive rods so far examined are either pathogenic, being involved in the formation of necrotic tissue lesions, or associated with pathological lesions.[20, 220] The obligate anaerobic, spore-forming bacteria classified in the genus *Clostridium* are noninvasive, that is, produce disease by the elaboration of potent exotoxins and enzymes. The basic pattern of these anaerobic diseases involves septicemia with metastic suppuration, intra-abdominal abscesses with peritonitis, as well as endotritis of the female genital tract.[183]

Members of the genus involved in exotoxin production which result in necrotic lesions include *C. tetani, C. perfingens, C. novyi, C. septicum, C. histolyticum*, and *C. botulinum*.

In contrast to the extreme pathogenicity and virulence of clostridial species, members of the genus *Bacillus* are for the most part saprophytes, being found predominantly in soil, water, and in similar environments. *Bacillus anthracis*, the causative agent of anthrax in cattle, is the only recognized bacillus considered to be pathogenic. Anthrax is transmitted to humans by way of animal contact or through contaminated feces or urine.

Vegetative cells of *B. anthracis* (Figure 4.6(b)), *B. macroides* (Figure 4.6(a)), *C. tetani* (Figure 4.6(c)), *C. sporogenes* (Figure 4.6(c)), and *C. botulinum* (Figure 4.6(d)) are representative of the morphology of the vegetative cells of the anaerobic clostridia and aerobic spore-forming bacilli. The vegetative cells vary in length from 3 to 8 μm and in width from 0·4 to 1·2 μm. The cell wall is approximately 4 μm thick and electron dense,[92, 93, 98] and where motile, cells contain numerous peritrichously arranged flagella (Figure 4.6(f),(g)). The cell walls of both the clostridia and bacilli are tripartite, with electron-dense outer and inner margins, enclosing a narrow space of medium electron density. The cell wall in *C. tetani* is thinner than that observed in the other clostridial vegetative cells.[94] In some instances, due to the high degree of staining, it is not possible to observe this layering. The cell wall of *C. botulinum* types A and E (Figure 4.6(d)) is similar to those of *C. perfringens* but an electron-dense irregular fuzz lines the cell wall surface[198] of *C. botulinum*.

Directly beneath the cell wall is the closely adhering unit cell or plasma membrane. In some cases it is possible to discern a periplasmic region[46] between the inner margin of the cell wall and the outer layer of the cell membrane. The cell membrane has electron-dense material adhering to it, making it

Figure 4.5 Comparison of mammalian, protozoan, and algal cell.
(c) *Porphyridium cruentum* alga showing large central pyrenoid (Pyr), large chloroplast (CHL), with lamellar associated phycobilisome granules (PyG), storage granules (SB), some endoplasmic reticulum (ER), and particle associated phycobilisomes. N: nucleus, NM: nuclear membrane, CM: cell membrane.

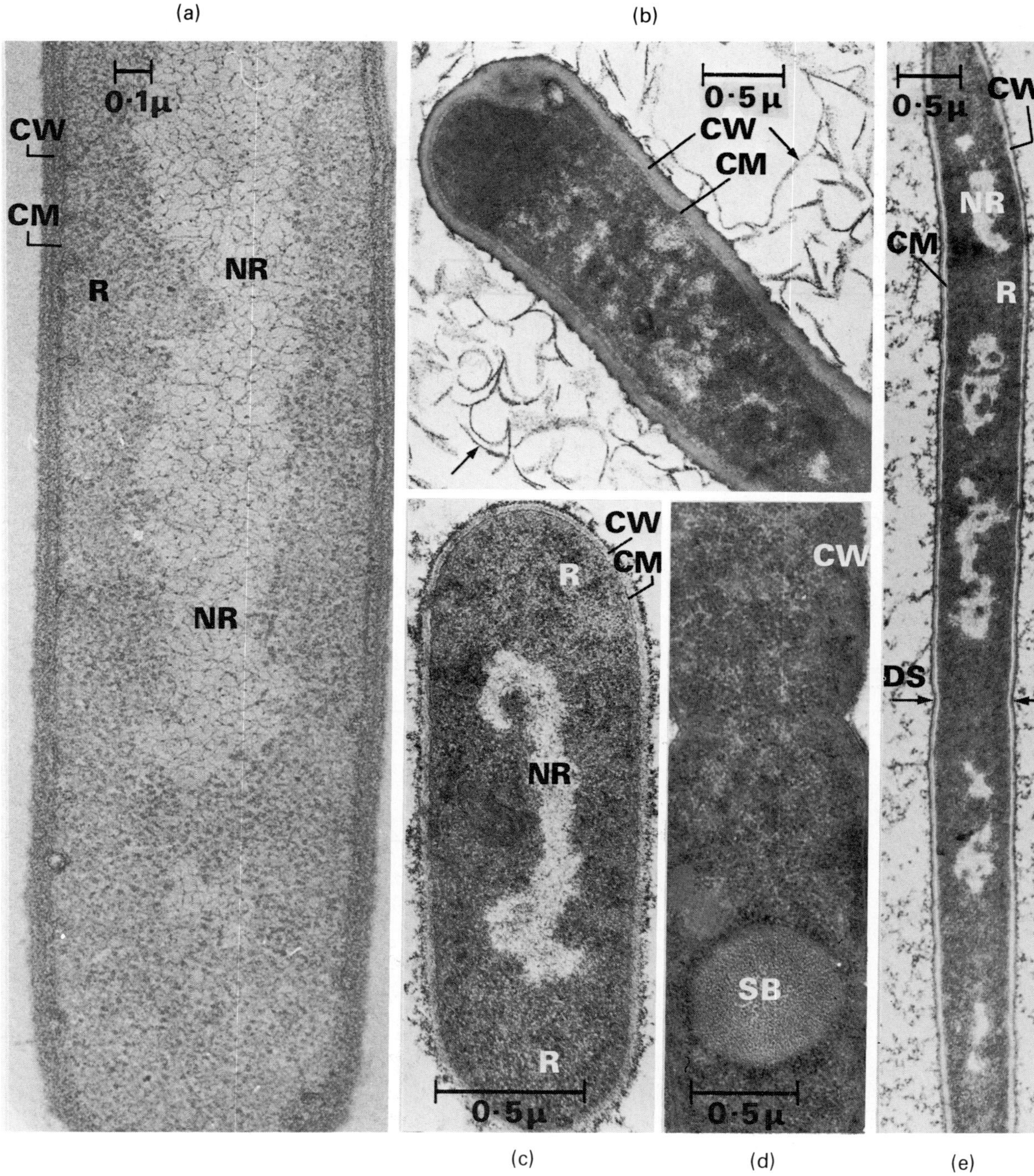

Figure 4.6. Representative bacilli and clostridia.
(a) *Bacillus macroides*, (b) *B. anthracis* strain *Sterne*, (c) *Clostridium sporogenes* ATCC8074, (d) *C. botulinum* type E, and (e) *C. tetani* (NCTC9575) seen in thin sections of chemically fixed cells, The gram-positive cell wall is uniformly thick and homogeneous. The underlying cell membrane (CM) adheres closely to the cell wall (CW) and encloses an electron-dense cytoplasm of ribosomes (R) and fibrous nucleoid. (NR). Note in (b) the extracellular polymer (arrows) produced by *B. anthracis.* DS: division septum.

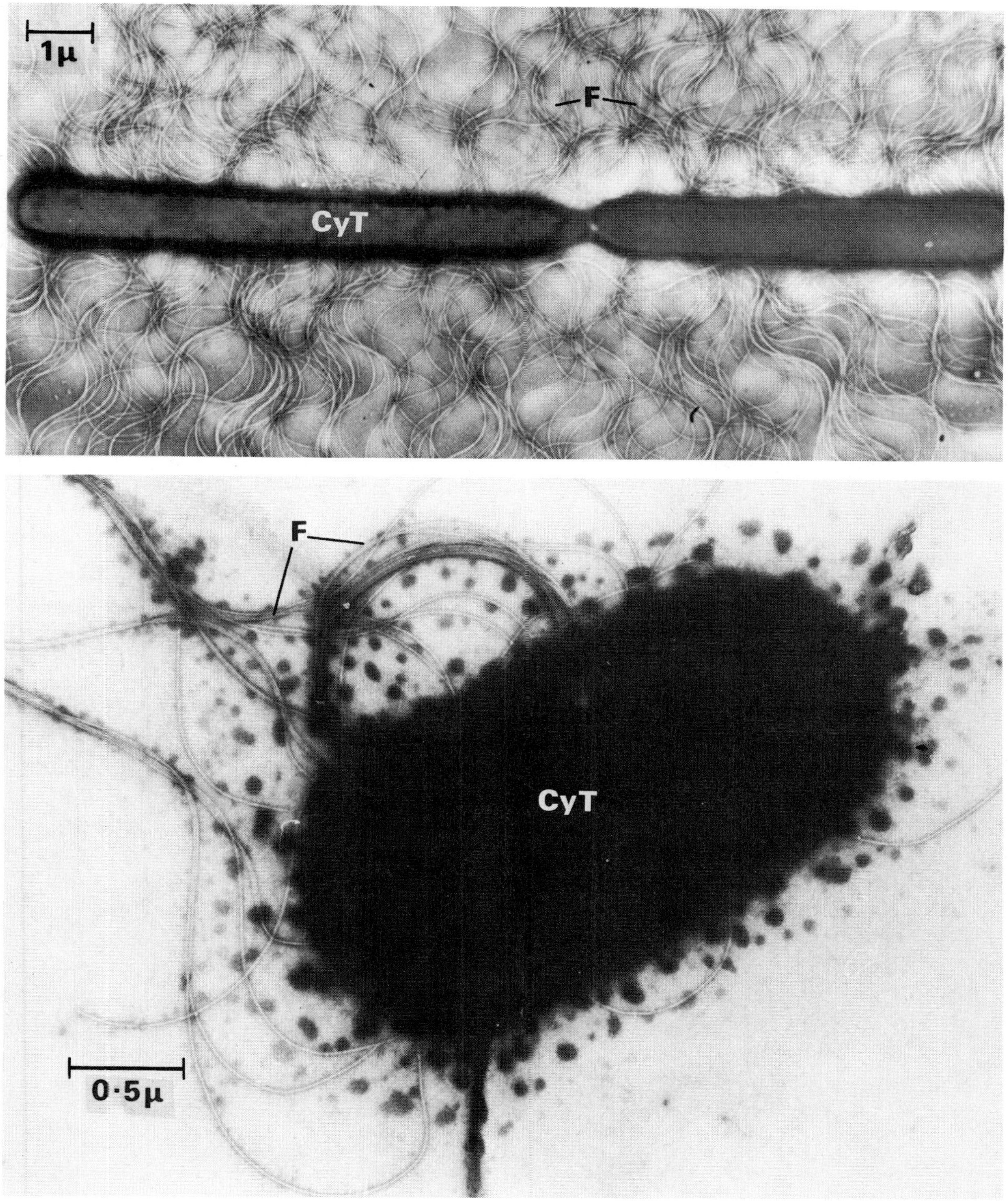

(f)

(g)

Figure 4.6. Representative bacilli and clostridia.
(f),(g) Negative staining of *C. tetani* NCT9575 and *B. fastidiosus* reveals a densely staining cytoplasm (CyT) (stains positively), with negatively stained peritrichously arranged flagella (F).

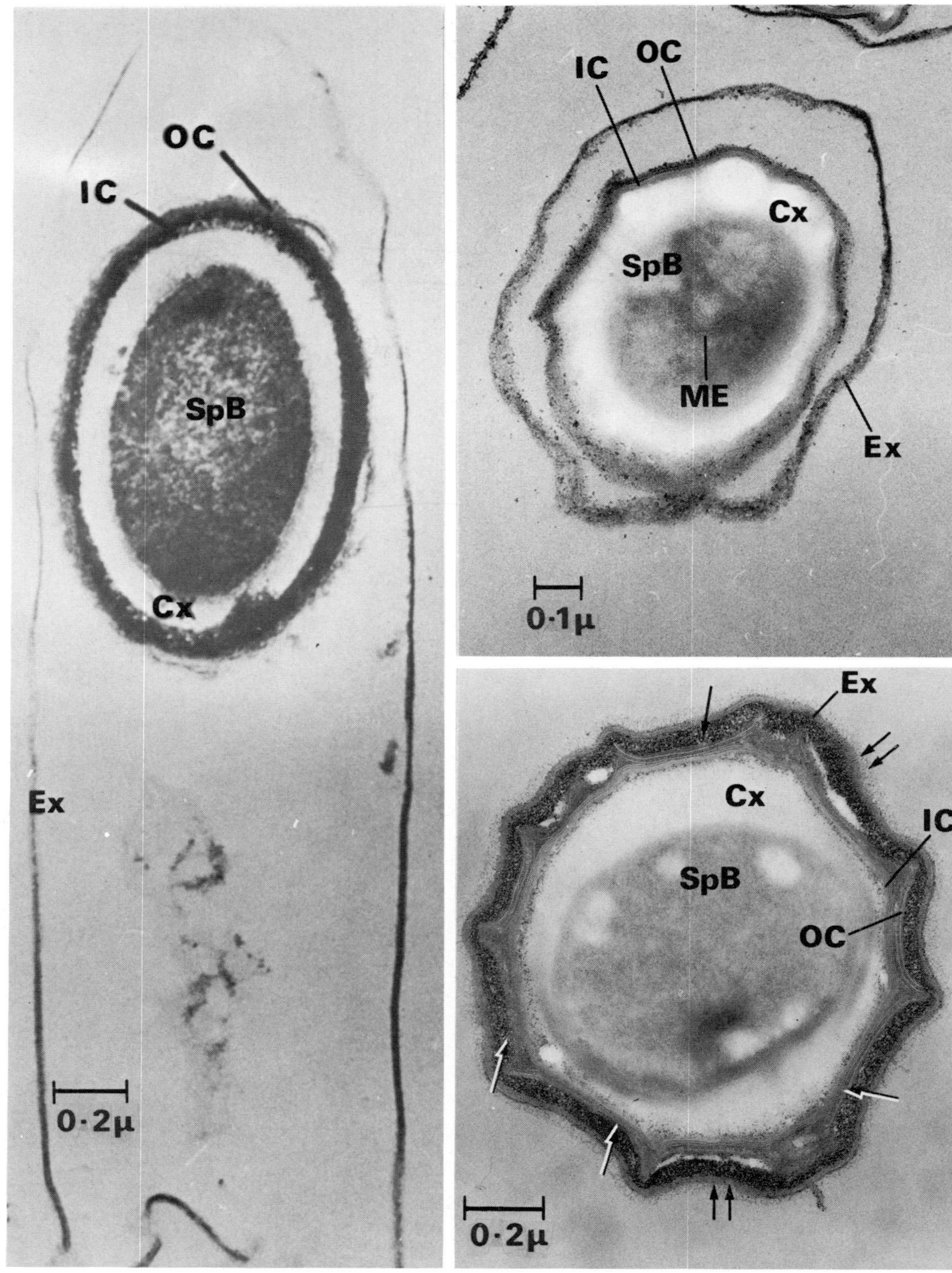

Figure 4.6. Representative bacilli and clostridia.
(h), (i), (j) Thin sections of spores of *C. botulinum* Holland 17B, *B. anthracis* Sterne, and *B. fastidiosus*, respectively. The outermost layer of the spore, the exoporium (Ex), surrounds the numerous lamellar spore coat layers consisting of outer coat (OC) and inner coat (IC), and an electron-transparent cortex (Cx). In (j) *B. fastidiosus* the region between the exosporium and lamellar coat (arrow) is filled with electron-dense particulate material. Note also the surface of the exosporium is covered with a hair-like fuzz (double-arrows) M: mesosome, SpB: spore body.

(k) (l) (m)

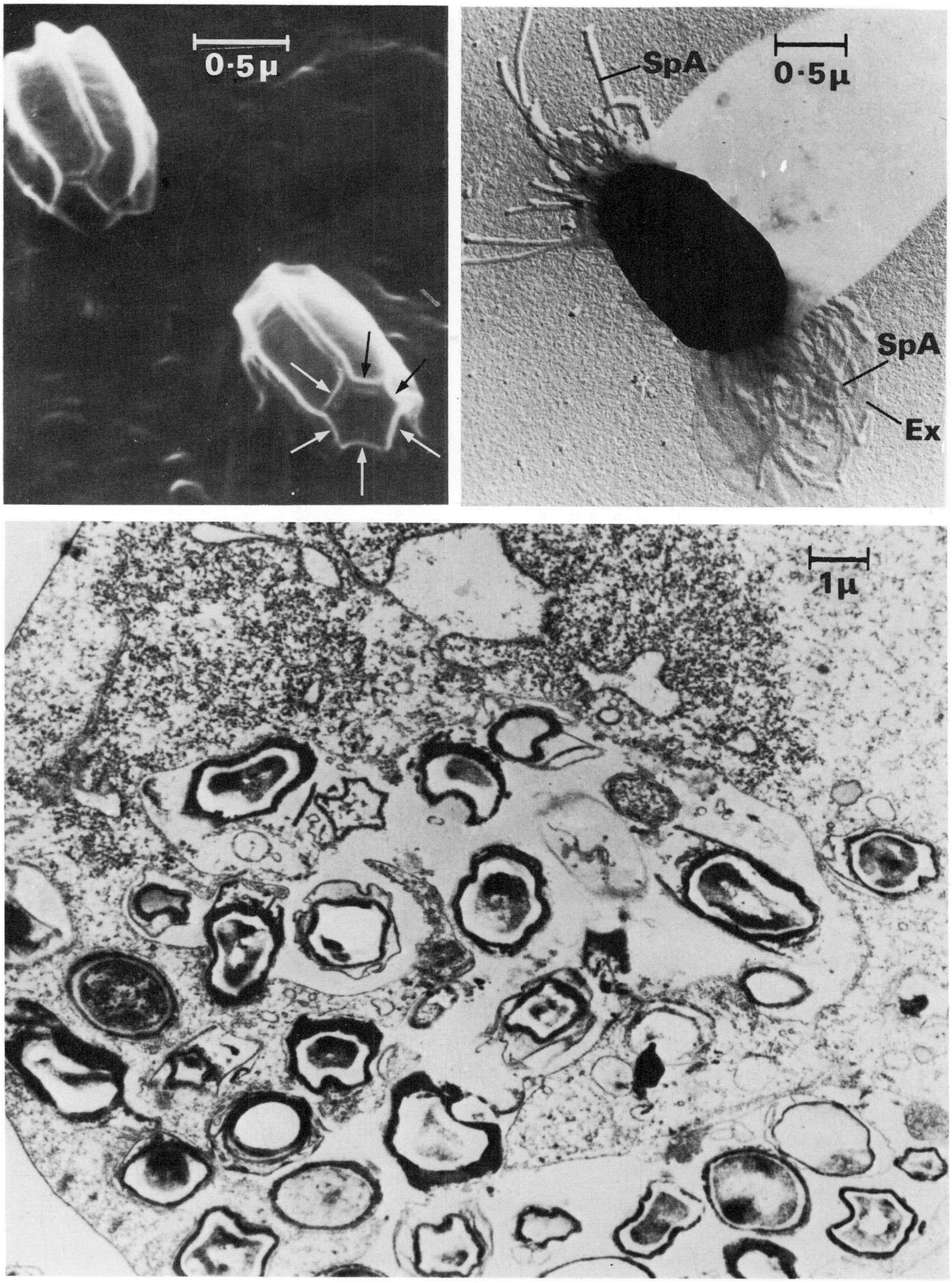

Figure 4.6. Representative bacilli and clostridia.
(k),(l) Surface views of *B. polymyxa* by SEM and *C. botulinum* type E by shadow-casting. The spore-coat ridges in SEM merge in a starlike pattern at the spore pole (arrows). In (l), the Au–Pd shadowed spore has a partially torn exosporium (Ex) which permits the enclosed spore appendages (SpA) to be seen. The long shadow projected by the spore indicates a tall structure.
In (m) *C. botulinum* Type 33A spores are enclosed in a human PMN-leukocyte. Several spores are in the process of germination within the large phagosome.

an asymmetric membrane. The membrane encloses an electron-dense particulate cytoplasm composed of ribosomes, as well as a centrally disposed nucleoid (Figure 4.6(a),(c),(e)). In some instances, lamellar mesosomes are in direct association with the cell membrane and the nuclear region (Figure 4.6(c)); however, the existence of, as well as the nature of, the mesosomes is in dispute (Higgins, personal communication). Note the large spherical inclusion body in the cytoplasmic region of *C. botulinum* (Figure 4.6(d)); the nature of this inclusion is unclear.

Endospores of the clostridia are usually terminally located within the vegetative cell (Figure 4.6(g))[92] and in most instances swell the vegetative cell presenting a drumstick-shaped appearance. In the bacilli, the endospore may be terminal to subterminal, but in most instances they do not swell the vegetative cell. A bacillus, similar to *B. macerans* has been isolated from the oral cavity of humans (Meyer, Holt, Leadbetter, unpublished communication). Its morphology is similar to that of other bacilli; however, the terminal endospore does swell the enveloping vegetative cell wall, and has an enveloping hair-like nap on its exosporial surface (Figure 4.1(b)).

The spores formed by several clostridia and bacilli are seen in Figure 4.6(h)–(m). The central region, or spore body, consists of an amorphous mass of ribosomes and nuclear material, both enclosed by a unit membrane. An electron-transparent space, the cortex, may be either very wide or narrow, depending upon the species. External to the cortex are the spore coat lamellae, which are concentrically arranged; their number varies from a few to 5–6 layers. The most external layer of spores consists of a loose-fitting membrane, the exosporium. In some of the bacilli, notably *B. polymyxa* and *B. fastidiosus*[121, 145] the spore coat lamellae are arranged in almost equidistant ridged layers around the spore proper (Figure 4.6(j),(k)). This ridging is clearly seen in the SEM of *B. polymyxa* spores (Figure 4.6(k)) and in spores prepared by surface replication (Figure 4.1(a),(b)). *Clostridium histolyticum* contains only 3 spore coat lamellae and is devoid of an exosporium.[12] The toxigenic strain of *C. botulinum* Type E possesses many (160–240) tubular appendages 24 nm in diameter with a central 8 nm lumen which protrudes from the spore surface Figure 4.6(l).[191] These appendages are enclosed in what appears to be a loosely fibrillar, delicate exosporium, approximately 6 nm thick. The tube appears to be sealed at its distal end and contains a hemispherical cap.

In a clinical isolate of *C. sporogenes*[92] the exosporium was seen to consist of at least two layers, an inner 8 nm thick layer composed of at least 10 laminae, as well as an outer layer with attached hair-like projections. These projections range from 25–125 nm in length.

In Figures 4.6(m), 4.7(a) spores of type E *Clostridium botulinum* are seen[193] which have been exposed to polymorphonuclear leukocytes. The majority of the engulfed spores have retained their characteristic geometric shape, while some are in the process of disintegration. When *C. perfringens* enters human tissue (Figure 4.7(b),(c)) it can establish itself in muscle fibers, resulting in progressive destruction of connecting collagen material. The integrity of the invading bacteria has been maintained within and peripheral to the tissue collagen.

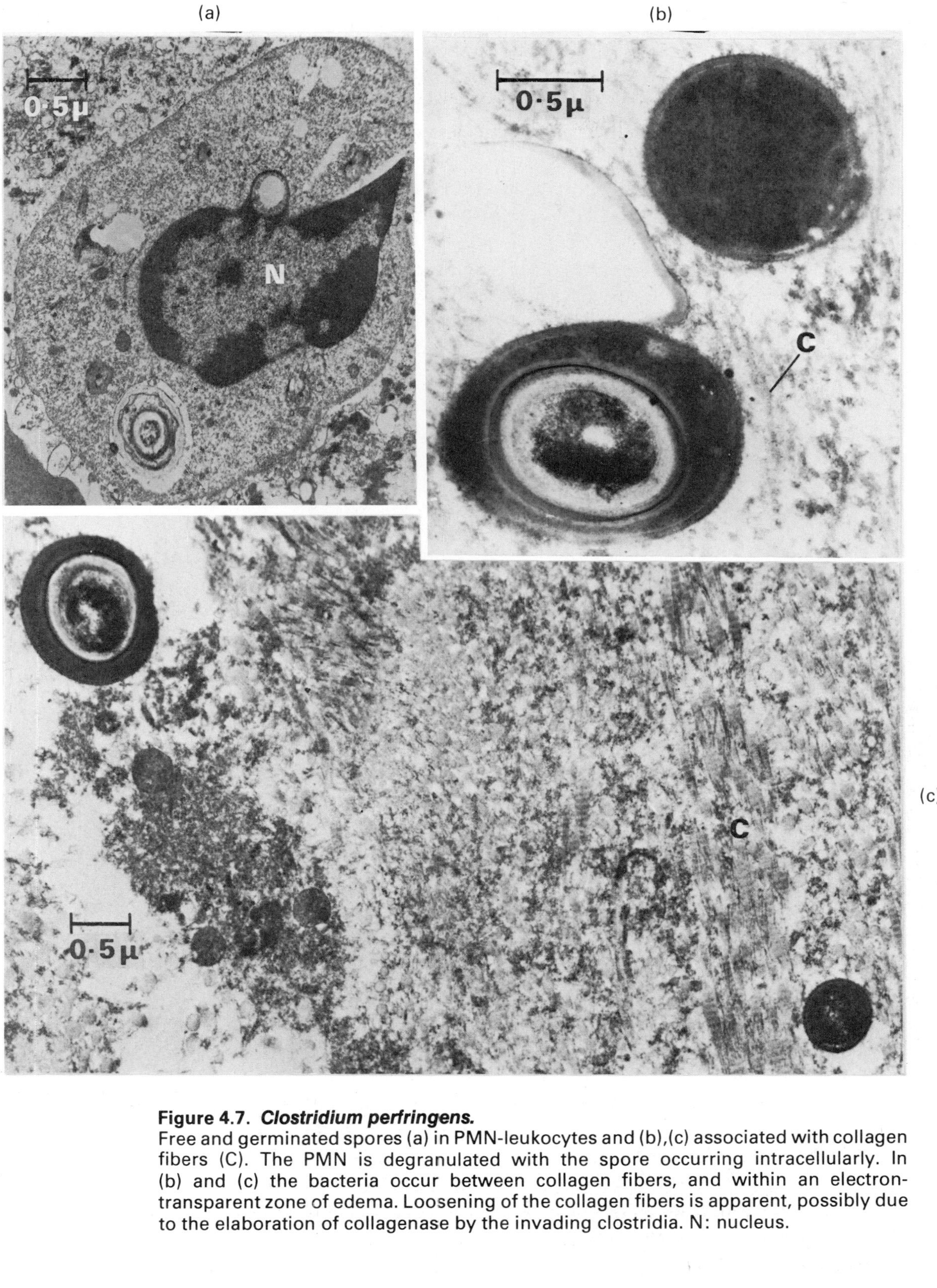

Figure 4.7. ***Clostridium perfringens.***
Free and germinated spores (a) in PMN-leukocytes and (b),(c) associated with collagen fibers (C). The PMN is degranulated with the spore occurring intracellularly. In (b) and (c) the bacteria occur between collagen fibers, and within an electron-transparent zone of edema. Loosening of the collagen fibers is apparent, possibly due to the elaboration of collagenase by the invading clostridia. N: nucleus.

4.2.2. *Listeria*

The type species of the genus *Listeria, Listeria monocytogenes*, is a ubiquitous gram-positive, motile coccobacillus. It infects humans and is found in a wide variety of animals where it causes listeriosis, encephalitis, or purulent meningitis.[20, 220] Fatal infection with *L. monocytogenes* results from granulomatosis, infantiseptica, and intrauterine infection, these latter infections being associated with widespread necrosis of internal organs and hemorrhagic areas of the skin.[183]

Ultrastructurally, *L. monocytogenes* is characteristically gram-positive (Figure 4.8(a)).[7, 68, 72] The thick cell wall is approximately 30 nm wide. The underlying cell membrane adheres very closely to the inner border of the cell wall and consists of two dense layers, each 2·5 nm wide, with two electron-transparent layers, each 3 nm wide. The cytoplasm contains electron-dense ribosomes which enclose a central region of fibrous nuclear material. Note in Figure 4.8(a) the large lamellar mesosome associated with both the division septum and nuclear region. In the negatively stained sample (Figure 4.8(b)), numerous mesosomal invaginations are evident.

4.2.3. Actinomycetes

The actinomycetes are gram-positive, predominantly aerobic procaryotes.[63] Included are both acid- and nonacid-fast organisms. Characteristic of this group of bacteria is their ability to form branching filaments, which in some of the families develop into a mycelial mass. The Euactinomycetes represent those cells which develop exclusively by mycelia and specialized spores. Included in this group are the genera *Streptomyces, Micromonospora, Actinoplanes, Streptosporangium*, and *Thermoactinomyces*. The Proactinomycetes, which are characterized by the formation of fragmentary mycelia, are also unicellular with no specialized spores being produced. Included in this group are the genera *Bifidobacterium, Actinomyces, Mycobacterium*, and *Nocardia*. The coryneform bacteria are included in the last group of Actinomycetes, in which the cells are unicellular and reproduce by binary fission. Members of this group include the genera *Corynebacterium, Arthrobacter*, and *Propionibacterium*.[186]

Among the actinomycotic diseases, actinomycoses of lower animals, especially cattle, are best recognized.[220] In some cases humans have been known to become infected by chance animal inoculation. These bacteria have also been found in some instances as "normal" inhabitants of the mouths of humans and cattle (see below). Characteristic of this group of organisms is their localization in pustular lesions or in the walls of abscesses,[20] where they occur as small, yellow granules, typical of sulfur granules.

A variety of proactinomycetes have been isolated from the human oral cavity, and have been implicated in several types of periodontal infections.[76] *Actinomyces viscosus, A. naeslundii, A. odontolyticus*, and *A. israelii* are representative of this group of human oral isolates.

In young cultures, the actinomyces present a characteristic cellular morphology consisting of branching filaments with thick cell walls (Figure 4.9(b)).

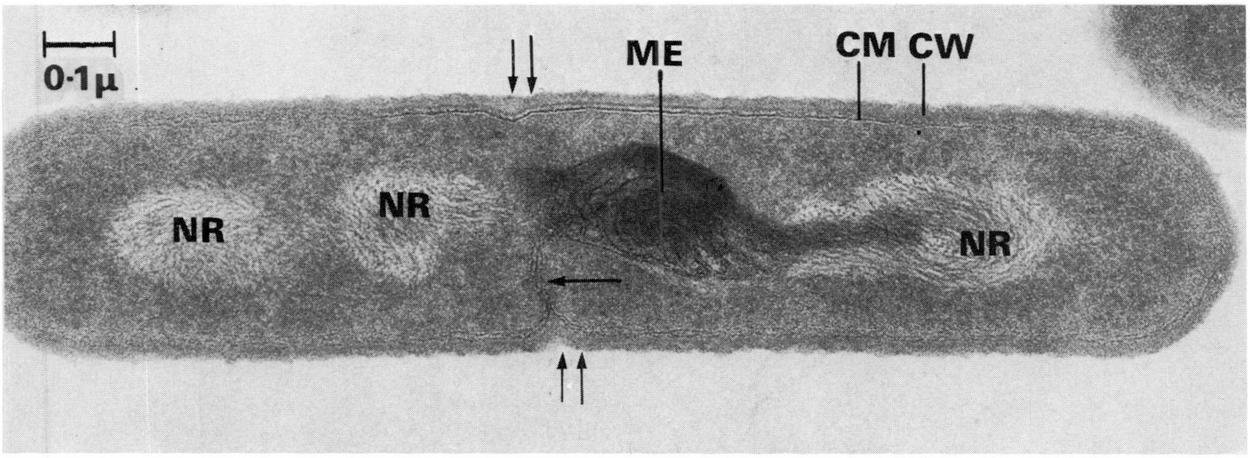

(a)

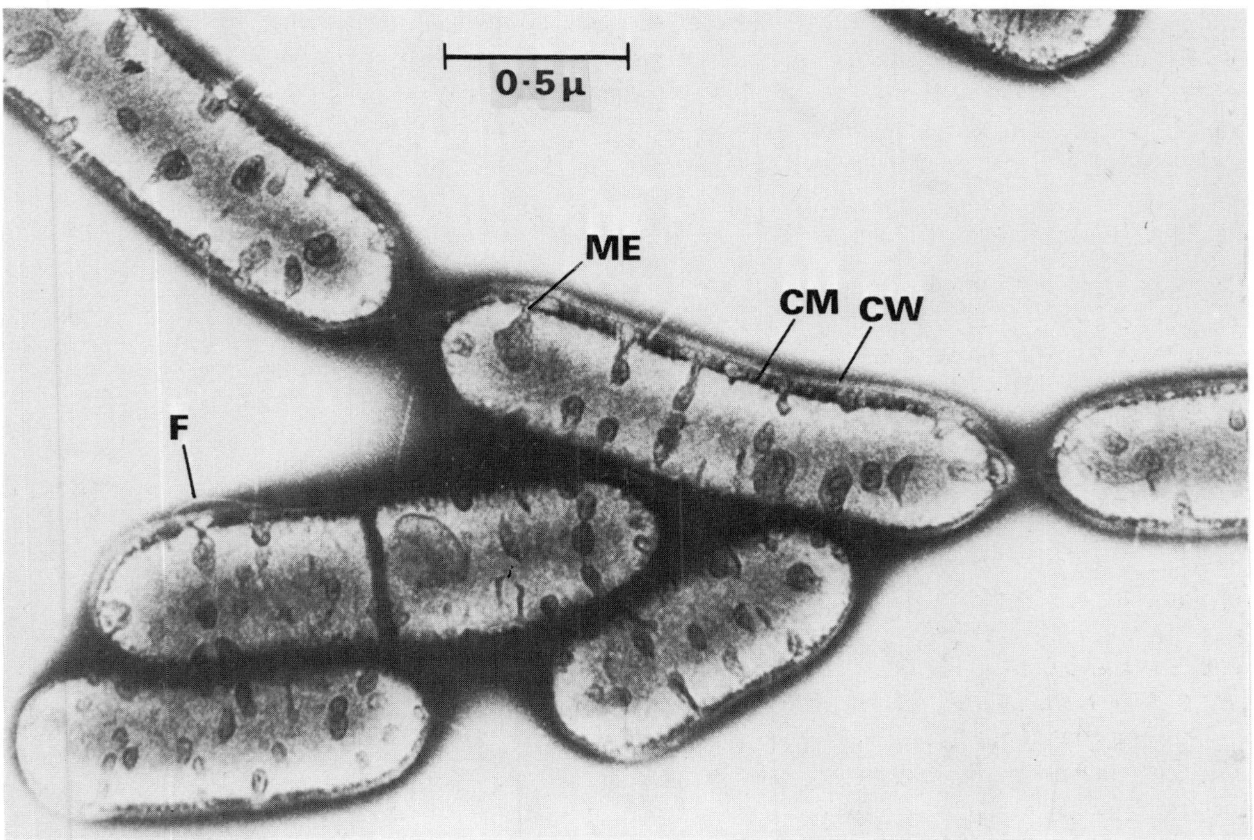

(b)

Figure 4.8. *Listeria monocytogenes* type 2.
(a) Thin section and (b) negatively stained cells. The thick and homogeneous cell wall (CW) surrounds the cell membrane (CM) and electron-dense cytoplasm. A mesosome (ME) is seen to originate from the inner portion of the cell membrane (arrow) and to be in close association with the nucleoid (NR) and developing cross-wall septum (double arrows). In (b) the negative stain (PTA) has penetrated the mesosome membranes revealing numerous invaginations.

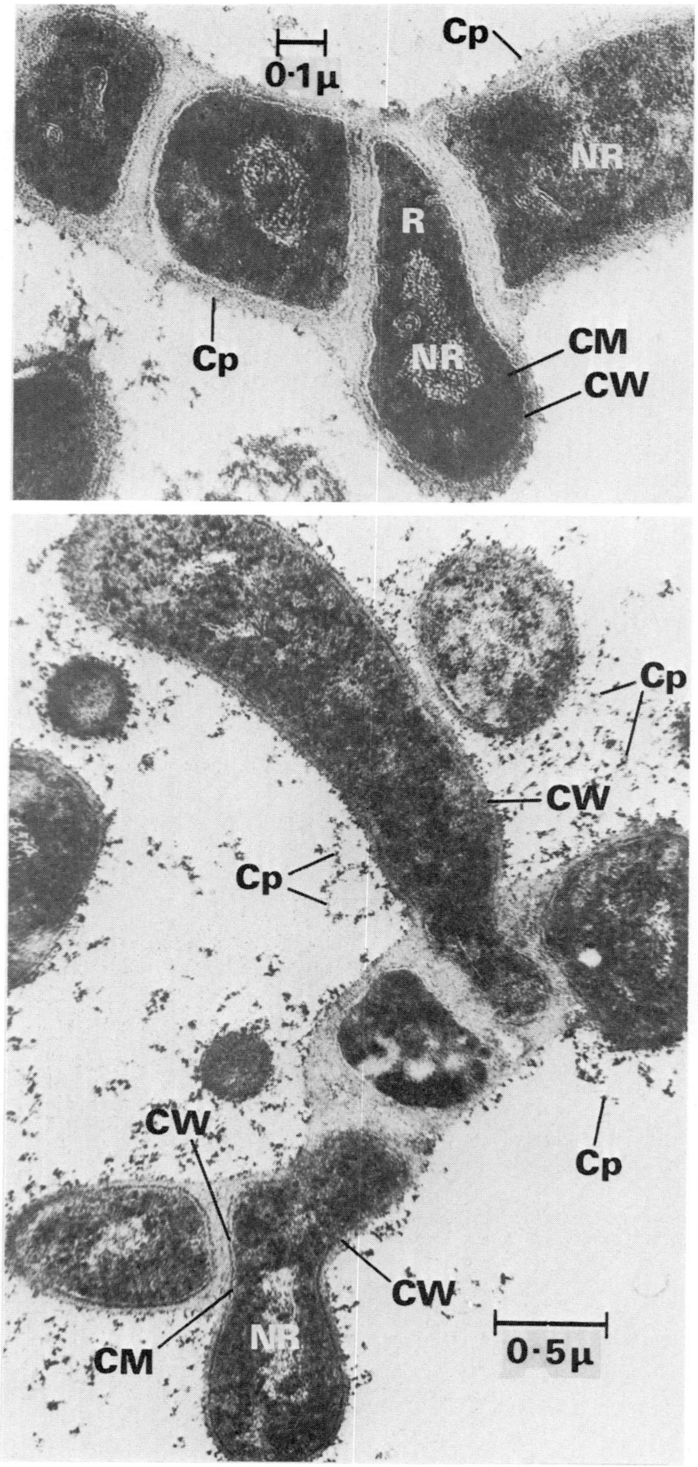

Figure 4.9. Representative *Actinomyces* species.
Cells prepared by chemical fixation and thin sectioning, SEM and freeze-fracture etching. Cells are surrounded by thick cell walls which enclose electron-dense cytoplasm. (a), (b) *A. odontolyticus* ATCC 17929 showing outer layer of cell wall (CW) covered with a loose fitting electron-dense capsule (Cp). The cells are pleomorphic, and in (b) appear to be derived from a central cell. R: ribosome, CM: cell membrane, NR: nuclear region.

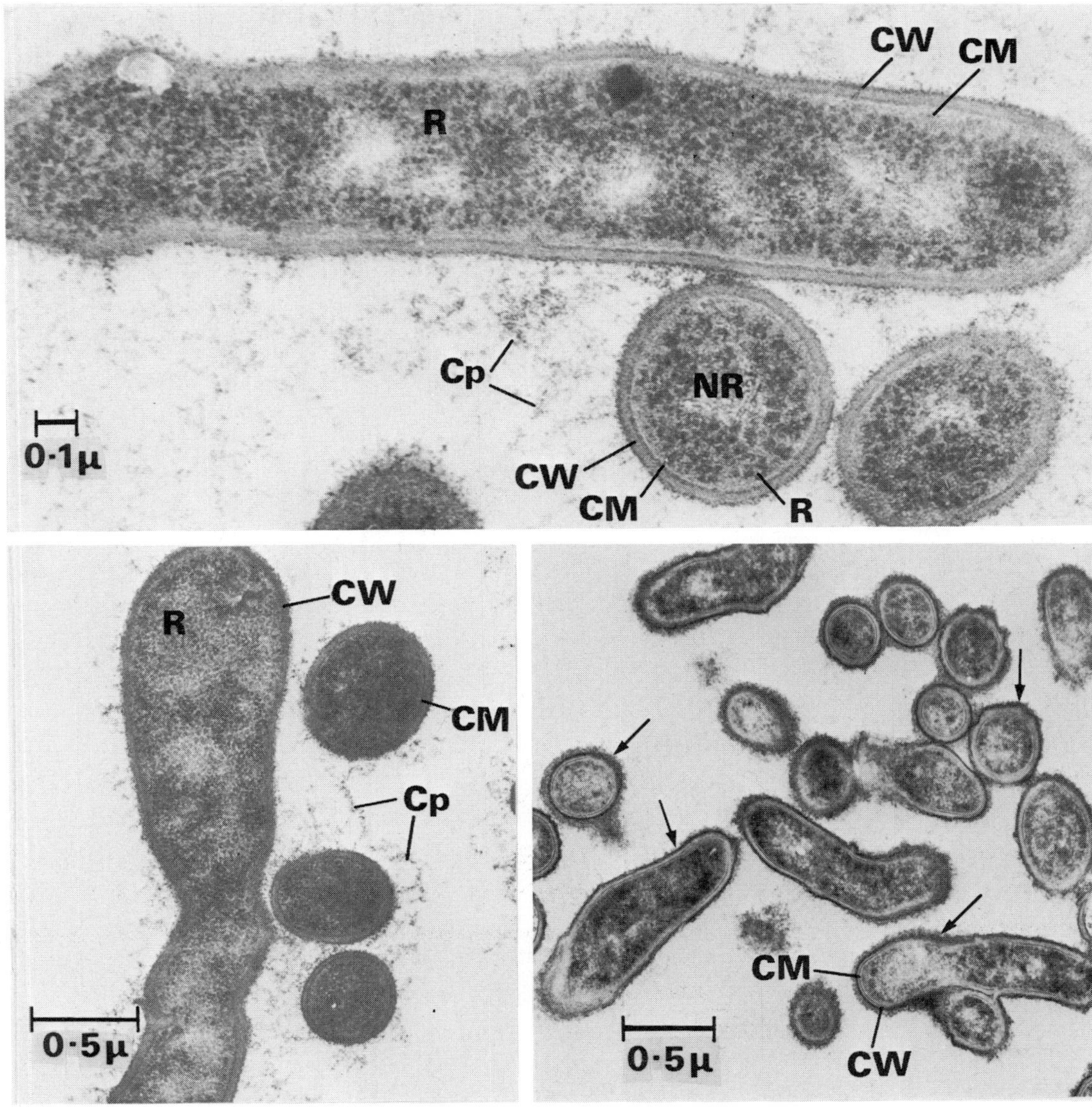

(c)

(d) (e)

Figure 4.9. Representative *Actinomyces* species.
(c), (d) (e) Human strains of *A. viscosus* W1053, *A. naeslundii* V16, and *A. israelii* W748, respectively, are morphologically similar to the other *Actinomyces*. *A. israelii* W748 has a smooth cell wall (CW), the outer surface of which is electron-dense (arrows). R: ribosome, CM: cell membrane, Cp: capsule, NR: nuclear region.

Figure 4.9. Representative *Actinomyces* species.
(f) *A.viscosus* fibrous capsule (Cp) radiating from the cell surface and connecting adjacent cells (arrows).
(g) By freeze-fracture-etching the fibrous capsular layer is evident on the surface of *A. viscosus* VP17281. CM: cell membrane, CW: cell wall, Cyt: cytoplasm.

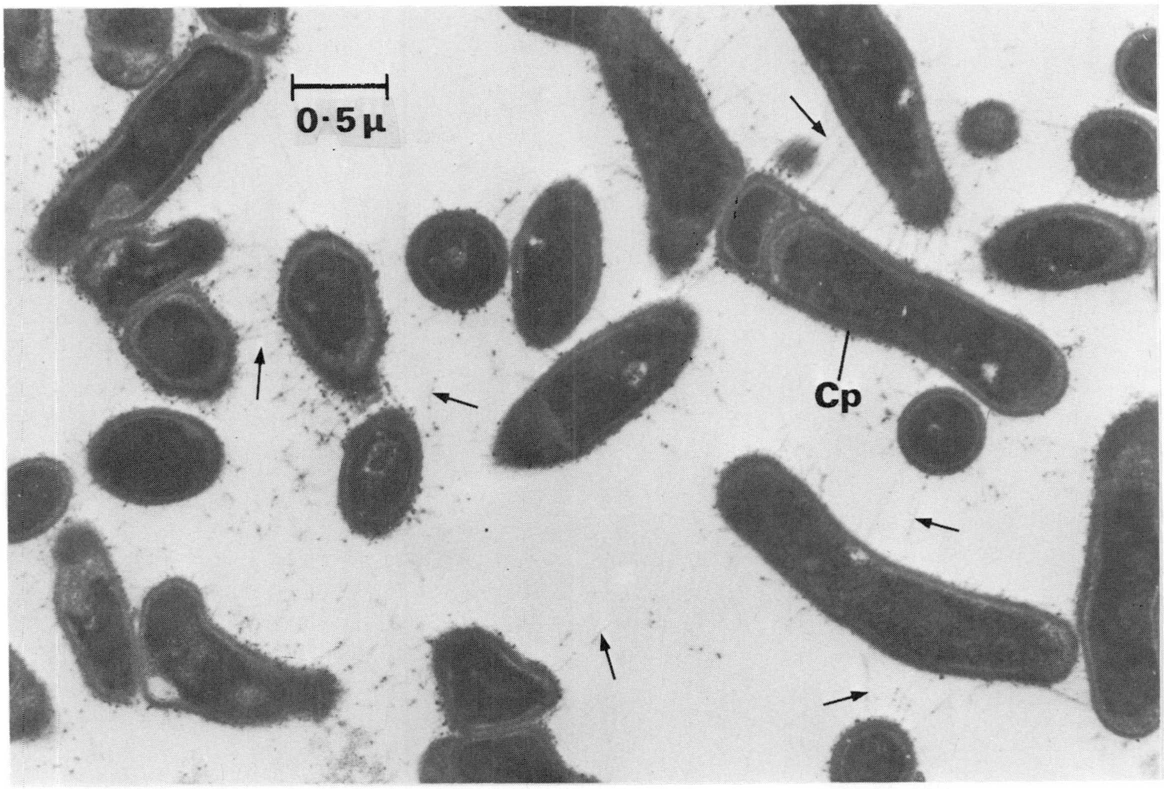

(f)

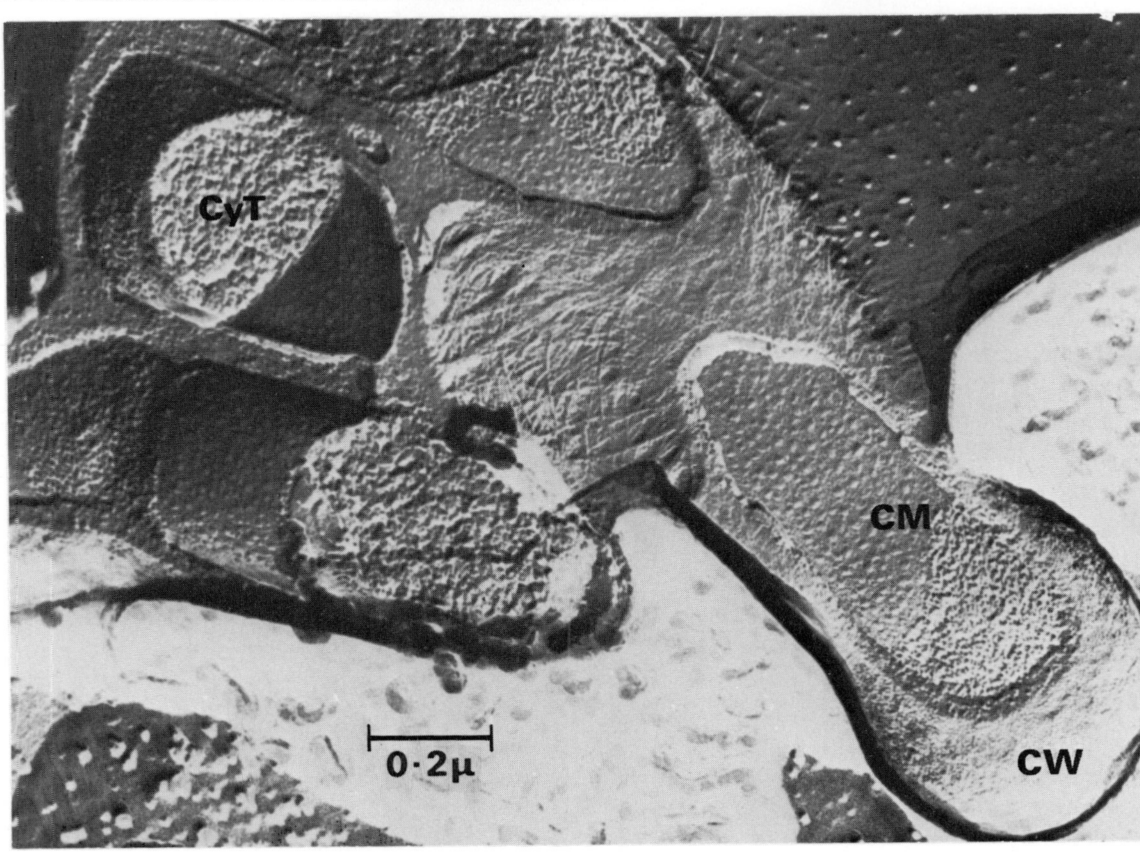

(g)

Figure 4.9. Representative *Actinomyces* species.
(h), (i) SEM of *A. viscosus* ATCC 15987 and *A. naeslundii* V16, respectively, showing the fibrous capsular layer on the surface. The fibers are arranged in a tight fitting mesh which most probably aggregates the cells into a tight mass.
(j), (k) *Actinomyces viscosus* WVU745 and WVU398B by SEM. Both filamentous and coccobacilli are apparent.

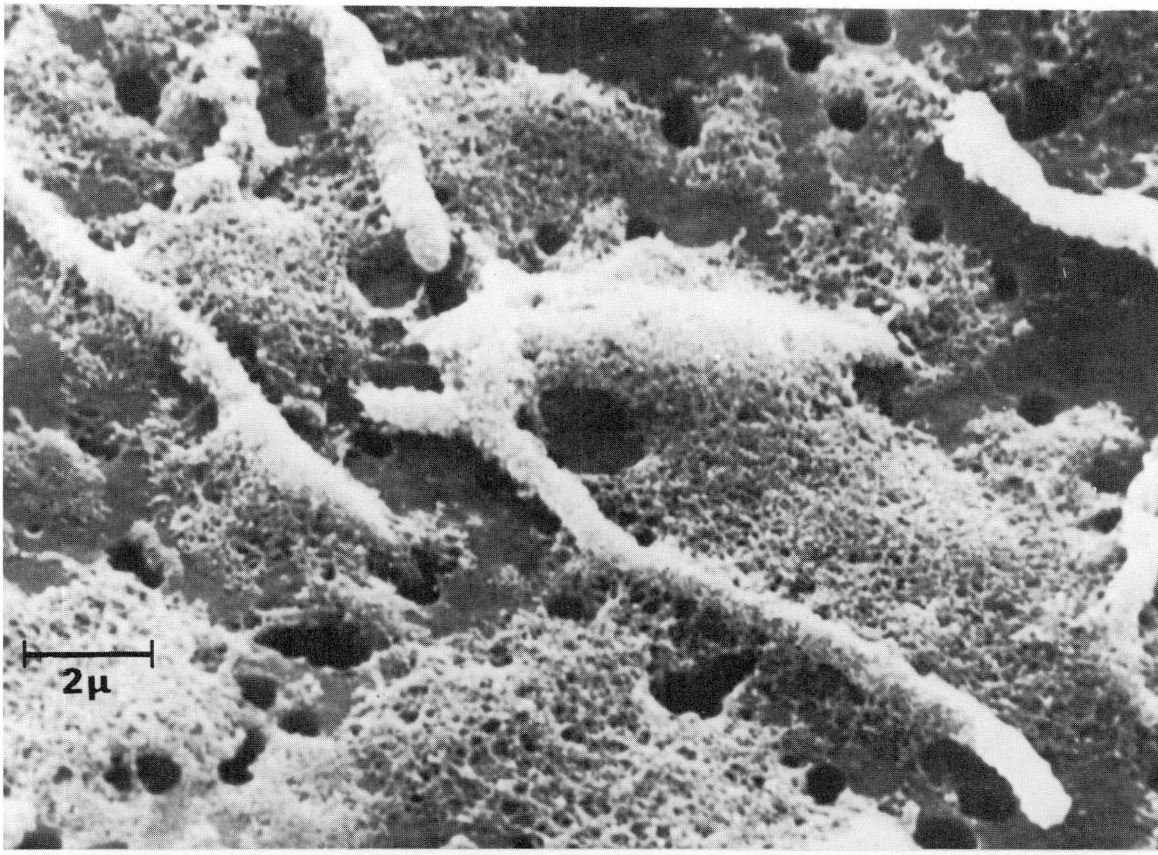

(h)

(i)

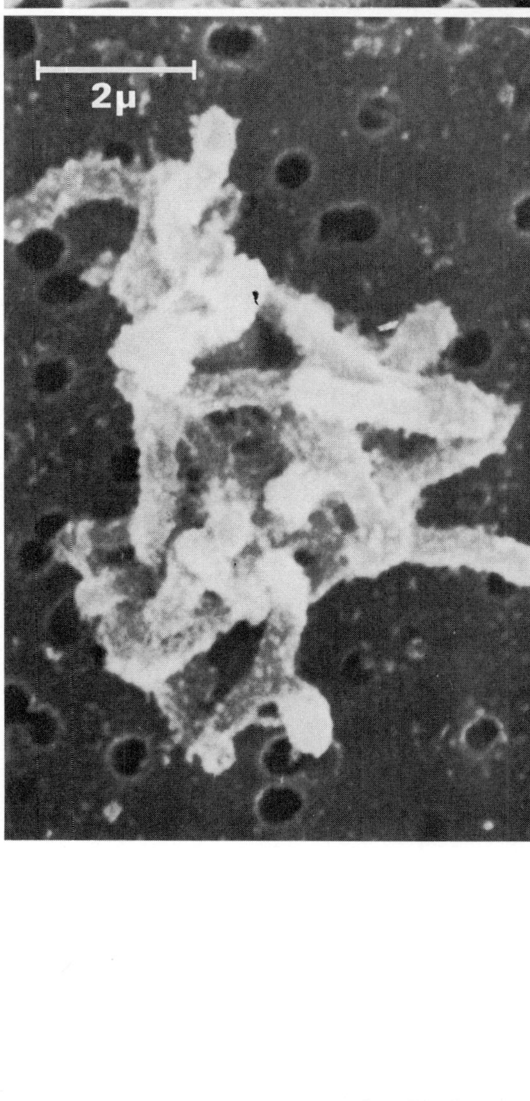

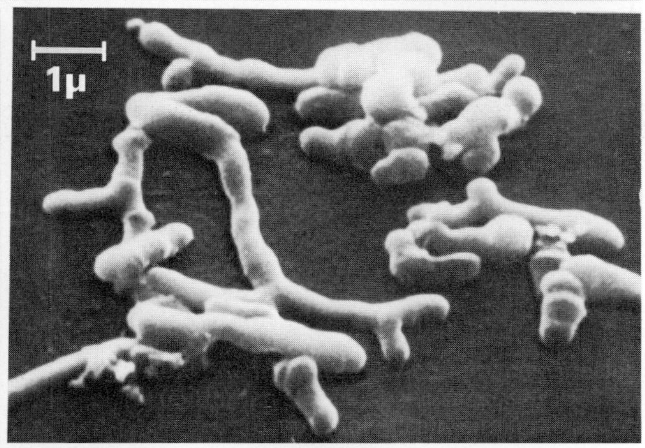

(j)

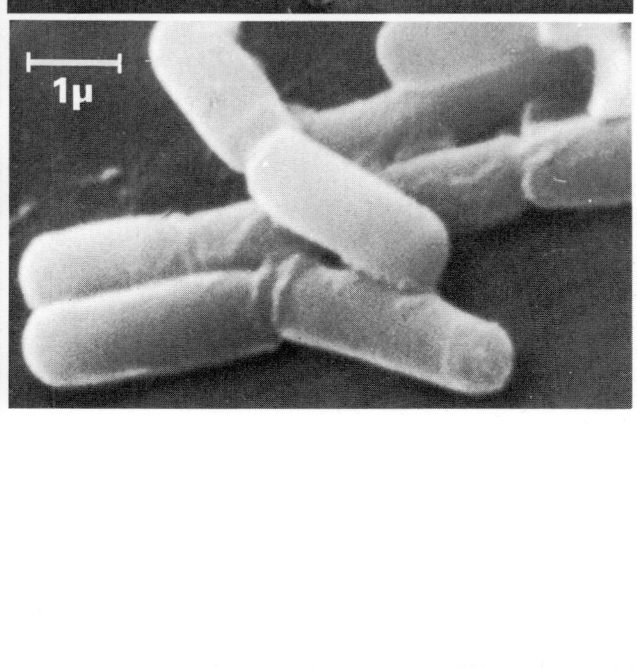

(k)

The filaments fragment into coccobacilli during the later stages of growth (Figure 4.9(a),(c)).

Scanning electron microscopy of *A. viscosus* reveals filamentous coccobacillus (Figure 4.9(j),(k)) forms as well as short rod forms. The cell surface is smooth with no evidence of a slime or capsular layer.

In contrast to the branching observed with *A. viscosus* and *A. naeslundii*, *A. bovis* (found predominantly in cattle) consists of long filaments which fragment into short rods.

Ultrastructurally, thin-section analysis of *A. naeslundii* and *A. odontolyticus* reveals the cell walls to be granular and approximately 45 and 31 nm thick, respectively (Figure 4.9(b),(c)). The most external layer of the wall is covered with an electron-dense adhering material – whether it represents cell wall polymer, comes from the medium, or is produced during fixation is unknown.

In *A. viscosus*, fibrils 4·8 nm in diameter emerge from the surface of the cell and extend for a considerable distance (Figure 4.9(f)). In many instances these fibrils extend between cells and may function to attach cells into a mesh or mat layer.[76] The fibrils are also present as a mat on the cell surface (Figure 4.9(g)). Large masses of extracellular material are also seen in SEM preparations (Figure 4.9(h),(i); Birdsell, unpublished communication) of several of the oral *Actinomyces*. Note in these figures how this extracellular matrix forms a meshed layer between cells, holding them in an aggregated mass.

Actinomyces naeslundii and *A. israelii* (Figure 4.9(e)) contain only short, hair-like structures on their outer surfaces. In addition, the cell wall is thicker, with some cells containing a slight cleft in the outer layer.

Cell division is by binary fission and septation, with the clustering of cells suggestive of right angle division planes. Thus, binary fission results in a straight row of cells which divide, but remain attached by septation. Subsequent to this first division, a second division occurs at right angles to it, resulting in a clustering of cells.

4.2.4. *Nocardia*

Members of the genus *Nocardia* are facultative intracellular parasites usually characterized as branching, filamentous organisms; however, the filamentous cells are seen to fragment into short rods or coccobacilli. Whether grown on tryptone, nutrient agar, or on defined media, the branched filamentous cells vary in length from 10–30 μm, with their width varying between 0·5–1·5 nm. In brain–heart infusion media *Nocardia* are usually club-shaped with long filaments that increase in size and elongate (Figure 4.10). Eventually, cells fragment into short rods and cocci. Similarly, on defined medium, *Nocardia* grow primarily as coccobacilli; as growth proceeds the cells elongate into slender, branched filaments; the filaments fragment into regularly shaped rods, which further fragment into coccoid cells. *Nocardia asteroides*, a free-living organism, is the best known member of the genus. It is known to cause a variety of infections in humans, typical of which is nocardiosis or Madura foot.[183] It is characterized by the production of a network of interlocking abscesses and granulomas in the soft tissues and bones of the foot. In addition, *Nocardia* may

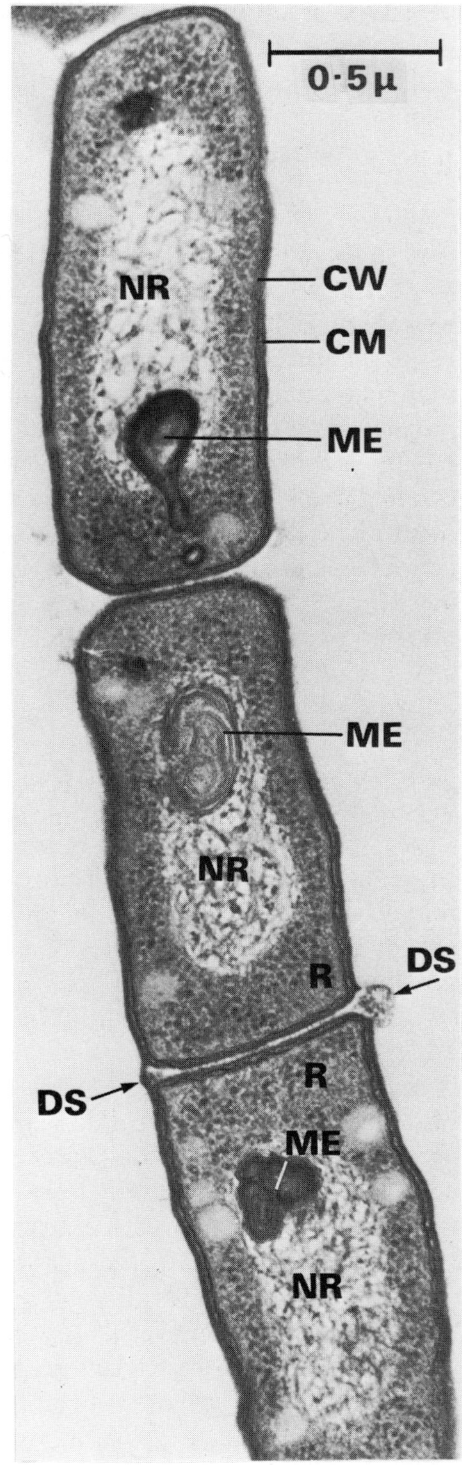

Figure 4.10. *Nocardia asteroides* 10905.
Thick electron-dense cell wall (CW) encloses the densely staining cytoplasm. Large vesicular mesosomes are associated with both the fibrous nucleoid and inner layer of the cell membrane (CM). The cells are connected by division septa (DS). NR: nuclear region, R: ribosome, ME: vesicular mesosome.

be responsible for infections of the respiratory tract, as well as involvement in brain abscesses.[20, 183] The organisms have also been found to accompany malignant diseases of the reticuloendothelial system, including Hodgkin's disease, lymphosarcoma, and leukemia.

Nocardia asteroides (Figure 4.10) possesses an electron-dense cell wall.[13] The external layer of the cell wall is covered with a slightly amorphous material, 1·5–3 nm thick. The cell wall and closely adhering cell membrane vary in thickness from 3–6 nm (characteristic of rod-shaped cells) to 50–130 nm (characteristic of coccoid cells). In some instances the cell wall is irregularly thickened, reaching thicknesses of 300–500 nm.

Depending upon the growth medium (i.e., brain–heart infusion broth), there is significant structural modification of the cell wall.[18] So, for example, in the early logarithmic growth phase, the cells are swollen and possess an enlarged outer layer of the cell wall, which is limited by a thin membrane. During mid-log growth, this outer layer thins and becomes uniform in diameter, resulting in the characteristic trilayered cell wall profile. Finally, in stationary growth, the envelope becomes asymmetric, with the inner layer of the cell wall increasing in thickness and staining density. Ultimately the cell wall develops a uniform density and texture, with a thickened diffuse outer layer.

4.2.5. Mycobacteria

Included in the genus *Mycobacterium* are acid-fast bacteria, resistant to decolorization with alcohol.[11] The cells are aerobic, either curved or straight rods measuring $3 \times 0{\cdot}5\ \mu m$. In some cases cells have been observed which show branching or irregular forms. The mycobacteria are ubiquitous, being found in soil, animals, and humans.

Several of the mycobacterial species are of medical interest and include *M. tuberculosis, M. bovis, M. leprae*, and *M. lepraemurium. Mycobacterium tuberculosis*, the causative agent of tuberculosis in humans, infects almost any tissue or organ of the host, with the lungs having the greatest involvement. *Mycobacterium bovis* causes tuberculosis in cattle, while *M. leprae* (Hansen's bacillus) has been found associated with human leprosy. *Mycobacterium lepraemurium*, an obligate, intracellular parasite, is the causative agent of rat leprosy, and has been experimentally transmitted to rats, mice, and hamsters.

Several mycobacterial-like organisms have been isolated from tuberculosis associated environments (i.e., TB hospitals) and are referred to as "atypical mycobacteria." Usually, these atypical acid-fast bacilli are found in the sputum of tuberculose patients, causing a mycobacteriosis, or pulmonary infection which mimics the classical tuberculosis by *M. tuberculosis.* The organisms have been divided into groups according to their ability to produce pigments.

The most common of the mycobacteria, *M. smegmatis* and *M. phlei*, are found almost ubiquitous in dust, soil, water, and as nonpathogenic forms in animals and humans. They are normal inhabitants of the prepuce and vulva where they can contaminate urine and can be mistaken for *M. tuberculosis.*

Mycobacterium tuberculosis, M. smegmatis, M. phlei, and *M. leprae* are seen in Figure 4.11.[11, 27, 29, 38, 61, 104, 161, 219] The cells contain a dense, homogeneous

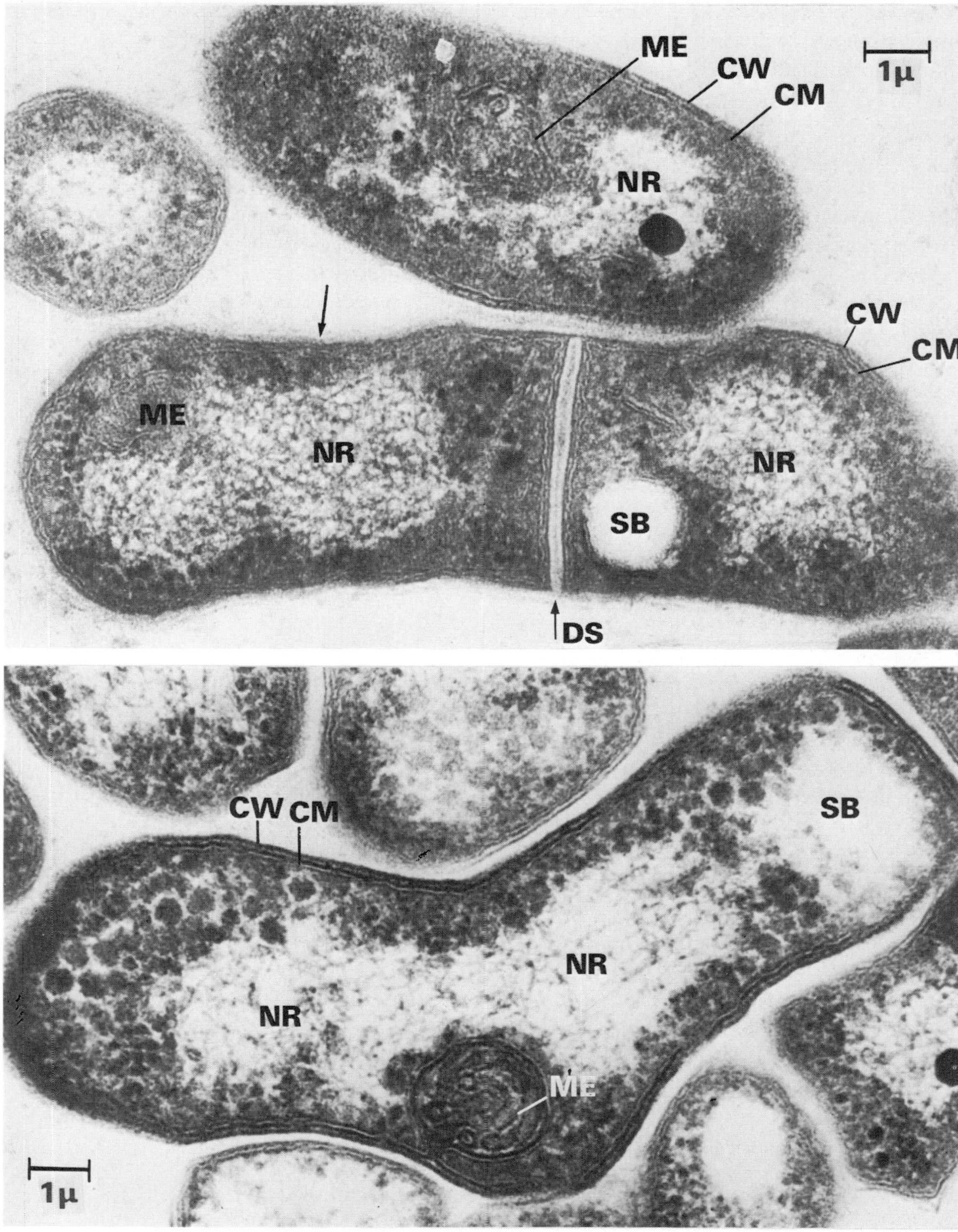

(a)

(b)

Figure 4.11. Acid-fast mycobacteria.
(a), (b) *Mycobacterium tuberculosis* H_{37} RV102. The cell wall (CW) stains poorly and its outermost layer is covered with a thin fibrous fuzz (arrow). The cytoplasmic region is filled with electron-transparent storage granules (SB) (vacuoles) and vesicular mesosomes (ME). DS: division septum. NR: nuclear region, CM: cell membranes.

cell wall approximately 15–20 nm thick (Figure 4.11(a)–(d),(g)) which, in most instances, is covered with an electron-dense fuzz, possibly similar to the mycosidic fibrils observed in *Mycobacterium* sp. strain NQ (Figure 4.11(e),(f)).[117, 118] These mycosidic fibrils consist of long parallel fibers, 5 nm in diameter. The cell surface of *M. smegmatis* is covered with fibrils which are 5–12 nm in diameter (Holt, unpublished communication). These fibrils may also correspond to the mycosidic fibril, or be a complex of lipopolysaccharide-mycolic acid. In some instances when non-acid-fast were compared with acid-fast *M. phlei* by electron microscopy, the non-acid-fast cells contained a thick cell wall, being almost three times as thick (38 nm) as the acid-fast *M. phlei.*[161] The non-acid-fast cells also contain large amounts of reserve compounds of irregular shape. The underlying cell membrane adheres tightly to the inner layer of the cell wall, and in some instances appears to be attached directly to it.

Mycobacterium leprae (Figure 4.11(g)) is found predominantly in altered mononuclear or epithelial cells. These procaryotes are sometimes referred to as "lepra cells" due to their close packing (in some cases the cells resembling packets of cigars) within the cells. In Figure 4.11(g) *M. leprae* is seen completely enclosed in a phagolysosome. The cell wall is similar to that found in the other mycobacteria. Possibly a mesosome is associated with the cell membrane, and the cytoplasmic region not only contains electron-dense ribosomes and fibrous nucleoid, but also an electron-lucent, irregularly shaped mass, possibly storage material.

4.2.6. Bifidobacteria

Lactobacillus bifidus, renamed *Bifidobacterium bifidum*, was first described by Tissier in 1889[201] as curved rods or rods which consisted of split ends giving a characteristic or bifid shape (Figure 4.12(a)). The organisms are gram-positive and tend toward some degree of pleomorphism with bizarre small branches and bulbous or swollen ends.[162] *Bifidobacterium bifidum* is readily isolated from the stools of breast-fed infants but the association between these infants and the bacteria is not clear. Recent investigations[59, 87, 221] have shown a large number of *E. coli* as well as *Bifidobacterium* in human feces. It is apparent that the latter is not the predominant microorganism in breast-fed human intestines.

The cell morphology as revealed by electron microscopy (Figure 4.12(a),(b)) is that of a typical gram-positive bacterium. The cell wall is thick and homogeneous, with a cell membrane which appears to adhere very tightly

---

Figure 4.11. Acid-fast mycobacteria.
(c), (d) *M. smegmatis* and *M. phlei.* Cells of *M. smegmatis* are juxtaposed in such a way as to indicate an intervening matrix (arrows). CW: cell wall, NR: nuclear region, ME: mesosome.
(e), (f) *Mycobacterium* sp. NQ. In the freeze-fracture etched replica (e) the cell surface is covered with a rope-like layer, consisting of ridges or channels, possibly mycoside fibrils (MF). CW: cell wall. The extent of the mycoside layer is evident when the species is examined by negative staining (f). The cells from the pellicle region of growth contain a thick layer of intertwining mycoside fibrils (MF).

(c) (d)

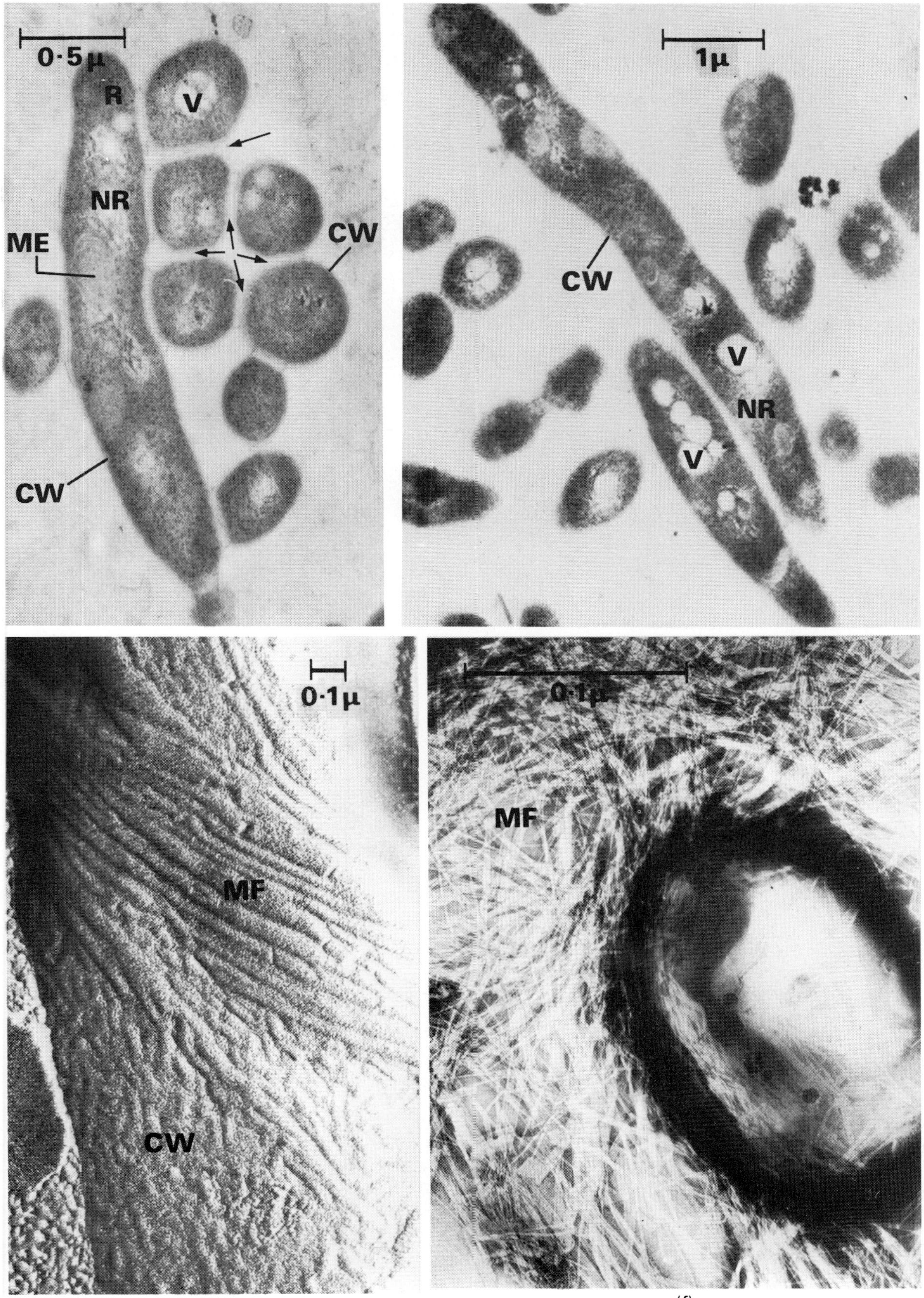

(e) (f)

Figure 4.11. Acid-fast mycobacteria.
(g) *Mycobacterium leprae* within a phagolysosome in a "histoid" nodule from a case of human leprosy. The bacterium is enclosed in a phagosome vacuole (PV). The thin cell wall (CW) is evident enclosing an electron-dense cytoplasmic region. R: ribosome, ER: endoplasmic reticulum, NR: nuclear region.

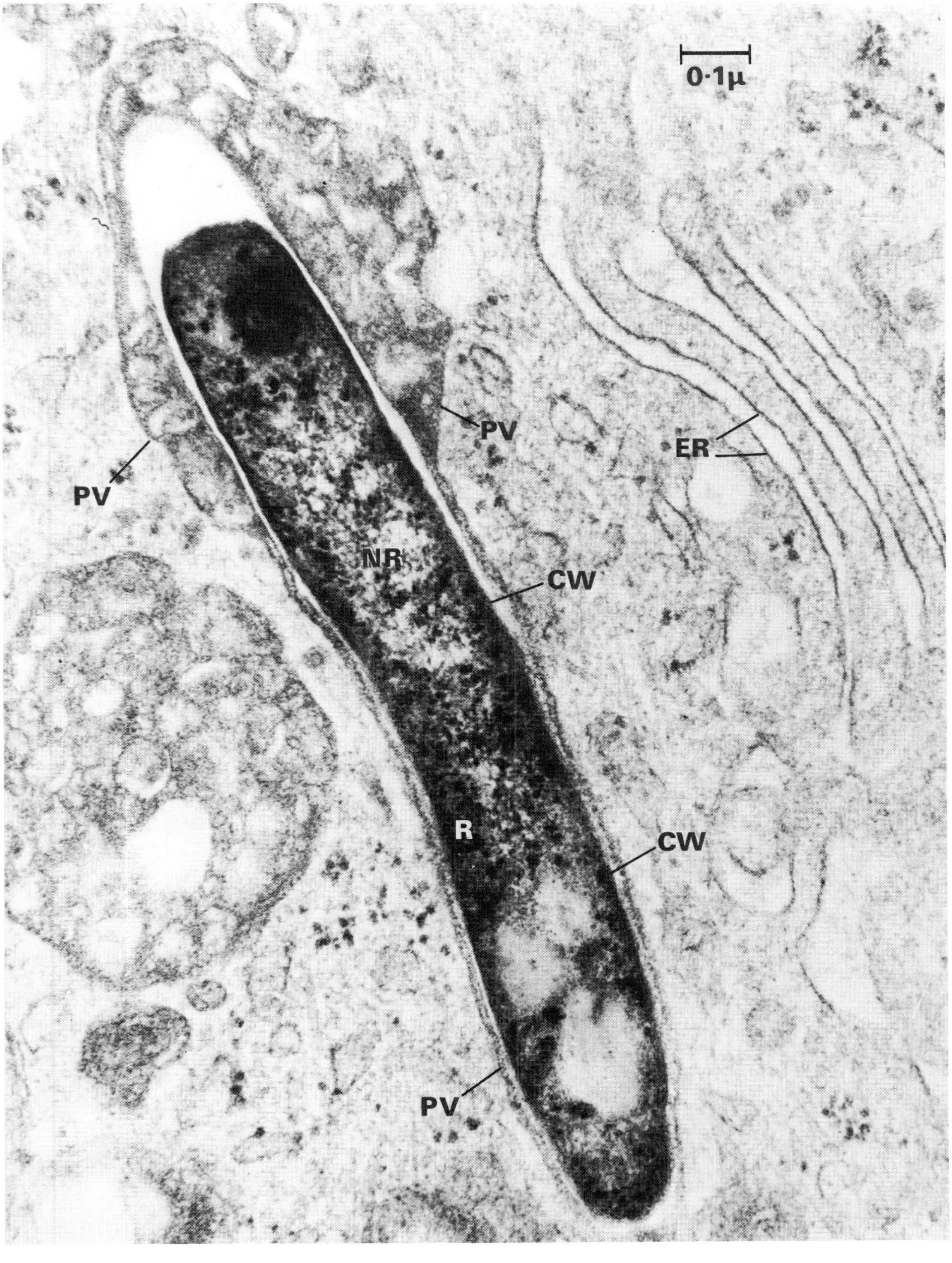
0·1μ
PV
PV
ER
NR
CW
R
CW
PV

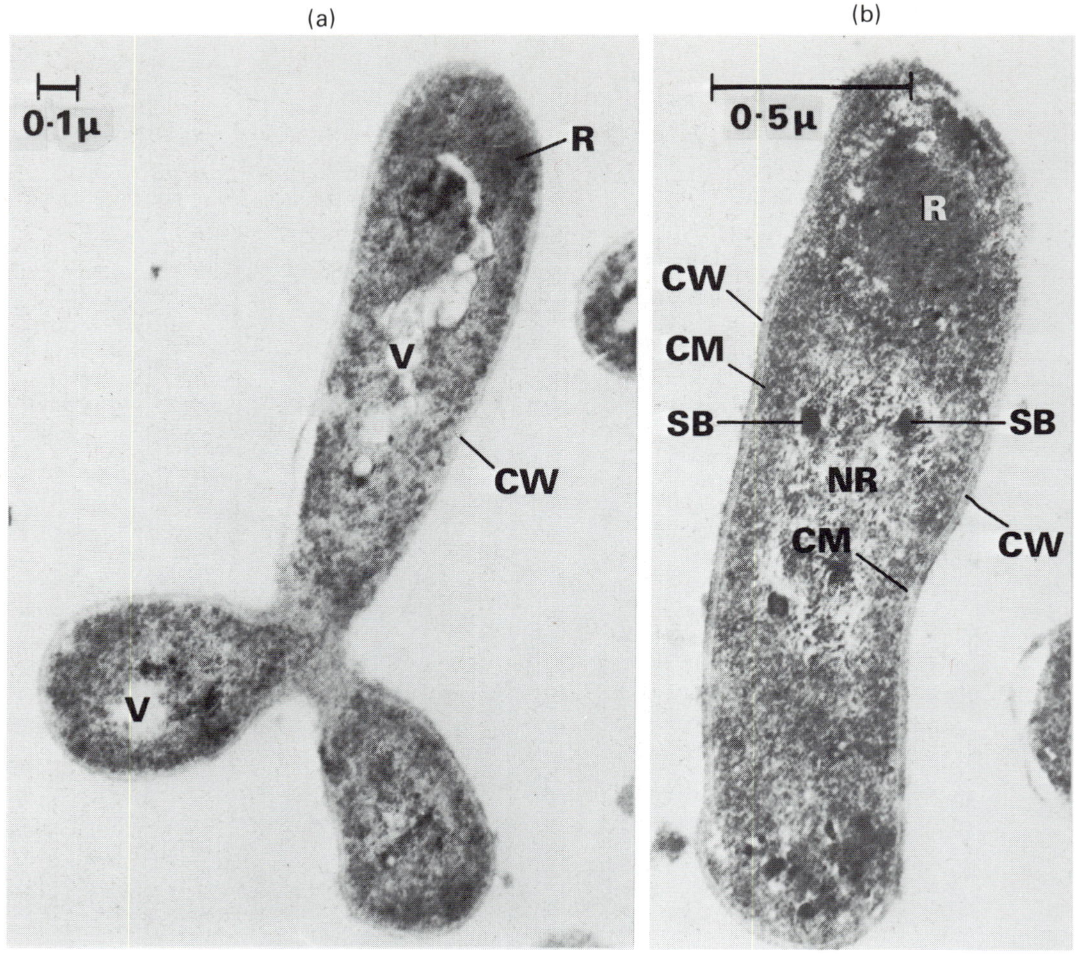

Figure 4.12. ***Bifidobacterium bifidum.***
(a) Characteristic branched or bifid cell. (b) Rod-shaped or unbranched form. The cell wall (CW) is thin and possibly multilayered. Electron-dense ribosomes (R) are interspersed with the fibrous nucleoid (NR) and electron-transparent vacuoles (V). SB: spore body, CM: cell membrane, NR: nuclear region.

to the inner layer of the cell wall. The nucleoid is centrally disposed and surrounded by a homogeneous layer of electron-dense ribosomes. In branched *B. bifidum*, the cell wall is thinner than that observed in the curved rods. It is apparent from the published literature that the growth medium has a dramatic effect on the morphology of these organisms.[162] Thus, great care should be exercised in interpreting their ultrastructure, especially as it relates to a "standard" form.

4.2.7. Corynebacteria

The corynebacteria (Klebs–Loffler bacillus) include those gram-positive to gram-variable, obligate aerobes which are characteristically club-shaped with pointed to rounded ends (Figure 4.13(a),(b)). The cells are nonmotile and

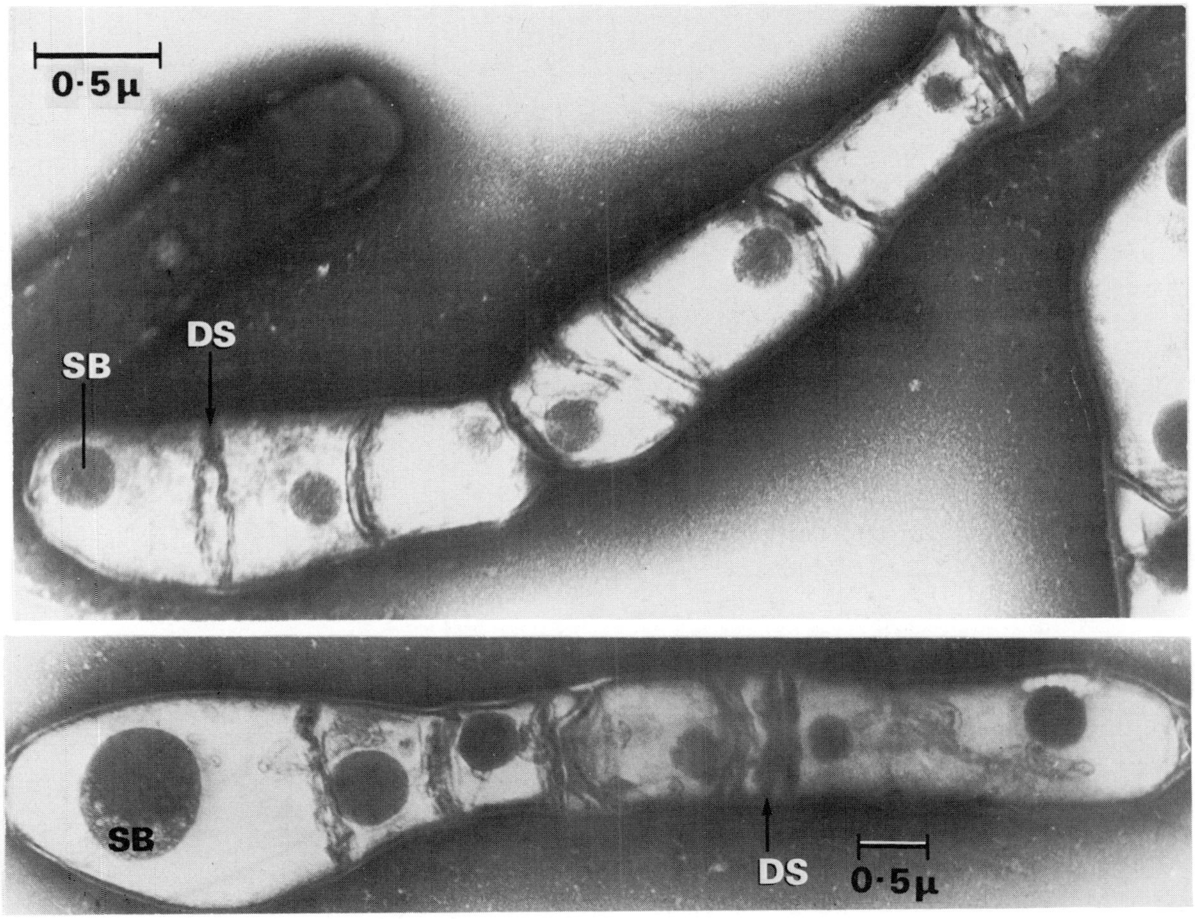

Figure 4.13. ***Corynebacterium.***
(a), (b) Negatively stained cells of *C. diphtheriae* reveal the characteristic club- or drum-shaped appearance, along with numerous spherical storage granules (SB), one per segmented cell. DS: division septum.

extremely pleomorphic. The predominant species within the genus, *Corynebacterium diphtheriae*[10] is the causative agent of membranous croup.[220] The designation of the epithet diphtheriae comes from the Greek word, leather, the result of the production by *C. diphtheriae* of a leathery diphtheritic membrane or pseudomembrane on the mucous membranes of the tonsils, pharynx, larynx, and nasal passages.

Corynebacterium diphtheriae, being distinctly pleomorphic, varies in both size and shape. Cells between 2–6 μm in length and 0·5–1 μm in diameter are observed; *C. diphtheriae* cells may be straight or curved, swollen in the middle region, or club-shaped at one or both ends. Usually, the cells vary in diameter between 0·5–1 μm for ovoid cells, and 1·5–5 μm for rod-shaped *C. diphtheriae*.

A characteristic cell division morphology distinguishes *C. diphtheriae* from other procaryotes. The cells divide by classical binary fission; however, during the division process a V-shaped arrangement of cells occurs. The cytoplasmic region usually contains metachromatic granules, sometimes referred to as polar bodies, Babes–Ernst bodies, or volutin granules.[111]

Depending upon the severity of infection, *C. diphtheriae* has been divided into three species: *C. gravis, C. intermedius*, and *C. mitis.* The pathogenicity and associated virulence of these corynebacteria is due to the excretion of an extracellular toxin which penetrates host tissue; there is little, if any, bacterial invasion. The toxicity of virulent *C. diphtheriae* is related to the presence of a temperate bacteriophage which carries the information necessary for toxin production.[44, 70]

In throat smears, it is sometimes possible to confuse a heterogeneous group of nonpathogenic, nontoxin producing diphtheroid bacilli with *C. diphtheriae.* These diphtheroid bacilli are almost ubiquitous, having been isolated from a wide range of host tissues, including the nose,throat, urethra, bladder, vagina and prostate gland, in addition to being found free in the environment, usually being isolated from soil and water.

Electron microscopic examination of thin sections of *C. diphtheriae* (Figure 4.13(c)–(f)) reveals the cell wall to be approximately 2·7–3·0 nm thick. The intermediate layer is extremely thick, sometimes approaching 2–2·3 nm in diameter. Lamellar mesosomes (Figure 4.13(c), (f)) and electron-transparent storage granules (Figure 4.13(c)) are seen in the electron-dense cytoplasm. The presence of a capsule in *C. diphtheriae* has not been demonstrated unequivocably; there is some evidence for capsule formation when the cells are grown in a glucose-yeast extract broth containing bovine serum. In some strains of *C. diphtheriae*, notably strain C4, an extracellular crystalline lattice which adheres to the outer layer of the cell wall has been observed (Figure 4.13(f)).[112] Two lattice types occur: a square lattice composed of parallel lines 312 nm in diameter with an average center-to-center spacing of 5·3 nm, and a lattice consisting of thinner parallel lines 2·2 nm thick with a much closer periodicity, averaging 3·5 nm. Of interest is the observation that the extracellular lattice is located at discrete sites on the cell surface; it is not diphtheria toxin *per se*, since it is found in both toxigenic and nontoxigenic strains.

Corynebacterium ovis (*C. pseudotuberculosis*), the causal agent of caseous lymphadenitis in sheep and ulcerative lymphangitis in horses, possesses a thick waxy material on its external surface which morphologically resembles a capsule. This lipoidal layer varies in thickness from 7 to 37 nm, and is readily found in both toxigenic and nontoxigenic strains; however, in attenuated or nontoxigenic strains, this layer is only 2·5–25 nm thick. Extraction and chemical analysis of the surface lipid reveals it to consist of sodium corynemycolic

Figure 4.13. *Corynebacterium.*
(c), (d), (e), (f) Thin sections of chemically fixed *C. diphtheriae* consist of a thick multilayered cell wall (CW) the outer surface of which is covered with a thin capsule (Cp). In some cells (f) a lattice structure (LS) is associated with the outer layer of the cell wall. *Corynebacterium diphtheriae* PW8 grown in low iron (d) reveals a metachromatic granule (MG), while a high iron phenotype ($PW8_r(P)^{tox+}$ (c) exhibits rudimentary branching and electron-transparent regions (storage granules) at the cell poles (arrows). CM: cell membrane, ME: mesosome, NR: nuclear region, R: ribosome.

(c)

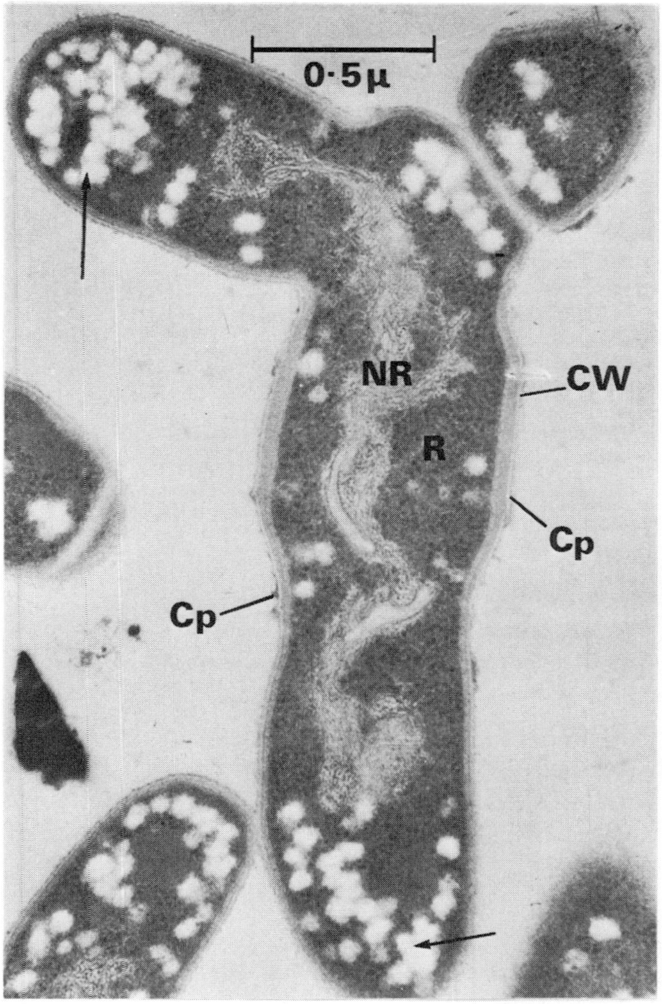
0·5μ
NR
CW
R
Cp
Cp

(d)

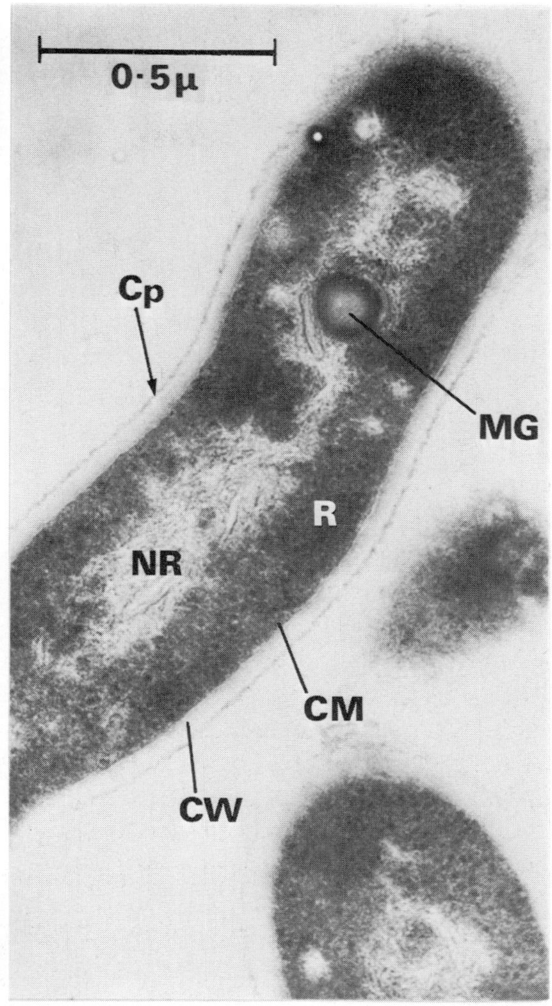
0·5μ
Cp
MG
R
NR
CM
CW

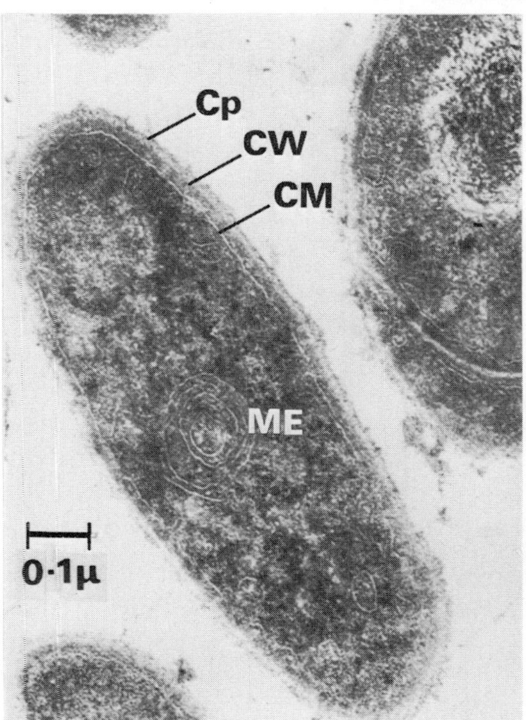
Cp
CW
CM
ME
0·1μ

(e)

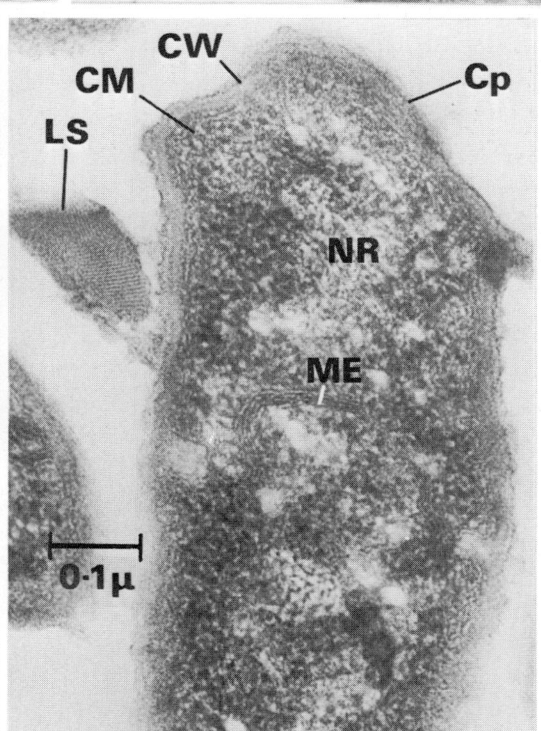
CW
CM
Cp
LS
NR
ME
0·1μ

(f)

acid, and the sodium salts of C14, C16, and C18 fatty acids. In liquid media *C. ovis* grows as a solid pellicle on the liquid surface.[83, 84]

The causative agent of erythrasma has been found to be the diphtheroid, *C. minutissimum* (Figure 4.13(g)–(i))[176] where it occurs on and in the horny layer of the stratum corneum. Cytologically, *C. minutissimum* is a rod-shaped gram-positive cell with rounded ends. The cells are morphologically very similar to *C. diphtheriae*, being surrounded by a thick cell wall composed of two electron-dense layers, each 3·5 nm wide, separated by an electron-lucent zone, 5·1 nm in width. In some instances *C. minutissimum* may adhere to host epithelial tissue by this extracellular material (Figure 4.13(h), (i)).[140, 141, 142] Directly beneath the inner layer of the cell wall and in intimate contact with it is the unit 7·5 nm cell membrane which encloses the electron-dense cytoplasm. Figure 4.13(g)–(i) shows the completed septa, characteristic of gram-positive cells in division.

4.2.8. *Rothia*

The single species of gram-positive cells which comprise this genus is the pleomorphic *Rothia dentocariosa*, previously known as *Actinomyces dentocariosus, Nocardia dentocariosus*, or *N. salivae*. Cells have been observed as cocci, filamentous, or diphtheroid in shape.

Rothia dentocariosa is readily isolated from the oral cavity, blood, and spinal fluid as well as from abscesses. Of interest is the fact that Roth and Flanagan[171] have been able to produce abscesses in experimentally infected mice with *R. dentocariosa.*

Culturing of *R. dentocariosa* in brain–heart infusion or in sheep blood results in the formation of two morphological cell types; coccobacillary cells after 24 hour growth (Figure 4.14(b)) and chained filaments after 48 hours (Figure 4.14(a)).[172] The cell wall of the coccoid forms is approximately 50 nm thick, the outer aspect of which is more electron dense and rugose. The filamentous cells (Figure 4.14(a)) vary in both shape and size; separating the individual cells within the filament are thick septal regions, approximately 100 nm thick. In these filamentous forms a division plane is apparent perpendicular to a completed septum, similar to that observed in the micrococci and sarcinae (Figure 4.17).

Rothia has also been observed to contain an external acid-mucopolysaccharide mucoid capsular or slime layer.[172] Employing an iron-binding technique, these investigators showed that coccoid cells of *R. dentocariosa* bound acidified, dialyzed iron over the entire outer surface of the cell. This Fe-capture may represent a conversion of a mucoid capsular or slime layer from an electron-transparent material to an electron-dense layer.

Figure 4.13. *Corynebacterium.*
(g), (h), (i) *Corynebacterium minutissimum* displays a thick electron-dense capsule (Cp) from both broth culture (g) or when viewed directly from a skin lesion (h), (i). In (i) the bacterium is associated with the surface of the superficial stratum corneum by fibers emerging from the electron-dense capsular layer (arrows). DS: division septum, CW: cell wall, CM: cell membrane, NR: nuclear region, Pg: Peptidoglycan layer.

(g)

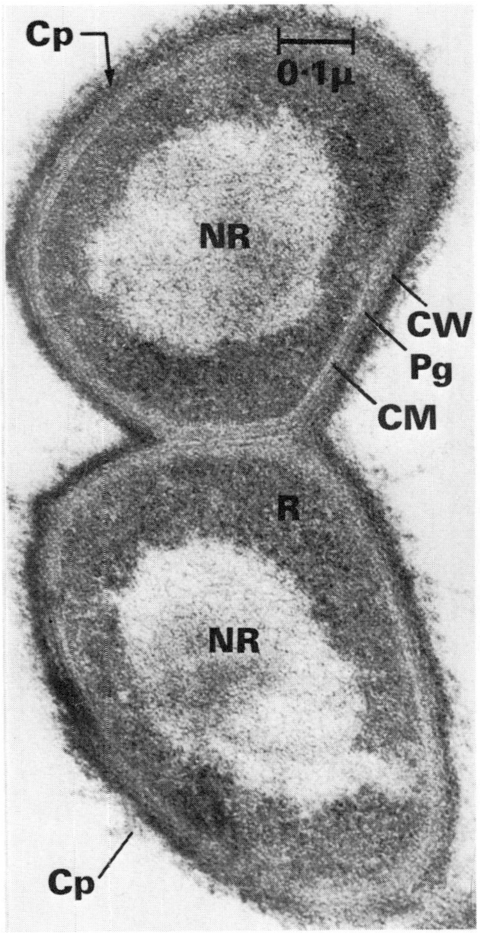

(h)

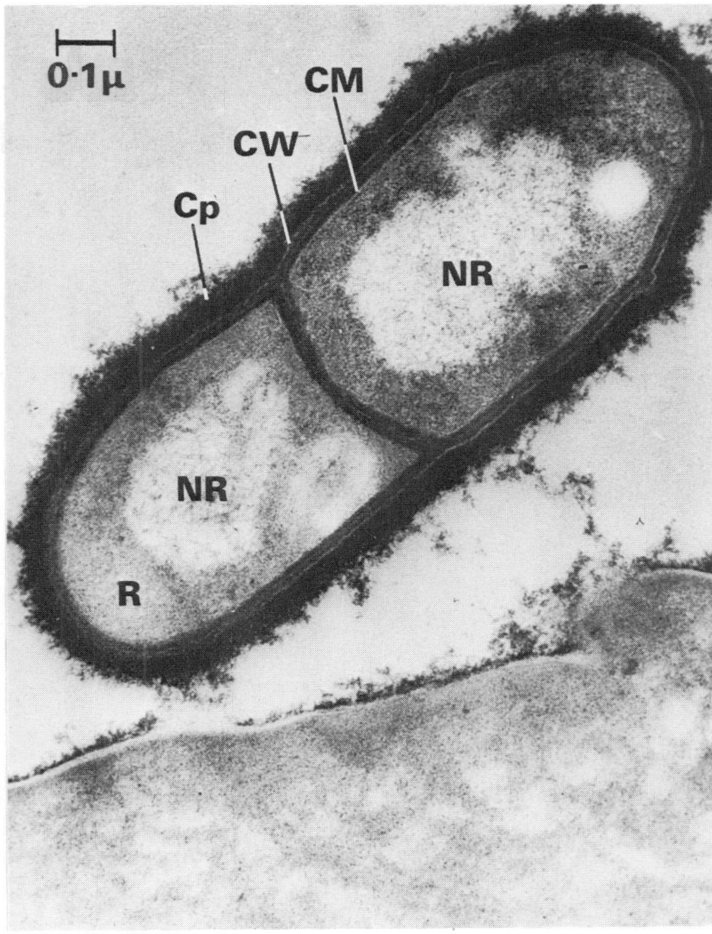

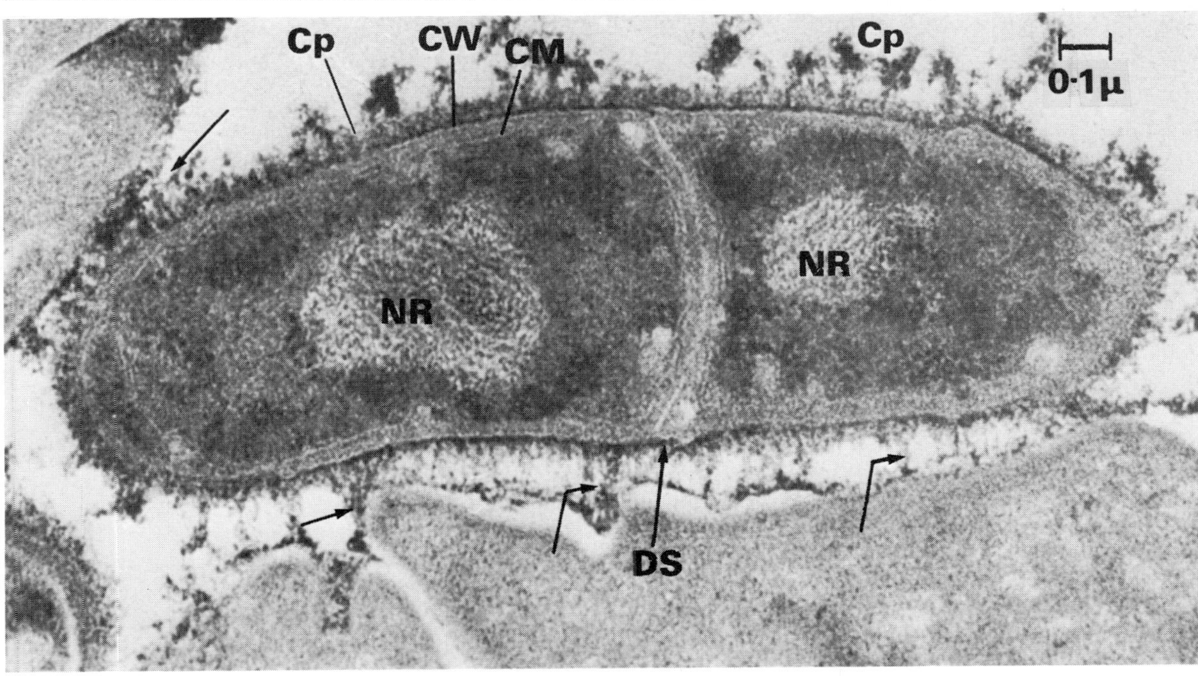

(i)

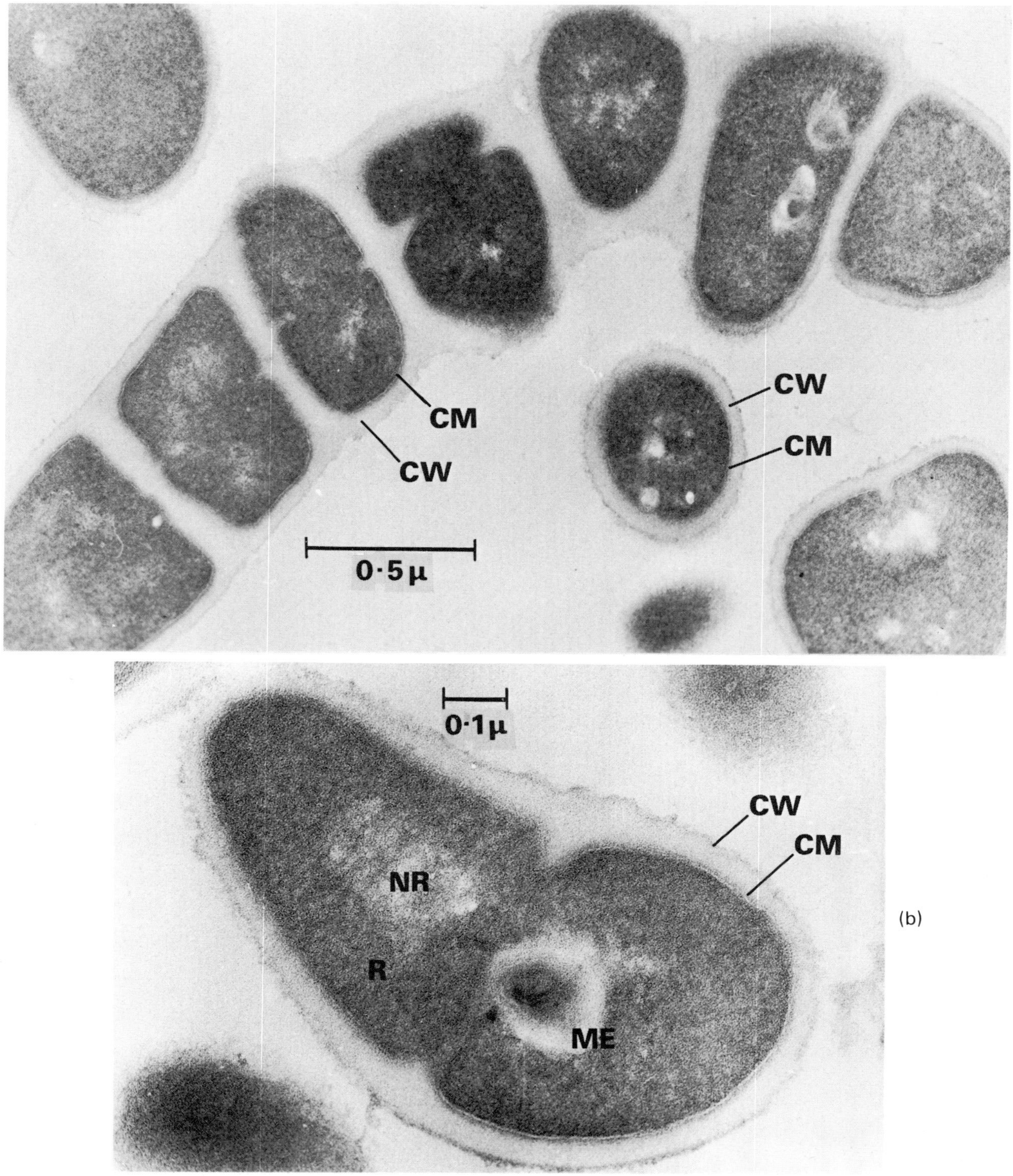

(a)

(b)

Figure 4.14. ***Rothia dentocariosa.***
These oral isolates occur in (a) chain-like filaments and (b) as swollen coccobacilli. A thick cell wall (CW) is characteristic of the genus. R: ribosome: CM: cell membrane, NR: nuclear region, ME: mesosome.

4.2.9. Lactobacilli

The lactobacilli are gram-positive, nonmotile, straight to curved rods. The members of the genus *Lactobacillus* are widely distributed in nature, as well as being associated with the human host as members of the oral cavity and alimentary tract. During a woman's active reproductive years a mixture of lactobacilli (sometimes referred to as Doderlein's bacillus) have been found to inhabit the vagina. In the oral cavity the lactobacilli are consistently isolated from carious lesions, where, because of their acidogenic and acidophilic capabilities and their ability to produce extracellular polymers (adherence), they may be involved with caries development.

A thin section of *L. acidophilus* is seen in Figure 4.15. The cell wall is thick and of variable electron density. The underlying cell membrane adheres very tightly to the inner layer of the cell wall. Associated with the cell membrane is a lamellar mesosome. Characteristic of many of the gram-positive bacilli is the electron-dense cytoplasm, the result of intense staining of ribosomal material. The nuclear fibrils are hidden amongst the electron-dense ribosomes. The initiation of the gram-positive mode of cell division, septum formation, usually with associated mesosomes is also seen in Figure 4.15.

4.2.10. Staphylococci

These gram-positive organisms usually occur in grape-like clusters, pairs, or as single cells in short chains. They are normal inhabitants of the skin, mouth, and nose.

The two principal members of the genus, *Staphylococcus aureus* and *S. epidermidis*, are best known for their involvement in skin and superficial tissue infections. They are found as the causative agents of boils, pimples, furuncles, abscesses, and wound infections. In addition, the staphylococci may also be found associated with systemic diseases such as pneumonia, emphysema, endocarditis, meningitis, and parotitis.[20, 183]

An electron micrograph typical of *S. aureus* is seen in Figure 4.16(a) (see also Suganuma[188, 189, 190, 191, 192]). The cells are similar to the other gram-positive bacteria, possessing a thick electron-dense cell wall, a closely adhering cell membrane which in some instances is separated from the wall by an electron-transparent space similar to the periplasmic space of gram-negative bacteria (see section 4.3.9). Of interest is the observation[40] that if *S. aureus* is not doubly fixed in glutaraldehyde-osmium, the periplasmic space is not apparent. Glutaraldehyde, which acts to stabilize proteins,[82] most probably stabilizes periplasmic proteins preventing them from being leached out during subsequent osmium fixation and ethanol or acetone dehydration. The thickness of the cell wall varies between 18 and 25 nm, with the outer dense layer being approximately 4–5 nm thick. The electron-dense inner layers of the cell wall including the periplasm and the outer lamina of the cell membrane are 7–8 nm wide. Cells of *S. aureus* strain R, fixed by the conventional Kellenberger–Ryter–Séchaud[113] osmium technique (Figure 4.16(b)), contain vesicular-mesosomal membranes in the central region of the cell, as well as a loose-fitting hazy

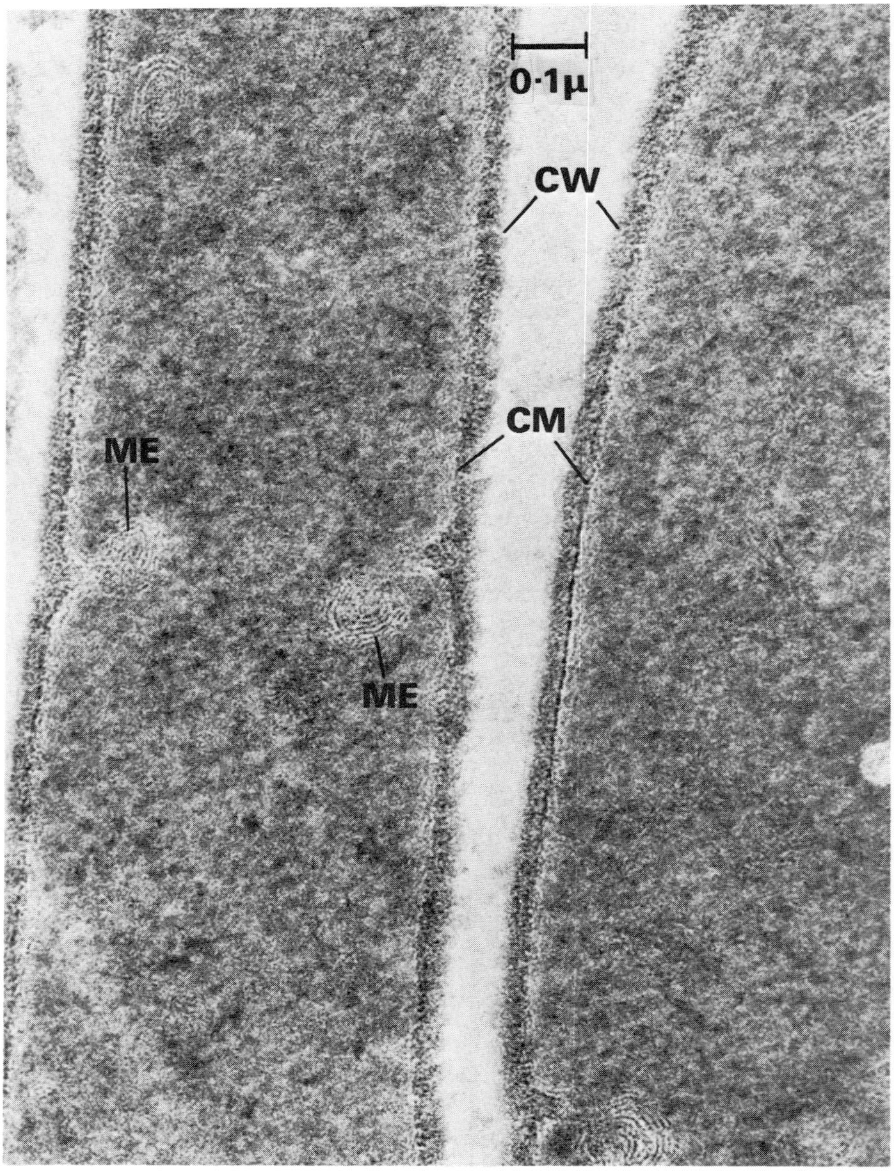

Figure 4.15. ***Lactobacillus acidophilus*** **NCTC4504.**
This gram-positive bacterium displays a thick electron-dense cell wall (CW) and tightly adhering cell membrane. (CM). The electron-dense cytoplasm contains several lamellar mesosomes (ME) originating from the inner portion of the cell membrane.

Figure 4.16. ***Staphylococcus–Micrococcus.***
Thin sections of chemically fixed (a), (b), (c), (d) *Staphylococcus aureus* and (e), (f) *S. epidermidis.* Compare with *Micrococcus* spp. (g) and (h). The cell walls (CW) of all organisms are thick structures which enclose a densely staining cytoplasmic region. A fine, hair-like fuzz (arrows) surrounds the cells in (b) compare with (h)). In (c), (d), (e), (f), in association with polymorphonuclear leukocytes(PMN), the staphylococci are initially observed (c) in a trough in the phagocytic membrane, then (d), (e) engulfed into phagosomes (PV), ultimately to be digested in later stages, with the loss of cell integrity (f). Note in (e) the loss of cell wall (CW) homogeneity with resulting fibrous formation of the wall (Fi). CM: cell membrane, V: vacuole, NR: nuclear region.

(a) (b)

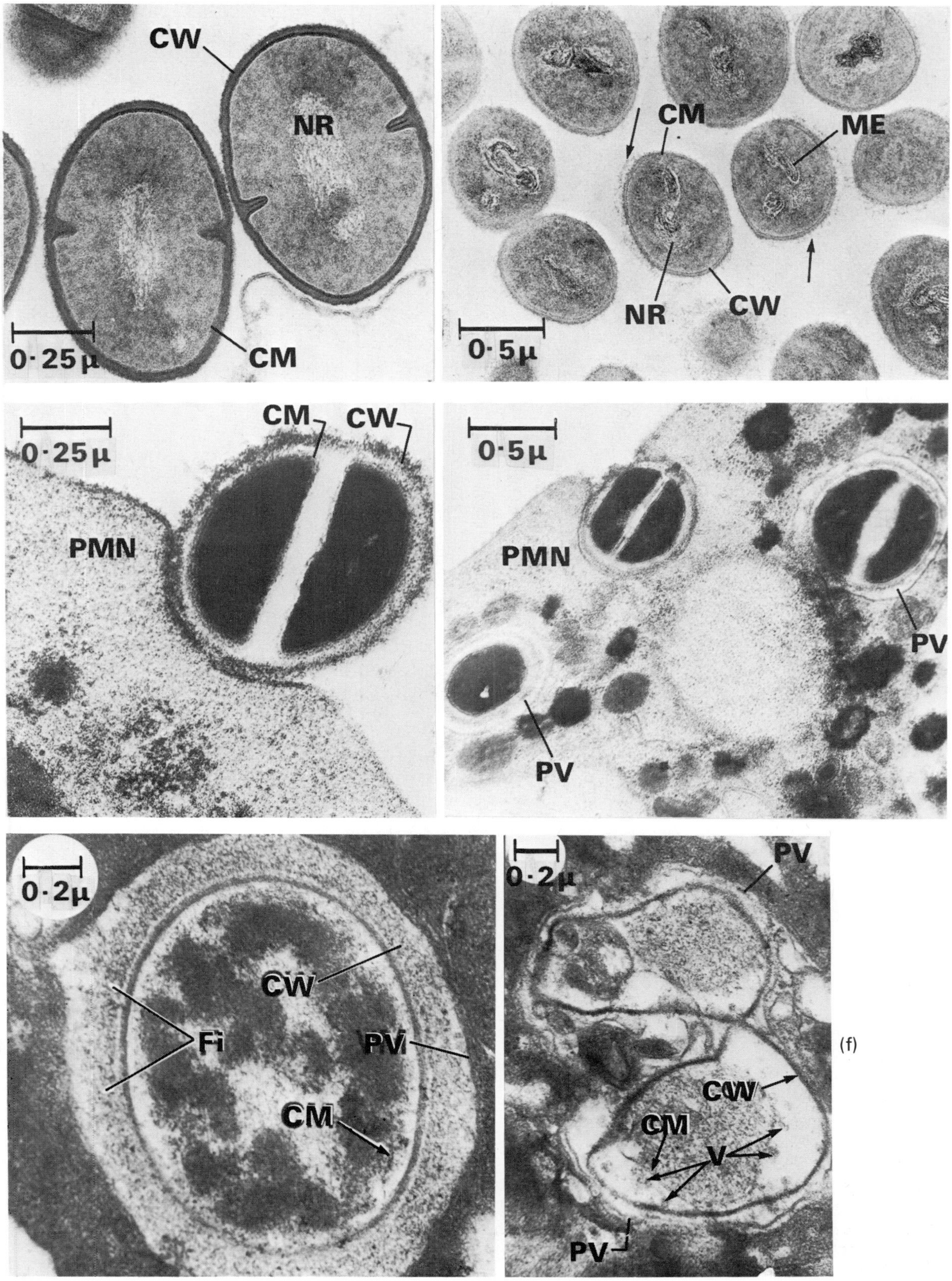
CW
NR
CM
0·25μ
CM
ME
NR
CW
0·5μ
CM
CW
0·25μ
PMN
0·5μ
PMN
PV
PV
0·2μ
CW
Fi
PV
CM
0·2μ
PV
CW
CM
V
PV

(c) (d) (e) (f)

layer associated with the outer surface of the cell wall. This loose-fitting layer could represent a capsular layer.

In examining the cell walls of bacteria, especially gram-positive cells, great care must be taken to determine the growth conditions from which the cells are taken for morphological examination. In a most enlightening study, Cole *et al.*[40] observed that cells of *S. aureus* strain H($S_m{}^R$) when grown in PyK broth to early or late logarithmic phase had a trilaminar cell wall, similar to that of *S. aureus* strain 209P. However, overnight growth of *S. aureus* H($S_m{}^R$) resulted in the loss of the trilaminar morphology of the cell wall, and the appearance of a very electron-dense and compact structure. In addition, the periplasm was also lost. Interestingly, growth of strain H in the presence of 1M $CaCl_2$ overnight yielded cells of comparable morphology to log-phase cells.

The cytoplasm is filled with electron-dense ribosomes which enclose a fibrillar nuclear region, and in some strains the cytoplasm also contains numerous mesosomal membranes.[74] Interesting evidence has been presented[67] that chemical fixation *may produce* mesosomes or "technikosomes." By freeze-etching, *S. epidermidis* shows the characteristic thick cell wall, underlying particle-studded cell membrane, as well as vesicular and tubular mesosomes.[75]

Staphylococcus aureus has been used for many years in model systems for the study of the phagocytic events involved in bacterial destruction.[26, 79] When a human isolate of *S. aureus* is mixed with human neutrophils, there is an almost immediate (within 5 min) uptake of staphylococci. Bacterial-neutrophil contact leads to an indentation of the neutrophil (Figure 4.16(c)), the development of pseudopodia (Figure 4.16(d)), with resulting rapid and complete bacterial engulfment (Figure 4.16(d), (e)). The pseudopodial cell membrane coalesces after engulfment, leaving a portion of the cell membrane surrounding the bacterial cell (Figure 4.16(d)). In addition to neutrophil engulfment, a vacuole is also formed around the bacterium – probably the result of an interaction between the bacterium and the neutrophil cytoplasm.

In well-preserved phagocytic vacuoles it is possible to observe an electron-dense fibrous material surrounding a staphylococcus within the vacuole (Figure 4.16(e)). This material could, as DeVoe *et al.*[52] suggest, represent degradative enzymes which are released into the vacuole from leukocytic granules; or it could represent debris from disintegrating *S. aureus.* Characteristic of the destruction of engulfed *S. aureus* is the almost simultaneous disintegration of both its cell wall and cell membrane (Figure 4.16(f)). In addition, there is an accompanying coagulation of the cytoplasm and its retraction from the disintegrating cell wall–cell membrane.

Figure 4.16. ***Staphylococcus–Micrococcus.***
(g) *Micrococcus radiodurans* and (h) *M. mucilaginosus.* The latter is surrounded by a thick electron-dense fibrous capsule (Cp) while the former is held together in tetrad arrangement but no capsule is apparent. CW: cell wall, CM: cell membrane, ME: mesosome.

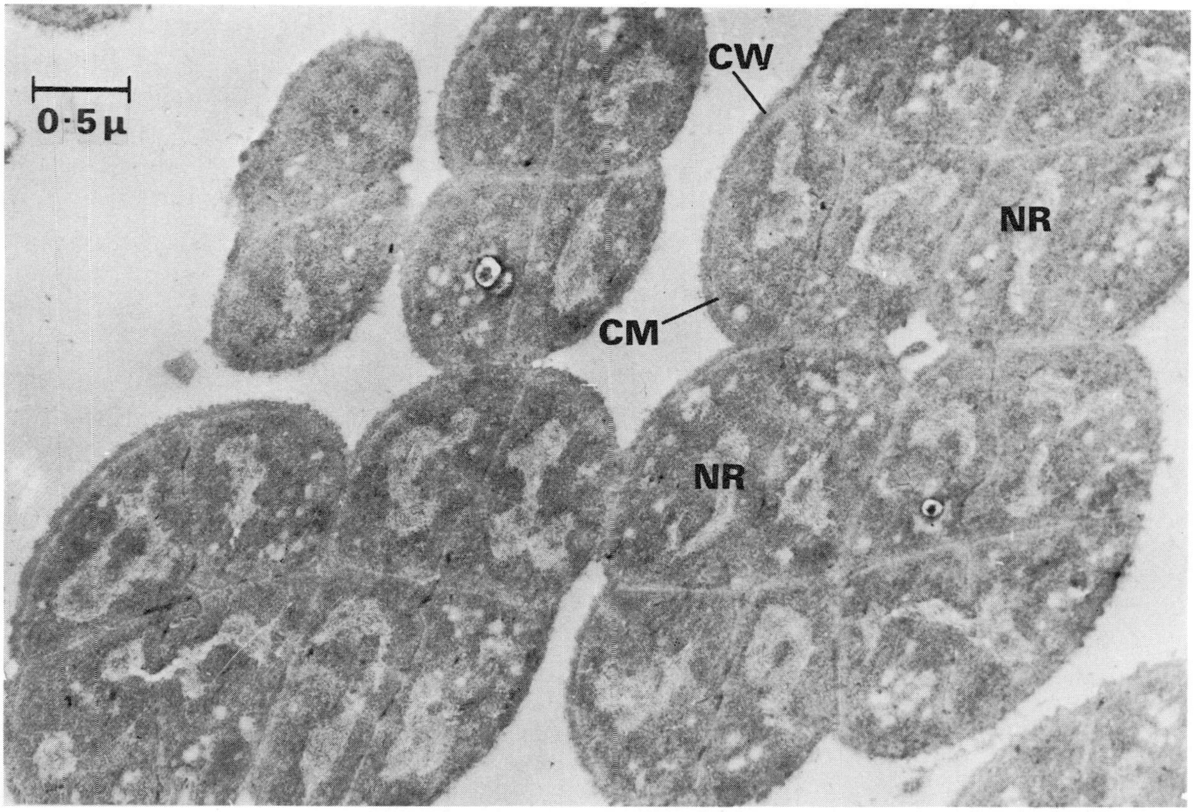

(g)

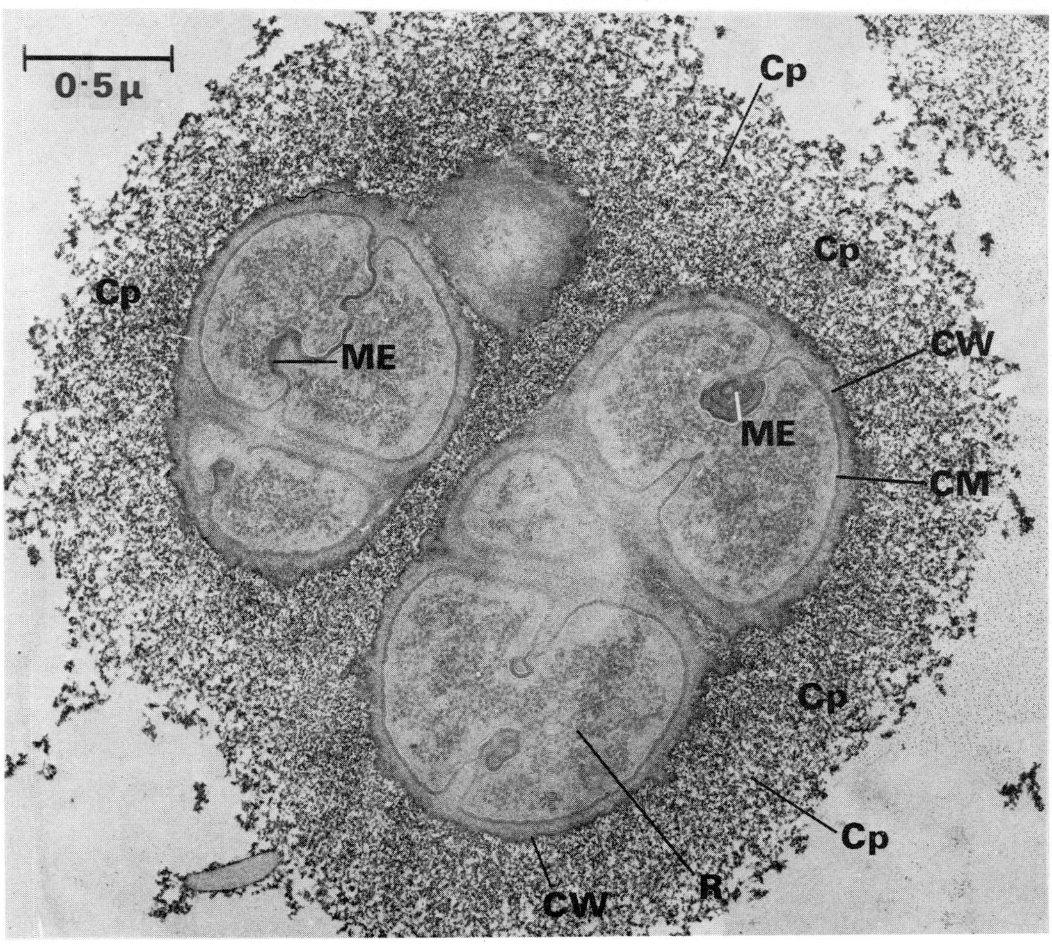

(h)

4.2.11. Micrococci: *Sarcina*, *Gaffkya*, *Micrococcus*

This group of bacteria is characterized by their ability to grow as packets. Morphologically they resemble in many respects the staphylococci. Where they have been found in association with humans, they probably occur as opportunists,[20] in most instances they are not considered to be a normal part of the human microflora. There are some reported cases of bacterial endocarditis due to *Gaffkya tetragena* but these reports are not of a regularity to justify consideration of *G. tetragena* as a human pathogen.

The two recognized species of anaerobic sarcinae, *S. ventriculi* and *S. maxima*, are representative of procaryotic cells which divide by a transverse binary fission involving three perpendicular division planes. The resulting cells are in packets (packages) hence sarcina of regular dimension.

Sarcina ventriculi has been isolated from the stomach of humans as well as from rabbit and guinea pig stomach contents. *Sarcina maxima* is found associated with oats and cereal grains as well as being found in soil. In addition to *S. ventriculi* and *S. maxima*, other sarcinae inhabiting soil, water, dust, and mud include the methanogenic and aerobic spore-forming sarcinae and several non-spore-forming aerobes. These species have been reclassified in the eighth edition of Bergey's Manual, and the reader is referred to the appropriate sections of the manual for discussion (see Diebel and Seeley[51]).

Light microscopic observations of *S. maxima* and *S. ventriculi* (Figure 4.17(a), (b))[98] reveal the cells to be arranged in packets, with usually 8 cells per packet in cubic configuration, hence sarcinae. Rectangular configurations of cells in multiples of 8 have also been observed. *Sarcina ventriculi* strain AL accumulates cellulose as a layer external to the cell wall and may account for its rough appearance in both light and electron photomicrographs. Thin sections of both the sarcinae (Figure 4.17(c), (d)) and micrococci (Figure 4.16(g), (h)) reveal cells in a tetrad configuration. The cell walls are of uniform electron density, ranging in thickness between 8 and 40 nm. In contrast to the smooth surfaced cell wall found in *S. maxima* (Figure 4.17(c), the cell walls of *S. ventriculi* (Figure 4.17(d)) and *M. mucilaginosus* (Figure 4.16(h)), the latter representative of packet-forming micrococci, are covered with a thick capsular or sheathed layer. *Micrococcus radiodurans* (Figure 4.16(g))[200] has a thinner extracellular layer. In *S. ventriculi* this layer may consist of cellulose fibrils. The rugose cellulose layer, and the three-dimensional arrangement of these packet-forming procaryotes is clearly seen in Figure 4.17(e), a platinum–carbon replica of *S. ventriculi*. Figure 4.17(e) shows how the individual cells of the packet are connected to adjacent cells by an almost continuous outer surface layer.

The cell membrane adheres very closely to the inner surface of the cell wall, and in most instances is difficult to observe. However, when apparent, it is approximately 7·5–10 nm in diameter.

The cytoplasmic region of these micrococci consists of closely packed electron-dense ribosomes, which tend to obscure the nuclear region. An occasional mesosome is also observed.

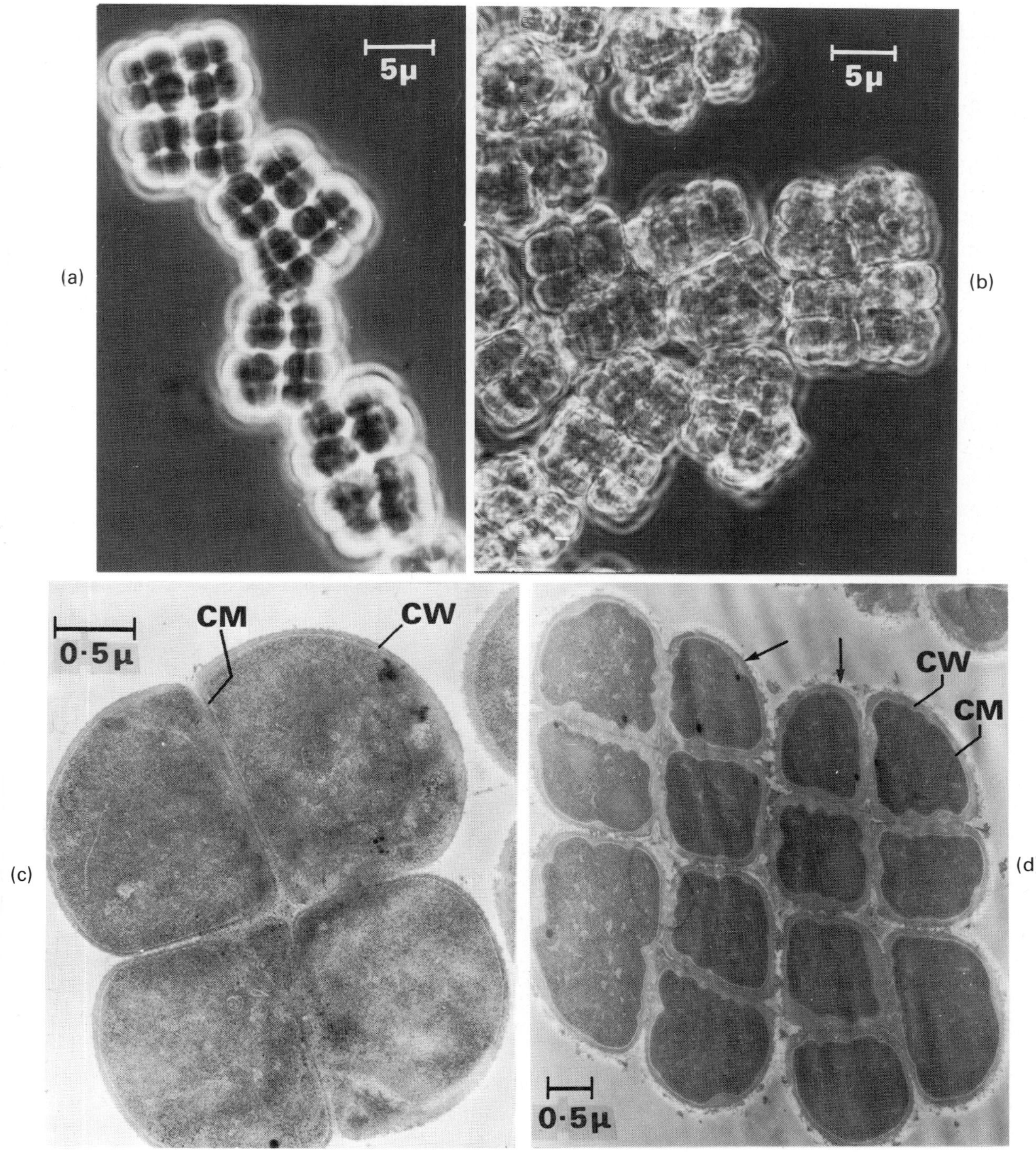

Figure 4.17. ***Sarcina.***
Phase contrast and electron micrographs of (a), (c) *Sarcina maxima* and (b), (d) *S. ventriculi*. The cells occur in tight packets. *Sarcina ventriculi* usually possesses more cells per packet than *S. maxima* and has a rough texture, probably due to extracellular cellulose. In thin section, *S. maxima* (c) is smooth surfaced while *S. ventriculi* (d) has a rough outer surface (arrows). CM: cell membrane, CW: cell wall.

(e)

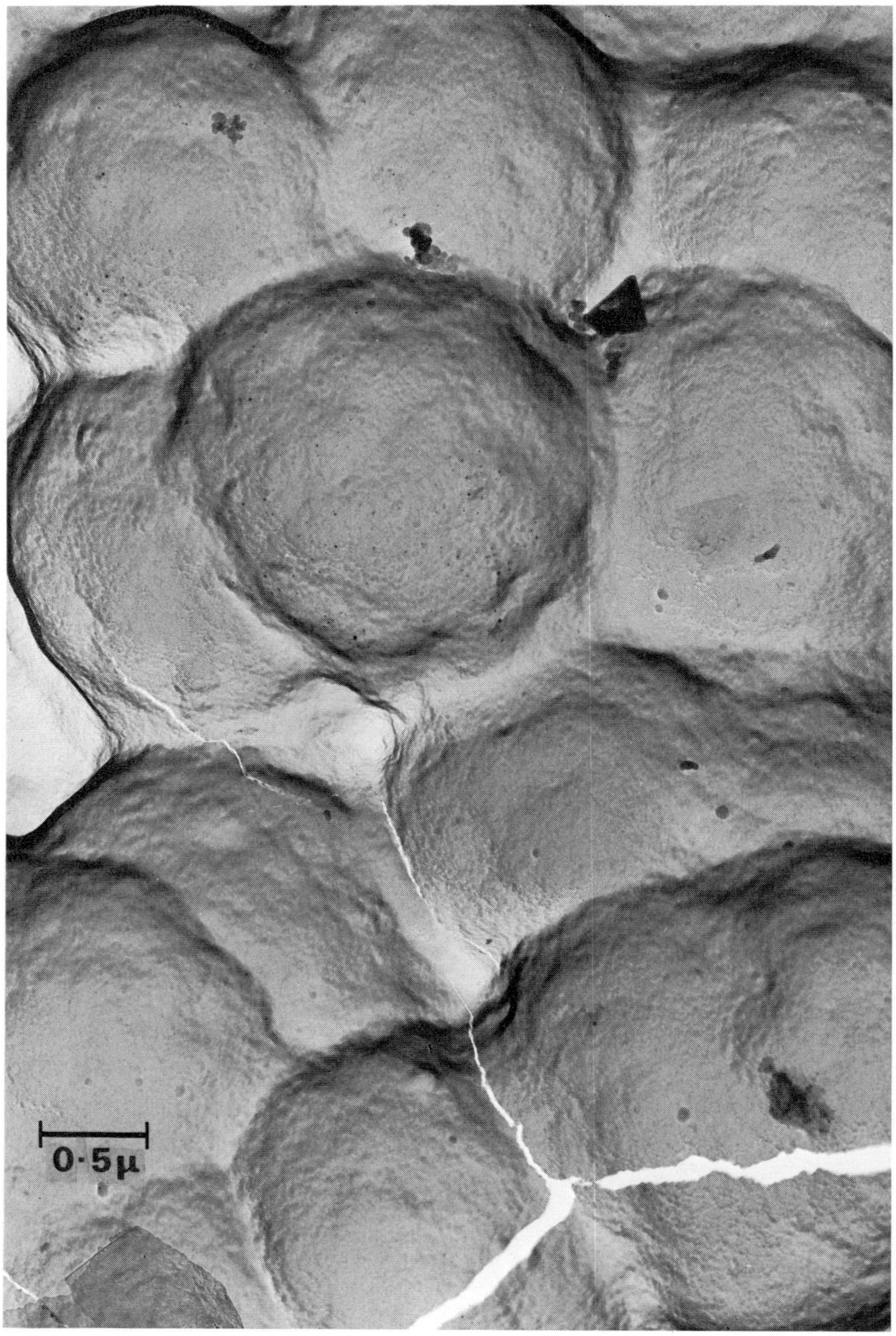

Figure 4.17. ***Sarcina.***
(e) *Sarcina ventriculi* after Pt–C surface replication, the rough outer surface being clearly evident.

4.2.12. Streptococci

These gram-positive cocci occur in pairs or chains, the chains sometimes reaching 50 or more cells. Within the chain the cells are arranged in pairs. Most species are facultative anaerobes.

In general, the streptococci are usually found associated with organic matter, especially where fermentable sugars are present. Several of the streptococci have been implicated in human diseases; *S. pyogenes, S. faecalis*, and *S. mutans* being implicated in the initiation of acute rheumatic fever, genitourinary tract infections, endocarditis and wound infections, and the destruction of tooth enamel with the production of carious lesions. The majority of human infections are the result of exposure to group A streptococci, while all of the other groups, with the exception of group N are indigenous for humans.

The anaerobic *Peptococcus* and *Peptostreptococcus* are of importance in infections of the female genitalia, as well as being implicated in infections such as puerperal sepsis, subacute bacterial endocarditis, and deep wound abscesses. In the oral cavity the peptostreptococci are found primarily in the gingival sulci. These anaerobic cocci are morphologically similar to the aerobic streptococci.

Figure 4.18(a)–(d) shows the basic morphology of the streptococci.[40, 89, 90, 136] The cell wall is composed of two thin electron-dense layers, each 5–7 nm thick, separated by a less dense layer, 14–17 nm thick. The underlying cell membrane is especially thick, being 12 nm in diameter, and adheres very closely to the inner layer of the cell wall over its entire length. The cell membrane is asymmetric in its staining characteristics; the outer dense layer staining more heavily than the inner layer. Characteristic of *S. faecalis* is the presence of raised cell wall bands, as well as mesosomal invaginations of the cell membrane (Figure 4.18(c), (d)).[89, 90]

In most instances the nuclear material is fibrous and dispersed throughout the cytoplasmic region, intermingled with electron-dense ribosomes.

The most external layer of the cell wall of the virulent streptococci (groups A, C, D, G, and H) is covered with a thin protein layer, the M-protein which varies both in thickness and electron density (Figure 4.18(e)–(g),(i)).[66, 68] Hyaluronic acid is considered to be a component of the M-protein layer, and in all likelihood also contains a variety of other antigens.

Group A streptococci are also found to contain hair-like fimbriae attached to their most external surface.[195, 197] Ferritin-conjugated antibodies to the M-protein revealed it to be located in these hair-like fimbriae (Figure 4.18(f)).

M-protein-associated adherence to host tissue in *S. pyogenes* was clearly demonstrated by Ellen and Gibbons.[66] These investigators showed that *S. pyogenes* STA628 and C203 was surrounded by a dense, fuzzy, fibrillar surface layer which bound the bacteria to epithelial cell membranes (Figure 4.18(i),(j)). Trypsin removed this layer along with the M-protein. Similar observations have been made with *S. salivarius* in which the bacteria adhere to epithelial cells by their surface fuzz or fimbriae. In addition, ruthenium red fixation and staining of *S. salivarius* clearly showed the presence of a distinct 85 nm diameter globular extracellular polymer. Of interest was the observation that in many instances where bacterial strains had lost their virulence, they also

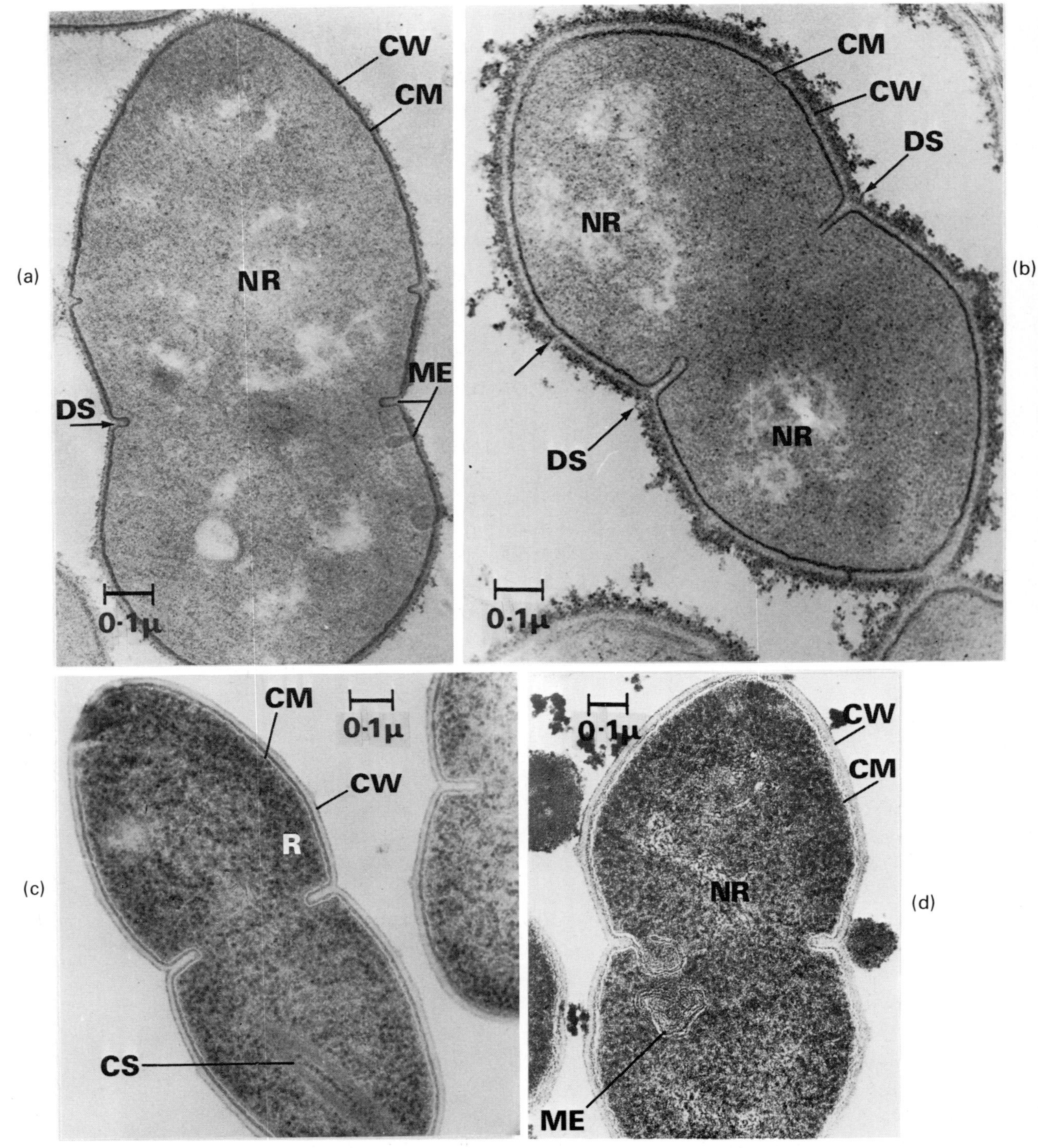

Figure 4.18. Representative streptococci.
Exponential-phase cells of (a), (b) group A *Streptococcus* and (c), (d) group D *Streptococcus.* The cell walls (CW) are thick and of uniform density. In some instances (b), an outer electron-dense layer (arrows) is associated with the cell wall (CW). Numerous division septa (DS) and mesosomes (ME) are visible. Note the core structure (CS) in the cytoplasm of *S. faecalis* (c). CM: cell membrane, NR: nuclear region.

lost their external M-protein layer. This result was most evident for the mucosal-associated pathogens, *Salmonella, Pneumococcus, Staphylococcus, Escherichia, Shigella*, and *Yersinia*.

The morphology of the oral streptococci has been examined by thin sectioning and freeze-etching[100, 106] and found to be similar to that of the other streptococci, as well as to gram-positive cells in general.

The cell wall of *S. mutans* serotype c and *S. salivarius* AHT is 11–35 nm thick and amorphous in texture (Figure 4.18(k),(l),(n)). The outermost layer of the wall is electron dense, and approximately 4–5 nm thick. This layer is covered with electron-dense fibrils or fuzz, which may represent glucan polymers. The intermediate layer of the wall is of low electron density and 7–16 nm thick. It is homogeneous in its staining properties and firmly apposed to the innermost, thin (3–11 nm), electron-dense layer of the cell wall. The cell membrane is 7 nm thick and of unit construction. In *S. salivarius* (Figure 4.18(l)), electron-transparent intracellular inclusions ring the periphery of the cytoplasm and may be similar to iodophilic granules.[57, 100, 208]

The outer surface of the cell wall of *S. mutans* after freeze-fracture etching is rough textured, with attached fibrous strands (Figure 4.18(m)) which may reflect extracellular polymer. In addition, the cell wall consists of two distinct layers, each 11–19 nm in diameter, which enclose a 6–14 nm particle covered cell membrane.

The relationship between extracellular polymer and cell clumping has been clearly demonstrated[106] with *S. mutans* GS–5 and two mutants GS–511 and GS–514. The mutants differ in the type of extracellular polymer produced. Wild type *S. mutans* GS–5 forms large clumps and aggregates of cells, with copious amounts of extracellular slime (Figure 4.18(p)), while the two mutants either produce very little extracellular slime (Figure 4.18(n)) or form a gel-like polysaccharide (Figure 4.18(o)) which is not as electron dense as the polymer found in wild-type cells.

The polymer consists of both fibrillar and globular polysaccharide at its most external border, with only the fibrillar polymer in the central region. The fibrillar layer consists of two parallel protofibrils, each 2 nm wide, to which an amorphous electron-dense material adheres. Johnson *et al.*[106] think that the extracellular polysaccharide may be involved in the adherence of *S. mutans* GS–5 to host cell tissue.

Streptococcus group A has also been examined after interaction with human monocytes (Figure 4.18(h)–(j)).[78] The ingestion process is very similar to that which occurs when human monocytes encounter other procaryotic cells. As is apparent in Figure 4.18(i) the monocyte surrounds the Group A streptococci by enveloping them with pseudopods. In Figure 4.18(h), a dividing representative of streptococcus has been enveloped in phagocytic vacuoles. There is a general disorganization of the enveloped bacterium, with the loss of cell wall–cell membrane definition. It is noticeable that the phagocytic vacuole is surrounded by an electron-dense material. Within 1 hour of streptococcal ingestion and envelopment by the phagocytic vacuole there is general bacterial degeneration. The cell wall and underlying cell membrane are partially fragmented, as well as a loss of ribosomal electron density. This loss of cytoplasmic

Figure 4.18. Representative streptococci.
The M-protein layer (MPL) of (e), (f) group A *Streptococcus* and (g) group A *Streptococcus* 543 MA(M) is visible as a thin hair-like fuzz on the outer surface of the cells. In (f), the cells were incubated with ferritin-conjugated type 5-antibody with the ferritin particles localized at the tips of the surface layer of the cell wall (CW) (arrows). (i) The M-protein layer attaches (arrows) *S. pyogenes* to the surface of a tongue epithelial cell. (h), (j) Ingested by a PMN-leukocyte, group A streptococci are enclosed in a phagocytic vacuole (PV) and in various stages of disintegration. Note the association (arrows) between the extracellular capsular layer of the bacterium and the surface of the monocyte membrane in (j). CM: cell membrane, NR: nuclear region, ME: mesosome.

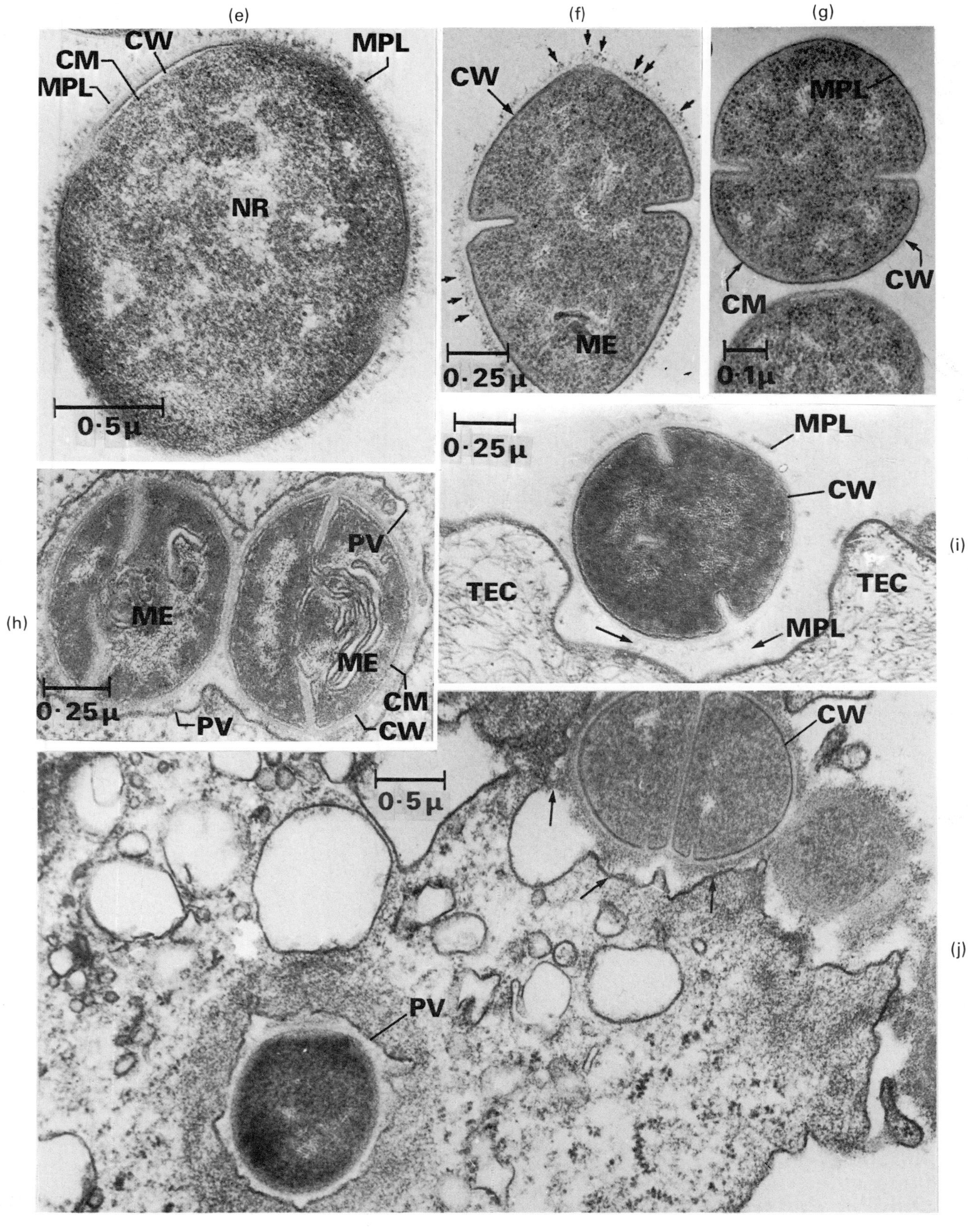
(e)
(f)
(g)
(h)
(i)
(j)
CW
CM
MPL
NR
0·5μ
ME
0·25μ
0·1μ
PV
TEC

Figure 4.18. Representative streptococci.
(k), (l), (m) Oral streptococci. Characteristic of these isolates is the extracellular fibrous polymer (EFP), either associated with the cell surface or free in the background (k). In (l) *S. salivarius* HHT electron-transparent iodophilic granules (IG) occur in the cell cytoplasm. CW: cell wall, CM: cell membrane, R: ribosome, NR: nuclear region.

(k) (l)

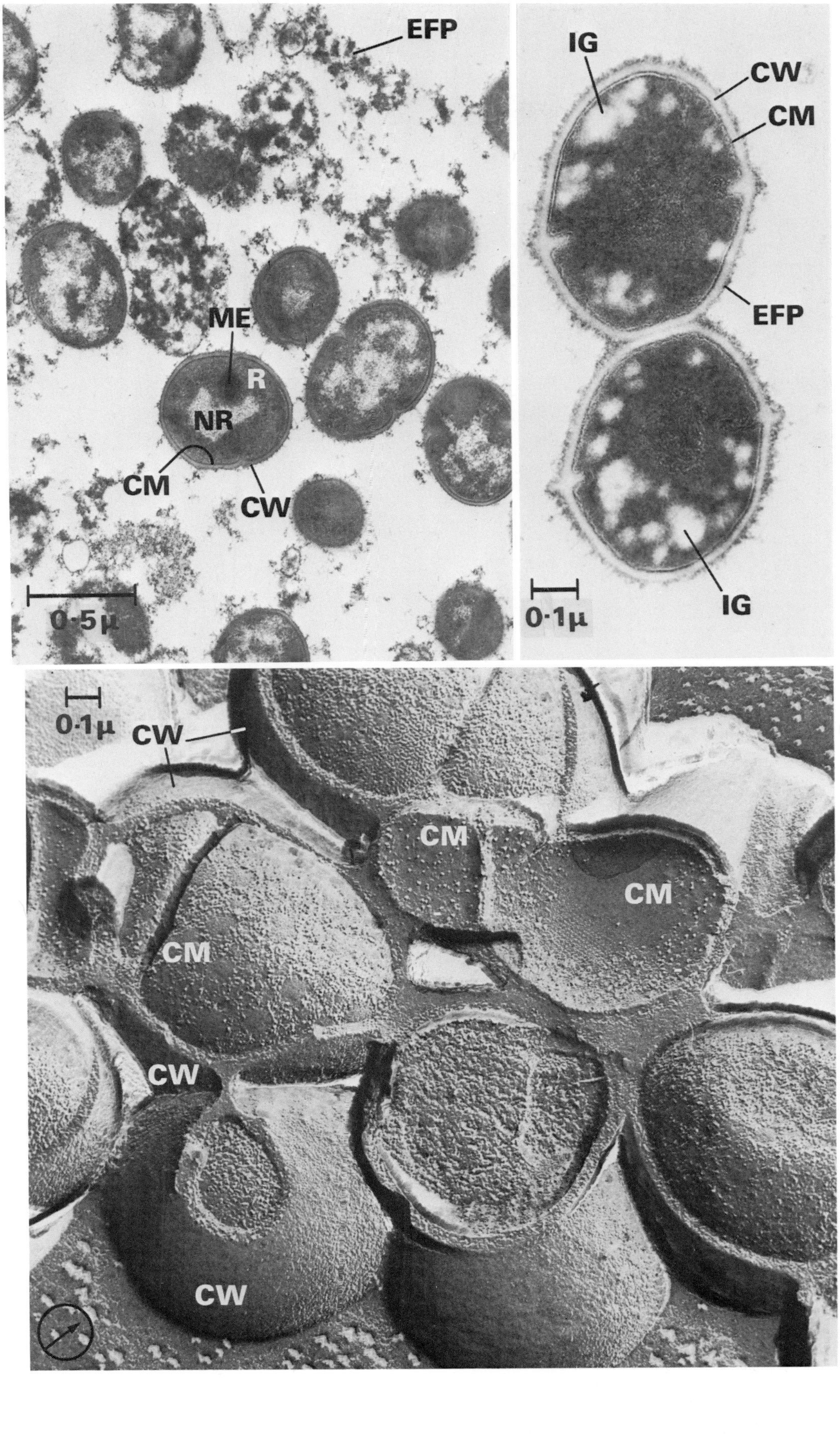

(m)

Figure 4.18. Representative streptococci.
In (m), the freeze-fracture-etched replica of *S. mutans* serotype a, the thick cell wall (CW) and particle-studded cell membrane (CM) is revealed. The cells are aggregated into a tight mass. R: ribosome, NR: nuclear region, ME: mesosome.
(n), (o), (p) Oral streptococci. In (o) the characteristic extracellular fibrous polymer (EFP) is visible. (n) *S. mutans* GS-5 shows electron-transparent iodophilic granules (IG) in the cell cytoplasm. In (p) the aggregation of *S. mutans* GS-5 by extracellular polymer is clearly visible. The electron-transparent space surrounding the cells is probably a preparative artifact. CW: cell wall, CM: cell membrane.

(n)

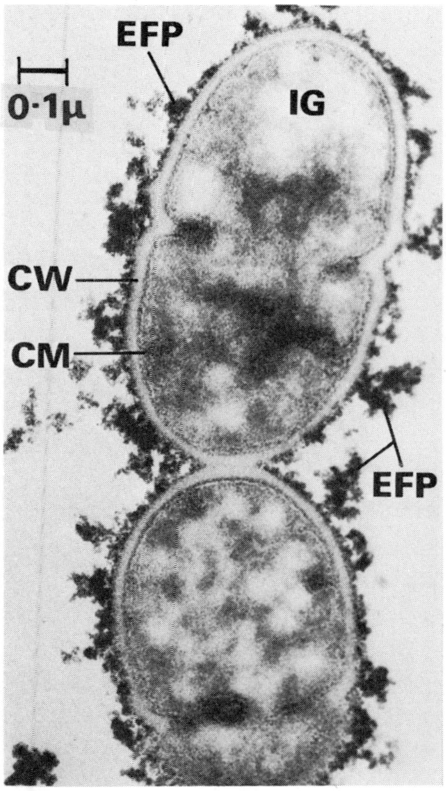

(o)

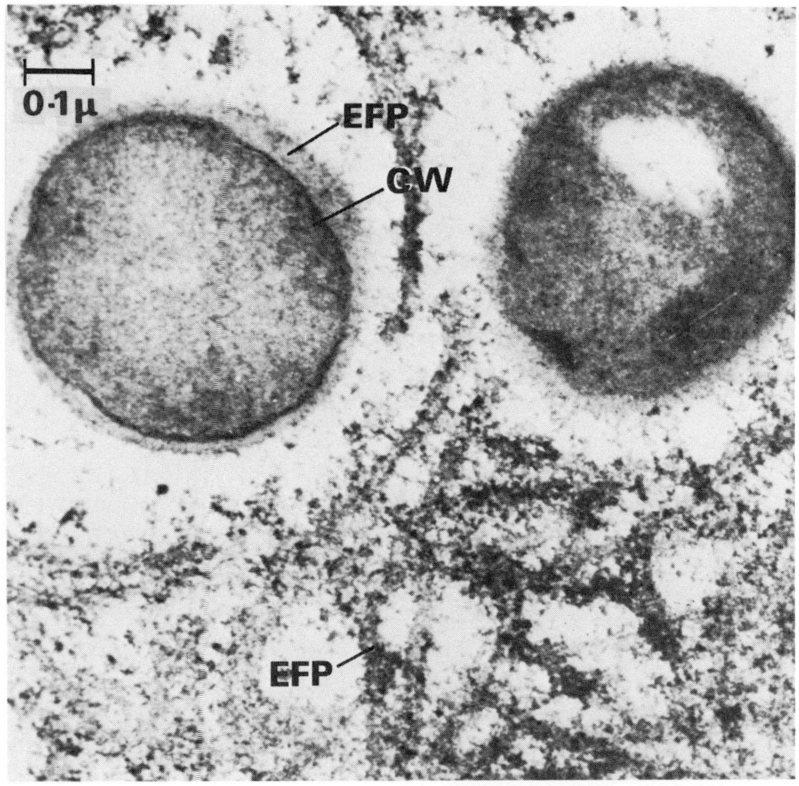

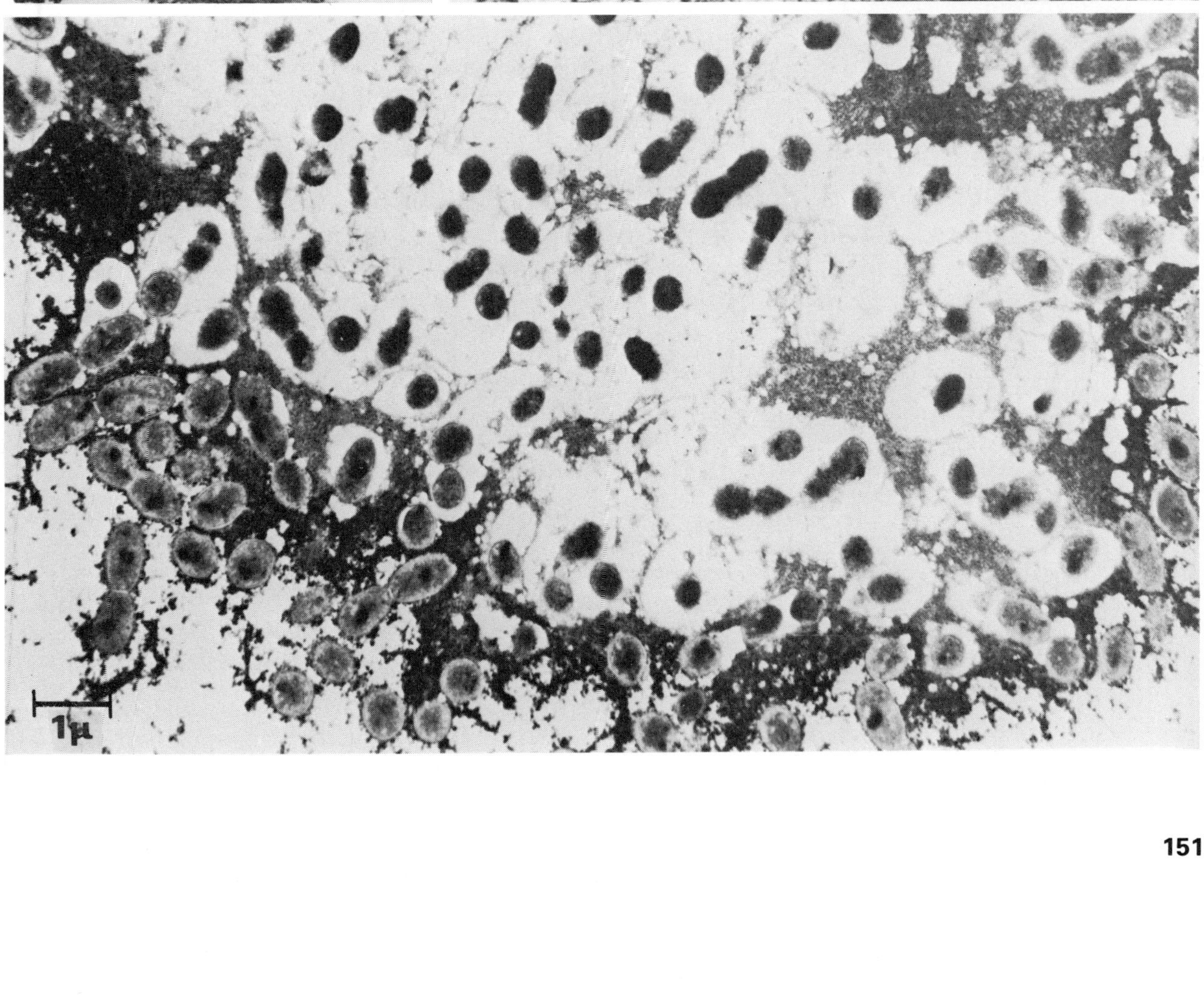

(p)

density reveals an extensive mesosomal membrane. Membrane-linked vesicles ring the degenerating cell wall.

Lancet to kidney-shaped diplococci were originally classified in the genus *Diplococcus*. These organisms are classified in the genus *Streptococcus* in the 8th Edition of Bergey's Manual of Determinative Bacteriology (see Deibel and Seeley[51]).

Streptococcus pneumoniae (*Diplococcus pneumoniae*), the best known of the original genus *Diplococcus*, causes both lobar and bronchial pneumonia, as well as emphysema, endocarditis, meningitis, otitis media, peritonitis, and corneal ulcers.[183] The cells are typically gram-positive, possessing a thick (200–250 nm) electron-dense cell wall (Figure 4.19(a)).[10, 202] The cell membrane adheres very closely to the cell wall, and the cytoplasmic region is electron dense.

The presence of a thick, dense fibrous capsule is characteristic of many of the virulent streptococci (diplococci). Ruthenium red positive staining (indicative of mucopolysaccharide material) of *S. pneumoniae* type III reveals an intensely stained capsule as well as a heavily stained cell membrane and peripheral parts of the cytoplasm (Figure 4.19(b)).[184] In contrast, the capsular layer when prepared by critical point drying (Figure 4.19(e))[34] is fibrous with numerous bulges along the fibrils.

In mouse lung tissue, *S. pneumoniae* is seen as early as 1·5 hours post-intranasal inoculation both extra- and intracellularly. The cell wall of these host-associated diplococci is electron dense, laminated, and approximately 17 nm thick. An electron-transparent artifactual space separates the external layer of the cell wall from the 35–40 nm thick capsule, which contains rough or string-like outer fringes. Staining of these cells with lead and uranyl acetate reveals the capsule to be of the same electron opacity as the underlying cell wall.

The extent of capsule thickness is most apparent when *S. pneumoniae* is reacted with type specific antiserum (Figure 4.19(c),(d))[9] where it is almost eight times as thick (35–280 nm) as control cells. The matrix of the swollen capsule is of a lower electron density than nonrelated diplococci but the capsule retains its reticular mosaic texture.

Figure 4.19. ***Streptococcus (Diplococcus) pneumoniae.***
(a), (b) Type III ATCC 6303; (c), (d) type III A-5, (e) type III OUS 2a. When fixed and stained with ruthenium red (b) the outer surface of the cells is covered with a thick fibrillar ruthenium red-positive capsule (Cp). The control cell (a) (minus ruthenium red) lacks this outer layer. After exposure to type specific antiserum (c), (d) the capsule (Cp) is swollen. Note the association of the capsule with the cell wall (CW) through thread-like connections (arrows). Addition of a ferritin conjugate to the type specific antiserum (d) localizes the ferritin (Fr) particles at the capsule surface. Critical point drying of OSU 2a (e) reveals the extracellular polymer to radiate from the cell surface as strands; vesicle or bulbous swellings occur along the strands. CM: cell membrane, NR: nuclear region, DS: division septum, ME: mesosome.

(a) (b)

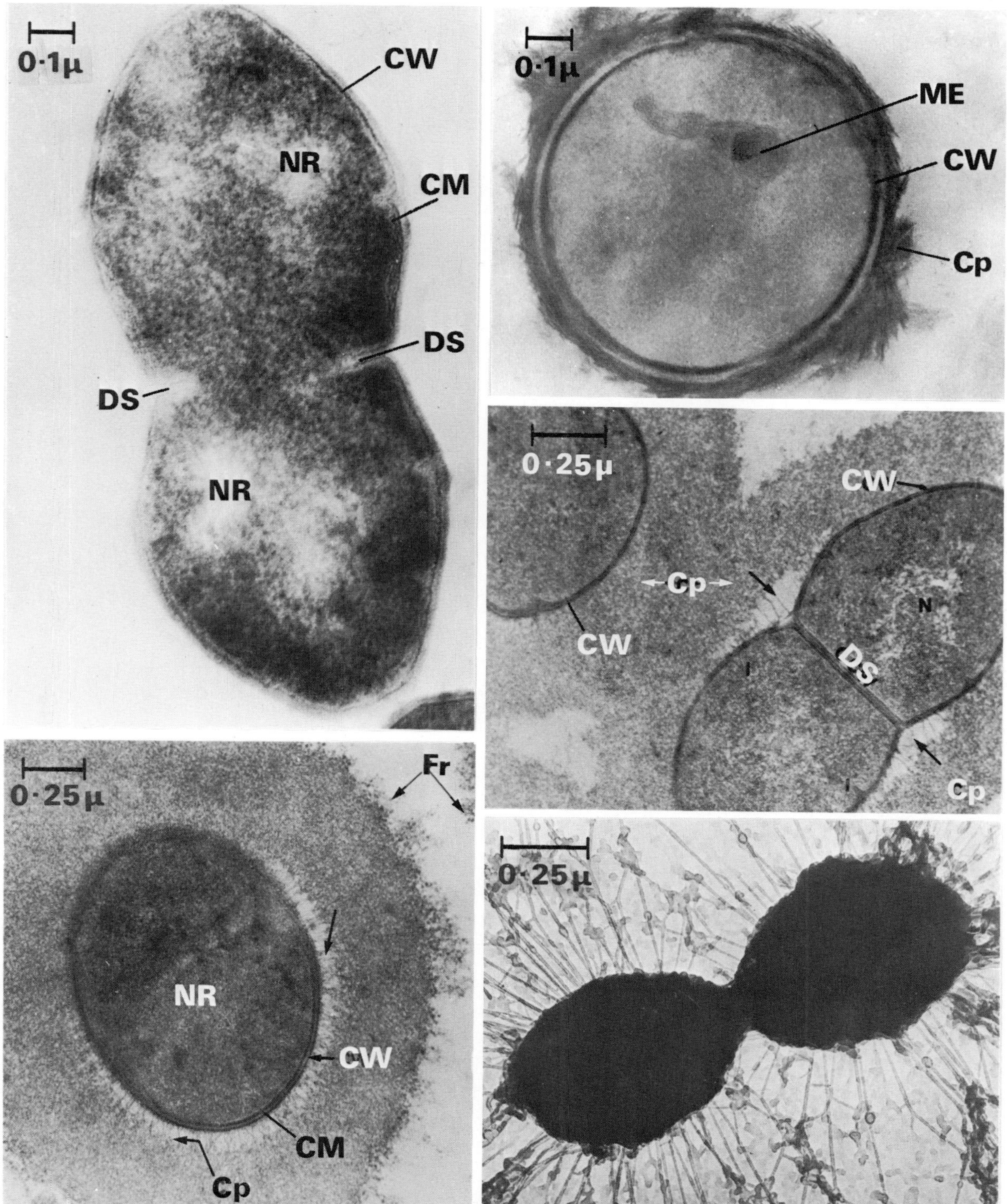
0·1μ
CW
NR
CM
DS
DS
NR
0·1μ
ME
CW
Cp
0·25μ
CW
Cp
CW
N
DS
Cp
0·25μ
Fr
NR
CW
CM
Cp
0·25μ

(c)

(d) (e)

4.3. Gram-negative procaryotes

4.3.1. *Brucella*

The genus *Brucella* includes the intracellular parasites, *B. melitensis, B. abortus, B. suis, B. neotomae, B. canis*, and *B. ovis*. These organisms are primarily pathogenic for animals. Infections of goats and sheep with *B. melitensis* (Malta fever), cattle with *B. abortus* (Bang's disease), and swine with *B. suis* results in significant economic loss. Transmission of the disease to humans occurs via infected animals, resulting in brucellosis, Malta and undulant fevers.[20, 183, 220]

The brucellae can invade nearly every human organ and tissue after exposure to animal excreta, blood, or tissues of infected animals. Primary infection occurs through the skin or the gastrointestinal tract.

Brucellae are small ($0{\cdot}5$–$0{\cdot}7 \times 0{\cdot}6$–$1{\cdot}6\ \mu m$), gram-negative, nonmotile coccobacilli. Both smooth and rough strains have been observed; the rough strains are the result of a thick capsule.[62, 209, 210]

The cell wall (Figure 4.20(a)–(f)) is multilayered, and along with the underlying cell membrane, it is similar to that seen in other gram-negative bacteria. In Figure 4.20(d), (e) the highly wrinkled outer membrane seen in transverse sections is probably an artifact of sample preparation. In Figure 4.20(a), (b), the typical constrictive cell division regarded as characteristic of gram-negative bacteria is apparent; although recent observations[32] reveal that as in gram-positive bacteria, gram-negative cells form a division septum prior to cell separation. The constrictions may be an apparent artifact of the preparative procedures for electron microscopy.

4.3.2. *Haemophilus–Bordetella*

Members are small, gram-negative, nonmotile strictly parasitic hemoglobin-requiring organisms. Included are *Bordetella* species (i.e., *B. pertussis*), *Haemophilus influenzae, H. ducreyi, H. vaginalis, H. aegyptius, H. suis*, and *H. parainfluenzae*.

Bordetella pertussis, or the Bordet–Gengot bacillus, is the causative agent of whooping cough, a catarrhal inflammation of the respiratory tract. The bacteria localize in the respiratory tract, where they aggregate into large masses among tracheal and bronchial cells and here elaborate their toxic products. They have never been observed to enter the blood stream.

Similar to the neisseriae, *B. pertussis* upon primary isolation is highly

---

Figure 4.20. *Brucella* species.
Thin sections of (a), (b) *Brucella canis* 317R and (c) *B. abortus* 544S. The cell wall is multilayered, consisting of an outer membrane (OM) and inner membrane (CM), which enclose a periplasmic space (PPS). A peptidoglycan (Pg) layer is also present between the outer and inner membrane. NR: nuclear region.

(a) (b)

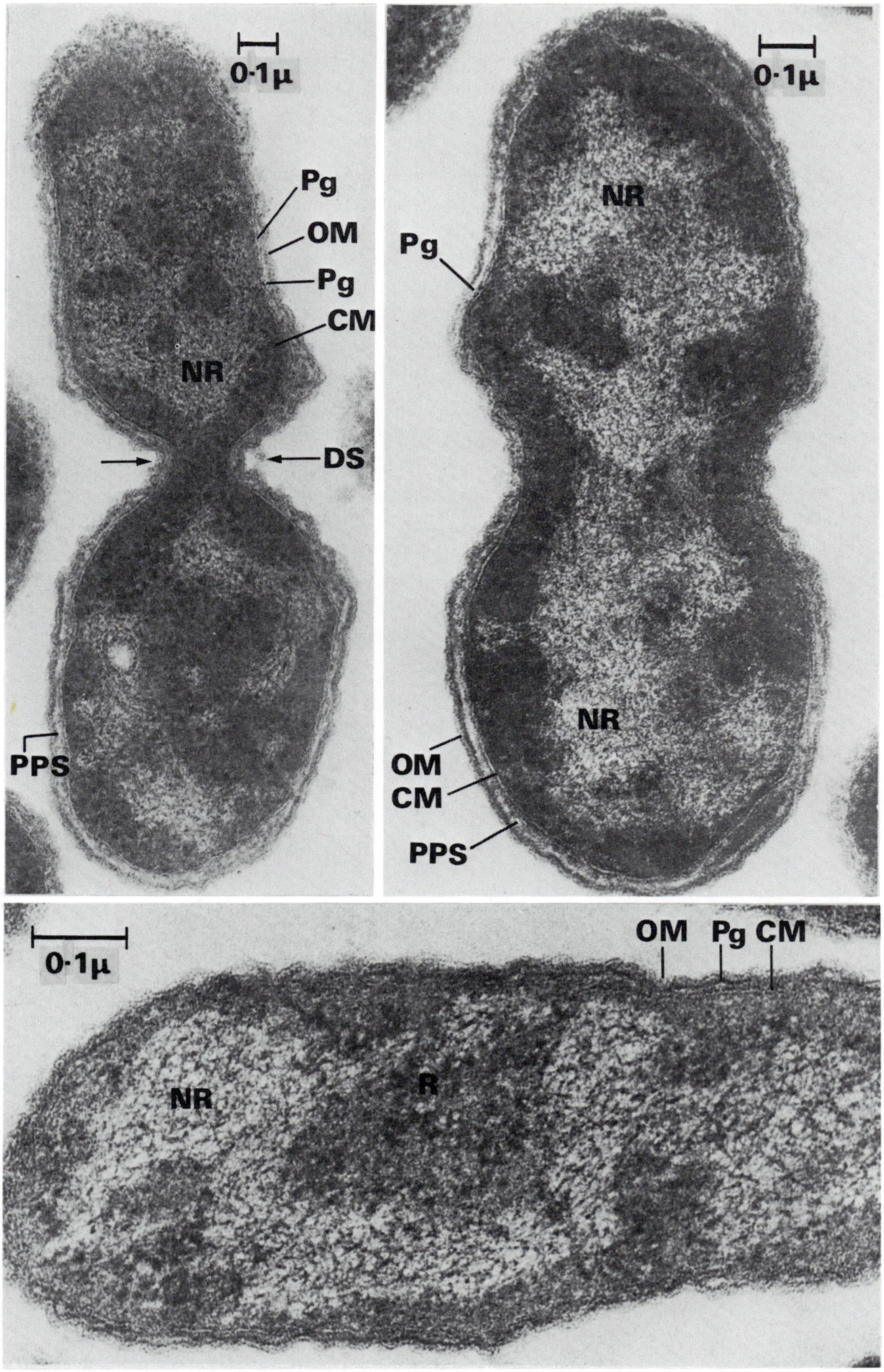

(c)

Figure 4.20. *Brucella* species.
(d) In transverse section *B. abortus* possesses a periplasmic space (PPS) filled with electron-dense material. (e) *B. melitensis* H38S. The very wrinkled nature of the outer membrane is apparent in both *B. abortus* and *B. meletensis.* In (f) a bacterial cell with a morphology similar to that of the brucellae is seen within a cytoplasmic vacuole of a hamster kidney cell 24 h post infection of a cell culture with virulent *B. abortus*. OM: outer cell membrane, CM: cell membrane.

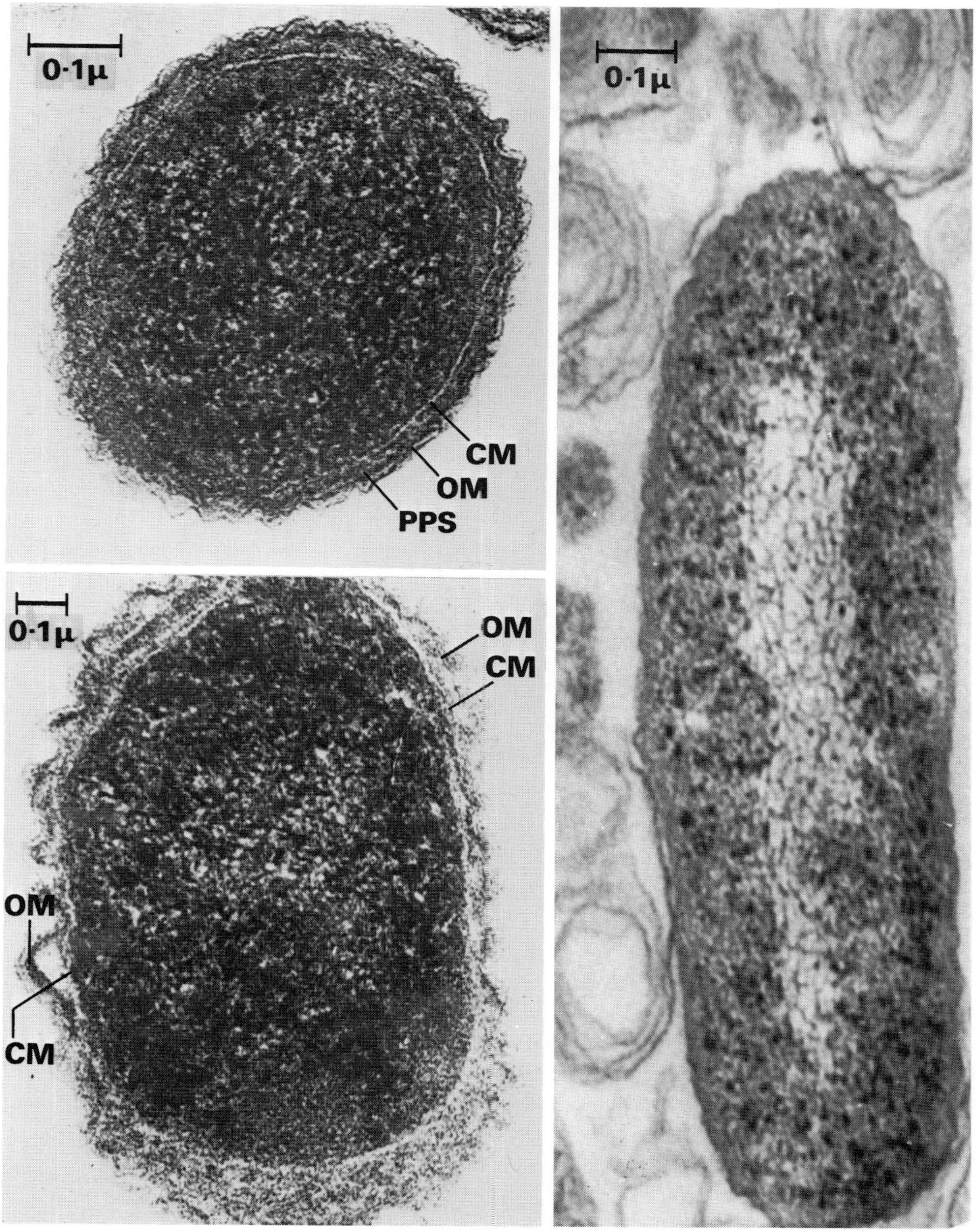

(d)

(e)

(f)

virulent; however, culturing on artificial media results in a dramatic loss of virulence. Four categories of *B. pertussis* virulence are recognized: the virulent encapsulated fresh isolate (phase I), cells of intermediate virulence (phases II and III), and nonvirulent, noncapsulated (phase IV) cells.

Bordetella bronchiseptica[167] (Figure 4.21(a)–(f)) vary in length from 0·5 to 3 μm and are approximately 0·5 μm wide. The cell surface as revealed by negative staining is rugose (Figure 4.21(a), (b)), typical of gram-negative cells. Numerous flagella are arranged peritrichously (Figure 4.21 (b)). Thin surface hairs or fimbriae are seen to emerge from the surface of the outer membrane in Figure 4.21(f). These surface hairs may constitute a primitive glycocalyx as described by Costerton *et al.*[45] and may be of importance in cell–cell or cell–surface interactions.

Haemophilus influenzae is the smallest known bacillus. It is an obligate parasite of the throats of approximately 30 percent of normal humans. Where it acts in a pathogenic state, it does so as a secondary invader, acting along with influenza virus in influenza infections and other respiratory diseases, including purulent meningitis and subacute bacterial endocarditis. It has also been implicated as the primary pathogen in mixed infections of the upper respiratory tract and bronchopneumonia.[20, 183] Its virulence also appears to be associated with the presence of a capsular layer. The encapsulated strains of *H. influenzae* are referred to as type b and are primary pathogens in the upper respiratory tract.

The haemophili (*H. influenzae, H. vaginalis, H. ducreyi*) are small (usually 0·3 × 1 μm), thin rods with rounded ends. Similar to other gram-negative bacteria, the cells contain a tripartite cell wall (Figure 4.22(a)–(d)).[48, 49, 58, 116, 165, 178] In some cases, when encapsulated *H. influenzae* are fixed in low concentrations of glutaraldehyde or by the R–K technique, both a periplasmic space filled with electron-dense material and an electron-dense capsule covering the outer membrane as a thin fibrous layer are seen (Figure 4.22(d)).

The capsule is best visualized when cells are exposed to type specific antiserum prior to preparative techniques for electron microscopy.[58] These authors found that type-specific antiserum is an effective capsular stabilizer, especially with *H. influenzae* type b cells. Further, the visualization of the capsule is influenced by the suspending buffer during antisera exposure. For example, tris-Ca^{2+} buffer and type specific antiserum results in the appearance of a broad, thick, homogeneous capsular layer (Figure 4.22(g)), while suspension of cells in 0·15 M NaCl and type b antiserum resulted in a more dispersed fibrous capsule (Figure 4.22(e), (f)), possibly due to a loss of associated capsular material by alteration or degradation of the outer membrane of the cell wall.

Figure 4.21. ***Bordetella bronchiseptica.***
(a), (b) Negative-stained and (c), (d), (e), (f) thin sections of *B. bronchiseptica.* The PTA negatively stained cells possess a wrinkled cell surface with numerous attached flagella (F). Electron-transparent, small, spherical structures (arrows) of unknown origin are present in (a). In thin sections, the central nucleoid is surrounded by the multilayered outer membrane (OM) and cell membrane (CM). In (d) the cell membrane (CM) has contracted around the cytoplasm clearly revealing the outer membrane (OM), a thin adhering peptidoglycan layer (Pg), and the inner cell membrane. Hair-like projections (arrows) are seen to emerge from the cell surface in (e), and in high magnification in (f). NR: nuclear region, Riribosomes.

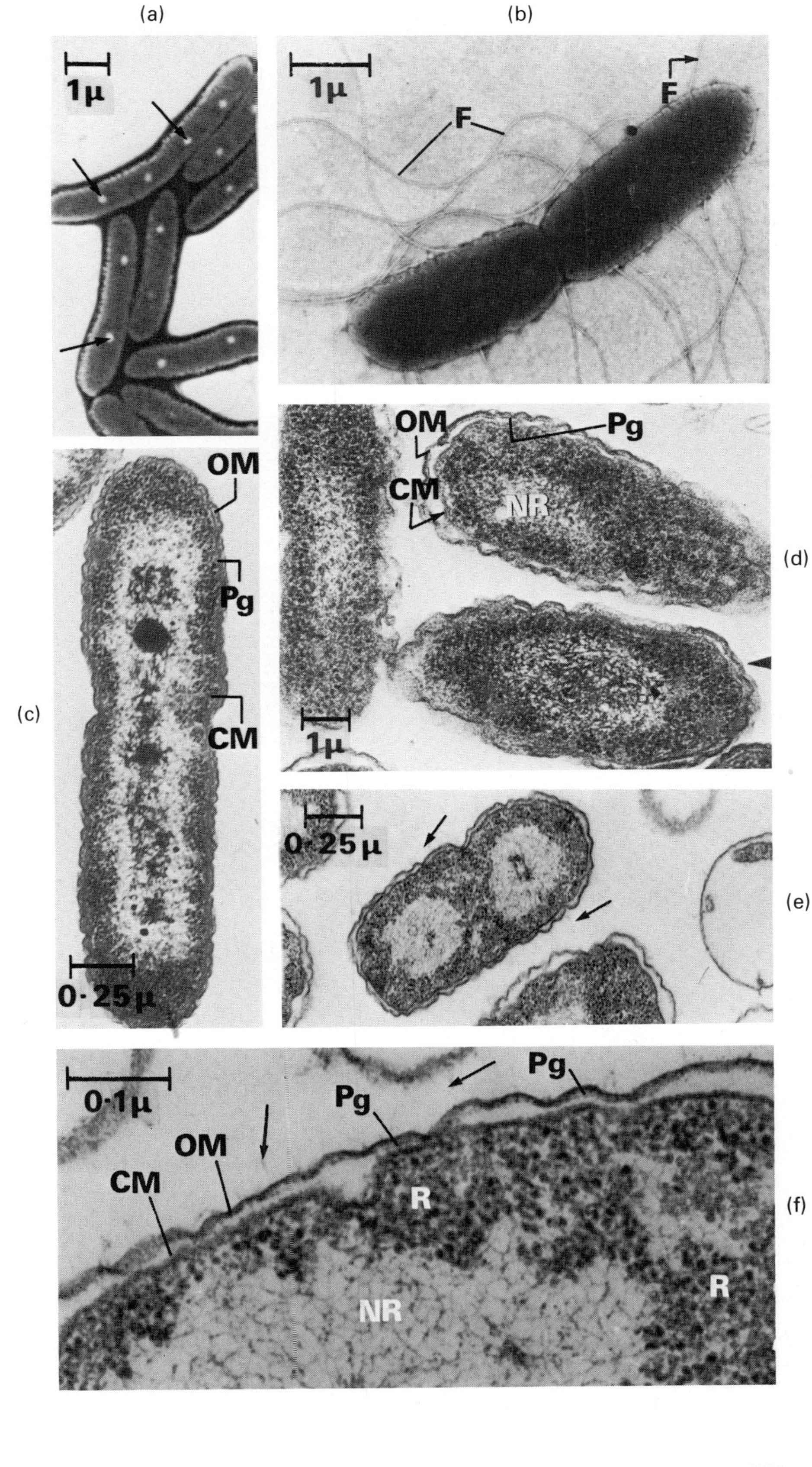
(a)
1μ
(b)
1μ
F
F
(c)
OM
Pg
CM
0·25μ
(d)
OM
Pg
CM
NR
1μ
(e)
0·25μ
(f)
0·1μ
Pg
Pg
OM
CM
R
NR
R

(a) (b) (c) (d)

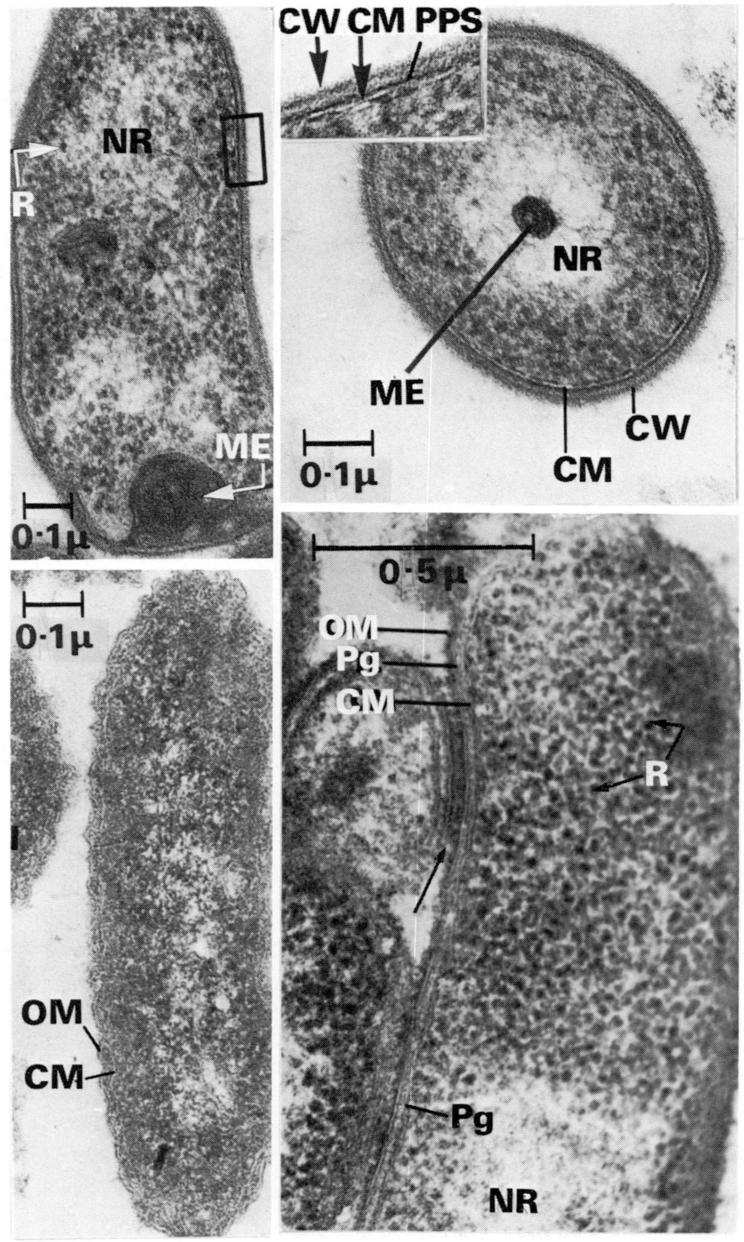

Figure 4.22. ***Haemophilus species.***
Thin sections of (a), (b) *H. vaginalis* 594, (c) *H. influenzae* NCTC 8476, and (d) *H. ducreyi* CIP 542. The cell wall (CW) consists of an outer membrane (OM) and inner membrane which enclose an electron-transparent periplasmic space (PPS) seen more clearly in the inset. Mesosomes (ME) originate at the cell membrane (CM). Where two cells (d) are adjacent to each other, they adhere by way of an electron-dense matrix (arrow). R: ribosome, NR: nuclear region, Pg: peptidoglycan.

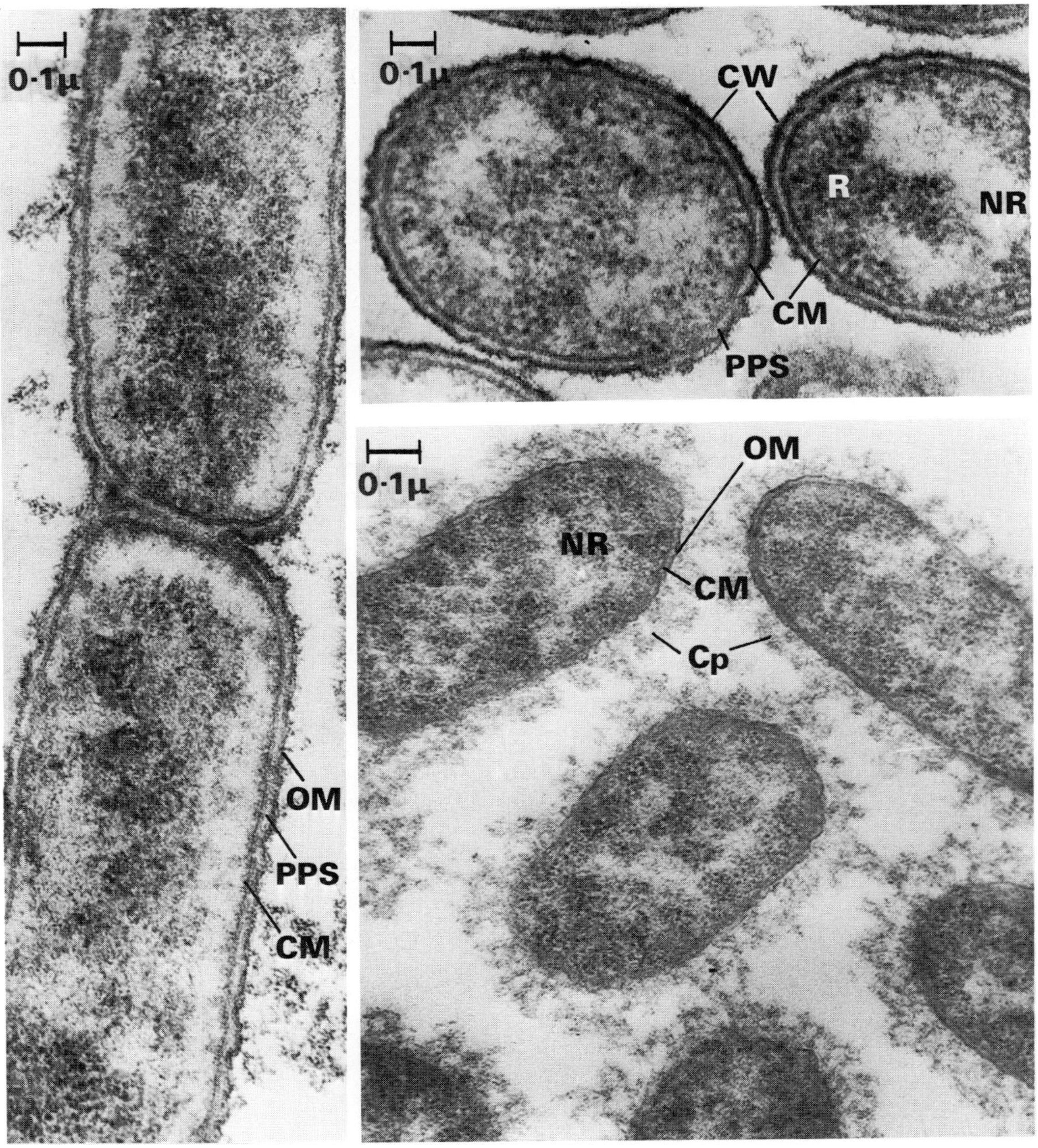

(e) (f) (g)

Figure 4.22. *Haemophilus species.*
(e), (f), (g). *H. Influenzae* type b, HIT6. When exposed to type b specific antiserum (g) the cell displays a fibrous electron-dense surface layer, which probably represents capsular antigen–antibody complexes (Cp). Control cells (e), (f) have a much thinner capsular layer. OM: outer cell membrane, CM: cell membrane, PPS: periplasmic space.

4.3.3. *Bacteroides*

These gram-negative, obligate anaerobes inhabit the oral cavity of humans, and the natural clefts and cavities of animals and insects. In humans they are the most numerous microorganisms recovered from the intestinal tract.[177] They are also found in the genital and upper respiratory tracts. In stool samples *Bacteroides* make up almost 95 percent of its bacterial content.

Their ability to produce hemolytic and necrotizing lesions (bacteroidosis) makes them of major importance in mucosal ulcers and abscesses, and as secondary invaders in surgery, alcoholic liver disease, diabetes, and in some malignancies.[121] *Bacteroides melaninogenicus* has been isolated with increasing frequency from the gingival crevice, where its incidence appears to be a function of the host's age.[114] It has also been suggested[131, 132] that *B. melaninogenicus* (*B. asnccharolyticus*) may be involved in some forms of peridontal disease.

The surface of negatively stained *B. asaccharolyticus* is wrinkled (Figure 4.23(a)), and in the lysed cell in Figure 4.23(b), the cell wall and underlying cell membrane are clearly seen. Thin fimbriae emerge from the outer layer of the cell wall.

These rod-shaped cells, typical of which are *Bacteroides fragilis, Bacteroides melaninogenicus* (and its subspecies), *B. oralis*, and *B. insolitus* (*B. fragilis*), possess a trilaminar cell wall composed of two electron-dense membranes; the outer membrane is approximately 9·5 nm thick, while the inner membrane is 3·5 nm thick (Figure 4.23(c)–(h))[205, 206, 207] (Holt and Leadbetter, unpublished communication). A thick capsule has not been observed but the surface of the outer membrane is covered with a fine hair-like nap, as well as long, thin, interconnecting fibers. In some instances it is possible to see bridging fibers joining the inner layer of the cell wall to the cell membrane, as well as an electron-dense periplasmic region.

The cytoplasmic region contains electron-dense, 5–10 nm diameter ribosomal particles which enclose the nuclear region. Numerous vesicles (LPS?) fill the background in Figure 4.23(g), (h). The function of the vesicles is unknown, but it is tempting to speculate that they are extruded endotoxin.

When examined by CPD–SEM an isolated colony of *B. asaccharolyticus* (Figure 4.23(i)) is covered with a thick crusted layer. Directly beneath this outer covering are the individual rod-shaped cells (Figure 4.23(m), (k)). In Figure 4.23(k) the adjacent cells are seen to be connected by thin fibrous strands emanating from the mucoid cell surface.

CPD–TEM of *B. asaccharolyticus* (Figure 4.23(l), (m)) reveals a radically different picture than that seen by CPD–SEM. The cells in Figure 4.23(l),(m)

Figure 4.23. *Bacteroides asaccharolyticus.*
(a), (b) Prepared by negative staining (PTA), (c), (d), (e), (f), (g) by fixation and thin sectioning. The wrinkled gram-negative surface is apparent (a). The lysed cell in (b) clearly reveals the outer (OM) and inner (cell) (CM) membranes. Thin sections reveal a tight-fitting outer membrane (OM) which, along with the cell membrane (CM) encloses an electron-dense periplasmic space (PPS). A thick, hair-like fuzz (arrow) covers the outer membrane and is characteristic of this species. R: ribosome, NR: nuclear region, C: collagen fiber, D: dictyosome.

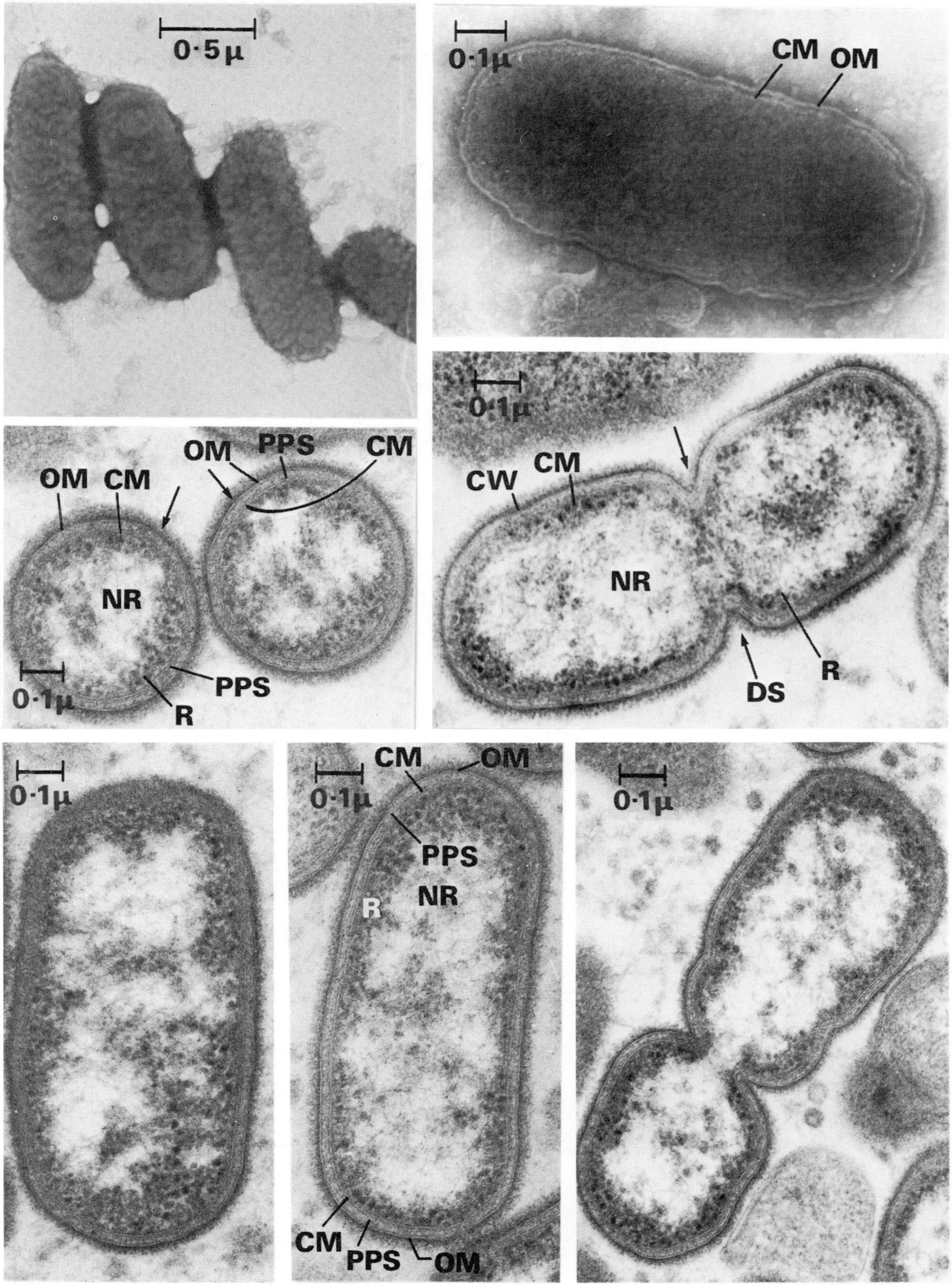
0·5μ
0·1μ
CM
OM
0·1μ
CW
CM
NR
DS
R
OM
PPS
CM
OM
CM
NR
PPS
R
0·1μ
0·1μ
CM
OM
0·1μ
PPS
R
NR
CM
PPS
OM
0·1μ

(a) (b) (c) (d) (e) (f) (g)

(h)

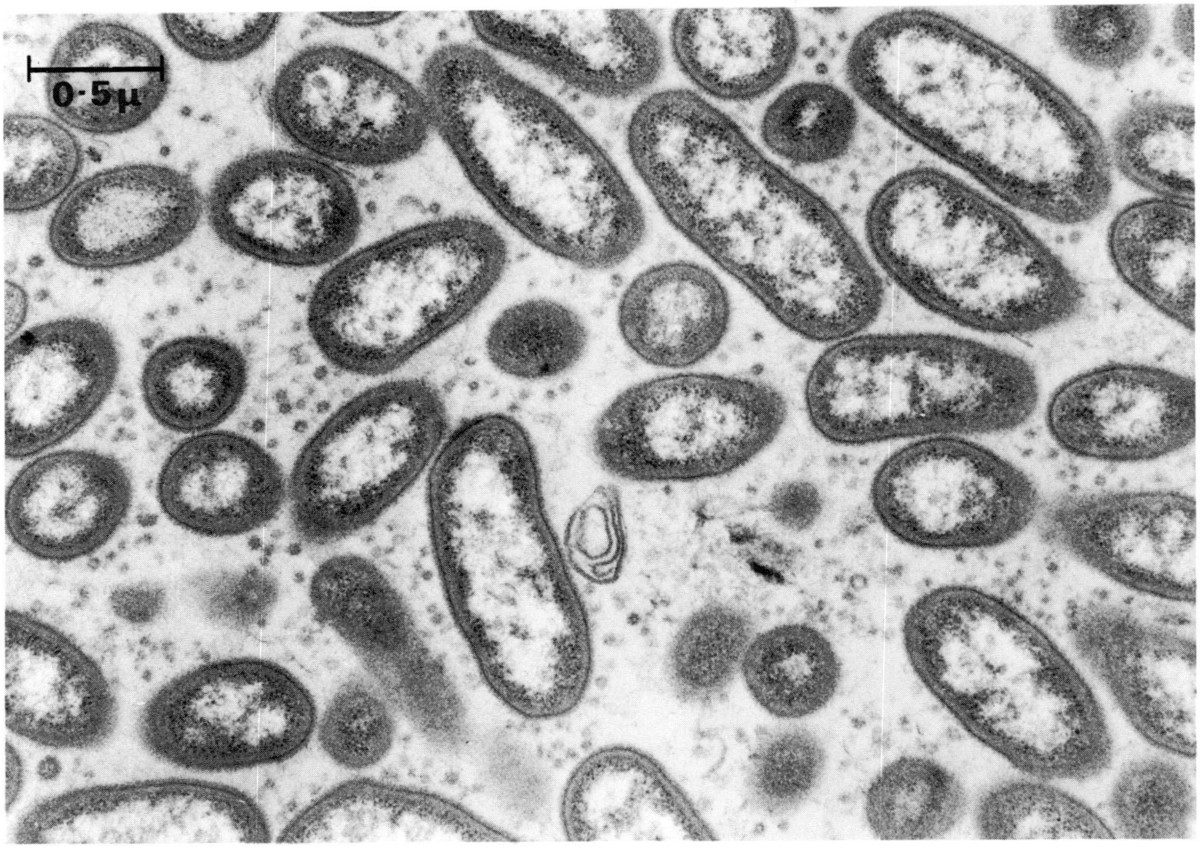

Figure 4.23. *Bacteroides asaccharolyticus.*
(h) At low magnification the background is shown to contain numerous small vesicles, the function of which is unknown.

are interconnected by long, thin fibers, which not only appear to connect adjacent cells, but possibly aggregate large numbers of cells into tightly packed units. Of interest is the long distances that these thin fibers traverse.

4.3.4. *Neisseria*

Neisseria, Branhamella, and *Moraxella* comprise small, gram-negative, aerobic, oxidase-positive cocci which are in some instances associated with human diseases. The anaerobic counterpart of *Neisseria, Veillonella*, is a common inhabitant of the oral cavity. They are morphologically very similar to *Neisseria*; however, a definitive morphological examination of the several *Veillonella* strains

Figure 4.23. *Bacteroides asaccharolyticus.*
(i), (j), (k) Prepared by CPD-SEM reveals the surface of the colony to be covered with a thick crust (i), with the individual cells (j), (k) clearly seen below this layer. Individual cells are connected by thin fibrils (arrows in (k)).

(i)

(j)

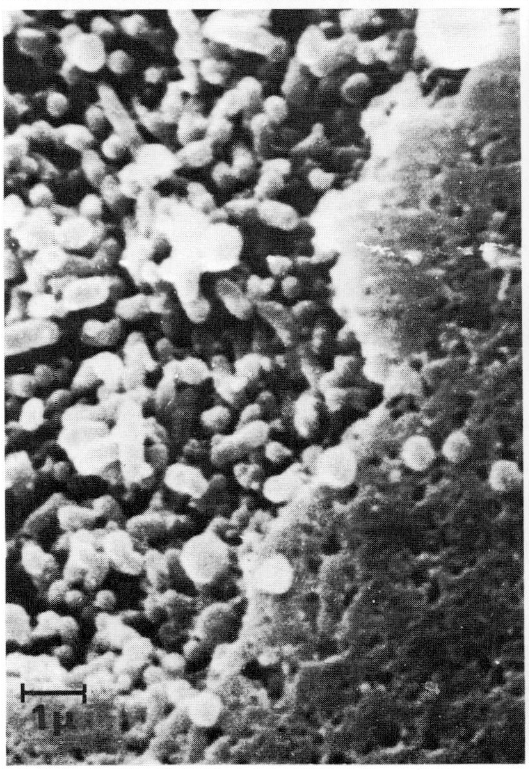

(k)

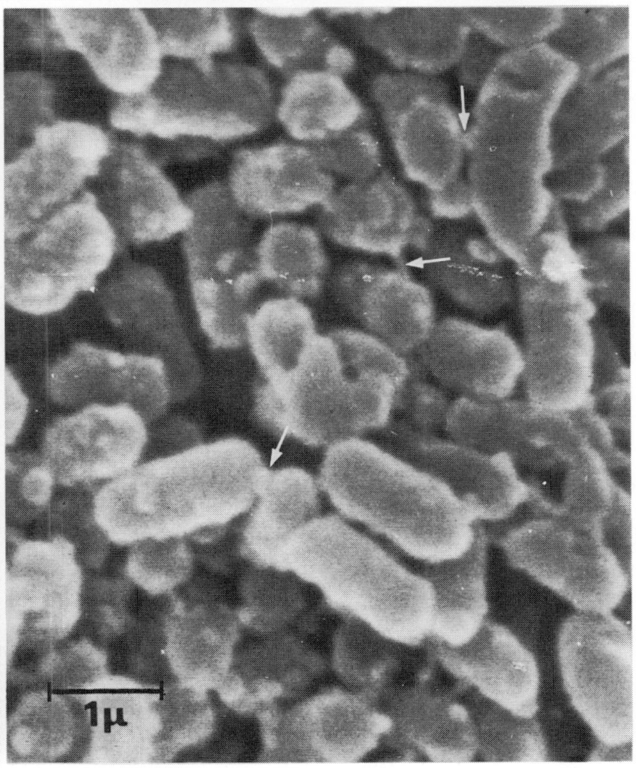

(l)

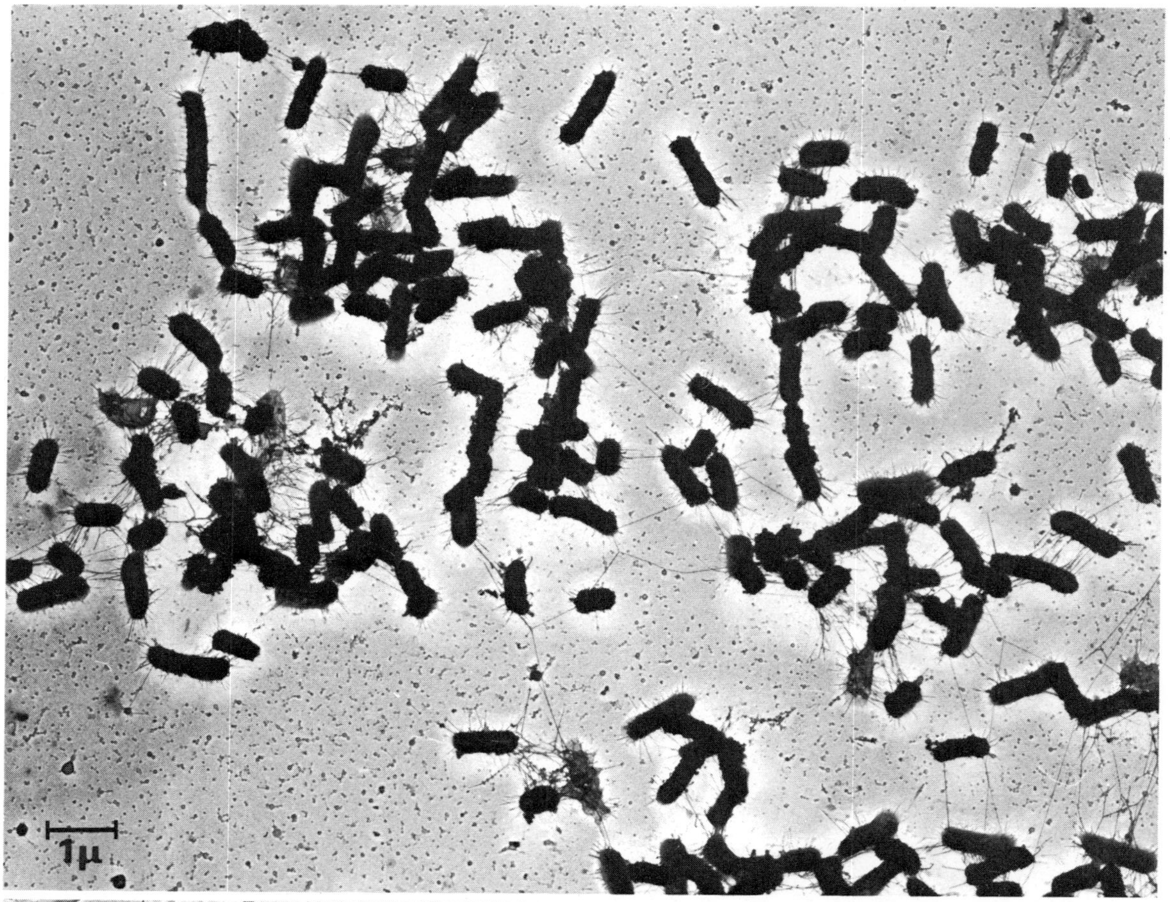

(m)

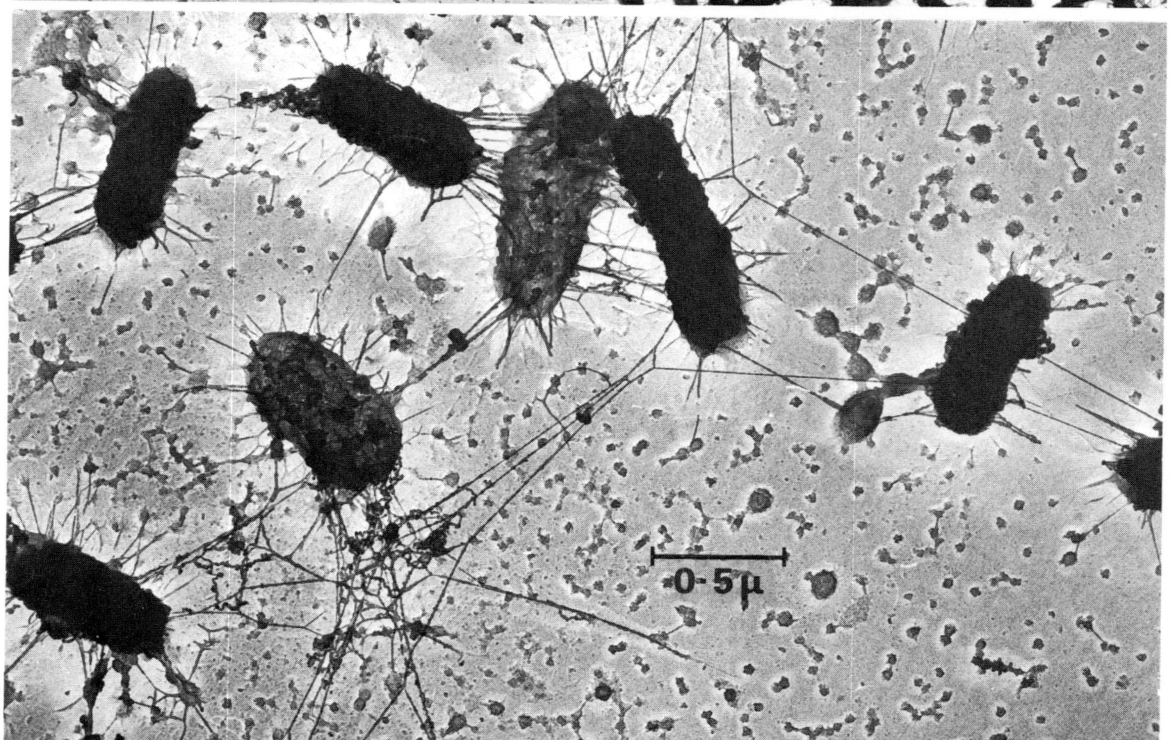

has not been undertaken, especially, for example, for the presence of surface-associated fimbriae or pili. See sections 4.3.5 *Branhamella*, 4.3.6 *Moraxella*, and 4.3.7 *Veillonella.*

Included in the genus *Neisseria* are two medically important species, *N. gonorrhoeae* and *N. meningitidis. Neisseria gonorrhoeae* is the causative agent of gonorrhea, one of the most prevalent diseases affecting humans, as well as being the most common venereal disease in the United States.[183] The primary sites of infection in humans are the urethra and cervix where the bacteria predominate as a superficial infection of the columnar epithelium of the lower urogenital tract. The gonococci often penetrate intercellular spaces where they produce an acute inflammatory response. Since the disseminated disease is the most common in individuals with asymptomatic anogenital infections, the acute inflammatory response may play some role in localizing the infection.

Neisseria meningitidis, the causative agent of meningococcal septicemia, and epidemic cerebrospinal meningitis as well as nasopharyngitis, is, by light microscopy, more irregular in size and shape than *N. gonorrhoeae.* In some cases, capsules have been observed in *N. meningitidis* by the India ink negative stain.

Electron microscopic examination of fresh clinical isolates of *N. gonorrhoeae*, gonococci grown in culture, and *N. meningitidis*, reveal bean-shaped cells, the flat sides of which are juxtaposed[196] (e.g., Figure 4.24(a)). The cells are similar in morphology to other gram-negative bacteria, consisting of a triple-layered cell wall approximately 8 nm thick and of unit construction (e.g., Figure 4.24(a),(e)–(g),(l)).

There is convincing evidence[164] that the virulence of *N. gonorrhoeae* is closely correlated with its colonial morphology, with four distinct colonial types being observed. The types are designated 1 to 4. Fresh isolates of *N. gonorrhoeae* are classified as type 1 and 2, while those gonococci which have undergone repeated transfer on artificial media are considered types 3 and 4. All types display the typical gram-negative morphology.

There are, however, clear differences between types 1, 2 and types 3, 4 gonococci, the most significant being that types 1, 2 organisms when examined by negative staining are always observed to contain pili on their external surface[138, 196] (Figure 4.24(c)), while types 3, 4 are never seen to contain pili (Figure 4.24(d)). The number of pili per cell varies between 25–50, with the average length of a pilus being 0·5–2 μm. Pili as long as 4 μm have been observed. The average pilus diameter is 8–8·5 nm. The pili exhibit a marked tendency toward lateral aggregation on the cell surface, but this may be an artifact of the negative-staining technique since uranyl acetate at pH 4 results in lateral aggregation, while phosphotungstic acid at pH 6·8 results in both dispersed and aggregated pili. Pili are not the only surface appendages found on gonococci. Grimble and Armitage[81] have observed large surface appendages, 8–40 nm in diameter which are distinct from pili on the surface of type 1

Figure 4.23. *Bacteroides asaccharolyticus.*
(l), (m) Prepared by CPD–TEM. Cells possess long, thin fibrils. In (l) cells are aggregated into clumps.

Figure 4.24. *Neisseria* species.
(a) *Neisseria gonorrhoeae* fixed *in situ* possesses a thin, multilayered outer membrane (OM), an underlying cell membrane, and a nuclear region (NR) dispersed among the electron-dense ribosomes (R). The cells are connected by "zones of adhesion" (large arrows), which in this grazing section appears as a fibrous network (double arrow). Pg: peptidoglycan, CW: cell wall. (b), (c) Type 2 cells are covered by a layer of thin pili (P) which in the freeze-etch replica (b) appear matted to the cell surface. In (b) numerous pits (circles) cover the cell surface. CM: cell membrane, OM: outer membrane. (d) Type 4 gonococci do not display pili.

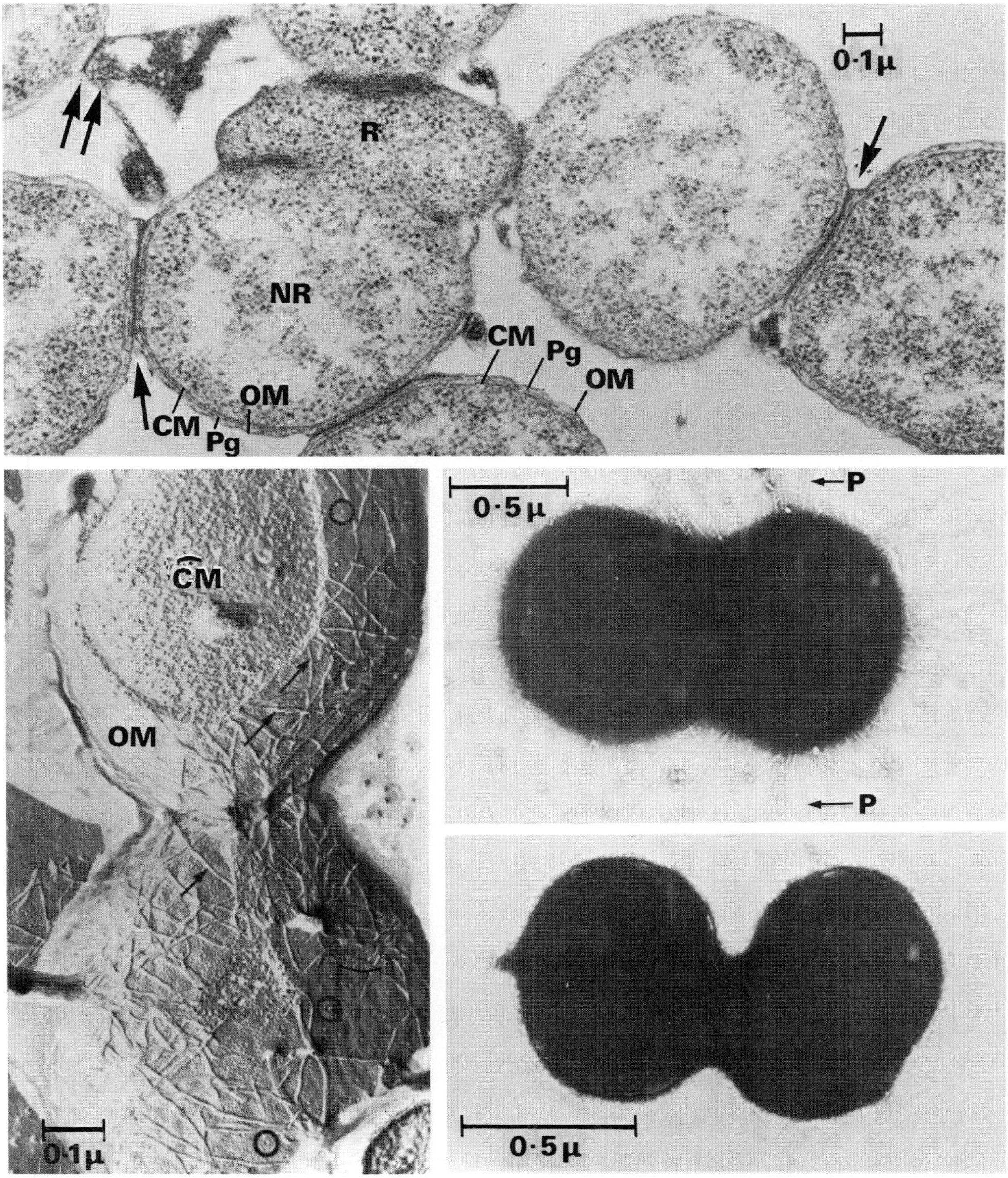

(a)

(b)

(c)

(d)

Figure 4.24. ***Neisseria*** **species.**
(e), (f), (g) Cells grown *in vitro* on an agar surface display a morphology similar to those fixed *in situ*. In (e) The cells are connected by "zones of adhesion" (see Figure 4.24(a)). In (f), the gonococcus is in the process of lysis. The cell membrane is bulged at the division tube or septum (arrow). R: ribosome, NR: nuclear region, Pg: peptidoglycan, CM: cell membrane, OM: outer cell membrane.

(e)

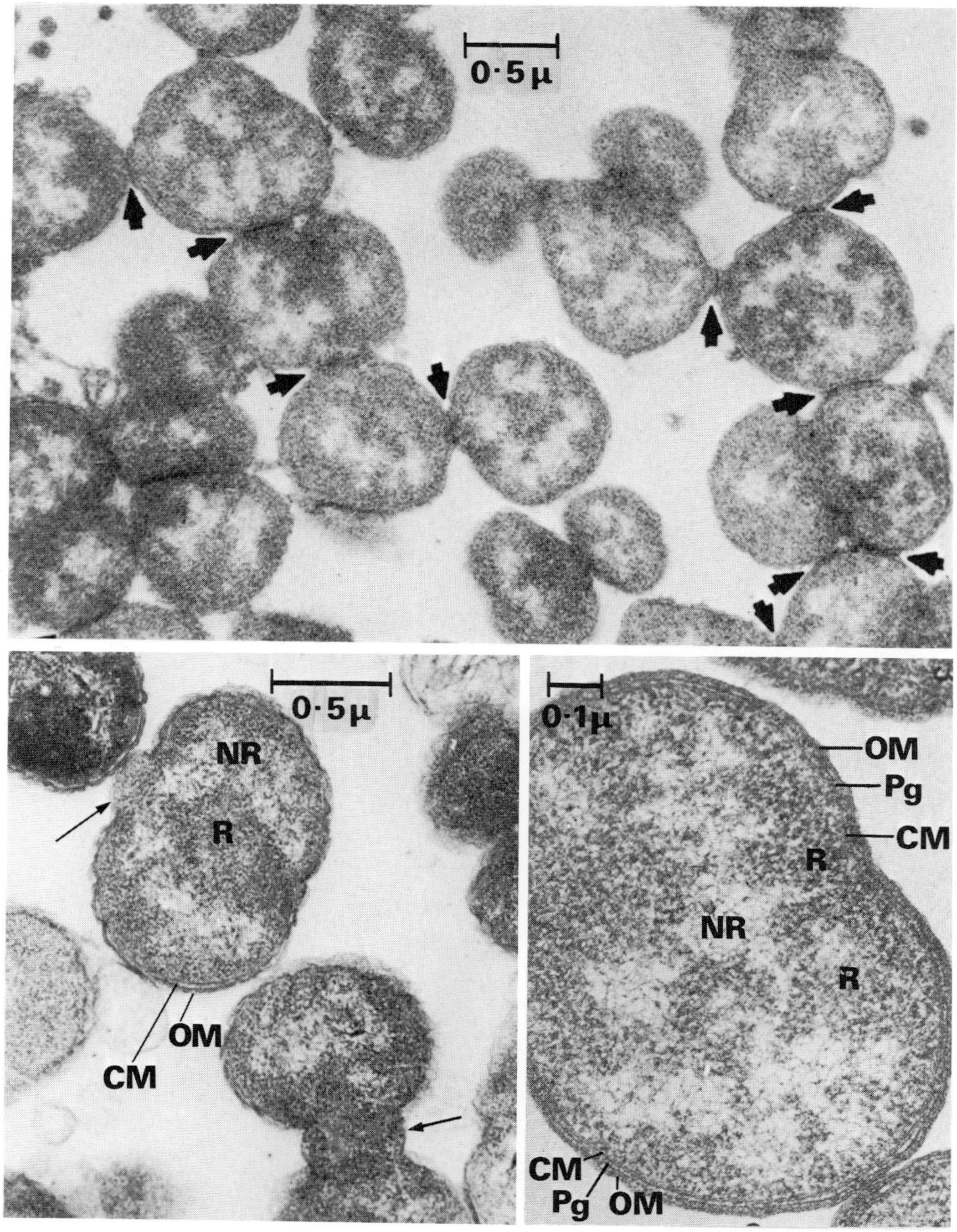

(f) (g)

N. gonorrhoeae; type 4 cells did not exhibit these structures. These appendages varied in length from 0·5 to 5 μm with occasional swellings of beads along their length, and usually occurring singly or in pairs, with as many as 5–10 being observed per cell. Whether these large appendages are real structures is unclear, since they are seen in distilled water suspended cells, and seldom in gonococci suspended in an isotonic environment. These large structures are, in addition, very reminiscent of that observed in electron micrographs of cell-associated outer membrane or lipopolysaccharide.

In addition to this major difference in the presence or absence of pili on the four gonococcal cell types, type 4 cells are also morphologically distinct from the other three types when grown on an agar surface. Types 1, 2, 3 gonococci when grown on an agar surface appear to be attached to adjacent gonococci through 2 nm "zones of adhesion" of their outer membrane (Figure 4.24(a), (e)); in contrast, type 4 cells occur either as single cells or as diplococci, with adhesion zones never being observed.[196]

Of interest has been the observation that approximately 90 percent of all gram-negative bacteria found associated with urinary tract infection are piliated. Further, only piliated *N. gonorrhoeae* appear to be capable of causing gonorrhea in humans; however, the role of pili in disease is not at all clear. Whether, for example, the pili hinder phagocytic activity, or are involved in adherence between the gonococci or the host epithelial cells is unknown.

Freeze-etched replicas of type 2 gonococci[194] (Figure 4.24(b)) reveal the pili to be in close contact with the cell surface and to wrap around the cell rather than extending out from the surface. In addition, the outer layer of the cell wall contains pits or holes on the raised portion of the wrinkled gonococcal surface, the function of which are unknown.

The associations that exist between *N. gonorrhoeae* and mammalian cells have also been examined with the electron microscope, employing cell culture techniques as well as the direct examination of host tissue exudates. As already mentioned, *N. gonorrhoeae* attaches to urethral epithelial cells (Figure 4.24(n)–(q)).[211, 212] This attachment may be a critical determinant in the dissemination of the gonococci to other tissues and locations in the host. For example, examination of urethral mucosa from individuals with gonococcal symptoms as early as 24 hours reveals numerous *N. gonorrhoeae* in close association with the urethral cell membrane as well as being lodged in microcontours of the host cells. In some cases the gonococci are seen within polymorphonuclear leukocytes where they appear to be undergoing phagocytic destruction (Figure 4.24(h),(i)).[211] With few exceptions, all the gonococci appear to be cell associated.

Neisseria meningitidis is ultrastructurally similar to *N. gonorrhoeae*, consisting of a typical gram-negative cell wall, 7–7·5 nm in diameter (Figure 4.24(l)).[56] Pili are observed on the surface of *N. meningitidis* group B, DS15, X and Y (Figure 4.24(e)).[53, 54, 55] Of interest is the observation that even in these piliated *N. meningitidis* strains only approximately 5 percent of the cell population possess visible pili, and the number of pili per cell is very low, approximately 5 pili/cell, with an average of only 1–2 per cell. By contrast, in group B *N. meningitidis* (ATCC 13090) almost 90 percent of the cell population is piliated.

The low number of pili per cell in some strains of *N. meningitidis* may be the result of *in vitro* growth or environment, similar to the loss of pili upon *in vitro* growth of *N. gonorrhoeae.* DeVoe and Gilchrist,[55] for example, observed that 80 percent of the cells of a fresh meningococcal isolate were piliated; in the first subculture the number of pili decreased to 50 percent of the initial level, and in the second subculture there was an almost complete loss of detectable pili; the latter similar to that found in meningococcal strains subcultured for long periods of time in the laboratory.

Characteristic of meningococcal infection is the presence of *N. meningitidis* within polymorphonuclear leukocytes (Figure 4.24(m).[56] Of considerable interest in establishing a role for cell associated endotoxic components was the very early observation by Murray[146] that the severity of meningococcal infection was inversely proportional to the severity of the disease; individuals with mild symptoms had numerous intracellular meningococci, while severe cases had few intracellular bacteria. In patients infected with many meningococci there was a marked destruction of both leukocytes and the invading meningococci, the latter destruction possibly accompanied by the release of bacterial endotoxic components such as lipopolysaccharide.

Gemella haemolysans (*Neisseria haemolysans*) (Figure 4.24(j))[165] has been isolated from mucous membranes and from the respiratory tract. It is similar to the other neisseria, with the exception that these cells possess a thick cell wall, morphologically similar to that of gram-positive bacteria. In addition, numerous lamellar mesosomes fill the cytoplasmic region. The outer layer of the cell wall is loose fitting, with a hair-like, electron-dense fuzz. It is not clear whether these cells possess a slime or capsular layer.

4.3.5. *Branhamella*

Branhamella catarrhalis, Moraxella lacunata, and *M. nonliquefaciens* are normal inhabitants of mucous membranes, especially of the respiratory tract. *Branhamella catarrhalis* has frequently been confused with *N. gonorrhoeae* and *N. meningitidis*; the only apparent morphological difference is that the *B. catarrhalis* cells are slightly biscuit-shaped rather than bean-shaped.

Branhamella catarrhalis, as well as *N. ovis, N. caviae*, and one strain of *N. cinerea*, are nonpathogenic and separated from *N. gonorrhoeae* and *N. meningitidis* by the presence of several distinctly different enzymes, as well as their distribution of fatty acids.

Ultrastructurally, the *Branhamellae* are similar to the pathogenic *Neisseria* in their cellular anatomical detail; the peptidoglycan layer, however, is almost twice as thick in *B. catarrhalis* as in *N. meningitidis* (3·5–4·5 nm compared with 1·5–2·5 nm).

4.3.6. *Moraxella*

Moraxella nonliquefaciens also possesses all of the ultrastructural characteristics typical of gram-negative bacteria. Members of this genus have been known to corrode and spread over agar surfaces. When grown on blood agar and

Figure 4.24. *Neisseria* species.
(h), (i) Polymorphonuclear leukocyte (PMN) – gonococci interaction. In (h) numerous gonococci (Gc) are within the urethral epithelial cell, while in (i) the gonococci are present intercellularly in PMN's from urethral pus. Note that the lysosome has not yet affected the integrity of the gonococcus in (i). MV: microvillus.

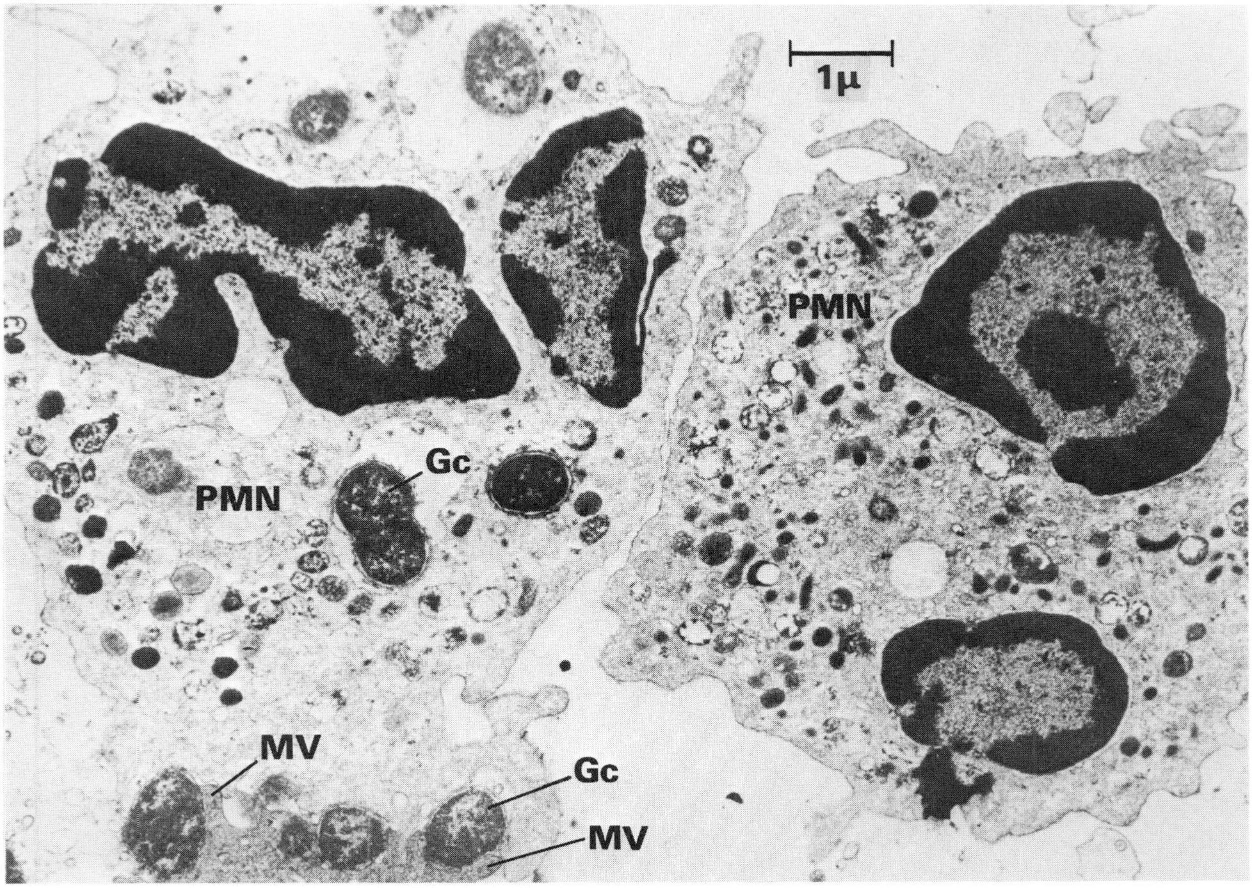

(h)

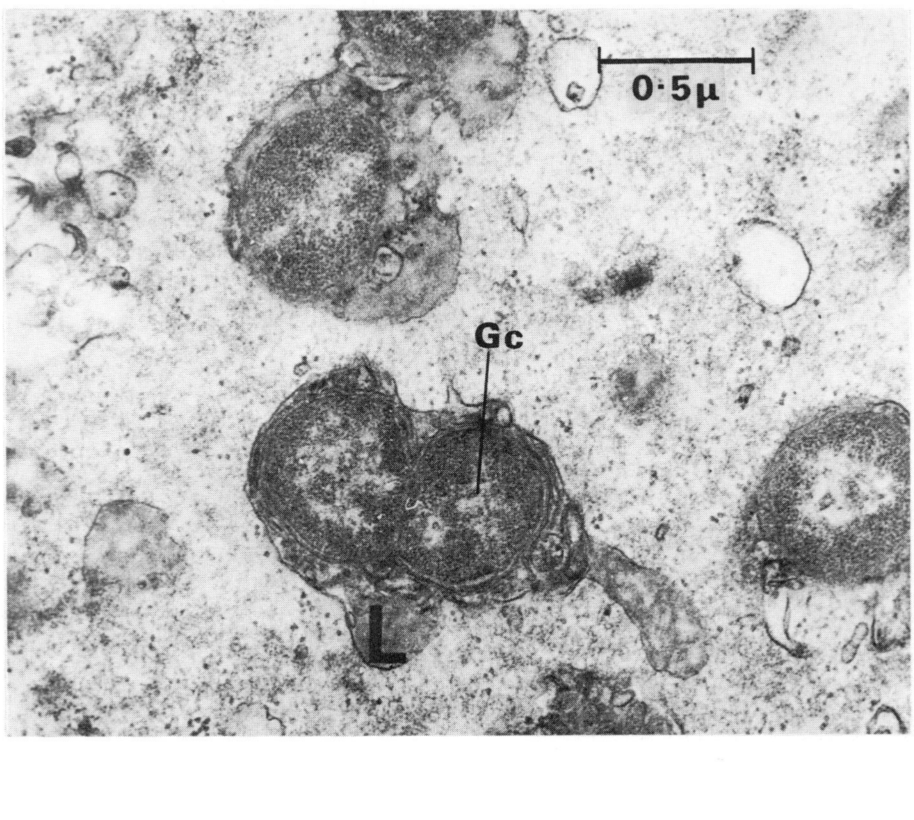

(i)

Figure 4.24. *Neisseria* species.
Thin sections of (j) *N. haemolysans (Gamella haemolysans)*, (l) *N. meningitidis* as well as (k) a negatively stained cell of *N. meningitidis* display similar morphology to the piliated gonococci. In addition to pili (P), the surface of *N. meningitidis* is covered with numerous membranous blebs (arrows in (k)) similar to LPS vesicles. (m) Within PMN leukocytes *N. meningitidis* is enclosed in phagocytic vacuoles, and several of them are in the process of disintegration. NR: nuclear region, CM: cell membrane, CW: cell wall, OM: outer cell membrane, PPS: periplasmic space, DS: division septum, MC: meningococcus, R: ribosome, Pg: peptidoglycan.

(j)

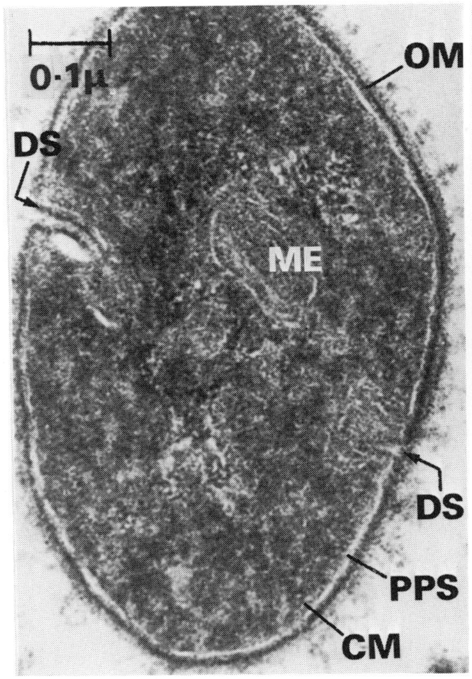
0·1μ
OM
DS
ME
DS
PPS
CM

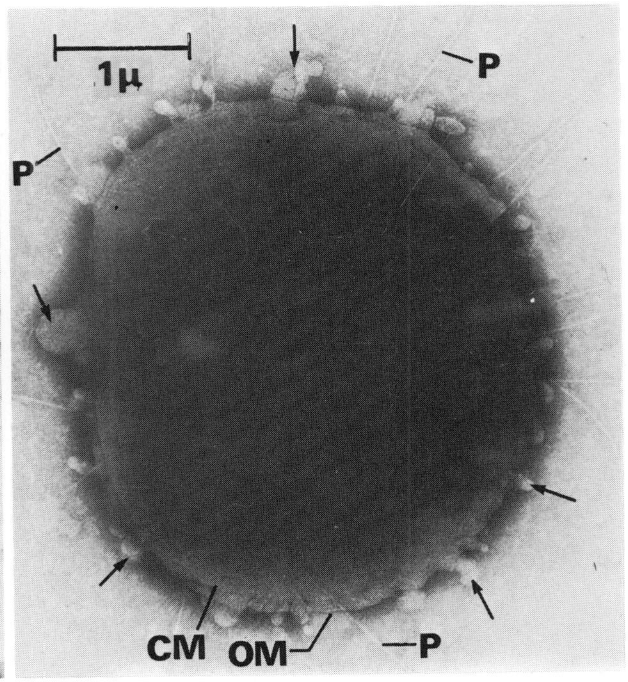
1μ
P
P
CM
OM
P

(k)

(l)

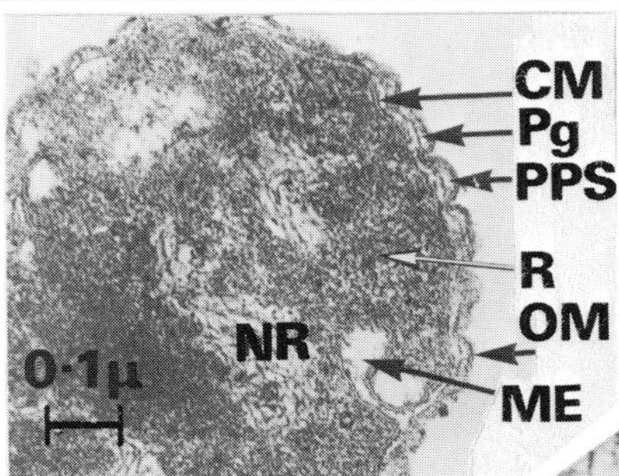
CM
Pg
PPS
R
OM
NR
0·1μ
ME

(m)

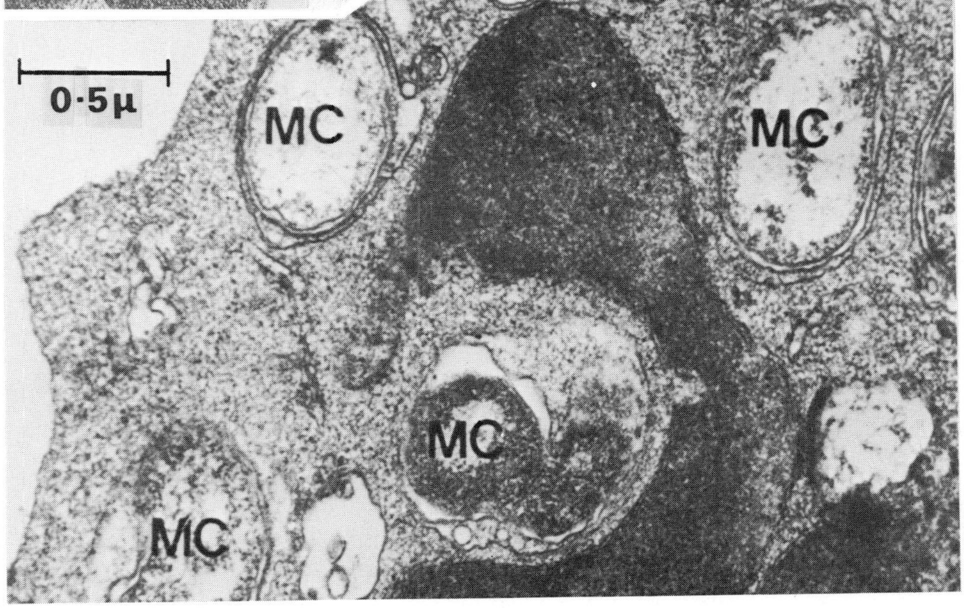
MC
0·5μ
MC
MC
MC
MC

Figure 4.24. *Neisseria* species.
In (n), (o), (p), (q) *N. gonorrhoeae* is seen to be localized on the surface of cervical epithelium (n), attached to the surface of urethral epithelial cells (arrows in o,p) and on the mucosal surface of a fallopian tube (q). Staining with colloidal thorium hydroxide outlines the acidic capsule of a commensal bacterium (clear arrow in (n)); the surface of gonococci (Gc) is essentially free of thorium (dark arrow). The epithelial cell coat and cervical mucosa are also thorium positive. On the surface of the urethral epithelial cell (o), the gonococci (Gc) are localized between adjoining cells. Numerous spherical vesicles cover the epithelial surface.

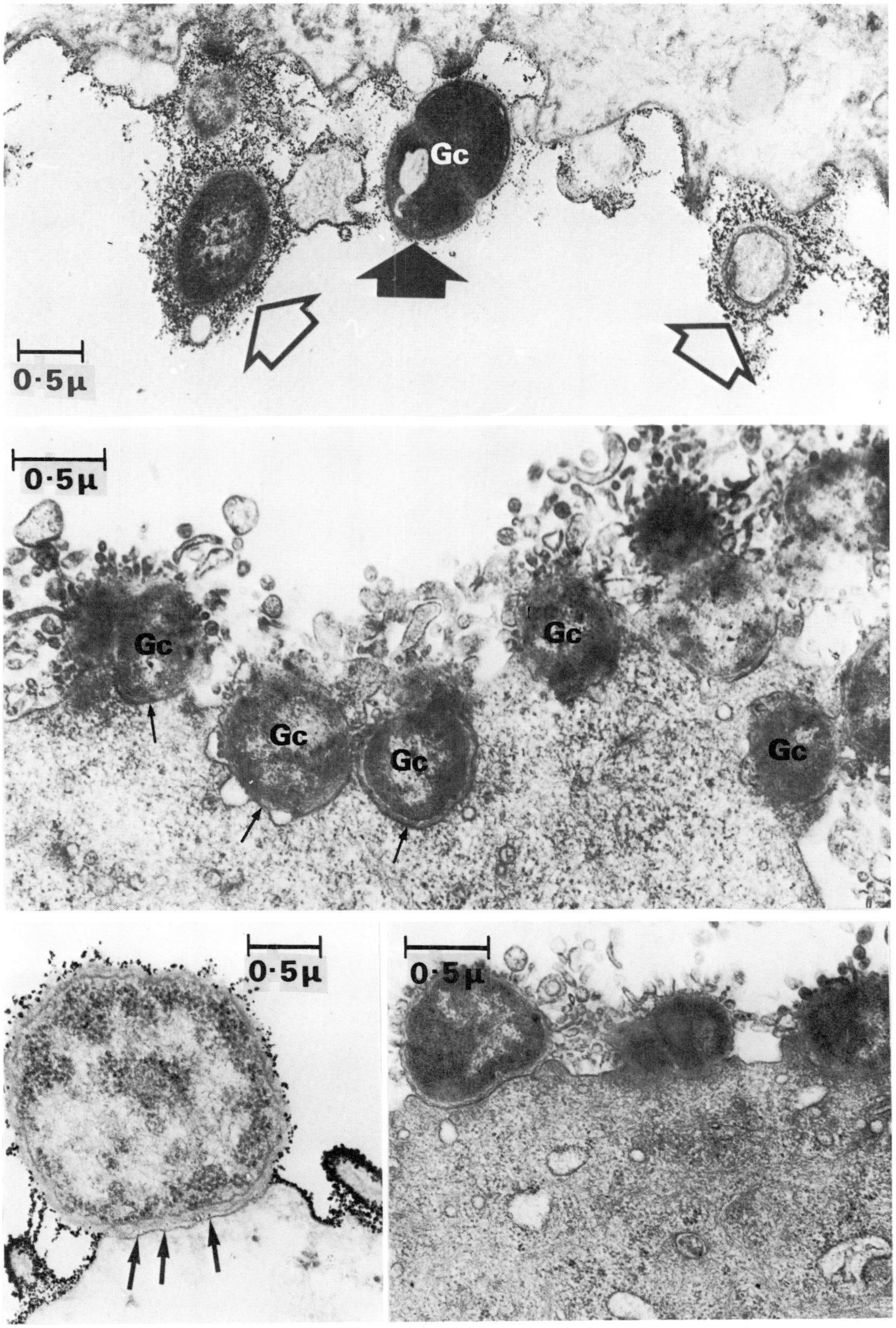

(n) (o) (p) (q)

examined by negative staining, numerous long filaments are seen attached to the outer cell wall membrane (Figure 4.25(a)). These long filaments are of uniform thickness (10 nm), and either straight or slightly curved. Nonspreading, noncorroding variants of *M. nonliquefaciens* have very few filaments on their outer surface. The filaments appear larger in diameter and straighter than those found in corroding, spreading cells.

Thin sections of *M. bovis* (Figure 4.25(b); Ritchie, unpublished communication), and negatively stained (Figure 4.25(c)) cells are similar to *M. nonliquefaciens*.

4.3.7. *Veillonella*

The genus *Veillonella* consists of small gram-negative anaerobic organisms which are most frequently found in the oral cavity. They do not appear to be active as primary pathogens, but may be associated with disease as secondary invaders. The veillonellae are found in dental plaque where they represent from 5 to 8 percent of the cultivable flora. Morphologically, the cells are very similar to *Neisseria*, and at one time were classified in the family Neisseriaceae.

Negative-stained cells of *Veillonella* (strain V2) and *V. alcalescens* strain CL (Figure 4.25(a),(b))[17, 217] are diplococcal in shape with the highly convoluted cell surface, characteristic of gram-negative bacteria examined by this staining technique.

The thin section (Figure 4.26(c)) of chemically fixed *Veillonella* (V2) is characteristic of gram-negative cells.

4.3.8. Fusobacteria

Members of the genus *Fusobacterium* are long, slender, gram-negative anaerobic bacilli which have been implicated in suppurative and gangrenous lesions. The genus includes *Fusobacterium nucleatum, F. necrophorum* (*Sphaerophorus necrophorus*), and *F. fusiforme* (*Leptotrichia buccalis*). From a morphological aspect, the fusobacteria are characterized by a thinness and delicateness of the cell wall layers as well as having either one or both ends of the cell tapered or pointed.

The leptotrichiae are obligate parasites commonly found in the human oral cavity.

Leptotrichia buccalis, the type species of the genus, is a relatively straight rod 4–15 μm in length and 0·6 μm wide. In Figure 4.27(a) it can be seen how one end of the cell is tapered and the other square.[126] The outer membrane is

Figure 4.25. *Moraxella* species.
(a) *Moraxella nonliquefaciens* 3067/66, as revealed by silicotungstic acid staining, possesses numerous long fimbriae emerging from the cell surface. (b), (c) *Moraxella bovis*, by thin sectioning and PT-acid staining, respectively, possesses a typical wrinkled gram-negative cell wall. Numerous delicate fimbriae completely cover the outer layer of *M. bovis* (c). CM: cell membrane, OM: outer cell membrane, R: ribosome, NR: nuclear region, Pg: peptidoglycan.

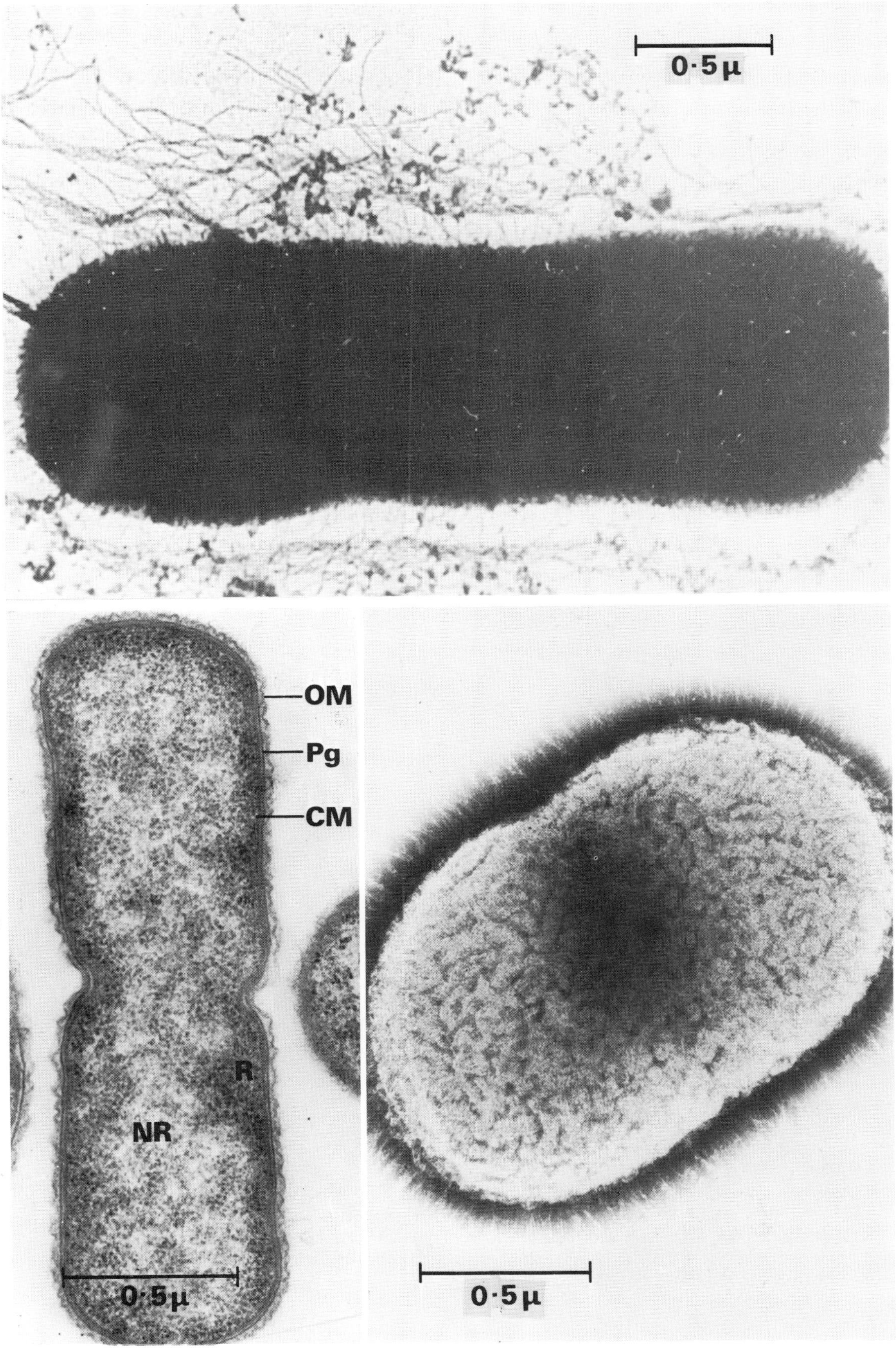

(a)

(b)

(c)

Figure 4.26. *Veillonella* species.
Dividing cells of (a) *V. alcalescens* and (b), (c) *Veillonella* strain V2. The negatively stained cells reveal the extremely wrinkled cell cytoplasm. The thin section (c) of a dividing cell shows the wrinkled outer cell wall membrane (OM), the intermediate peptidoglycan layer (Pg) and inner cell membrane (CM). Electron-dense ribosomes (R) limit the coagulated nucleoid. NR: nuclear region.

(a)

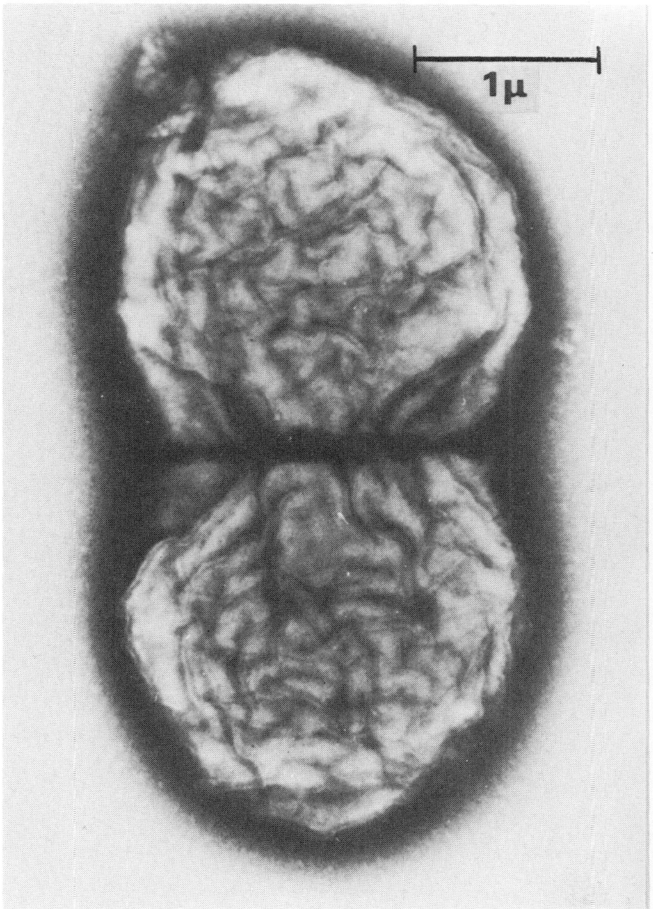

(b)

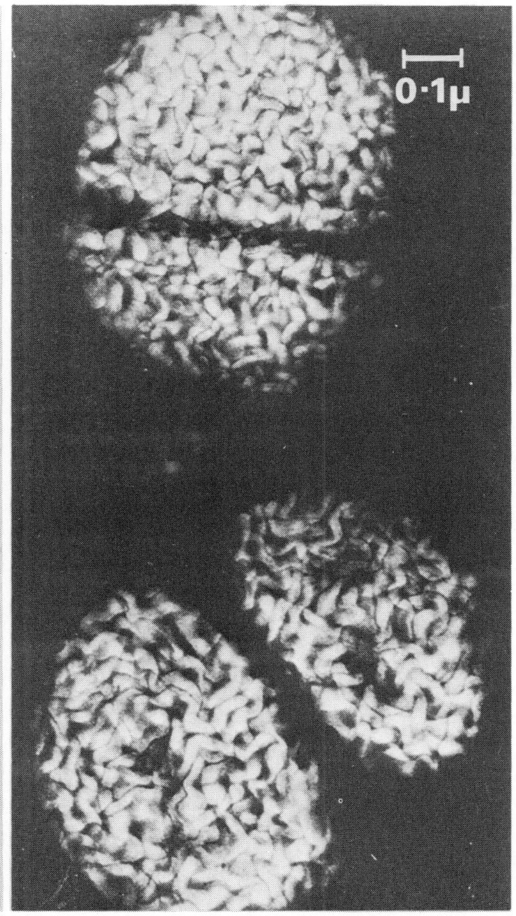

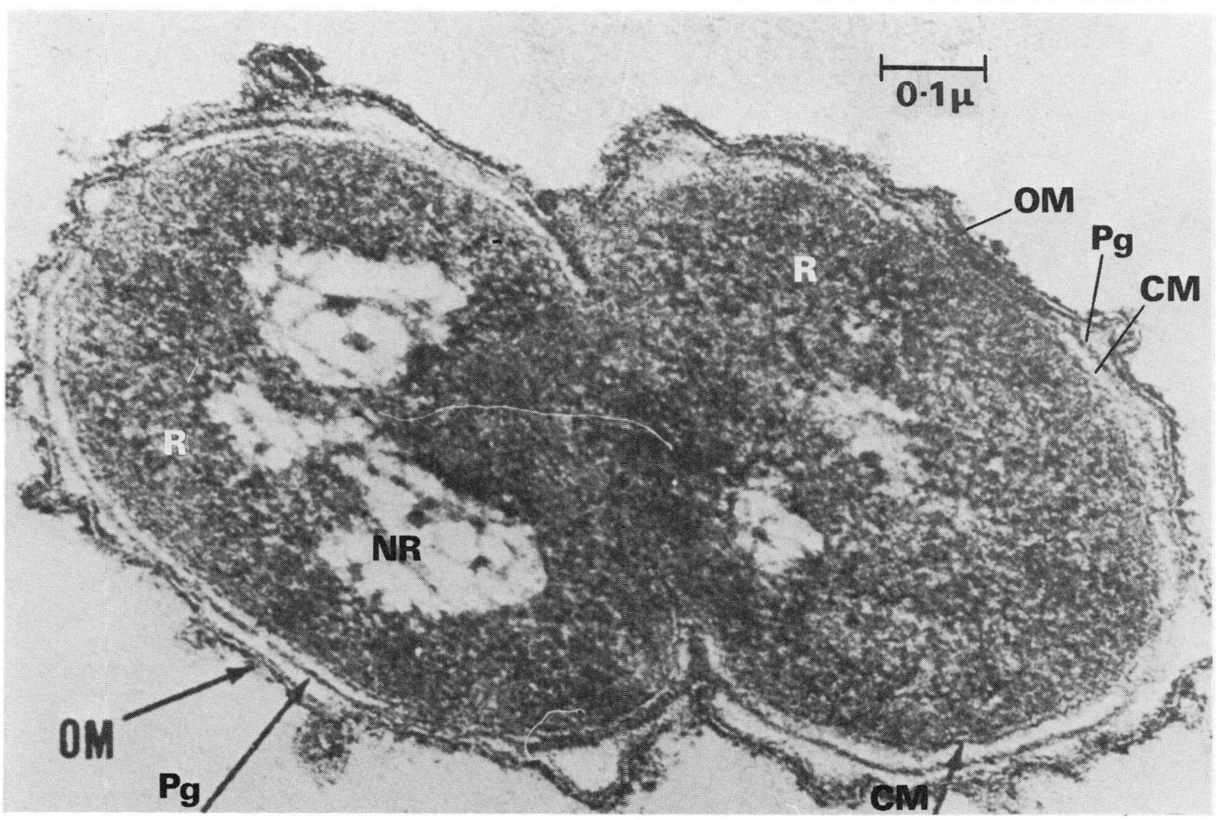

(c)

approximately 7 nm wide with the middle, electron-dense, single membrane (5 nm) representing the peptidoglycan layer. The underlying cell membrane follows the track of the peptidoglycan and is separated from it by a 5 nm space which is possibly the result of the preparative techniques for electron microscopy. Several strains (ATCC 14201, 19616, CS1/AC) of *L. buccalis* possess scale-like fimbriae on the outer layer of the outer membrane (Figure 4.27(c),(d)). These surface fimbriae when examined in tangential section consist of membranous folds in a scalloped arrangement. Several investigators[96] have not observed these fimbriae in their strains of *L. buccalis* (Figure 4.27(b)). The difference between the cell surfaces (fimbriated compared to smooth) may be strain differences.

4.3.9. *Capnocytophaga*

The members of this group comprise gram-negative fusiform-shaped bacteria (Figure 4.28(a)–(d)) found in the human oral cavity. These organisms, placed in the genus *Capnocytophaga*, have been isolated from both supragingival dental plaque and from periodontal lesions,[123, 149] and are distinguished by colonial and morphological characteristics from other fusiform-shaped bacteria by their ability to "glide"[86] over solid surfaces. There is a growing body of evidence[105, 149, 150] implicating these organisms in some forms of periodontal disease.

When examined by electron microscopy (Figure 4.28(d)–(m))[101] the capnobacteria are similar in morphology to other gram-negative bacteria. Several of the strains, however, possess an electron-dense periplasmic space, as well as electron-dense hair or "fuzz" on the surface of the outer membrane (Figure 4.28(h),(j),(k)).

The cell wall consists of three layers, an outer and inner double-track membrane, approximately 7·4 and 6·6 nm thick, respectively. Depending upon the strain, the outer layer of the cell wall is either taut fitting (Figure 4.28(e)) or loose (Figure 4.28(d),(f),(g),(l)). These two unit membranes enclose the thin, 3·3 nm wide peptidoglycan. Several of the capnobacteria have a thick cell wall, approximately 50 nm in diameter (Figure 4.28(h),(i)). These thick cell walls usually contain a hair-like fuzz on their outer membrane surface. Ruthenium red fixation and staining clearly reveals this external layer to vary in thickness between 3·2 and 8 nm, with an average thickness of 4·8 nm (Figure 4.29(j),(k)). In many instances when the cells are fixed directly on an agar surface they appear to be arranged in both longitudinal and transverse rows or

---

Figure 4.27. *Leptotrichia buccalis*
(a) *L. buccalis* ATCC 14201, (b) *L. buccalis* L-11, and (c), (d) human oral isolate. The cell in (a) possesses characteristic tapered (arrow) and blunt ends (double arrow). Visible in (c) is the scalloped cell wall, the presence of which has been confirmed by freeze-fracture etching (arrow in (d)). In (b) strain II has a loose fitting, smooth, outer cell wall membrane (OM), enclosing the thin peptidoglycan layer (Pg) and cell membrane (CM). Note the absence of the scalloped layer in this strain. DS: division septum.

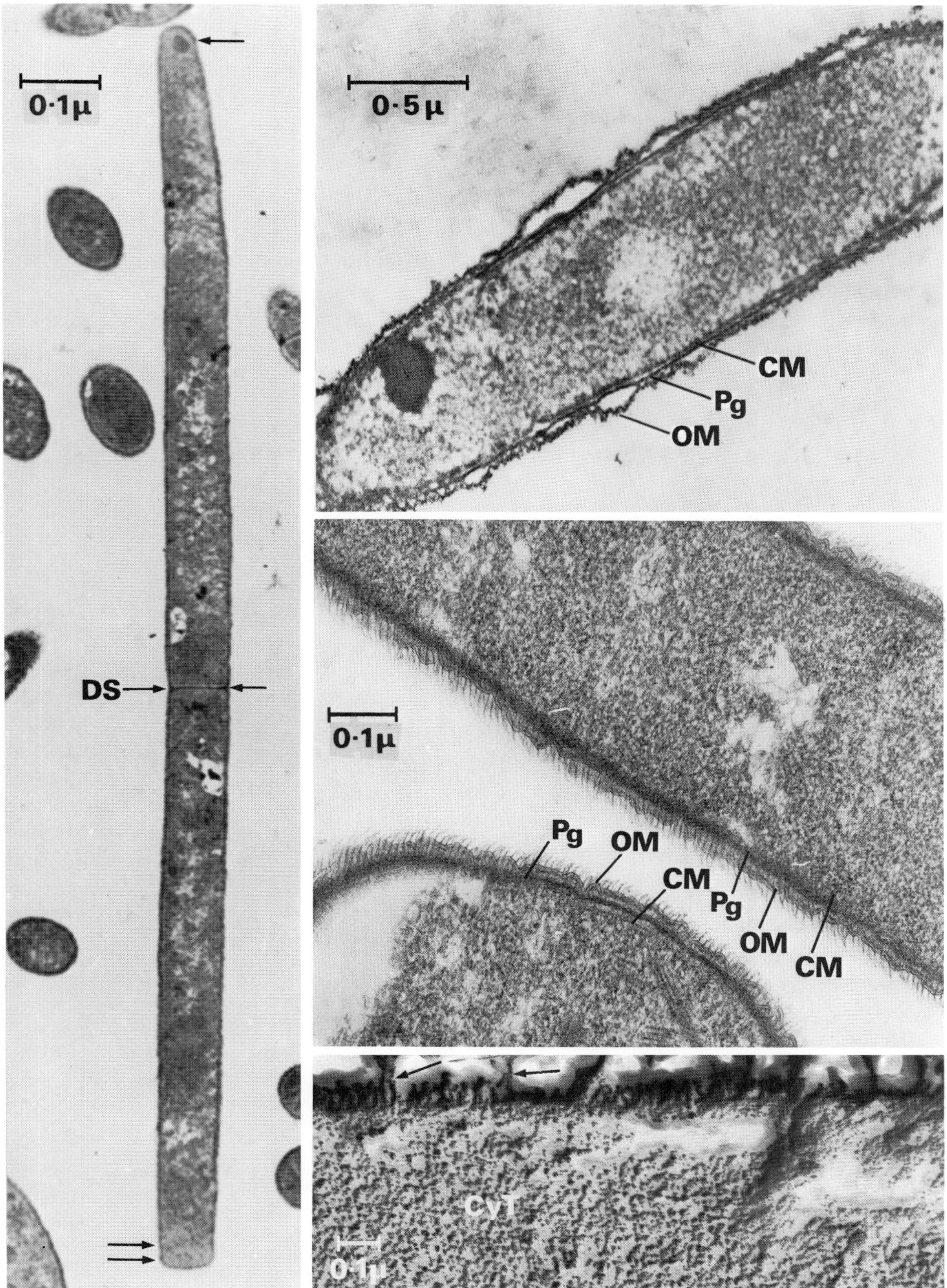

(a)

(b)

(c)

(d)

sheets of tightly enmeshed cells (Figure 4.28(m)). These rows and sheets of cells may be held together by the hair-like outer membrane fuzz.

Numerous vesicles and cell surface-associated membrane blebs are present in older cultures (Figure 4.28(l)), and these may be lipopolysaccharide released during the later stages of cell growth.

4.3.10. *Escherichia* and related genera

Escherichia coli and its "relatives," commonly referred to as the coliform-enteric bacilli, are gram-negative, short rods. These coliform-enteric bacilli are represented by the genera *Shigella, Yersinia, Proteus, Erwinia, Enterobacter*, and *Serratia*, as well as *Vibrio, Aeromonas, Beneckea*, and the *Photobacterium.* Also included are the *Klebsiella, Edwardsiella, Citrobacter*, and *Providencia.* Without exception, all the Enterobacteriaceae are gram-negative, aerobic to facultative anaerobes, and vary in length between 1–8 μm.

Escherichia, Klebsiella, Serratia, Arizona, Edwardsiella, Citrobacter, as well as the *Providencia*, are normal inhabitants of the human intestinal tract, but usually do not penetrate the intestinal wall. However, when the intestinal wall becomes pathologic with a lowering of the resistance of the host, or when there is an alteration in the "normal" intestinal flora, there usually results a penetration of coliforms (i.e., *Salmonella, Shigella*) through the intestinal canal into the bloodstream with resultant pyelonephritis, cystitis, cholecystitis, abscesses, peritonitis, or meningitis.[20, 183]

Escherichia coli

The "work horse" of procaryotic cells, *E. coli*, is a permanent resident of the intestinal tract of humans and animals, and is representative of the morphology of the gram-negative coliforms. In negative-staining (Figure 4.29(e)) and SEM (Figure 4.29(b)) the surface of *E. coli* is typically rugose. Thin sections of chemically fixed cells (Figure 4.29(a),(c),(d)) reveal the cell wall to consist of several layers; a triple-layered 5–5·5 nm outer membrane, at times resolvable into globular subunits, and an electron-transparent space, 4–4·5 nm thick. The inner layer, the cell membrane, is also triple layered, and approximately the same width as the outer membrane of the cell wall.

Between the outer and inner membranes of the cell wall is an intermediate layer, consisting of globular macromolecules; the distance between the globules being approximately 12–14 nm. This layer has been referred to as the G–layer, is 5–5·5 nm thick, and represents the peptidoglycan component.

In some *E. coli* strains the surface of the outer membrane is covered with a

Figure 4.28. *Capnocytophaga* species.
Capnobacterial isolates from the human oral cavity. (a), (b), (c) Phase contrast photomicrographs of representative gliding isolates. In (c) the cells have aggregated and are attached to each other. (d), (e), (f), (g) Electron micrographs of representative gliding strains of capnobacteria, showing characteristic rounded to tapered ends ((d), (e), (f)). CM: cell membrane, OM: outer cell membrane, PPS: periplasmic space.

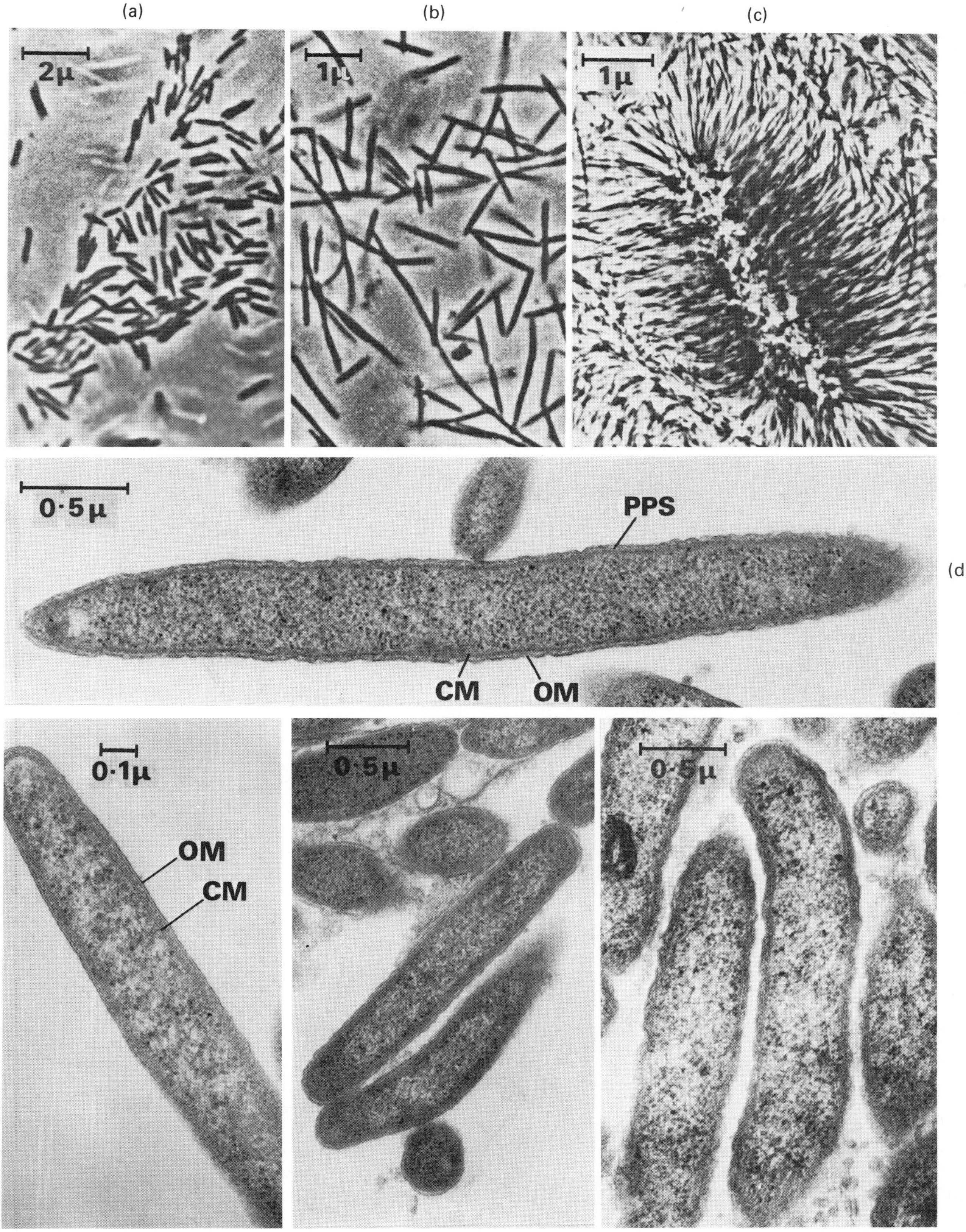
(a)
2μ
(b)
1μ
(c)
1μ
0·5μ
PPS
(d)
CM
OM
0·1μ
OM
CM
(e)
0·5μ
(f)
0·5μ
(g)

Figure 4.28. ***Capnocytophaga*** **species.**
(h), (i), (j), (k), (l), (m) Electron micrographs of representative gliding strains of capnobacteria. Loose-fitting and wavy cell walls with electron-dense periplasmic spaces (PPS) are visible in (h), (i), (j). Ruthenium red fixed and stained cells ((j), (k)) reveal the outer layer of the cell wall to be ruthenium red positive (arrows), indicative of acidic mucopolysaccharide polymers. In (l) numerous small vesicles (VB) (membrane blebs) emerge from the outer membrane of this strain. Cells fixed directly on an agar surface (m), and sectioned parallel to the cell mass, reveal the cells adhering tightly to each other and arranged in a longitudinal and transvers plane. Pg: peptidoglycan, NR: nuclear region, R: ribosome, CM: cell membrane, OM: outer cell membrane.

(i) (j) (k)

(h)

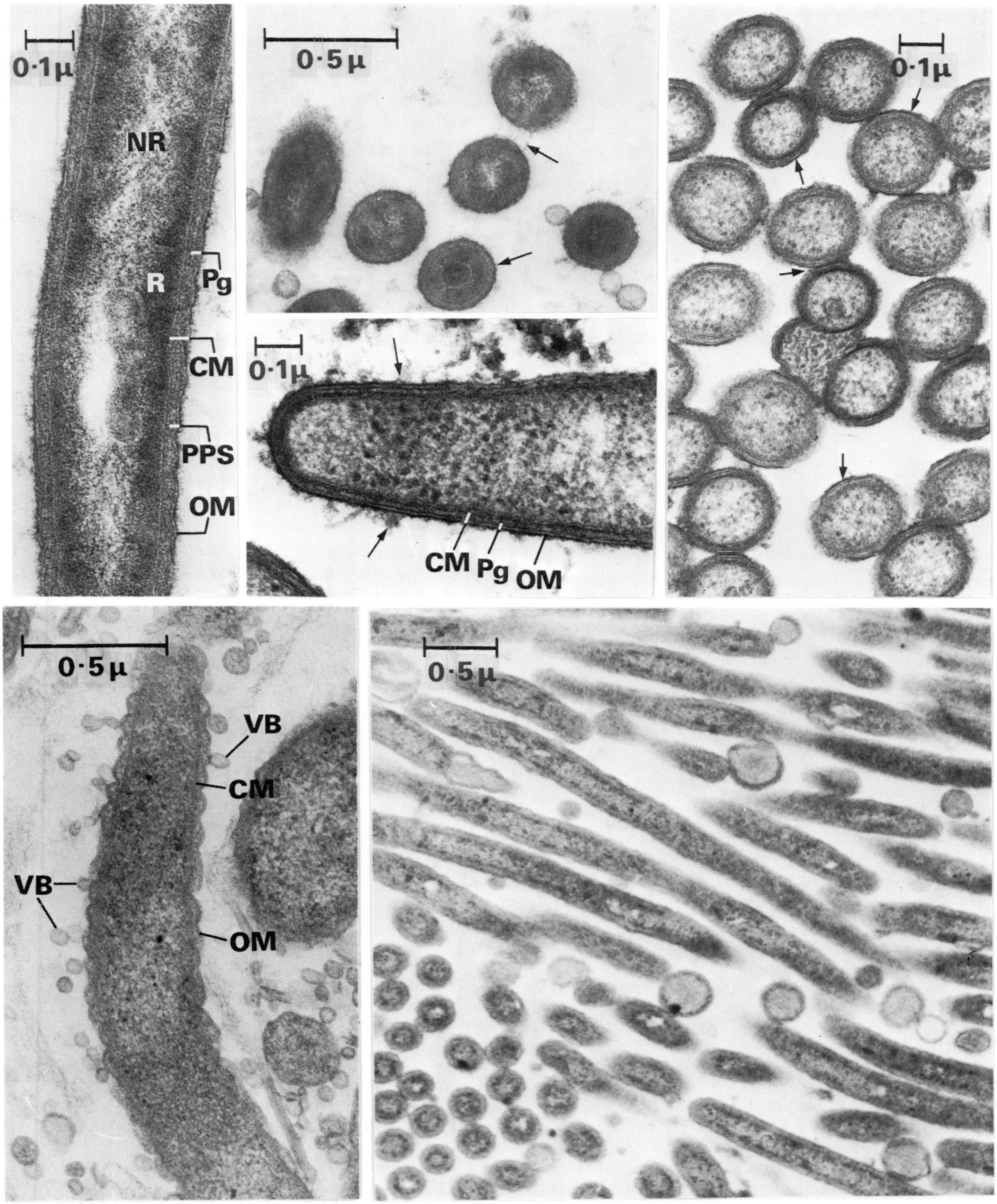

(l) (m)

"soft" lipoprotein coat as well as long, thin fimbriae or pili (Figure 4.29(f)). Acrolein or acrolein-glutaraldehyde fixation of *E. coli* B results in the formation of a smooth, 7·5 nm thick outer membrane, which is in close apposition to the peptidoglycan layer. The cell membrane is 7·3–9·3 nm thick and asymmetric, with the outer face more electron dense than the inner face. Acrolein-glutaraldehyde results in the formation of a dispersed nucleoid, rather than the tight, packed fibrous nucleoid seen after the classical Ryter–Kellenberger fixation.[113]

The actual appearance of the cell wall is dependent upon the fixation technique as well as on the "age" and physiological condition of the culture at the time of fixation. For example, when an exponential phase culture of *E. coli* is fixed in osmium, the cell wall appears straight, with an inner dense line 2–3 nm thick and an outer triple-layered membrane, 5·5–6 nm thick. The outer surface of the outer membrane of *E. coli* fixed in glutaraldehyde-osmium consists of 15–36 nm diameter vesicles. These vesicles are in intimate contact with, and probably are an outgrowth of the outer membrane. They may also represent lipopolysaccharide which is being lost from the outer membrane. Further, the cation (i.e., Ca^{2+}) concentration in both the growth medium and in the fixation buffer has a dramatic effect on the morphology of the outer layer of the cell wall; low levels of Ca^{2+} result in a loss of the structured outer layer while high concentrations bind or link this outer layer in a stable hexagonal fashion.[14, 15, 30]

Salmonella typhimurium

Salmonella typhimurium (Figure 4.30(a),(b)), *Proteus vulgaris* (Figure 4.30(c),(d)), and *P. mirabilis* (Figure 4.30(e)) are rod-shaped and in all respects morphologically identical to *E. coli.* The salmonellae are involved in two gastrointestinal diseases, the enteric fevers and gastroenteritis.

Klebsiella pneumoniae

Known as Friedlander's bacillus, this is closely related to the coliform group in morphology, and in its biochemical characteristics but it bears some resemblance to *Streptococcus pneumoniae* in its respiratory involvement, producing lobar or lobular pneumonia, and in its ability to form a capsular layer.[184] It is distinguished from the other members of the enteric bacilli in that it is nonmotile.

Figure 4.29. *Escherichia coli.*
(a) Strain W1655^{+}, (c) strain B, and (d) prepared by chemical fixation and thin-sectioning; (b) strain 6 by SEM; (e) strain B by negative staining; and (f) strain 6 by freeze-fracture etching. The cells exhibit a loose fitting cell wall and electron-dense cytoplasmic region. By SEM and silicotungstic acid negative staining the cell wall is highly convoluted and irregular in contour. Numerous long fimbriae similar to pili emerge from the outer layer of the cell wall in the freeze-fracture etched replica. Note the unequal division in (c). R: ribosome, NR: nuclear region, CM: cell membrane, DS: division septum, OMI and OM2 outer cell membranes.

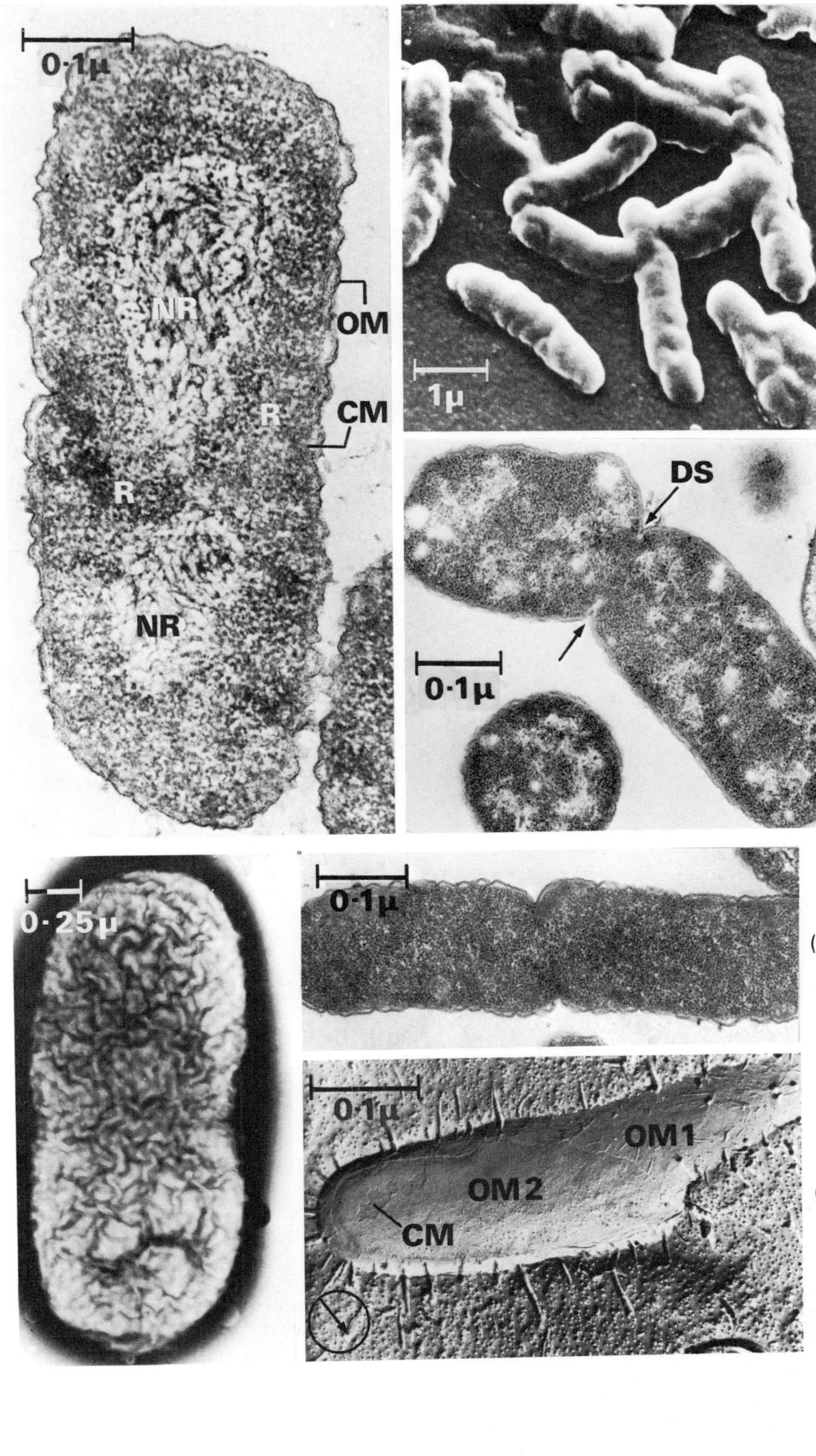

(a) (b) (c) (d) (e) (f)

Figure 4.30. ***Salmonella typhimurium* and *Proteus vulgaris.***
(a) The thin section of *S. typhimurium* is representative of these enteric microorganisms. The cell wall consists of the wrinkled outer membrane (OM), the surface of which is covered with an electron-dense fuzz (arrow). The underlying cell membrane (CM) follows the contour of the outer membrane. (b) Negative stained (PTA) *S. typhimurium* SW 1061 reveals a wrinkled and convoluted surface as well as numerous long flagella (F). (c), (d) similarly, *P. vulgaris* possess long, thin flagella (F), as well as numerous straight fimbriae, or pili (P). (e) Pt–Ir shadowcasting of *P. mirabilis* shows the surface to be covered with numerous spherical vesicles, short thin fimbriae or pili (P) and several thick flagella (F). NR: nuclear region.

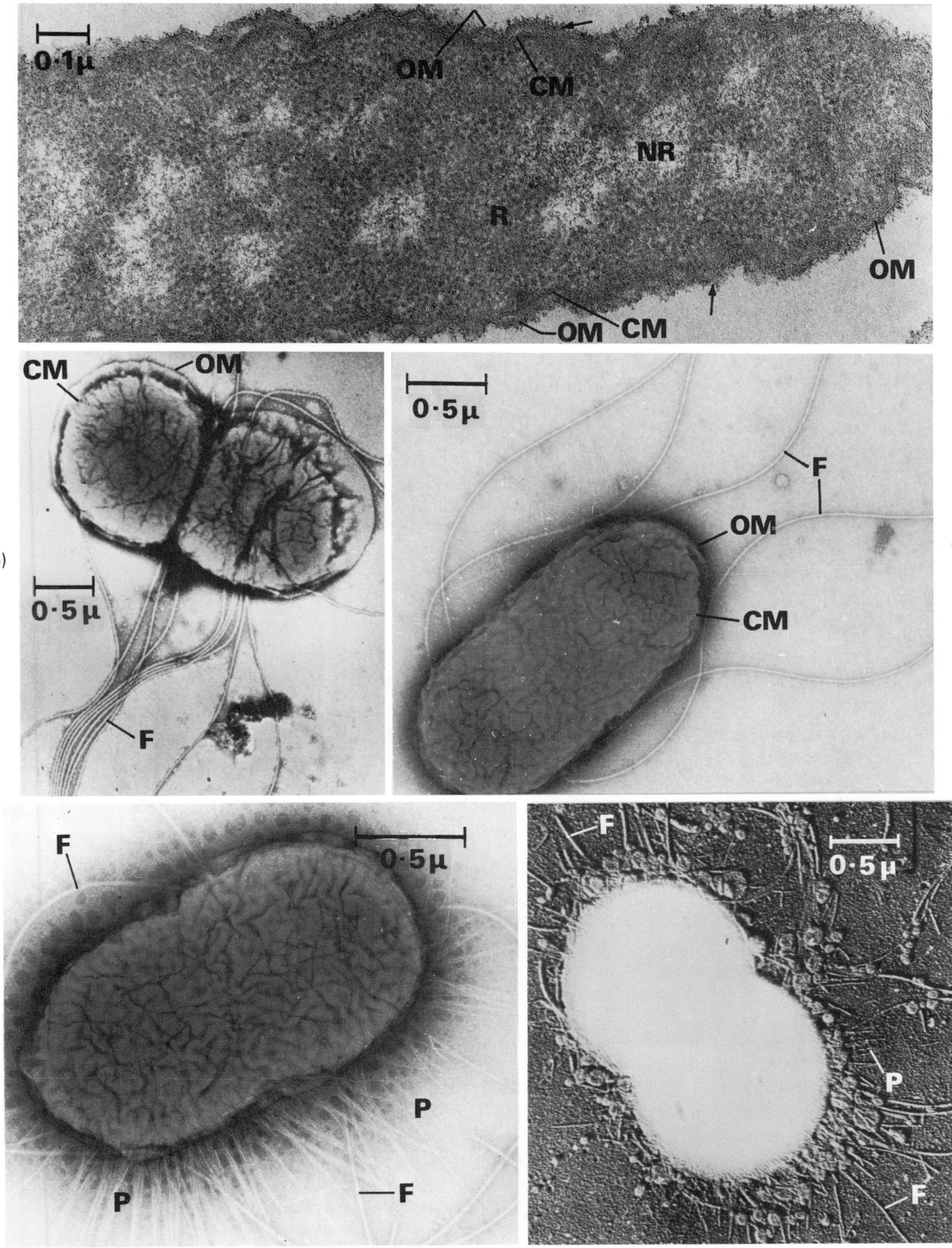

(a) (b) (c) (d) (e)

Klebsiella pneumoniae and *K. aerogenes*, strain GSCH (*K. pneumoniae*) are seen in Figure 4.31(a)–(d)[184] (Holt, unpublished communication). The capsule of *K. aerogenes* GSCH is uniformly distributed around the cell as an electron-dense fibrous layer. Note the electron-density of the cell wall layers. The reason for this concentration of electron-dense stain in the layers of the cell wall is unclear; however, ruthenium red staining and fixation did not increase the staining properties of the capsule, possibly indicating a chemically different capsule. In contrast to encapsulated *K. aerogenes* (Figures 4.31(c)), noncapsulated strains are essentially smooth-surfaced, with only a slight fuzz being apparent (Figure 4.31(d)).

After fixation of *K. pneumoniae* in osmium, and ruthenium red staining, dehydration in alcohol, and embedding in Epon, several morphologically distinct capsules are apparent;[184] these include spike-like capsules (Figure 4.31(a)) as well as net-like and fibrillar cross-linking structures (Figure 4.31(b)). These various capsule forms may be a result of dehydration effects on the capsule polymer.

Proteus

The members of this genus are frequently found as normal inhabitants of fecal material, water, soil, and sewage.

Several species within the genus are of medical importance, and include *P. vulgaris, P. morganii*, and *P. mirabilis.*

Proteus vulgaris ranks with *E. coli* in importance as a cause of urinary tract infections. In several rare instances, *P. vulgaris* has also been involved in peritonitis. *Proteus morganii* has been implicated in infectious (summer) diarrhoea in infants.

The cell wall of *P. vulgaris* is identical to *S. typhimurium* (Figure 4.30(a)), consisting of an outer membrane approximately 10–12 nm in diameter, and morphologically identical to the inner 8 nm cell membrane. In some strains mesosomes are also seen. These include fused, vesicular, and straight or slightly curved membranes.

In negatively stained and shadowcast preparations *P. vulgaris* is surrounded by numerous straight filaments (Figure 4.30(d),(e)), morphologically similar to pili. Large filaments, which represent fragments of thin flagella, are also seen. Of interest in Figure 4.30(d) is that these "pili" appear to be concentrated along one side of the cell, with the opposite side being devoid of them.

Figure 4.31. *Klebsiella* species.
(a) and (b) *K. pneumoniae* cells, fixed and stained with ruthenium red and either dehydrated in ethanol and embedded in Epon 812 or dehydrated in Durcupan and embedded in Araldite 502. In both cases the ruthenium red positive capsular layer appears as long threads emerging at regular intervals along the cell wall. CM: cell membrane, OM: outer cell membane.

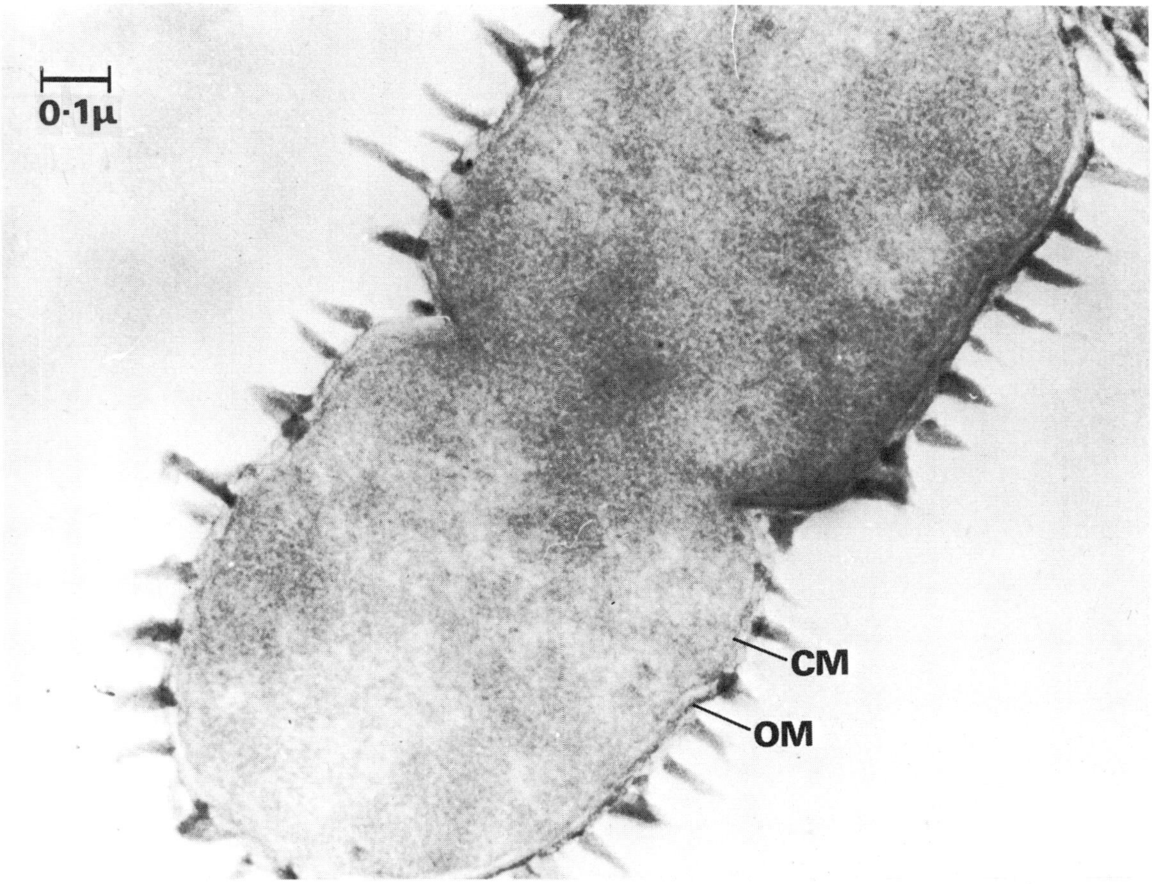

(a)

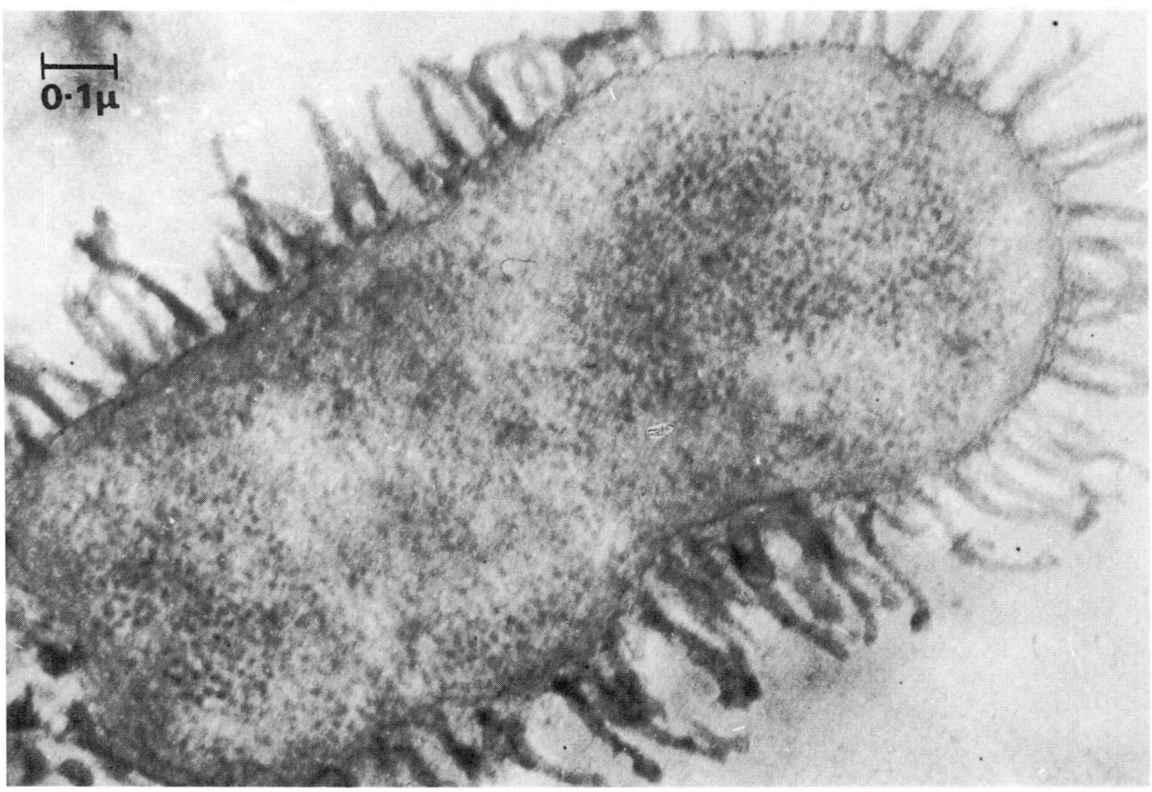

(b)

Figure 4.31. *Klebsiella* species.
(c) An encapsulated strain of *K. aerogenes* covered with a thick, electron-dense, fibrous capsule (Cp). The underlying outer (OM) and cell membrane (CM) are intensely stained.
(d) A noncapsulated *K. aerogenes* devoid of the thick, fibrous, electron-dense capsule but showing a tightly adhering electron-dense layer (arrow) associated with the surface of the outer membrane. R: ribosome, NR: nuclear region.

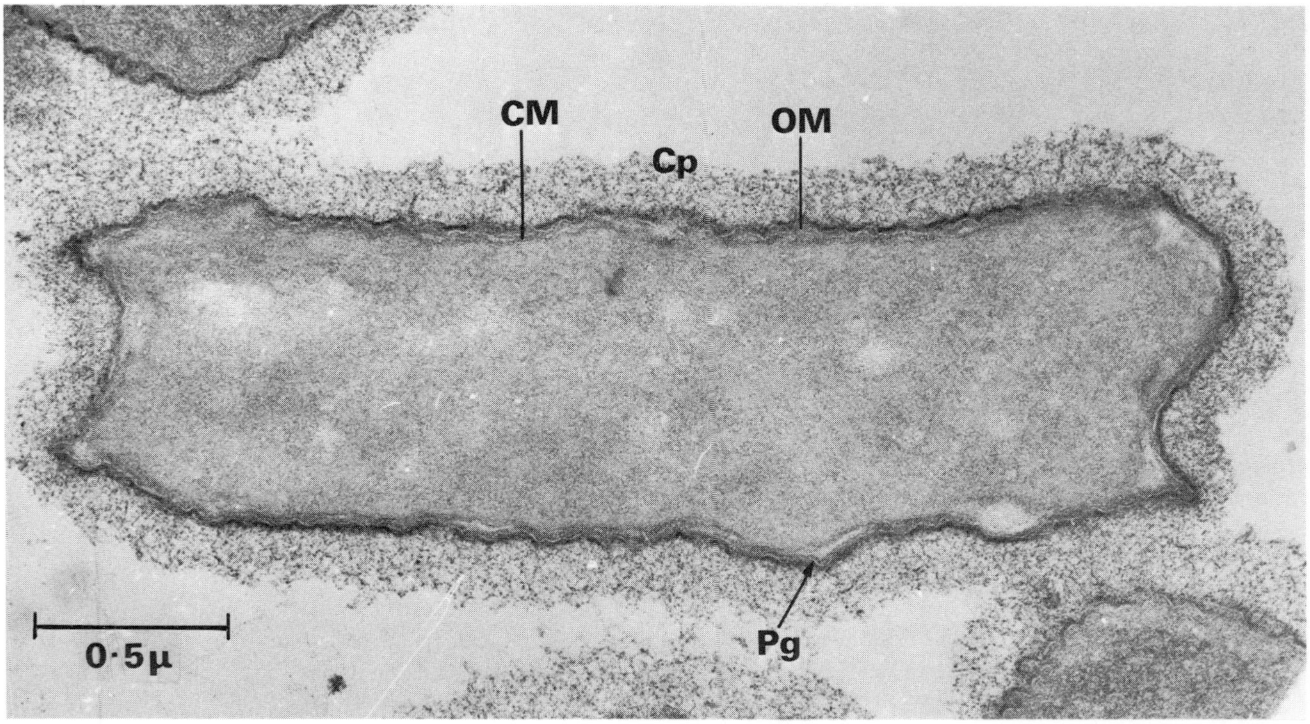

(c)

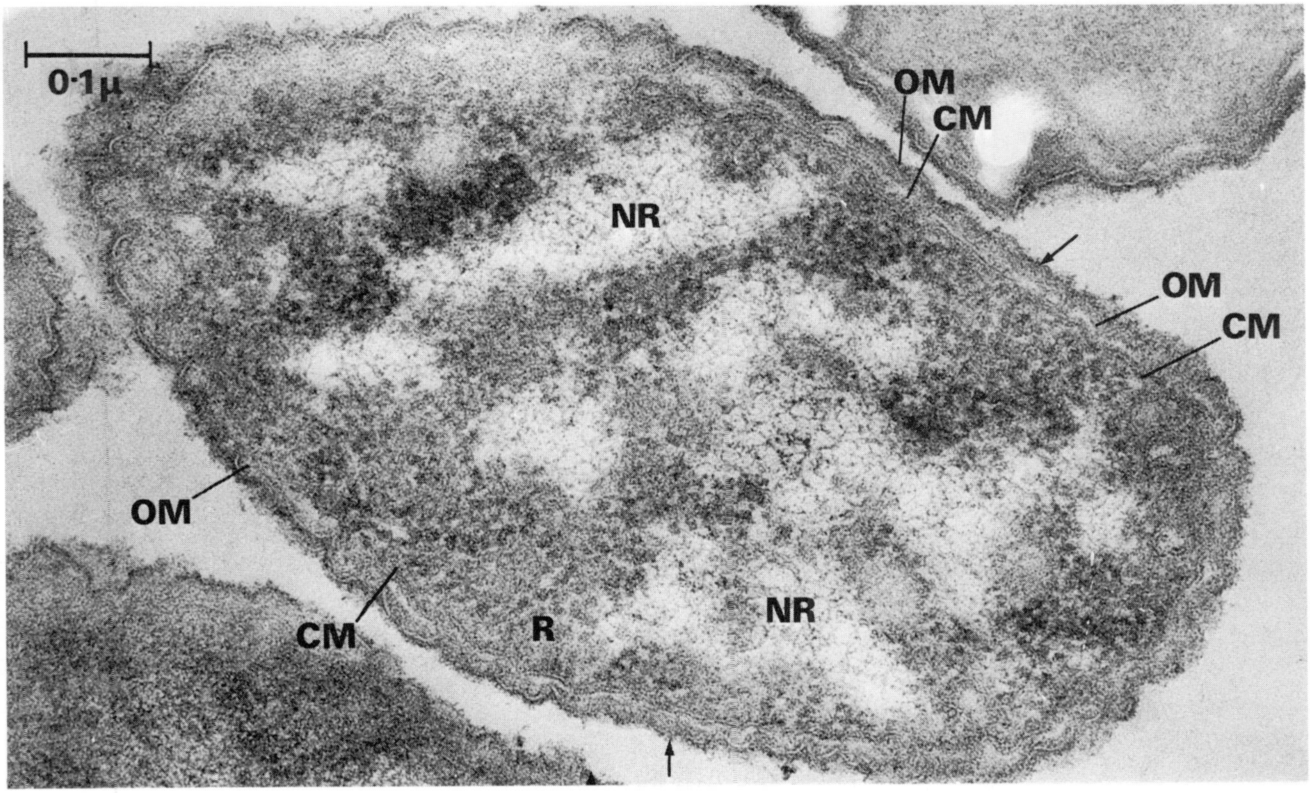

(d)

Yersinia

Members of this genus included *Yersinia pestis* (*Pasteurella pestis*), the causative agent of bubonic plague, *Y. enterocolitica*, and *Y. pseudotuberculosis. Francisella tularensis*, originally classified as *Pasteurella* or *Yersinia tularensis* is morphologically identical to *Y. pestis.* The zoonotic disease produced by *F. tularensis*, deerfly or rabbit-fever, is transmitted to humans by insect vectors from infected animals. *Yersinia enterocolitica* and *Y. pseudotuberculosis* are responsible for yersiniosis, a zoonotic disease of which human infection is usually acquired accidentally. The primary lesion in yersiniosis involves an invasion of the small intestine in the region of the ileum, ultimately invading the mesenteric nodes, where there is extensive involvement.

Yersinia are ovoid to rod-shaped gram-negative bacteria, are relatively small, being 0·5–1 × 1–2 μm. Both *Y. enterocolitica* and *Y. pseudotuberculosis* are motile; *Y. pestis* is nonmotile. Fresh isolates of these coccobacilli possess a capsule, which, like that of many encapsulated bacteria isolated from human or animal hosts, disappears after several generations of *in vitro* growth.

As is apparent in Figure 4.32(a)–(c), *Y. pestis* and *Y. enterocolitica* are typically gram negative, possessing a wrinkled outer cell membrane, a thin, and partially visible, intermediate peptidoglycan layer, and an inner cell membrane. Note the division tube in Figure 4.32(a) (arrow), similar to that observed in several dividing gonococci (see Figure 4.24(e),(f)). The outer membrane of *Y. enterocolitica* (Figure 4.32(c)) is covered with a slightly electron-dense, fuzz-like layer. Whether this material is representative of a capsule or lipopolysaccharide is unclear.[2]

Also in Figure 4.32(c)[3] fimbriae of varying width are apparent. The significance of the fimbriae, especially in relation to virulence, is unknown.

An early report by Crocker *et al.*[50] indicates that *Y. pestis* possesses an extracellular layer, similar to a capsule, but recent studies (Brubaker, unpublished communication) have not confirmed this.

Serratia marcescens

These gram-negative, pink-red pigmented, motile, peritrichously flagellated rods have been found free in nature living as saprophytes in soil and water. They have also been found as common inhabitants of hospital environments, where for many years they were readily controlled by antibiotic therapy. Prolonged use of the more common antibiotics has resulted in an increased number of resistant *S. marcescens* strains, with considerable concern in nosocomial hospital infections.

Of interest was the observation by Acker[1] that the cell surface of *S. marcescens* is relatively smooth with only a slight degree of granularity (Figure 4.33(a)). Treatment of cell cultures with polymyxin B (2·5–5 μg/ml) for as little as 10 min resulted in a general disorganization of the cell surface components into tubular, sausage-like structures (Figure 4.33(b)). Similarly, thin sections of chemically fixed cells (Figure 4.33(c)) revealed the classical gram-negative morphology; however, polymyxin B treatment appeared to have "loosened"

(a)

(b)

(c)

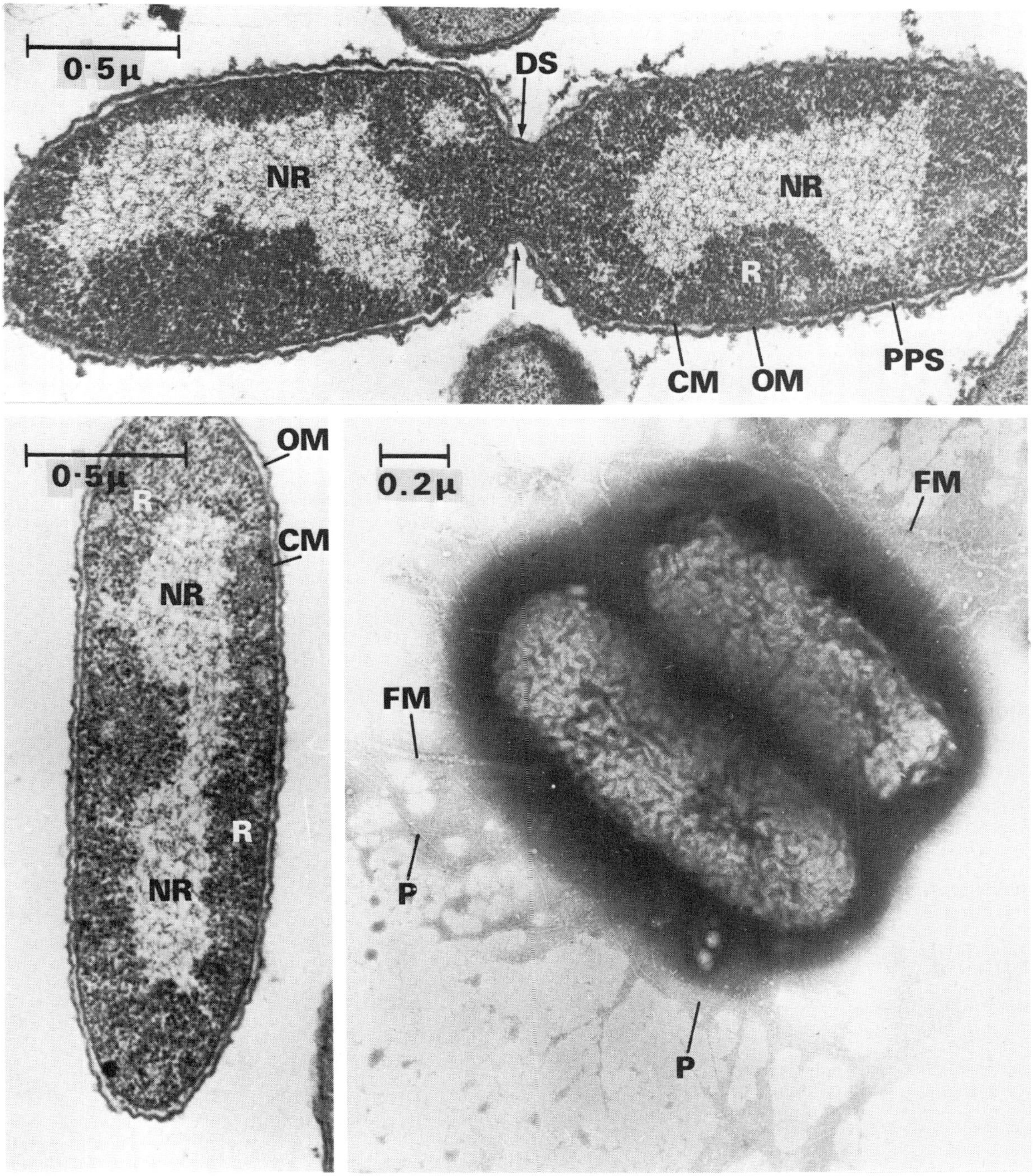

Figure 4.32. *Yersinia* spp.
(a), (b) *Y. pestis* EV76, and (c) *Y. enterocolitica.* In thin section *Y. pestis* is typically gram-negative in its morphology, possessing a wrinkled outer membrane (OM), an underlying cell membrane (CM), and an electron-dense ribsome (R)-packed cytoplasm. In the dividing cell in (a) the nucleoid has been equally divided between the two daughter cells. Note that constrictive division has given rise to a "division tube" at the central region of the dividing cell. NR: nuclear region, PPS: periplasmic space, DS: division tube.
(c) Uranyl acetate negatively stained *Y. enterocolitica* displays a wrinkled cytoplasmic region as well as thin pili-like structures (P) and possibly thicker fimbriae (FM).

the outer (LPS?) layer of the outer membrane, resulting in the appearance of long strands emerging from the surface. The periplasmic space is electron dense, as are the ribosomes and central nuclear material.

4.3.11. Selenomonas–*Alysiella–Simonsiella*

Characteristic of this group of gram-negative bacteria is their "crescent moon" shape and tufts of "slime-appendages" emerging from the middle of the concave side of the cell.[107, 187]

Three species of these disc-shaped bacteria have been described by Simons[180] and placed in the genus *Simonsiella* Schmid. *Simonsiella muelleri* and *S. crassa* are found in the oral cavities of human and domestic animals. *Simonsiella filiformis*, because of its disc-shaped morphology, is placed in a separate genus, *Alysiella* (see Steed[187] for a differentiation of the two genera).

Muller[144] described disc-shaped bacteria in the saliva of humans. Similar organisms have been observed in the throats of chickens.

The cells vary in dimension between 3·0 and 5·5 μm in length and from 0·9 to 1·1 μm in width, and their morphology as revealed by electron microscopy is represented by *A. filiformis* (Figure 4.34(a),(b)). Thin sections of chemically fixed cells reveal a 15–22 nm thick relatively electron-dense, gram-negative cell wall. The outer layer is covered with a loose fitting electron-dense material.

When isolated directly from its host the cells contain setae, or slime appendages emerging from the surface of the flat edge of the ribbon of bacterial cells (Figure 4.34(b)). These slime appendages may be involved in adherence, e.g., attaching the bacteria to the host epithelial cells. Cells of similar morphology to *A. filiformis* have been isolated from the oral cavity (Figure 4.34(c)) (Holt and Leadbetter, unpublished communication).

4.3.12. *Vibrio–Spirilla*, and related genera

Vibrio and *Spirilla*

These species are usually found in both fresh water and marine environments. Included in the genus *Vibrio* are *Vibrio cholerae* (*V. comma*), *V. cholerae* biotype eltor, and *V. parahaemolyticus*. The spirillaceae include two genera, *Spirillum* and *Campylobacter*. All are gram-negative, aerobic, actively motile

---

Figure 4.33. *Serratia marcescens.*
(a) Portion of a negative stained control cell of *S. marcescens* No. Se 143–b reveals a nonstructured, granular, outer cell surface. The outer limit of the cell is clearly delineated as a smooth layer.
(b) Treatment of *S. marcescens* with 2·5–5 μg polymyxin B/ml for 10 min resulted in a disorganization of cell surface components into tubular or sausage-like structures.
(c) Similarly treated cells in thin section exhibited a normal gram-negative morphology, with the exception that long strands, presumably LPS, emerged from the outer layer of the outer membrane. R: ribosome, NR: nuclear region, CW: cell wall, CM: cell membrane, OM: outer cell membrane, F: flagellum.

(a) (b)

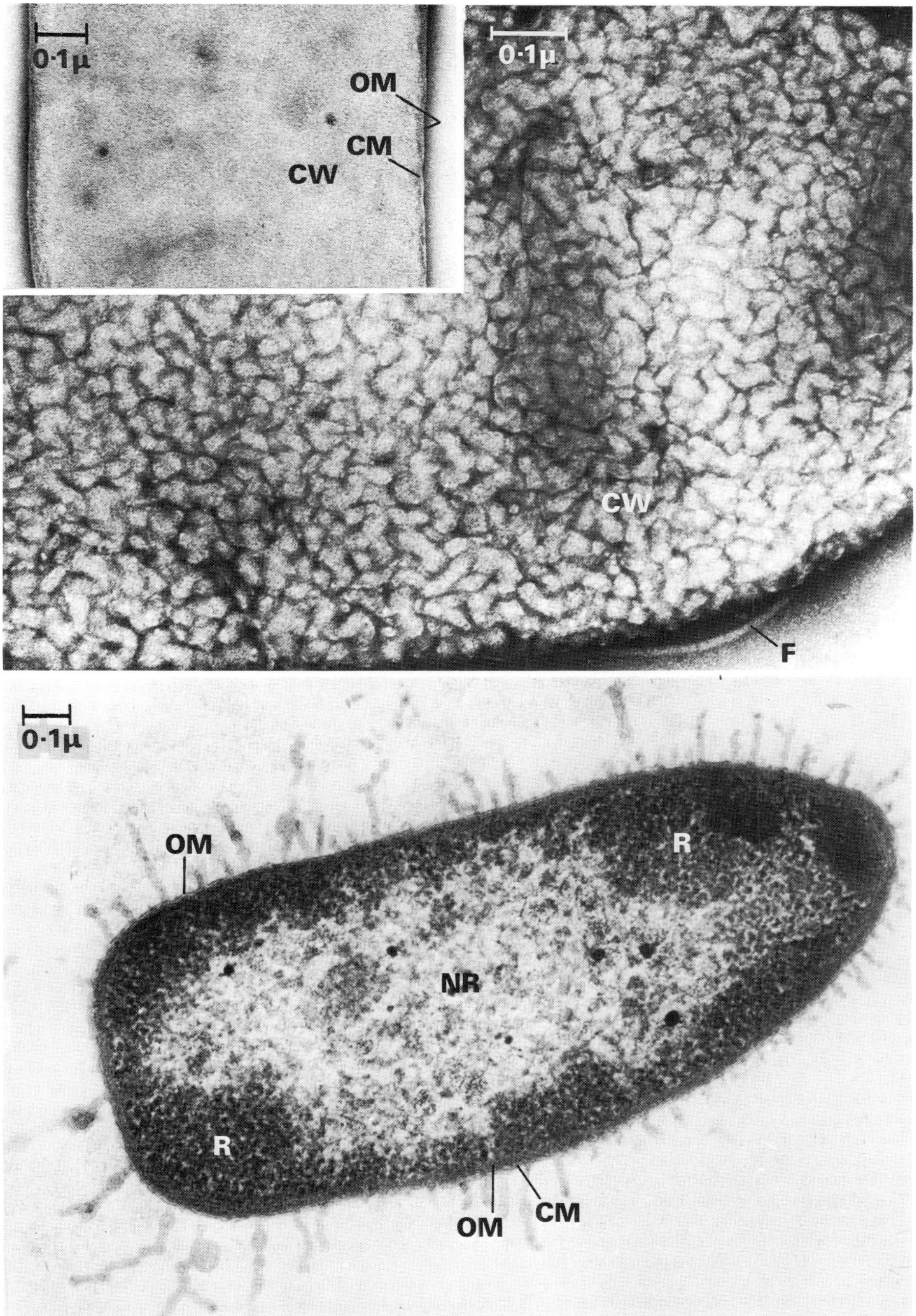

(c)

(a)

(b)

(c)

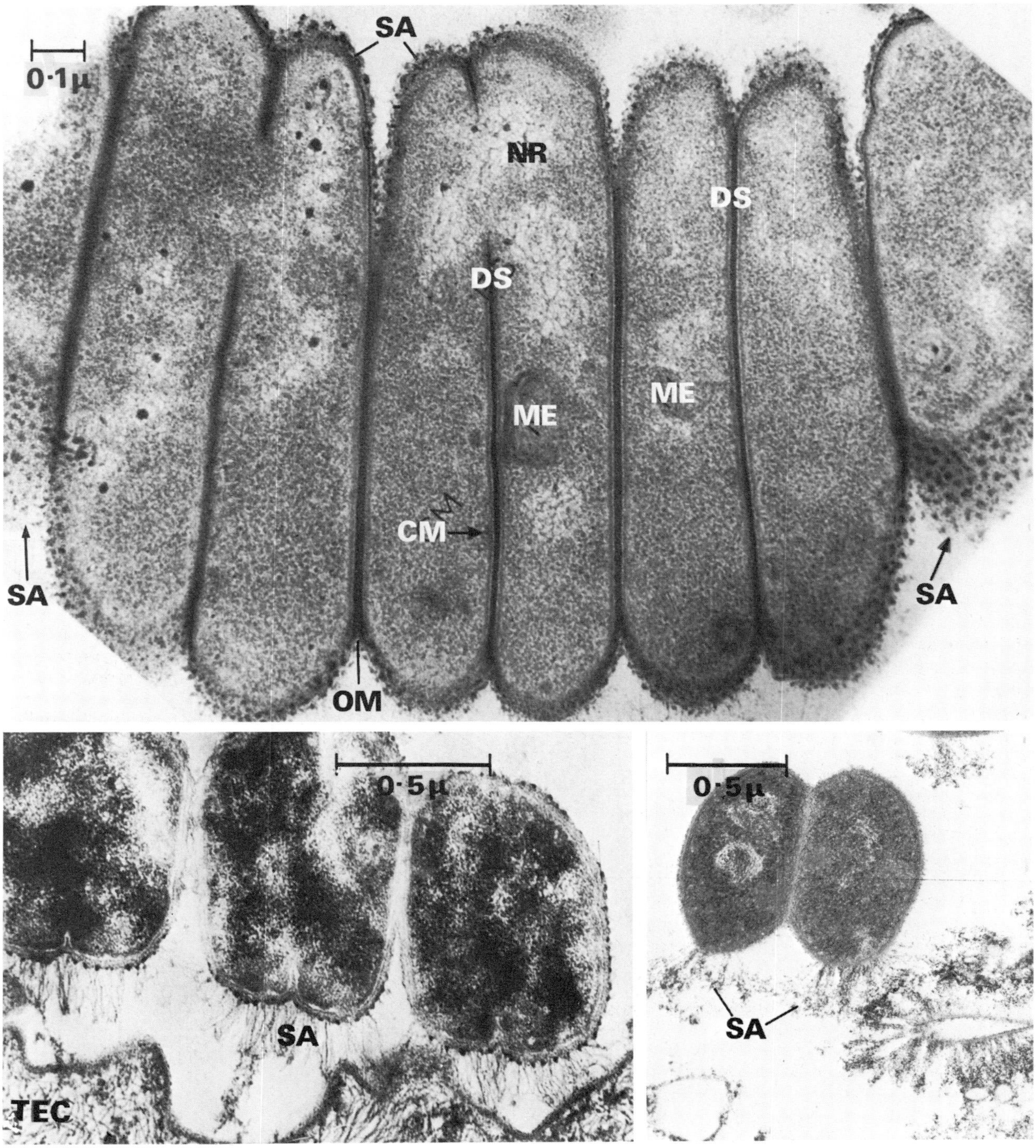

Figure 4.34. *Alysiella* species.
(a), (b) Thin sections of *A. filiformis*, and (c) human oral isolate of similar morphology. In (a), a longitudinal section of dividing cells, the multilayered cell wall encloses the unit cell membrane (CM). Slime appendages (SA) consisting of thin fibrils characteristic of this genus exist at the cell poles, and may act in attachment of the bacterium to host tissue, as seen in (b), where there is a very close if not actual contact between the slime appendages and an oral (tongue) epithelial cell (TEC). NR: nuclear region, DS: division septum, ME: mesosome, OM: outer cell membrane.

straight or curved rods. Humans are the primary hosts for *V. cholerae*, while *V. cholerae eltor* is the major etiological agent of Asiatic cholera.

Morphologically, *V. cholerae* (Figure 4.35(a),(b)),[114] is a small comma-shaped rod, similar to the other gram-negative bacteria. The peptidoglycan layer is not clearly visible in the micrographs presented. The electron-dense layers of the outer and cell membrane appear to be slightly rough or granular in appearance, the granules being 4–5 nm in diameter. In *Vibrio* species (Figure 4.35(g)) a polar flagellum is clearly seen.

The outer membrane of the cell wall has been implicated in cholera toxin production, specifically the toxin being involved in vesiculation or alteration of host-membrane permeability.[115] Negative-stained preparations of *V. cholerae* in the process of maximum toxin production possess an increased amount of cell-associated surface material (Figure 4.35(c)). Of interest was the fact that toxin production was also associated with the appearance of extracellular lipopolysaccharide.

In addition to this cell-associated surface material, *V. cholerae, V. cholerae eltor*, and other *Vibrio* species possess fimbriae or pili associated with their cell surface (Figure 4.35(d)).[203] Approximately 10 percent of a cell population of *V. cholerae* contained these short, curved fimbriae, each of which is approximately 7·2 nm in diameter. The fibrils have been observed to reach lengths of 1 μm. The number of fimbriae on the surface of *V. cholerae* is small (approximately 9 fimbriae per cell) while *V. cholerae eltor* may contain as many as 50 fimbriae per cell. Approximately 50 percent of the *V. cholerae eltor* contain long tapered fimbriae, sometimes reaching lengths of 7–9 μm. *Vibrio* species (Figure 4.35(g)) on the other hand contain far fewer fimbriae, between 1 and 3 fimbriae per cell.

In addition to containing pili, the surface of several of the *Vibrio* species are covered with a layer of branching strands, each strand being approximately 6–8 nm in diameter (Figure 4.35(e)). These strands are randomly distributed over the cell surface, and could represent a slime or capsule layer.

In order to cause disease, *V. cholerae* must enter the lumen of the small bowel, a process which usually occurs by the consumption of tainted water. The ingested *V. cholerae* produce a potent exo- or enterotoxin which acts on the lining of the bowel resulting in the loss of large amounts of water and essential salts. Unlike *V. cholerae, V. cholerae eltor* produces a hemolysin which lyses red blood cells of lower animals, especially sheep and goats. *Vibrio parahaemolyticus*, a marine vibrio, is the major cause of the gastric food poisoning gastroenteritis or "summer diarrhoea" and, owing to its marine habitat, is the major cause of seafood poisoning. *Vibrio parahaemolyticus* has been shown to be capable of swarming over a solid surface, as well as possessing large numbers of flagella (Figure 4.35(f),(h)).[204, 218]

Recent studies[148] have revealed that after invasion of the intestinal tract by *V. cholerae*, there develops a very close association of the vibrios with the intestinal villi (Figure 4.35(e),(j)). This association or adherence of *V. cholerae* to the intestinal villi may be of considerable importance in disease production, with those vibrios free in the lumen not involved in toxin or disease production. So, for example, when Nelson *et al.*[148] examined *V. cholerae* adherence and

Figure 4.35. Representative *Vibrio* species.
(a), (b), (c) *Vibrio cholerae* 569B, and (d) *V. eltor* NCTC5395. The cell wall of these gram-negative cells is relatively smooth and separated from the underlying cell membrane (CM) by an electron-transparent space. In (b), some of the cells have plasmolyzed, clearly revealing the layers of the cell wall. Cells in maximum toxin production (c) when examined by negative-staining exhibit large patches of extracellular material, morphologically similar to out membrane (OM) fragments. The vibrios in (d) are also covered with thin pili (P) or fimbriae as well as larger flagella (F). NR: nuclear region, PPS: periplasmic space.

(a)

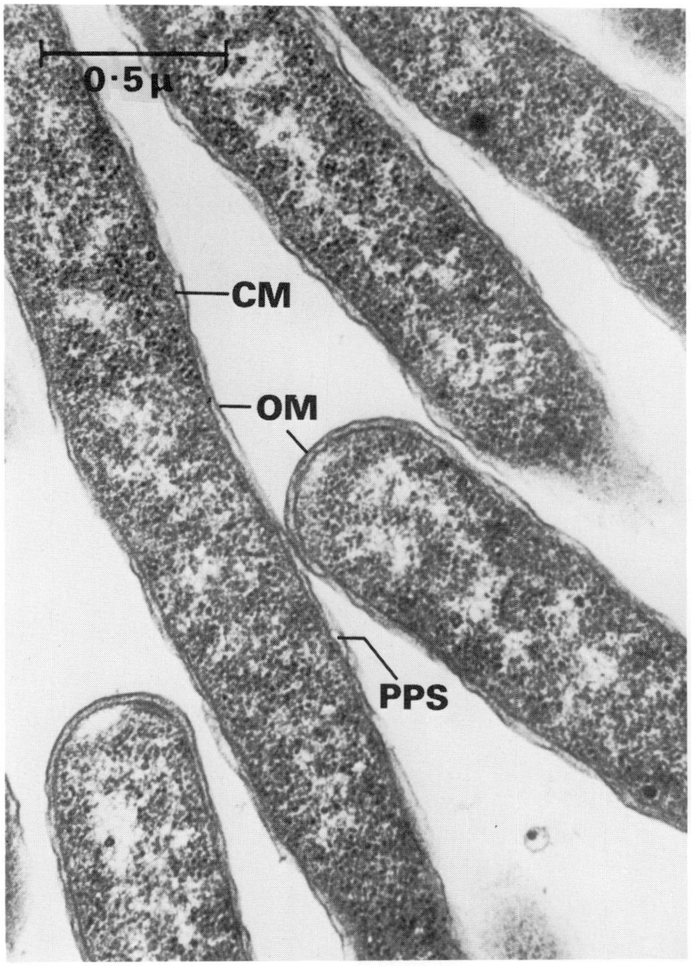

(b)

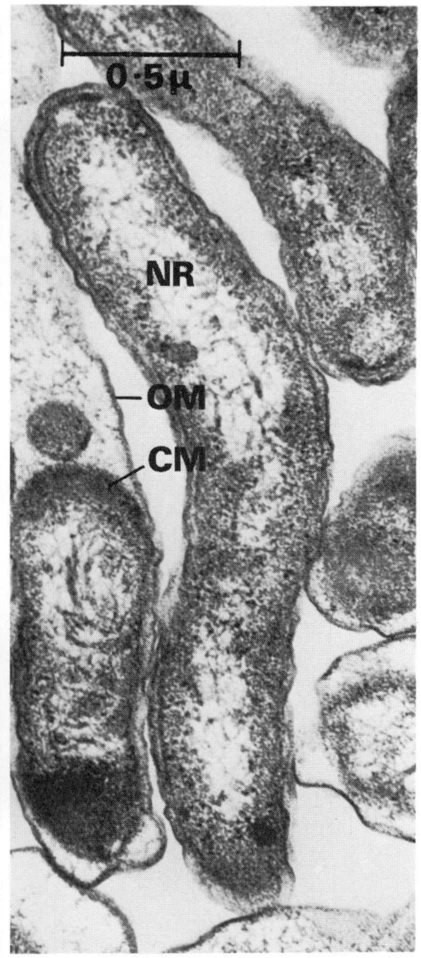

(c)

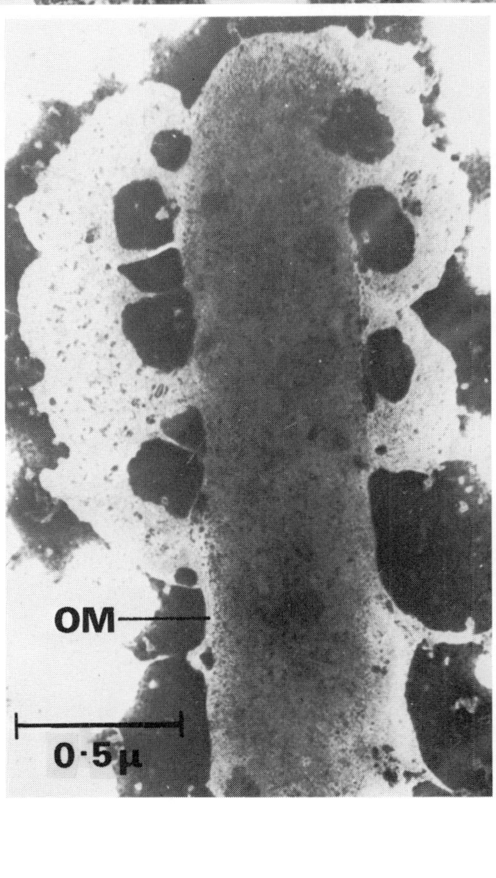

(d)

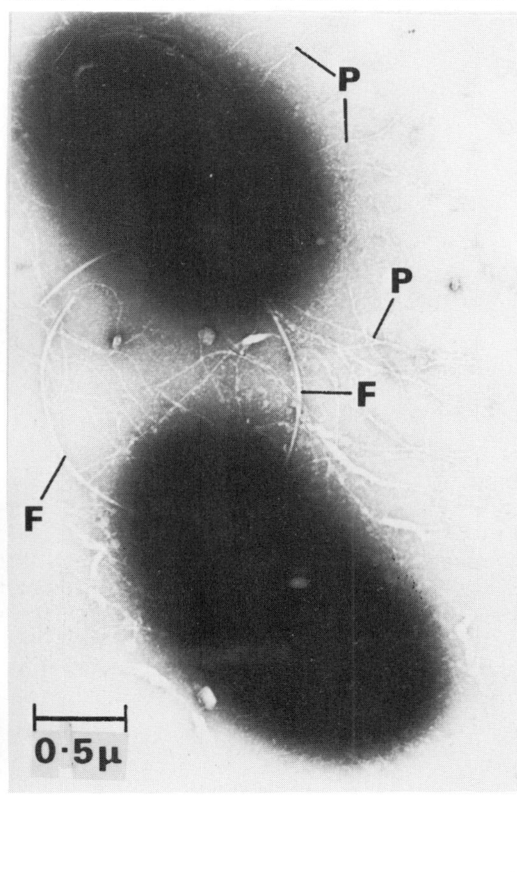

Figure 4.35. Representative *Vibrio* species.
(e) *Vibrio* sp. 4716 is enmeshed in a dense network of fibrous strands which are in close apposition to the bacterial cell surface. (f) *Vibrio parahaemolyticus* has numerous flagella (F) emerging from the surface, while (g) *V. cholerae* NCTC 8021 when shadow-cast with Au–Pd are seen to possess one long, polar flagellum (F). Short, thin fimbriae or pili (P) can also be seen in (g). CM: cell membrane, OM: outer cell membrane.

(e)

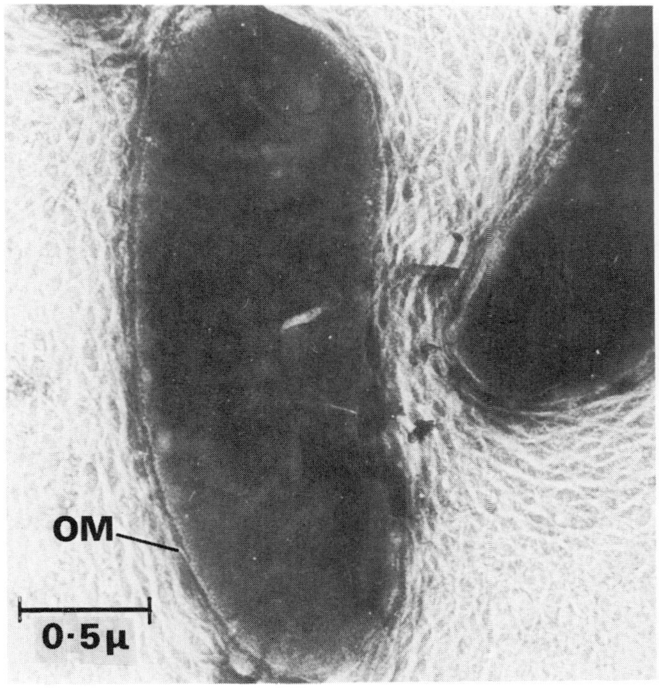

(f)

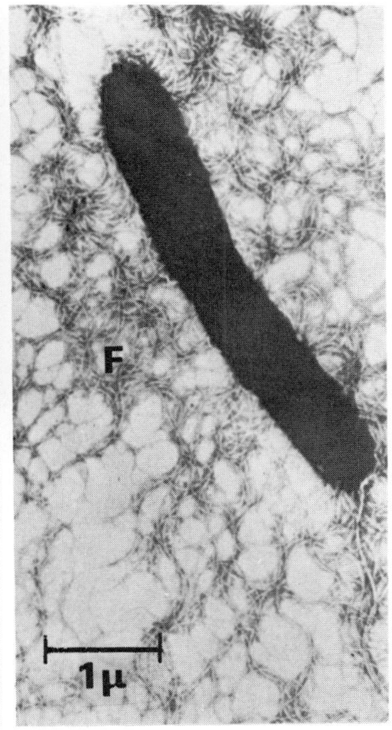

(g)

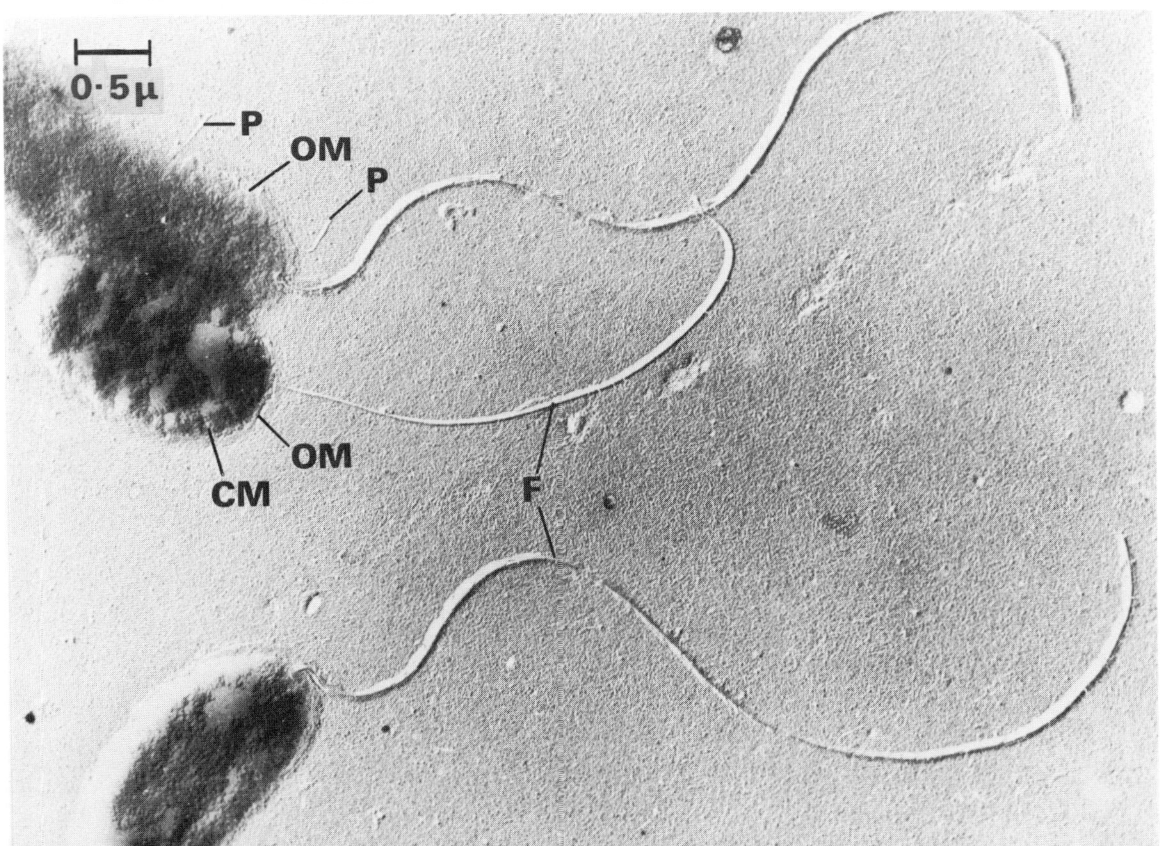

Figure 4.35. Representative *Vibrio* species.
(h) *Vibrio parahaemolyticus* showing numerous flagella (F) emerging from the surface.
CyT: cytoplasm, OM: outer cell membrane.

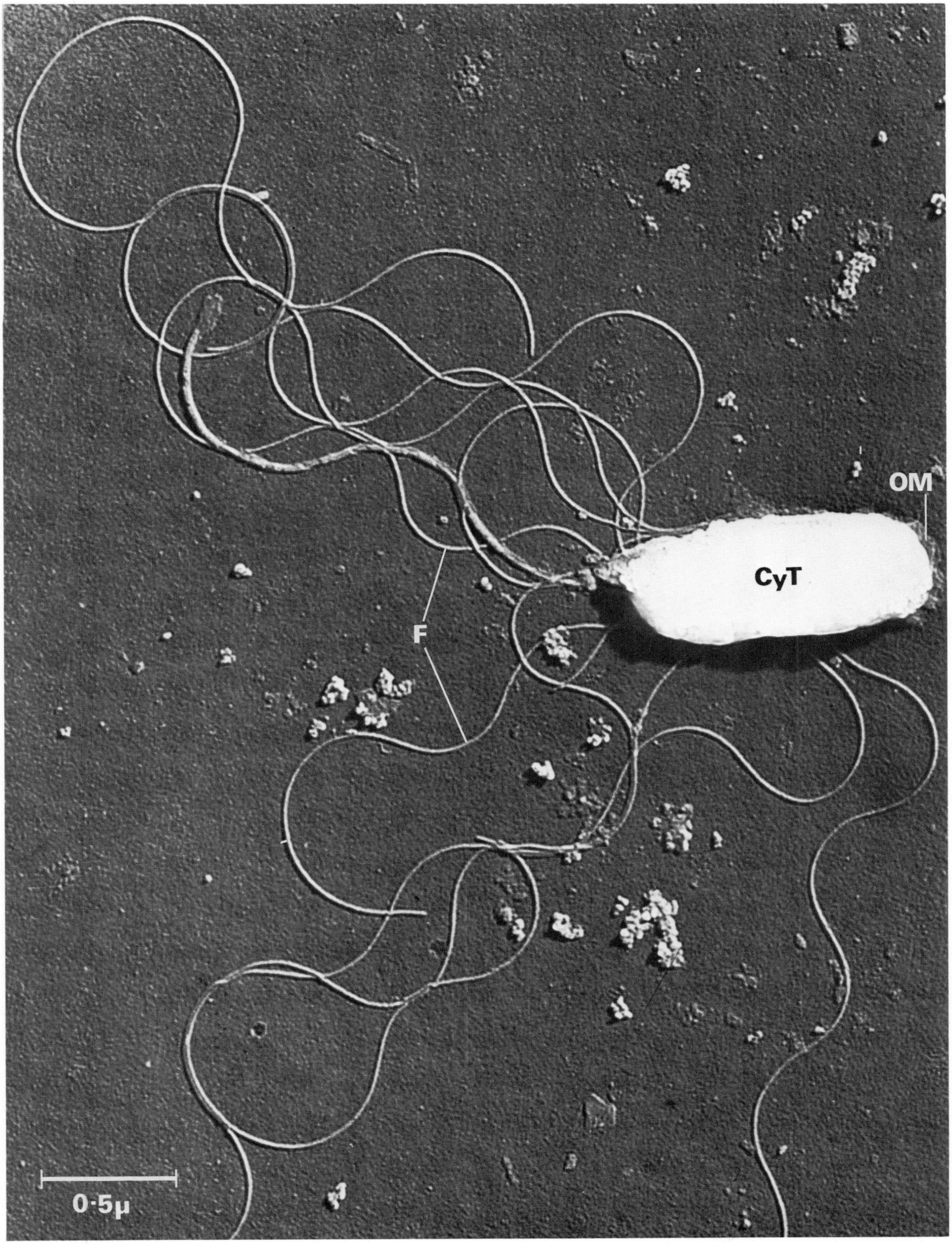

(h)

Figure 4.35. Representative *Vibrio* species.
(i), (j) The association of *V. eltor* 3083 with the brush border of a rabbit villus. The outer membrane (OM) of the bacterial cell wall is in very close association or contact with the surface of microvilli (MV) (arrows). Note both longitudinal and transverse attachment of the vibrio to the cell surface. CM: cell membrame, NR: nuclear regions, PPS: periplasmic space.

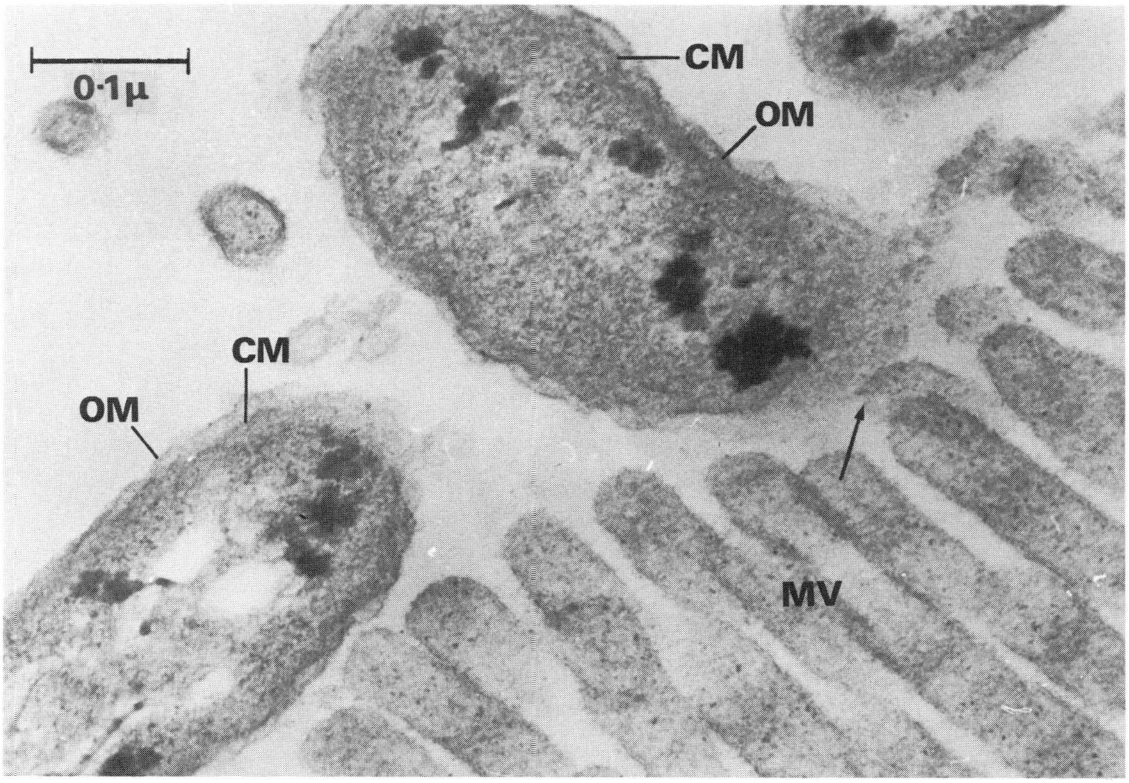

(i)

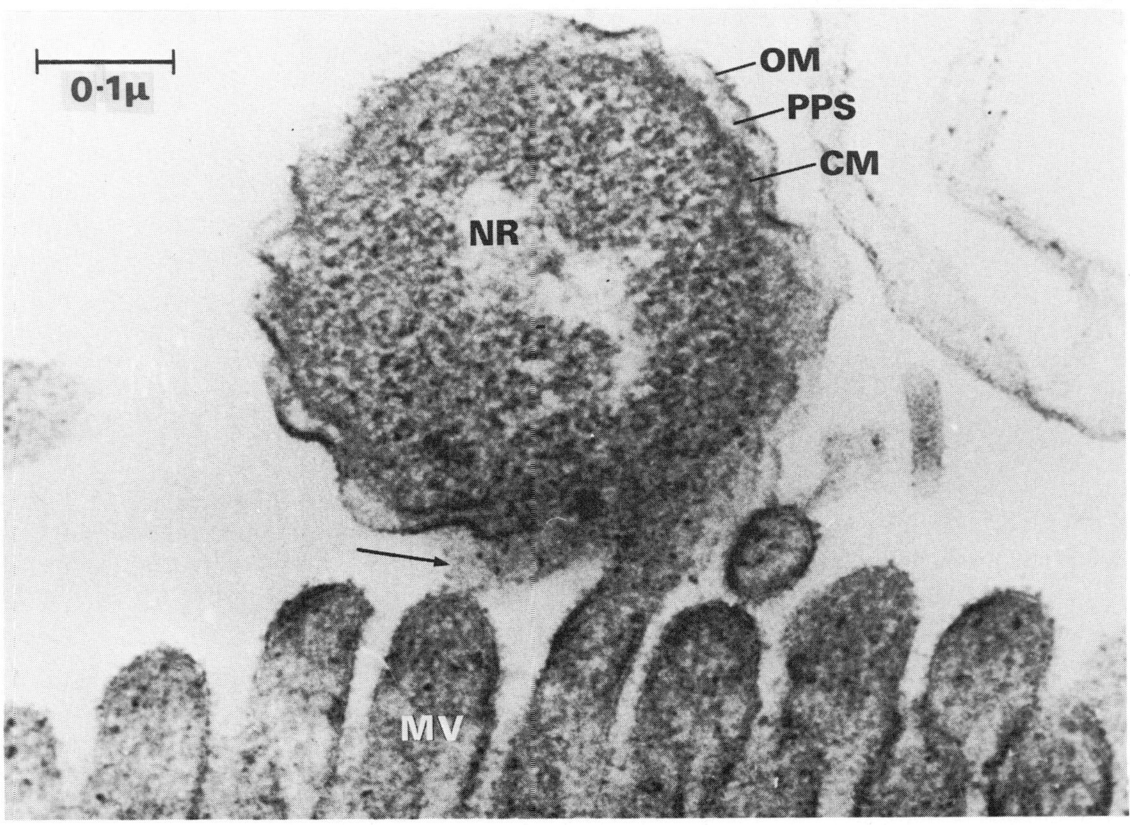

(j)

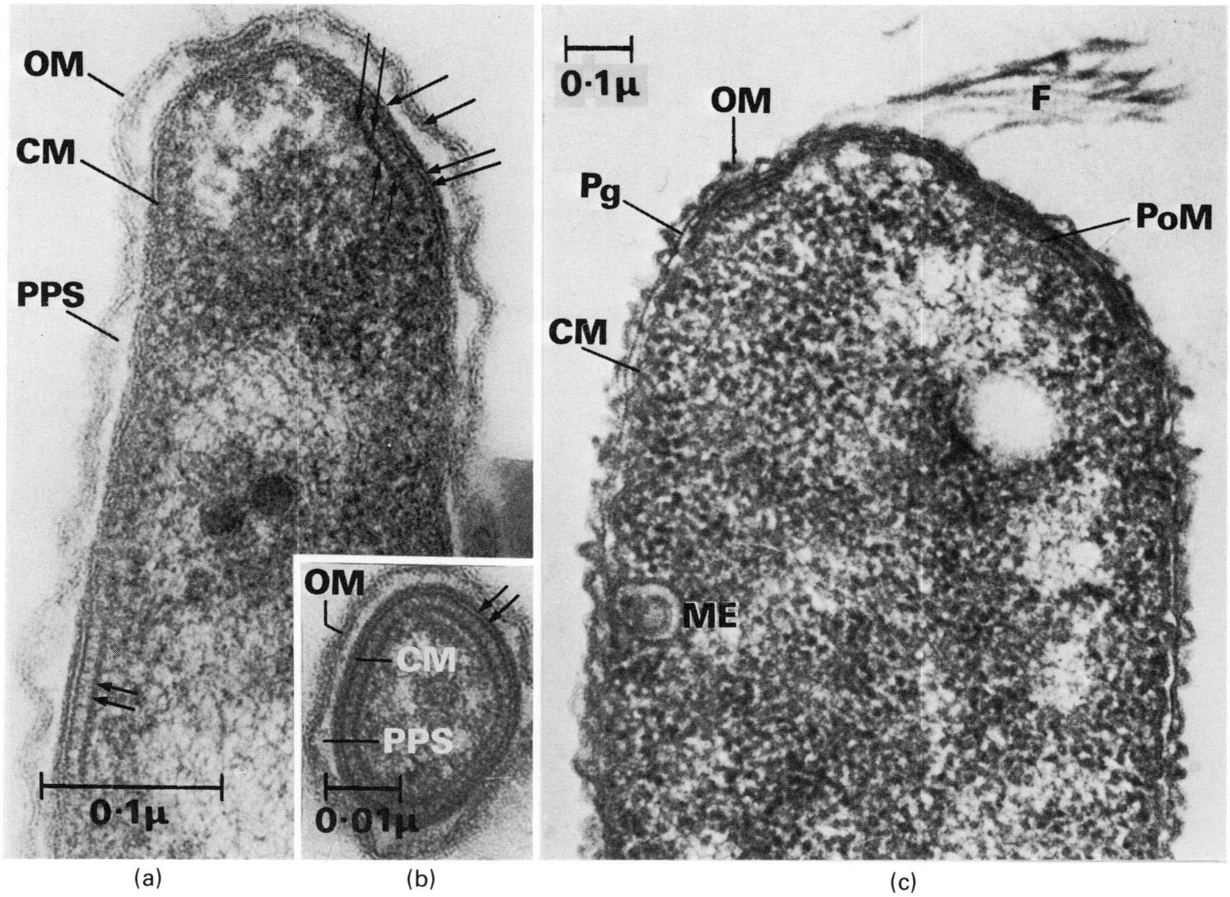

Figure 4.36. *Vibrio–Spirillum.*
Thin sections of chemically fixed cells of (a), (b) *Vibrio fetus* (*Campylobacter*) and (c) *Spirillum serpens.* In *V. fetus* a wide electron-transparent (periplasmic) space (PPS) separates the outer (OM) and cell membrane (CM). Surfaces (outer and inner) of these membranes are covered with electron-dense material (arrows). At a polar location and internal to the cell membrane is a ribbed layer (double arrow), seen more clearly in (b). *Spirillum serpens* (c) possesses the classical wrinkled outer membrane, typical of gram-negative cells, an intermediate peptidoglycan layer (Pg), and inner cell membrane (CM). A tuft of flagella (F) emerges from the cell pole. Note the polar membrane (PoM) in (c) and its morphological similarity with that observed in *V. fetus*. ME: mesosome.

colonization in the rabbit intestine (which mimicked human gut infection with *V. cholerae*), they found that early in infection (10–20 min postinfection) there was virtually no attachment of vibrios to the microvilli. At 1 hour postinfection, vibrios were found distributed with a fair degree of regularity along the intestinal epithelium. Of interest in this study was the observation that one villus had several hundred organisms attached, while an adjacent villus showed no

---

Figure 4.36. *Vibrio–Spirillum.*
By negative staining and shadowcasting (d), (f) *V. fetus* (*Campylobacter*), (e) *Oceanospirillum atlanticum*, and (g) *S. serpens* are spiral shaped, and where flagella (F) are present, they exist in a polar arrangement.

(d)

(e)

(f)

(g)

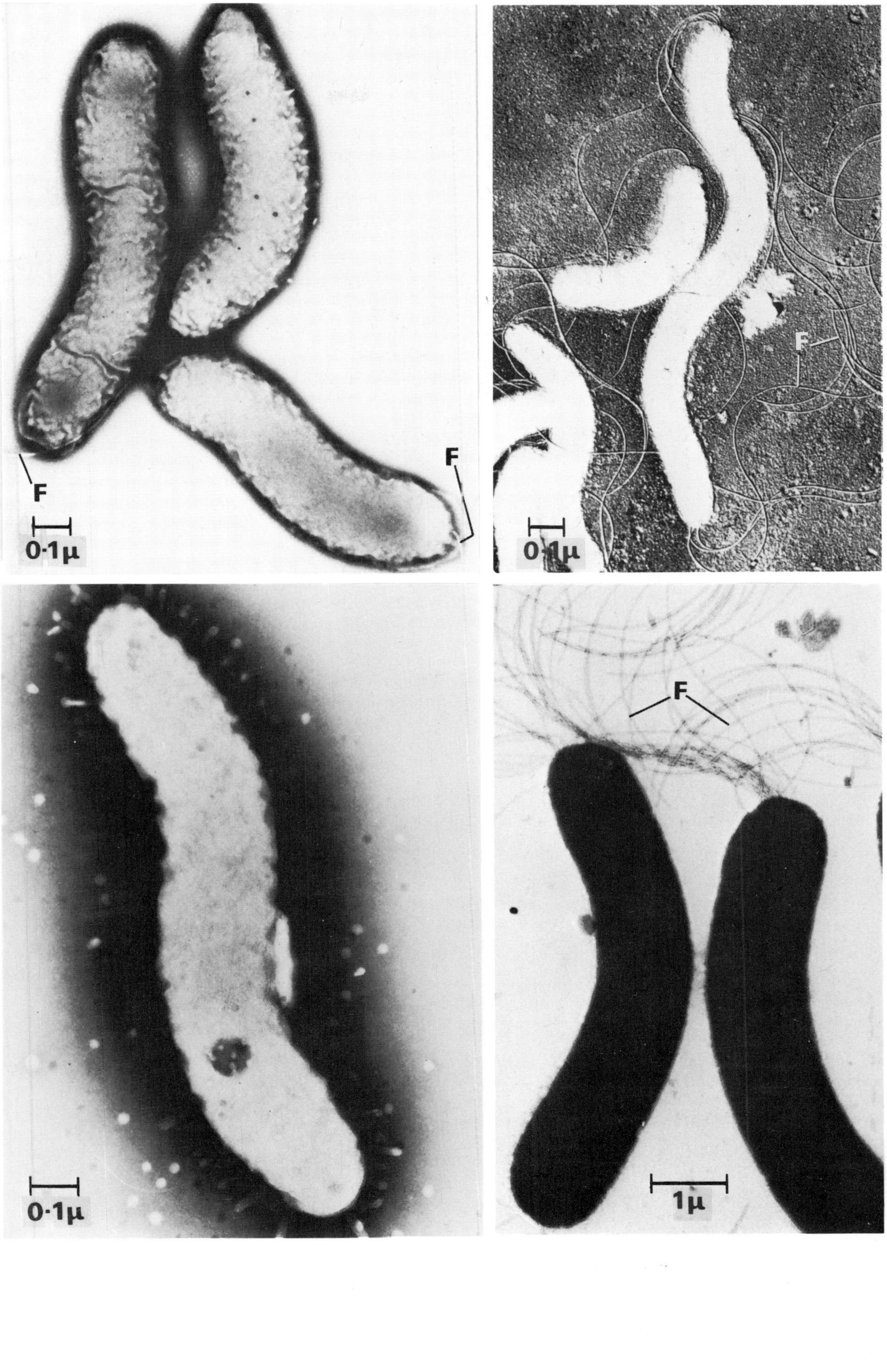

attachment. The vibrios tend to show a predilection for attachment along transverse furrows or grooves of the villi, with the bacteria usually attached end-on with the flagellated pole extending out into the intestinal lumen. By 7 hours postinfection, the vibrios started to disappear from the intestinal surface.

The cells were attached directly to the brush border of the intestine (Figure 4.35(i),(j)); interestingly, they never penetrated the microvilli to its base. Very late in infection, the cytology of the microvilli changed to reveal a fragmentation, elongation, and vesiculization.

Campylobacter

A poorly studied genus, *Campylobacter* is closely related to members of the genus *Spirillum*, the latter represented here by *S. serpens* (Figure 4.36(c),(g)) and *Oceanospirillum atlanticum* (Figure 4.36(e)). *Campylobacter fetus*, an animal parasite which may play some role in cattle abortion, consists of curved rods (Figure 4.36(d),(f))[168] approximately 0·3–0·8 × 0·5–5 nm, with a single polar flagellum. In Figure 4.36(f) the cell surface is wrinkled, with numerous surface blebs. The blebs may represent lipopolysaccharide (endotoxin) which is being shed from the cell. A loose-fitting cell wall (Figure 4.36(a)) encloses the asymmetrically electron-dense membrane. The outer layer (closest to the cell wall) of the cell membrane is almost twice as thick as the inner layer. This extra thickness may represent a tightly adhering peptidoglycan layer. A series of ribs are seen below the cell membrane (Figure 4.36(a),(b)) which may be continuous with an inner membrane layer.

Spirillum serpens

This is also characteristically gram-negative in construction (Figure 4.36(c)). Characteristic of *S. serpens* is the presence of a polar membrane (Figure 4.36(c)) and a tuft of polar flagella (Figure 4.36(c),(g)) which can occur in both mono- and bipolar arrangements.

4.3.13. *Pseudomonas*

These gram-negative, actively motile, oxidase-positive, short rods are morphologically similar to the other gram-negative rods discussed in this chapter. The majority of pseudomonads are saprophytes found in water, soil, and sewage.

Pseudomonads of medical importance include *Pseudomonas aeruginosa,*

Figure 4.37. *Pseudomonas* species.
(a), (c) *P. aeruginosa*, (b) *P. cepacia*, and (d) *P. multivorans* possess all of the morphological characteristics of gram-negative bacteria – a wrinkled outer membrane (OM), intermediate thin peptidoglycan (Pg), and inner smooth cell membrane (CM). (c) Uranyl acetate staining and (d) shadowcasting reveal thin fimbriae or pili emerging from the cell surface. Note in (d) the two size classes of fimbriae: long (arrow) and short (double arrow). DS: division septum, NR: nuclear region.

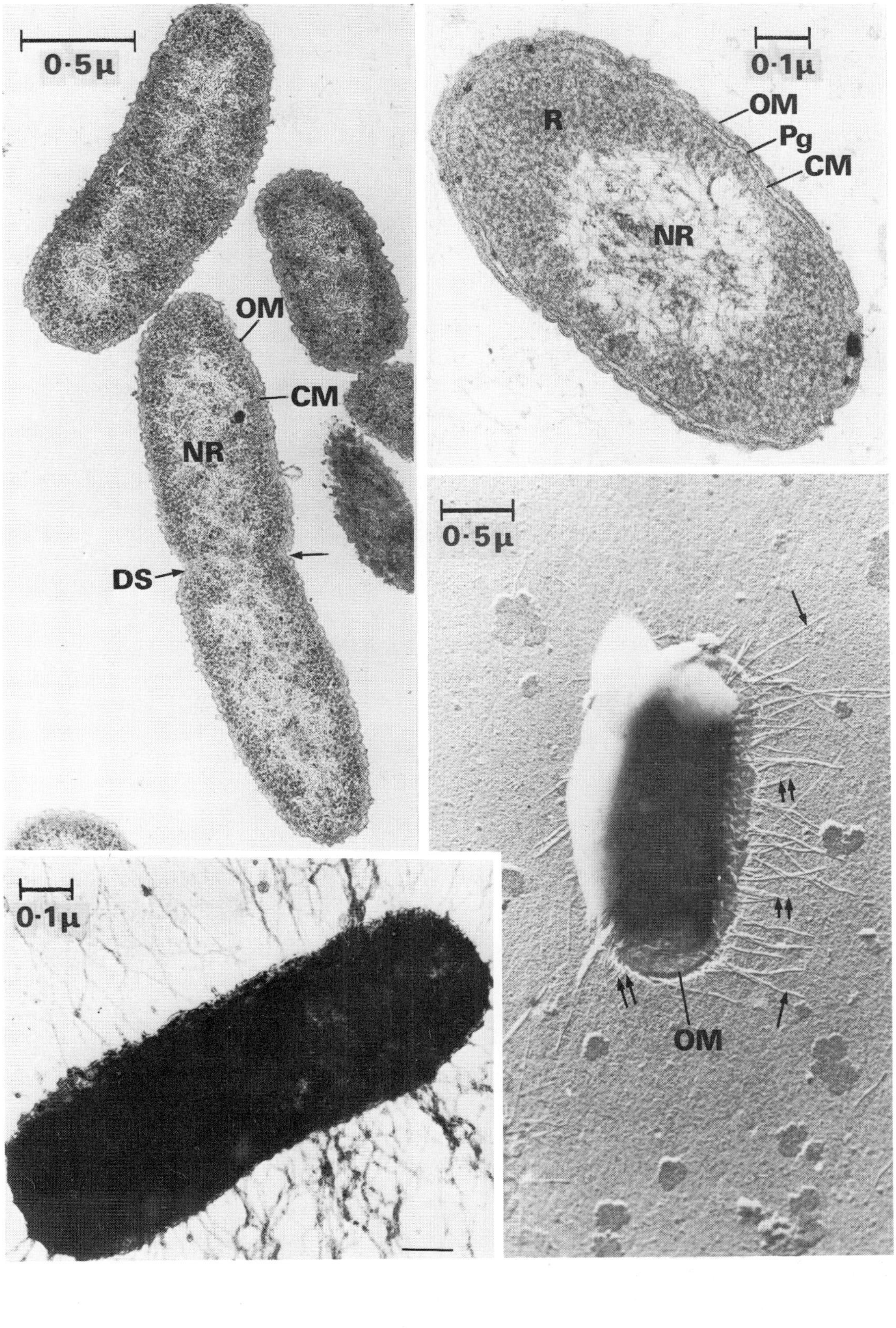

(a) (b) (c) (d)

P. pseudomallei, P. mallei, and *P. cepacia*. *P. aeruginosa*, the organism of blue-pus, is widely distributed in soil and water and is often found as a normal inhabitant of the skin and intestinal tract.[20, 183]

Characteristic of *P. aeruginosa* is its participation as a secondary or opportunistic invader, being important in otitis media, abscesses, wound and sinus infections, as well as in bronchopneumonia.[183] It is also important in gram-negative shock, and sepsis complicating severe burns.

Pseudomonas aeruginosa infections are usually characterized by tissue necrosis, the result of active proteolytic enzyme production by the bacterium, especially the production of extracellular collagenases, lecithinases, lipases, and hemolysins.

Pseudomonas pseudomallei, the causative agent of meliodosis, is responsible for septicemia with widespread lesions over the body. The disease is endemic in Southeast Asia, known as the "Vietnamese time bomb" because of its explosive and unexpected appearance.

Pseudomonas mallei, the causative organism of Glanders (Farcy), is a disease of horses, mules, and donkeys. The organism is readily transmitted to humans through wounds, scratches, or abrasions of the skin, and produces ulcerating, tubercle-like nodules in the lungs, superficial lymph nodes, and mucous membranes. Involvement of the lymph nodes results in Glanders, while involvement of the mucous membranes produces Farcy.

Ultrastructurally, the cell wall of the pseudomonads is trilaminar (Figure 4.37(a),(b)), composed of an outer and an inner membrane. A distinct and slightly electron-dense periplasmic space separates the inner and outer membranes (Figure 4.37(b)).[94] The cell membrane limits the cytoplasm and its enclosed electron-dense ribosomes and nuclear region. *Pseudomonas aeruginosa* (Figure 4.37(c),(e)) contains an external fibrillar slime layer, approximately 13 nm in diameter and of variable length, sometimes approaching 3 μm; of interest was the observation that when these strains were passed in animals they formed a distinct capsule as well as long, thin pili (Figure 4.37(d)). These extracellular components are lost during *in vitro* culture on artificial media. Both outer and inner membranes are asymmetric and approximately 5 nm wide.

In some strains of *P. aeruginosa* the cytoplasm contains numerous mesosomes, the appearance of which is dependent upon the method of fixation for electron microscopy. For example, Hoffman *et al.*[95] found that by standard R–K fixation mesosomes were not visible, while glutaraldehyde, osmium and phosphotungstic acid fixation resulted in the visualization of mesosomes. Those seen in *P. aeruginosa* strain Cl-3 consisted of trilaminar membranes,

Figure 4.37. *Pseudomonas* species.
(e), (f), (g) *Ps. aeruginosa* showing the morphological characteristics described for (a), (b), (c), (d). In addition to surface fimbriae, large outer-membrane blebs can be seen (f). In (e) a thick, interconnected, extracellular meshed layer radiates from the cell surface after critical point drying. In (g) uranyl acetate staining reveals rhapidosomes (Rhp) within the cytoplasmic region. OM: outer cell membrane.

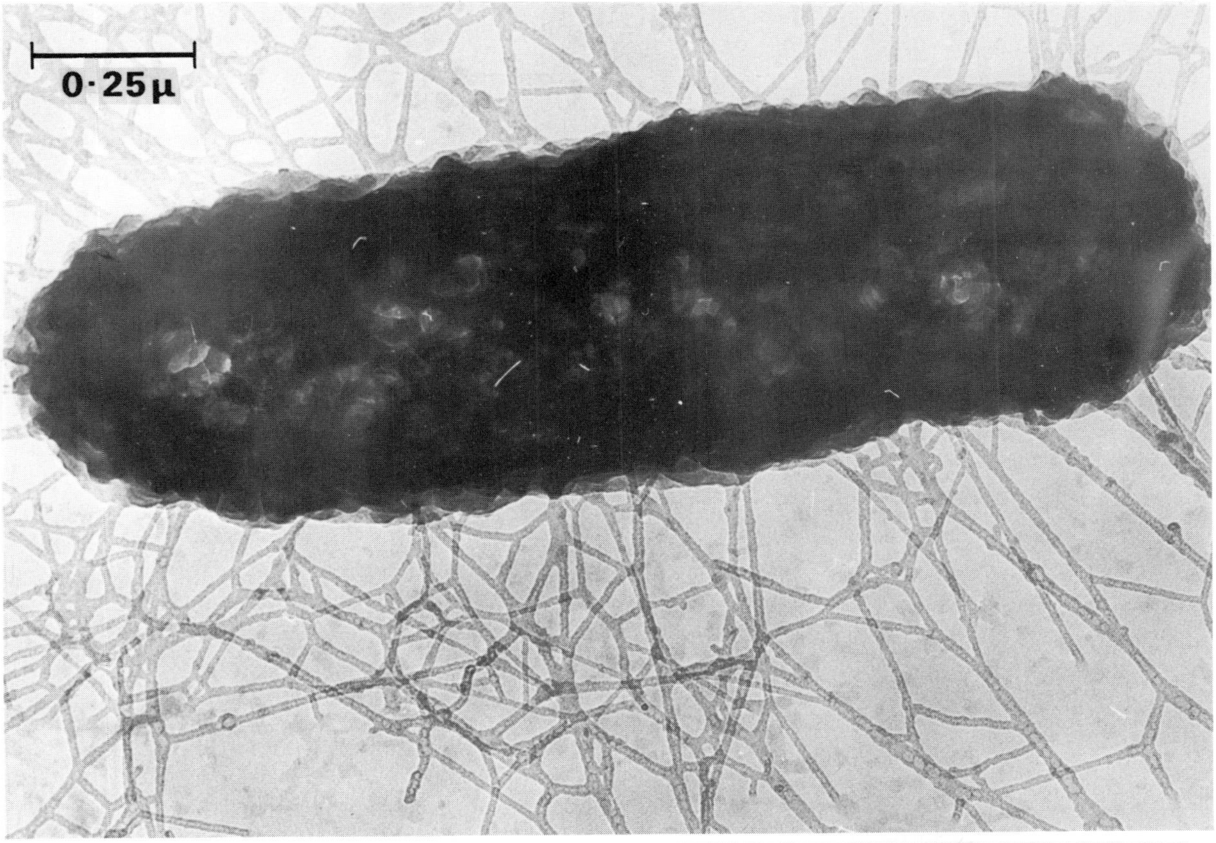

(e)

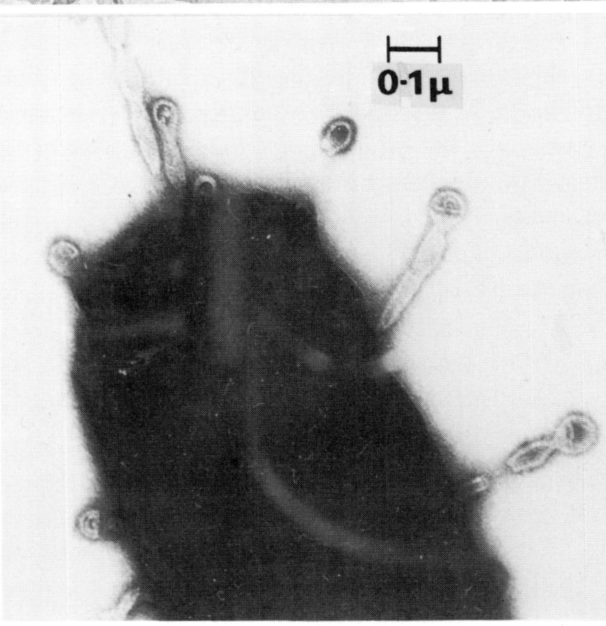

(f)

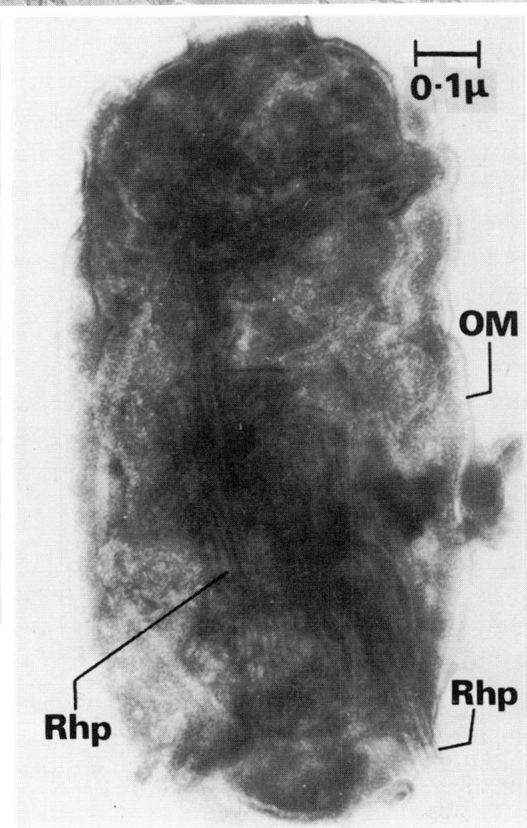

(g)

approximatley 5 nm thick. In addition to mesosomes, some strains of *P. aeruginosa* contain within their cytoplasmic region fimbriae or rhapidosomes (Figure 4.37(g)).[8] These structures are similar to pyocin polysheaths,[88, 125] and their numbers appear to be a function of culture age; older cultures having approximately 17 rhapidosomes per cell, while young, rapidly growing exponential cultures have 1–2 per cell. The mean width of the individual rhapidosome is 24 nm. Older cultures of *P. aeruginosa* also contain tubules or membrane extensions of the outer membrane of the cell wall (Figure 4.37(f)). These tubules can reach lengths of 135 nm, and may represent lipopolysaccharide or lipoprotein in the process of being lost from the cell wall.

4.3.14. Spirochetes

These gram-negative, flexous, thin bacteria are found associated with a wide variety of hosts including termites, clams, shrimps, pigs, and humans, as well as living free in the environment. The order, Spirochaetales, includes five genera: *Treponema*, the causative agent of such diseases as syphilis and yaws (*T. pallidum, T. pertenue*); *Borrelia*, of which *B. hermsii* and *B. recurrentis* are associated with relapsing fevers; and *Leptospira* (*L. interrogans*), the causative agent of infectious jaundice or Weil's disease. The majority of the nonpathogenic spirochetes are included in the fourth genus, *Spirochaeta* (e.g., *S. aurantia, S. stenostrepta*) while spirochetes found associated with sea molluscs are placed in the genus *Cristispira.*

Treponema pallidum is usually found in very large numbers in chancres and mucous patches in the human host, seldom being found free in the environment. Similar to *T. pallidum* in morphology is *T. refringens*, but this treponeme occurs as a nonpathogenic inhabitant of the normal flora of the human male and female genitalia, as well as being associated with *T. pallidum* in syphilitic lesions. Other treponemes often found associated with the human host include *T. buccale, T. denticola*, and *T. microdentium*, occurring as nonpathogenic saprophytes in the oral cavity. *Treponema pertenue*, isolated from yaws-associated skin lesions is morphologically identical to *T. pallidum.*

The various members of the genus *Borrelia* are transmitted by different

Figure 4.38. Spirochetes.
Representative free-living and host-associated spirochetes. (a), (b), (c), (d) Phase contrast photomicrographs of representative living spirochetes. In (a) the regularly coiled *Spirochaeta stenostrepta* is seen, (b) the small *Treponema denticola* with its tight coiling. *Borrelia anserina*, isolated from chicken blood is seen in (d). The cell coiling in *Borrelia* species tends towards a slight irregularity with pointed ends. *Leptospira interrogans icterohaemorrhagiae* with its characteristic hooked ends is seen in (c). Negative-stains of (e) *B. recurrentis*, (f) *T. pallidum* Nichols, (g) *T. microdentium* (i) *T. calligyrum* (j) *B. hermsii*, (k) *T. pertenue* Gauthier, (l) *L. pomona*, and (n) *S aurantia* show the basic morphology of spirochetes: an outer sheath (OS), cell wall–cell membrane complex (CWCM), and periplasmic fibrils (PF). In (l) *L. pomona* the periplasmic fibril is sheathed (arrow). Thin sections of chemically fixed *T. pallidum* Nichols and *B. recurrentis* are seen in (h) and (m), respectively. CM: cell membrane, OM: outercell membrane, PC: protoplasmic cylinder, ME: mesosome.

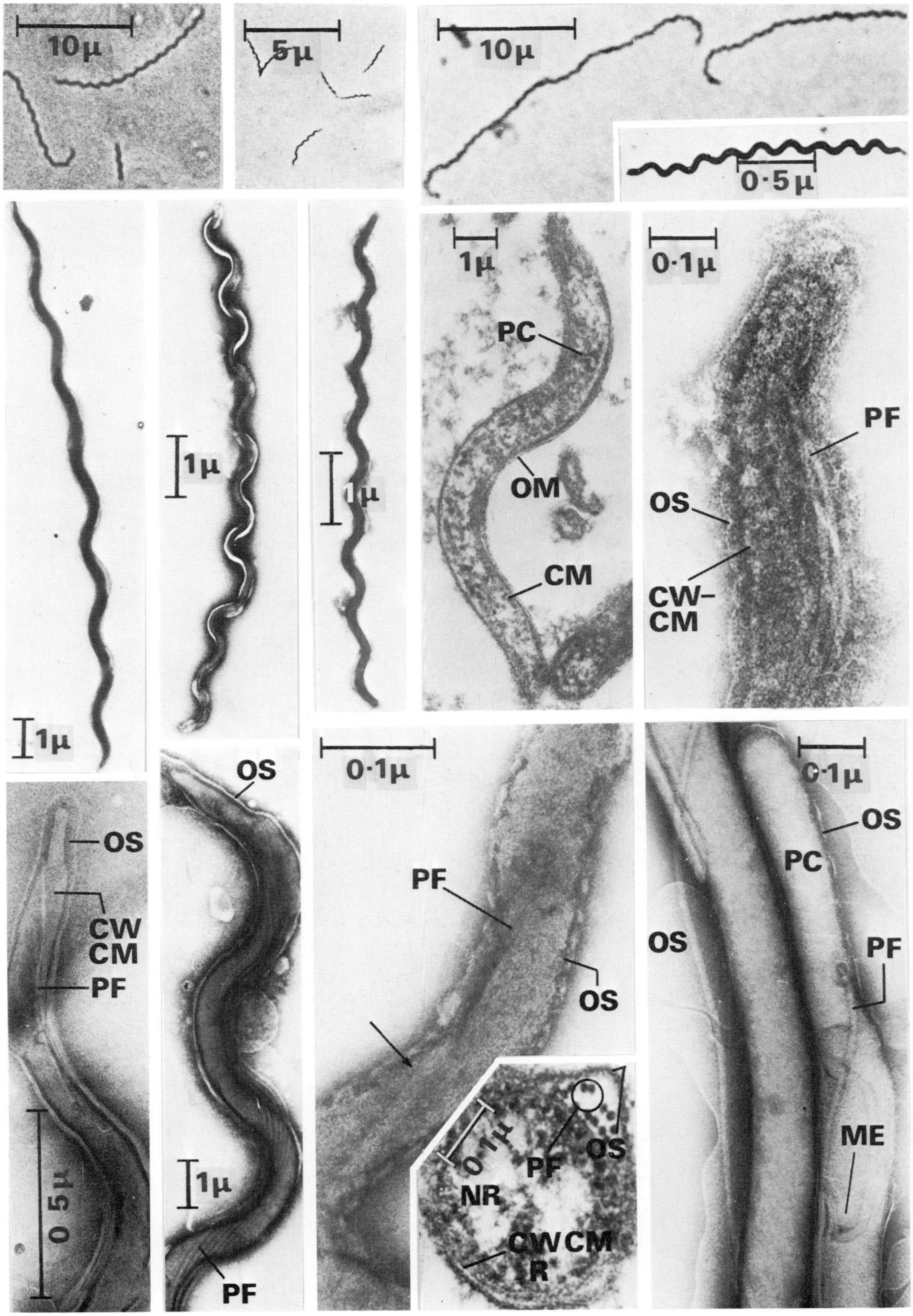

(a)
(b)
(c)
(d)
(e)
(f)
(g)
(h)
(i)
(j)
(k)
(l)
(m)
(n)
10μ
5μ
10μ
0·5μ
1μ
1μ
1μ
1μ
PC
OM
CM
0·1μ
PF
OS
CW–CM
OS
CW
CM
PF
0 5μ
OS
1μ
PF
0·1μ
PF
OS
0·1μ
PF
OS
NR
CW CM
R
0·1μ
OS
PC
OS
PF
ME

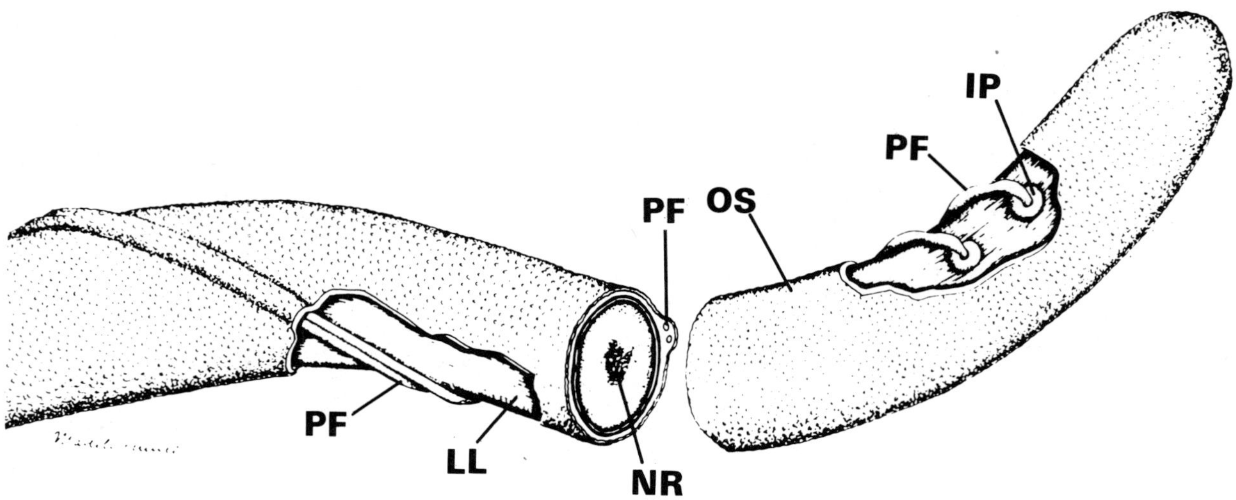

Figure 4.38. Spirochetes.
(o) Diagramatic representation of a typical spirochete as interpreted from electron micrographs. An outer sheath (OS) envelopes the cell with the periplasmic fibrils (PF) between the outer sheath and the layers of the protoplasmic cylinder (depicted here as a lipoprotein layer (LL), as found in *Spirochaeta stenostrepta*). The periplasmic fibrils insert into the cylinder by way of an insertion pore (IP). NR: nuclear region.

arthropod vectors; those associated with relapsing fever of humans (*B. recurrentis*) are louse borne, being transmitted by *Pediculous humanus* subspecies *humanus*, while rodent-human transmission occurs through the tick vector *Ornithodorus*.

Leptospirosis, an acute febrile disease caused by *L. interrogans* (*L. icterohemorrhagiae*) is also referred to as swamp fever, swineheads disease, infectious jaundice, and Weil's disease. These spirochetes are found in both domestic and wild animals, and the disease is usually zoonotic, occurring through accidental contact with infected animals or soil contaminated with urine from infected animals.

Two interesting fusospirochaetal infections, Vincent's angina and acute necrotizing ulcerative gingivitis, are of microbial origin and found in the oral cavity of humans. The lesions are produced as grayish-white pseudomembranes in the throat and mouth, and may be the result of an association of two bacteria, a

---

Figure 4.38. Spirochetes.
(p), (q), (r), (s), Host-associated spirochetes. Thin-sections of *T. pallidum* in rabbit tissue and endoperineurium and epineurium, and in a peripheral nerve fiber, are seen in (p), (q). The treponemes (Tp) have displaced a portion of the collagen fibers (C) and lie in an electron-transparent space. In the nerve fiber cross section (q), at least five treponemes are seen; note the massive destruction of the collagen fibers (C). *Treponema pallidum* in a primary chancre of the human penis is seen in (r). The treponemes (Tp) have invaded intercellular spaces with some resultant collagen destruction; similarly in a biopsy of secondary syphilis (s) the treponemes (Tp) are in the intercellular space. CWCM: cell wall–cell membrane complex.

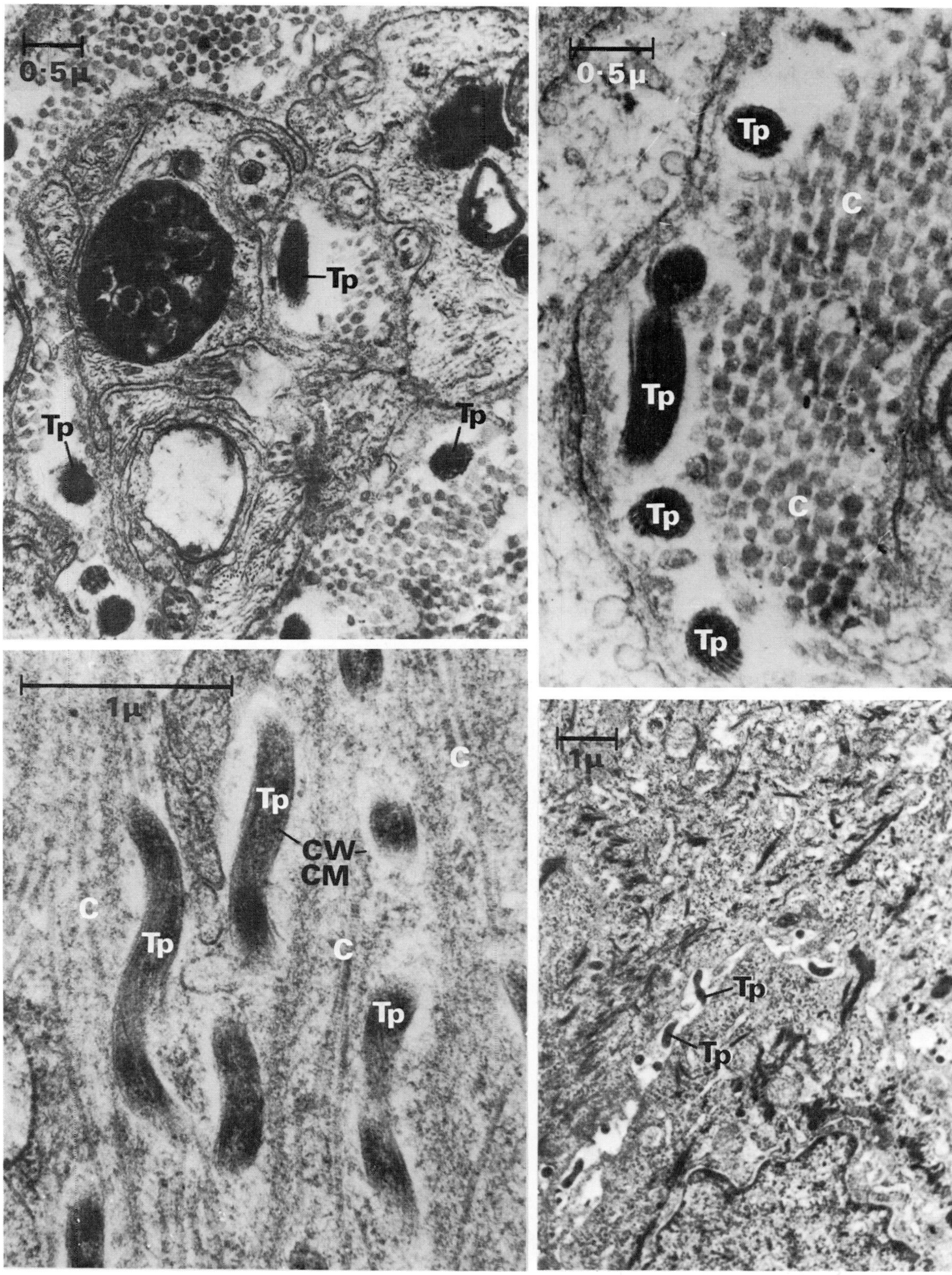

(p) (q) (r) (s)

gram-negative, cigar-shaped organism, similar in many respects to *Bacteroides melaninogenicus*, and a spirochete, *Treponema vincentii*.

The localization of these two organisms within the gums and mouth results in acute necrotizing ulcerative gingivitis, or "trench mouth". When found in the throat and tonsils, the disease is referred to as Vincent's angina.

The various genera of spirochetes are distinguished by their dimensions and degree of coiling. Thus, the genus *Spirochaeta* consists of cells which are 0·2–0·75 × 5–500 μm in length with wide, regular, stable coils and rounded ends (Figure 4.38(a). The genus *Treponema* consists of cells which are usually tightly coiled with an average of 8–14 regular, rigid coils (Figure 4.38(b),(f),(g)). The treponemes range in size between 5–20 × 0·09–5 μm, with round to tapered cell poles. *Borrelia* (Figure 4.38(d),(e)) are large cells (in comparison to spirochaeta, treponema, and leptospira), ranging in size between 10–30 μm, and 0·3–0·7 μm in width. The coils of the *Borrelia* are larger, longer, and looser than those of the treponemes and spirochaeta and tend to be irregularly spaced. *Leptospira* (Figure 4.38(c)), ranges in size between 6–20 × 0·11 μm. Characteristic of leptospira strains are the regular tight coiled spirals in the central portion of the cell, with a more regular, looser coiling in the more distal portions. The cell poles, either one or both, are bent into characteristic hooks.

Electron microscopic examination of the members of the order Spirochaetales reveals an ultrastructural anatomy similar in general to that found in other gram-negative bacteria (see Holt[97] for a review of the subject). However, spirochetes possess three major structural components which tends to distinguish them from other procaryotes:

1. An outer sheath (Figure 4.38(h)) composed of two electron-dense, 5–8 nm layers which enclose an electron-transparent layer of similar dimension which surrounds the cell and is its most external layer (Figure 4.38(i)).
2. A protoplasmic cylinder (Figure 4.8(h),(j),(n)) which consists of a cell wall–cell membrane complex which encloses the cytoplasmic region composed of ribosomes and centrally disposed nucleus. The protoplasmic cylinder consists of two electron-dense layers, separated by an electron-transparent space. This cell wall–cell membrane complex is 9–12 nm thick. In some instances, for example in *T. genitalis* and *S. stenostrepta*, the outer electron-dense layer of the protoplasmic cylinder is asymmetric, being almost twice as thick as the inner electron-dense layer.
3. The third cell component, the periplasmic fibril (Figure 4.38(i),(j),(n)), taxonomically distinguishes all spirochetes from other bacteria. Thus, a procaryotic cell is considered to be a spirochete if it has an periplasmic fibril.

Figure 4.38. Spirochetes.
(t), (u) Sections from the colonic epithelium from a rectal biopsy from a patient with systemic lupus erythematosus and a portion of an epithelial cell of a normal human colon, respectively. Both reveal spirochetes (Sp) either free in the intestinal lumen or in close association with the epithelial lumen or in close association with the epithelial microvilli (MV). RD: residual desmosome, Cyb: cytoplasmic membrane bleb.

(t)

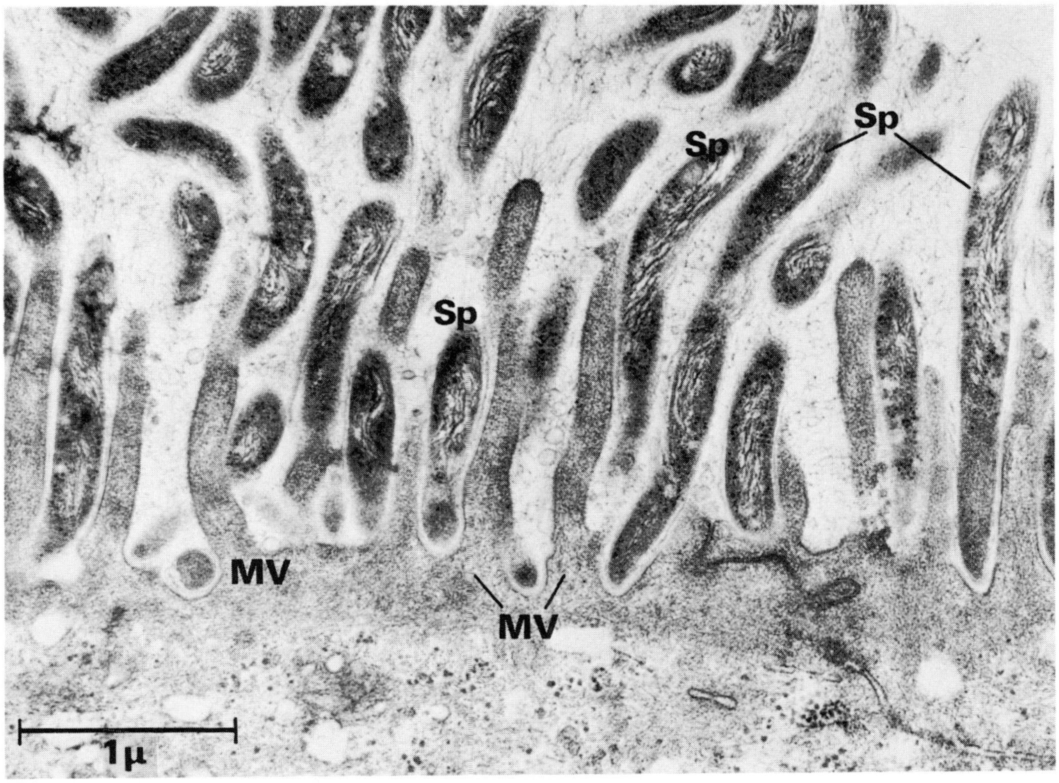

(u)

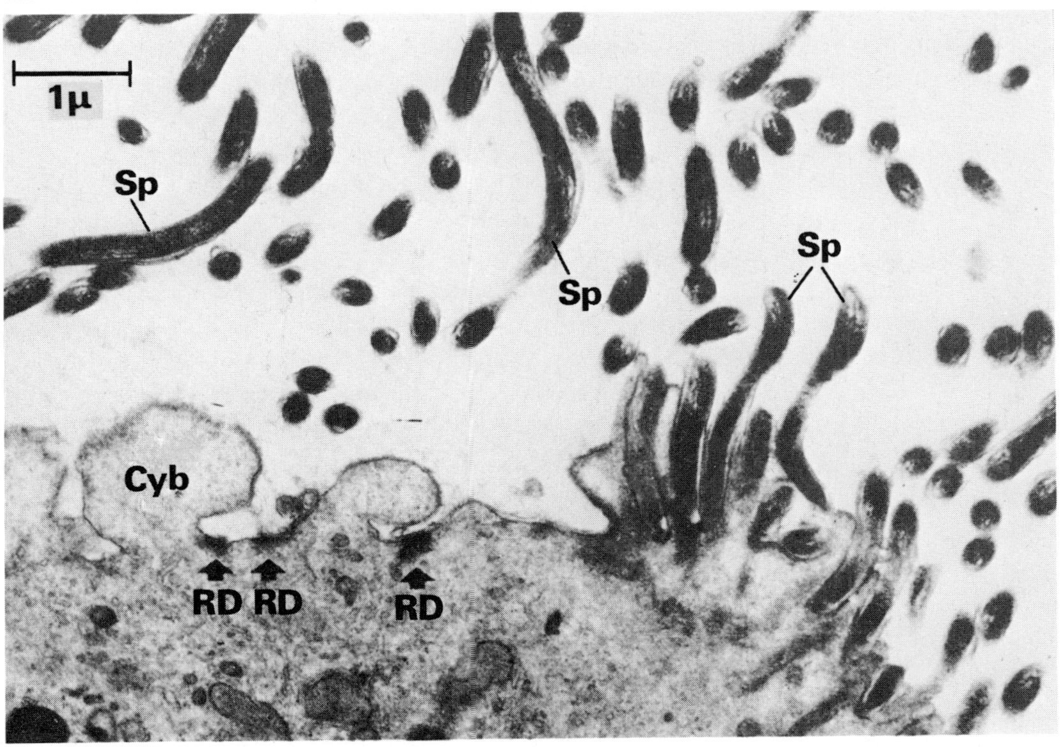

The periplasmic fibril is ultra-structurally similar to the bacterial flagellum but, in contrast to the external location of the flagellum, the periplasmic fibril is located internally, winding around the clylinder, and lying between the inner layer of the outer sheath, and the outermost layer of the protoplasmic cylinder; the flagellum by contrast is located exclusively external to the outer membrane.

Figure 4.38(o) diagrammatically depicts the major anatomical features deduced from representative electron photomicrographs of the order Spirochaetales.

Taxonomically, the number of periplasmic fibrils can be used in the identification of the various genera of spirochetes. *Borrelia* species, for example (Figure 4.38(j),(m)), have many periplasmic fibrils (usually more than 10) attached at both ends of the cell, while *Leptospira* (Figure 4.38(l) and the majority of the *Spirochaeta* (Figure 4.38(n)) have only one periplasmic fibril at each end of the cell; treponemes (Figure 4.38(i),(k)) have anywhere from 1–10 periplasmic fibrils inserted into the cell poles. *Cristispira* have periplasmic fibrils which may number in the hundreds.[173]

The localization of several of the host-associated disease-producing spirochetes has only been carried out over the past several years. Examination of testicular tissue infected with *T. pallidum* reveals the spirochetes to be located almost exclusively in extracellular spaces (Figure 4.38(p),(q),(s))[151, 152, 153, 154, 155, 156] as well as being associated with collagen fibrils (Figure 4.38(q),(r)). The treponemes were also seen to penetrate and localize in fibroblast and Leydig cells of monocytes, as well as within mesenchymal cells. In experimental syphilomas of rabbits, *T. pallidum* has been observed in the endo-, peri- and epineurium of nerve fibers (Figure 4.38(p),(q)). *Treponema pertenue* (Gauthier) is found within intercellular spaces in the stratum spinosum and stratum basale (Figure 4.38(s);[103] they have never been seen within (intracellularly) or between cells of the dermal layer.

In cases of human spirochetosis, spirochetes have been observed to attach directly to the surface of the colonic epithelium (Figure 4.38(t); however, they do not appear to enter cells. In some cases of intestinal spirochetosis, spirochetes similar in morphology to *Borrelia* sp. have been observed to be attached between microvilli at the apex of colonic epithelial cells.

Spirochetes have also been observed to inhabit the healthy human intestine (Figure 4.38(u)[139], morphologically distinguished as slender bacteria with narrow tips attached to superficial and cryptic cells of the colonic mucosa. Spirochetes with a similar morphology have also been observed between microvilli in contact with the epithelial plasmalemma, within patent intracellular spaces, and in close proximity to where these spaces open into the intestinal

---

Figure 4.39. *Mycoplasma* species.
(a), (b), (c) *Mycoplasma orale* showing cells both short, coccal-shaped to long or branching filaments. A unit cell membrane (CM) is the outermost layer of the cells in this genus. Some filaments contain numerous blebs or evaginations (arrows). (d), (e) *A. laidlawii* occur as constricted filaments with bleb-like evaginations (arrow), or as pleomorphic forms or coccobacilli. CyT: cytoplasm, NR: nuclear region, R: ribosomes, SB: storage body.

(a) (b) (c) (d) (e)

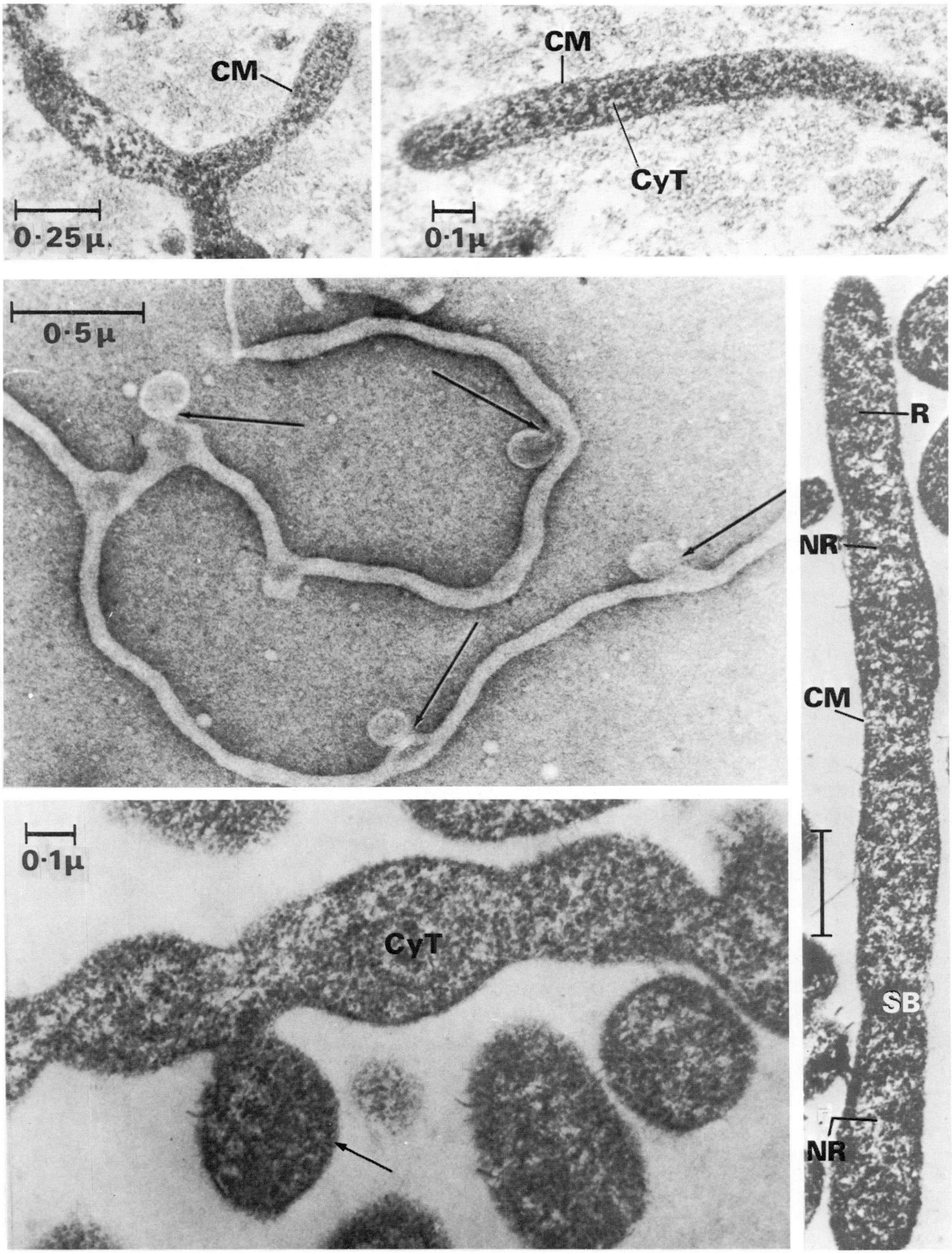

lumen. A small number of spirochetes are also found in the lamina propria and in the lumen of crypts.

4.3.15. Mycoplasmas

These procaryotes were originally described as pleuropneumonia organisms (PPLO) because they were isolated from the lungs of cattle in which large amounts of serous fluid were found. These original isolates are now firmly established as bacteria, and bacteria morphologically similar to them have been isolated from a wide variety of environments. Only in recent years that the PPLOs have been reclassified and place into the order *Mycoplasmatales* which consists of four genera: the sterol-requiring *Mycoplasma*, the *Acholeplasma* (capable of growth without sterols), the *Ureaplasma* or *T-Mycoplasmas*, and *Spiroplasma*, the latter being predominantly plant pathogens.

Two important characteristics of the genera separate them from other procaryotes: their extreme pleomorphism (e.g., Figure 4.39(a)-(c)), [4] and the absence of a cell wall (e.g., Figure 4.39(d),(e)).[124] Thus, the most external layer (excluding a capsule when it occurs) of these bacteria is a unit membrane comparable to the cell membrane of walled bacteria and similar to that of *Spiroplasma* (see Section 4.3.13), and halobacteria. The mycoplasmas are gram-negative and are among the smallest known free-living bacteria, being approximately 200 nm in diameter.

Many mycoplasmas are not pathogenic; however, a high proprotion of them are pathogenic for vertebrates, insects, and plants. The known human pathogenic species are *M. hominis* and *M. pneumoniae*. Its pathogenicity is thought to be due to the ability to necrotize respiratory tissue.

Several of the other mycoplasmas which may be of medical importance include *M. fermentans* and the *Ureaplasma* (T- or "tiny" mycoplasmas). However, the exact relationship between these latter mycoplasmas and human infections has not been conclusively proved. It is known, however, that the majority of these mycoplasmas are inhabitants of the normal flora of both humans and animals. The saprophytic mycoplasma, *A. laidlawii*, originally isolated from sewage, has, due to its ease of growth, been most extensively characterized.

In exponential growth phase, *M. laidlawii* is filamentous (Figure 4.39(d),(e)), being approximately 0·4–0·5 μm wide with filaments which can reach lengths of 2 μm.[134] Each cell consists of an external 7 nm thick cell membrane which encloses the fibrillar nucleoid and 20 nm diameter electron-dense ribosomes. In some cells, electron-dense granules (Figure 4.39(e)) are found, and these appear to partition the cytoplasm into a granular region enclosing a filamentous nuclear region.

As *M. laidlawii* enters the late exponential or stationary growth phase, the cells tend to constrict and branch. Some cells in this stage of growth are swollen and most probably undergoing disintegration. Incubation for extended periods (well into the stationary growth phase) leads to the formation of round bodies, 0·5–1 μm in diameter. Similarly, when cells of *M. pneumoniae* enter the late stationary growth phase, they tend to become vacuolated, with many

(f)

(g)

(h)

(i)

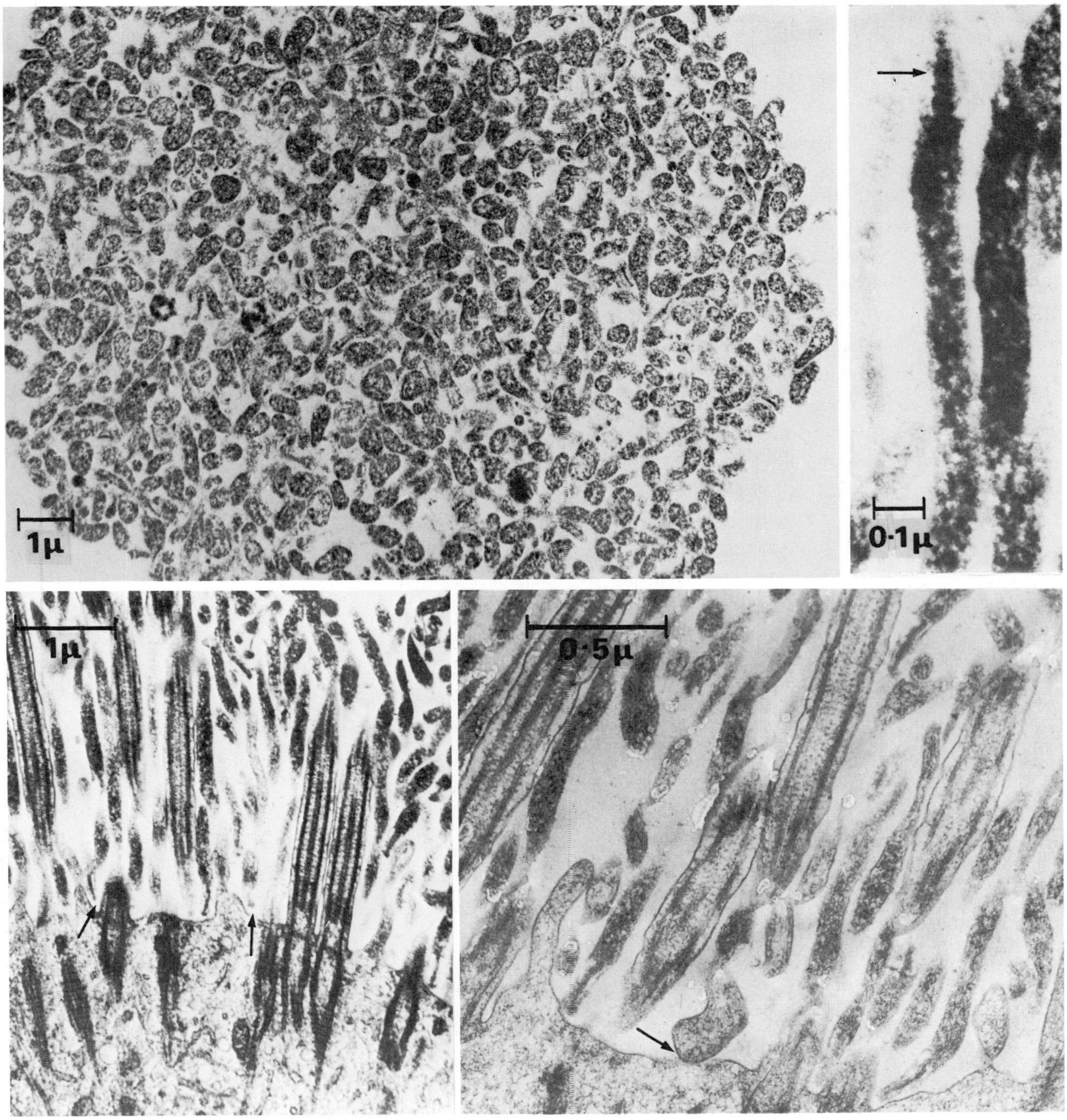

Figure 4.39. ***Mycoplasma* species.**
(f), (g), (h), (i) *Mycoplasma pneumoniae* is highly pleomorphic, with most cells varying from round to pear-shaped; elongate and irregular forms are also present. *M. pneumoniae* infected tracheal tissue organ culture, (h), (i), reveals the cells in close association with the organ cell cilia (arrows). Note in (g) that the cell pole is tapered (arrow) so as to insert into the cilial cell membrane.

pleomorphic and empty cells (Figure 4.39(f)) being found; these bodies are representative of nonviable cells.

The interaction between *M. pneumoniae* and human respiratory tissue has been examined extensively by Collier and Clyde[43] and Powell *et al.*[163] Of interest in the observations of Collier and Clyde was that 48 h post inoculation of organ cultures with *M. pneumoniae* resulted in extensive infiltration of the cilia, with many mycoplasmas at the base of cilia and extending across the width of a single epithelial cell (Figure 4.39(h),(i)). When associated with cultured cells, *M. pneumoniae* (Figure 4.39(g)) has a distinct morphology, consisting of tapered terminal ends in close proximity to the host cell membrane. The tapered ends consist of an electron-dense core surrounded by an electron-lucent zone. The attachment of *M. pneumoniae* to the respiratory epithelium results in both morphological and physiological alterations of the host tissue which ultimately lead to ciliostasis and host cell death. The loss of the epithelial layers may be a result of peroxide production.

Ureaplasma (human *T-Mycoplasma*) vary in morphology from spherical to ovoid, each cell being approximately 140–840 nm in diameter (Figure 4.39(j)). Branching filaments, 2 μm in length, have also been observed (Figure 4.39(y)).[16] The cells are surrounded by an 8–9 μm unit cell membrane, the outer surface of which is covered with a dense hair-like coat 9–11 nm thick. This hair-like layer is also apparent in thin-sectioned cells (Figure 4.39(x),(z)) as well as after ruthenium red fixation.[169] Green and Hanson[80] employing specific antibody binding, along with ruthenium red fixation and staining, also showed the presence of a thick extracellular matrix surrounding *M. meleagridis* (Figure 4.39(k). This extracellular layer (most likely a heterocarbohydrate polymer such as polygalacturonic acid), may be involved in cell–cell attachment or cell association.

Mycoplasma hominis (Figure 4.39(s),(t)) and *M. neurolyticum* are short filaments or rods, approximately 0·25 μm wide.[60] Again, the external surface is covered by an 8–9·5 nm thick unit membrane. The cytoplasm is filled with granules of various sizes and electron densities, as well as possessing a centrally disposed nuclear region.

Dividing cells of *M. hominis*[133] and *M. meleagridis* are connected by thin tubules, 17–19 nm thick (Figure 4.39(t),(u)).[80] These tubules, which morphologically appear to be back to back unit membranes, may be involved in cell division, the mother cell dividing by transverse binary fission and being held together by the tubules. As the daughter cells move further apart the tubules may break and reseal.

Mycoplasma orale was originally isolated from the tonsils of two patients with scarlatina. This species has also been found as a normal inhabitant of the

---

Figure 4.39. *Mycoplasma* species.
(j), (k) Human *T-Mycoplasma* (*Ureaplasma urealyticum*) and *M. meleagridis*, respectively, fixed and stained with ruthenium red possess a thick ruthenium red-positive capsule (Cp). CM: cell membrane.

(j)

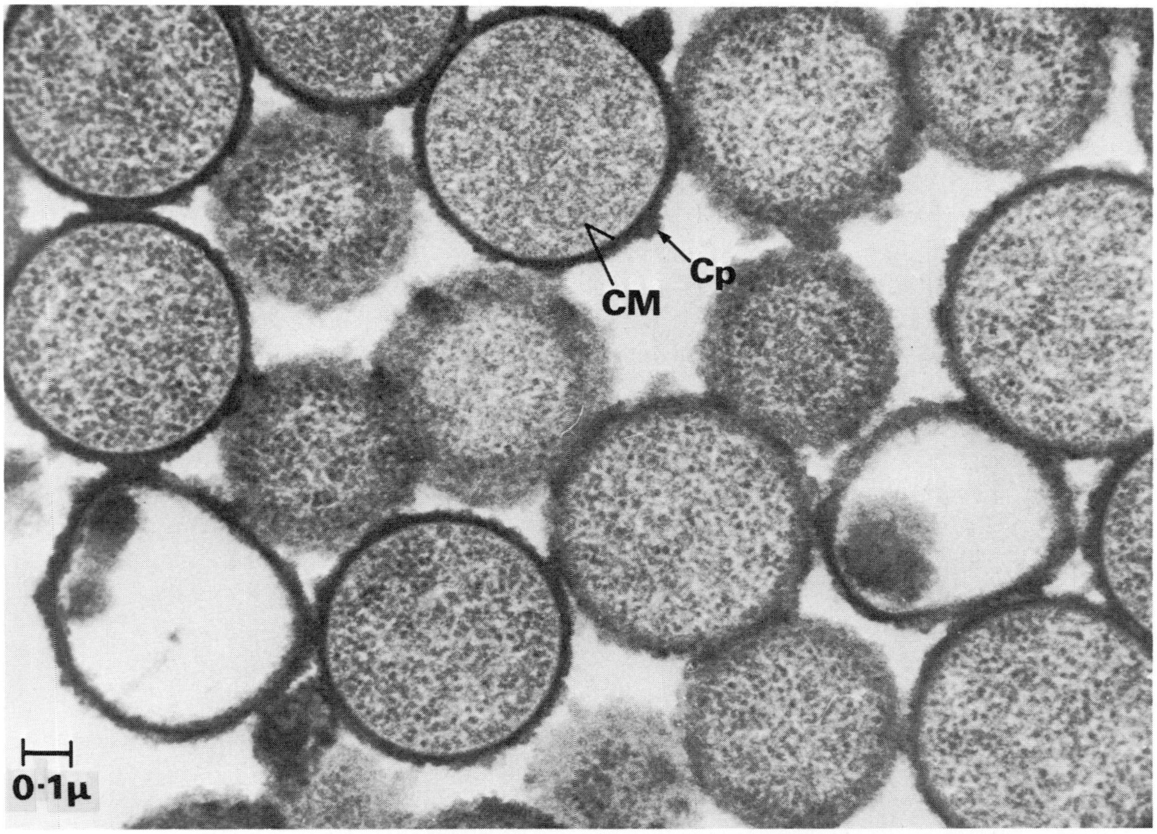

(k)

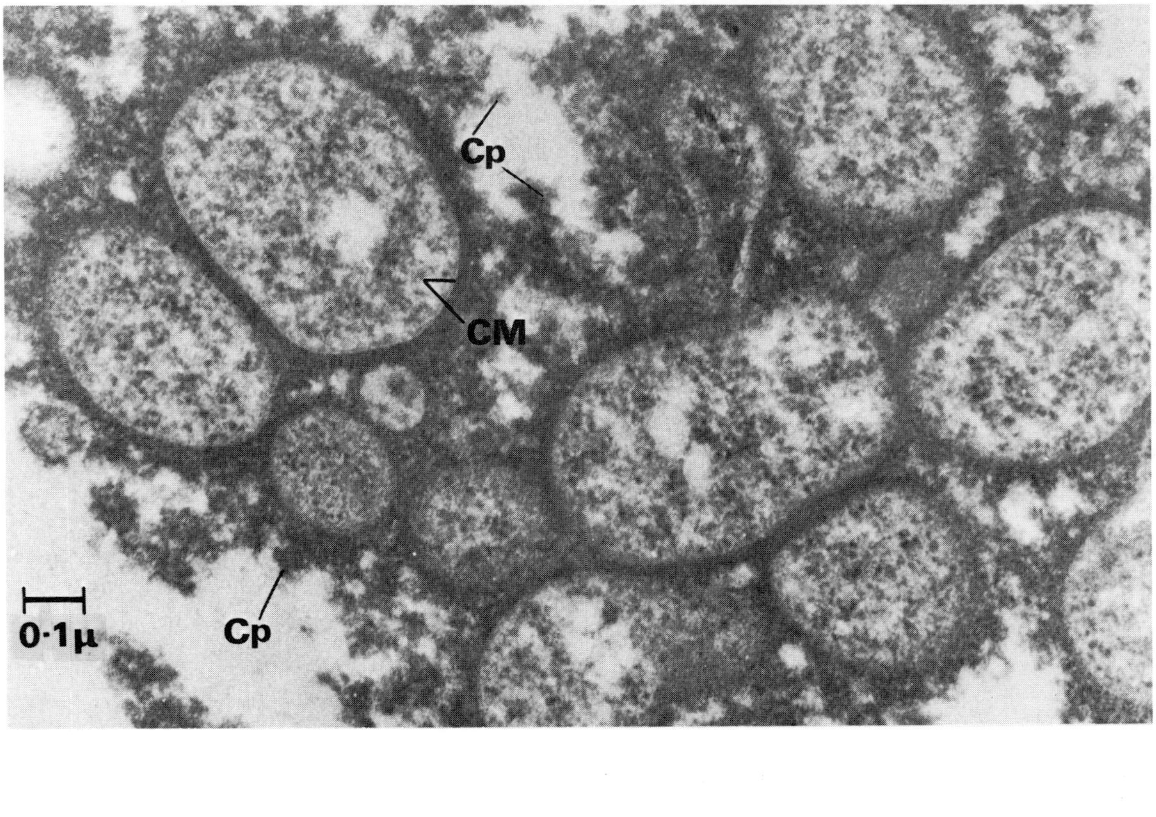

oral cavity. Anderson and Barile[5] were also able to isolate *M. orale* from patients with acute myelocytic and lymphocytic leukemia.

Nakamura and Kawaguchi,[147] in a very interesting series of experiments, examined various regions of plate-grown *M. orale* colonies by electron microscopy. They found that the central region of *M. orale* colonies consisted of, for the most part, large empty bodies, 1·8–2·3 μm in diameter. Morphologically, these cells appeared nonviable. Where the large cells appeared intact, they had dimensions of 1·6–1·9 μm and contained within their cytoplasmic region ribosome-like granules and DNA fibrils. Some of these large cells also contained small (67–100 nm), electron-dense bodies, sometimes referred to as elementary bodies, as well as filamentous structures, the latter being present in a helical arrangement.

The central region of the *M. orale* colonies also contained long filamentous cells, 2·7–45 nm in diameter, while the external portion of *M. orale* 1 colonies consisted of numerous elementary and electron-dense bodies (Figure 4.39(v),(w)). These latter bodies were larger than the elementary bodies, and, by negative-staining, filaments 50–100 μm wide and up to several micromillimeters in length appeared to emerge from the cell surface. In some cases the filaments were branched, swollen, or had blebs emerging from their surface.

Domermuth *et al.*[60] examined some nineteen *Mycoplasma* strains and showed that the basic cell morphologies were almost identical; all cells consisted of a limiting outer unit membrane, approximately 7·5 nm thick. There was some variation in the diameter of the cell membrane depending upon the strain examined. *Mycoplasma neurolyticum*, for example, had a cell membrane which was 10 nm thick; however, the majority of the other mycoplasma strains had a cell membrane that was consistently close to 7·5 nm. By electron microscopy, cells of *M. arthriditis* (Figure 4.39(l)), *M. hominis* (Figure 4.39(s),(t)), *M. pulmonis* (Figure 4.39(m)), and *M. salivarium* (Figure 4.39(n)) varied from spherical, filamentous to pleomorphic, and consisted of small to large cells. The cytoplasmic regions are electron dense, characteristic of ribosomes, and contain fibrillar nuclear regions. In some mycoplasmas, for example *M. gallisepticum*, the ribosomes are seen to be arranged in square groups of four, while *M. pneumoniae* contain rod-shaped structures in their cytoplasm.

In a most interesting study, Lemcke[124] examined the effects of the suspending buffer on the ultrastructure of mycoplasma species. When, for example, *M. mycoides*, which is filamentous in culture, is washed and suspended in a hypertonic buffer, the cells assume a spherical shape. Addition of sucrose to this hypertonic buffer to render it hypotonic, results in a reversal of these spherical cells to their original filamentous forms. *Mycoplasma laidlawii*, strain *A*, or *M.*

Figure 4.39. *Mycoplasma* species.
(l) *M. arthritidis*, (m) *M. pulmonis*, (n) *M. salivarium*, all display similar morphology. (o), (p), (q) *M. laidlawii* occur as constricted filaments with bleb-like evaginations or as pleomorphic forms or coccobacilli. (r) *M. orale* showing variation in shape from short coccal-shaped to long or branching filaments. R: ribosome, NR: nuclear region, OM: outer cell membrane. Many are probably nonviable forms.

(l) (m) (n)

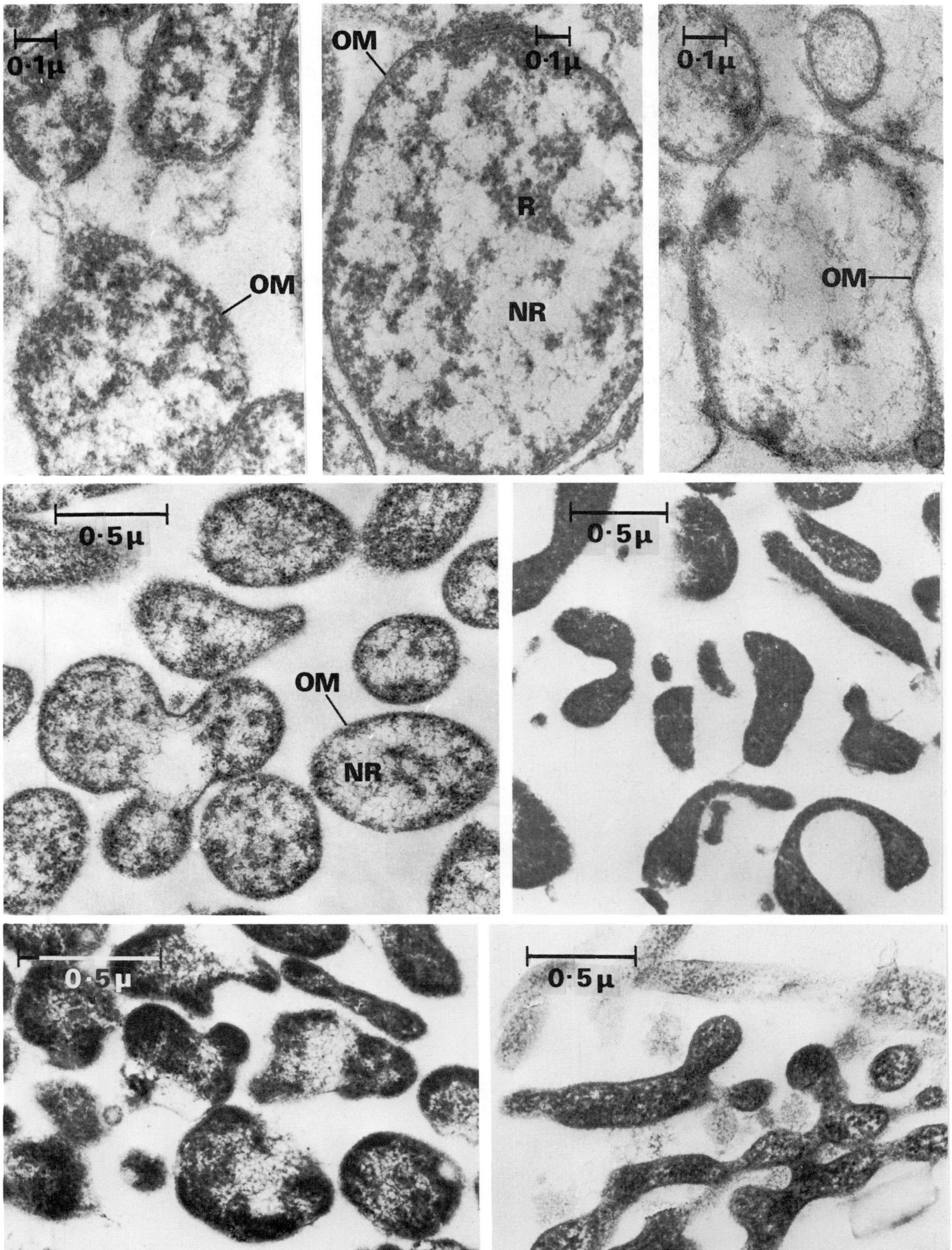

(o) (p)

(q) (r)

orale (filamentous form) suspended in either 0·001 M or 0·01 M sodium cacodylate or phosphate buffer and fixed with low concentrations (1 percent) of glutaraldehyde produced coccoid, coccobacillary, rod-shaped, or filamentous cells (Figure 4.39(o),(r)). Increasing the sodium cacodylate concentration to 0·1 M resulted in an hourglass to sinous cell morphology (Figure 4.39(p)). In addition, numerous membrane invaginations and membrane-limited vesicles become apparent at this high buffer osmolality. In overall appearance the cells were collapsed with concave or biconcave (hourglass) forms. There is the general appearance of membrane damage and leakage of internal material at this high osmolality (Figure 4.39(g)). These morphological forms were also apparent when glutaraldehyde concentrations greater than 1 percent were used.

4.3.16. *Spiroplasma*

These mycoplasmas have been isolated from "stubborn" disease of orange and other citrus trees,[19, 41, 42, 175] and have been observed in diseased plant tissue, as well as in infections of insect vectors.[213, 216]

Electron microscopic examination of negatively stained cells of *Spiroplasma citri* (Figure 4.40(a),(f)) reveals helical filaments, 100 nm in diameter. The amplitude of the helix is 200–300 nm. Many of the filaments have associated with them blebs, which most probably represent outpockets of the limiting surface membrane.

Of interest was the observation by Cole *et al.*[41] that staining *S. citri* with phosphotungstic acid (PTA) results in the loss of the helical filaments, while ammonium molybdate (NH_4MoO_4) preserved them. This effect of the two stains may be related to the low osmolality of PTA (54–105 m osm/kg) compared to an osmolality of 330–700 m osm/kg, depending upon the concentration of NH_4MoO_4. This latter osmolality is very similar to that of the *S. citri* growth medium (400–800 m osm/kg). Thin sections of chemically fixed cells (Figure 4.40(c),(e)) reveal the 120 nm diameter filaments to consist of a limiting unit membrane, 7·2 nm wide. Note in Figure 4.40(d), that the external surface of this membrane may be covered with an electron-dense fuzz. The cytoplasm is both electron dense and compact.

When *S. citri* is fixed in glutaraldehyde in growth medium in the absence of cacodylate buffer the helical nature of the cells is preserved; inclusion of cacodylate in the fixation protocol resulted in the formation of irregular filaments, and the loss of the helical configuration. This loss of helical shape may be due to osmotic effects, similar to that observed for PTA and NH_4MoO_4.

4.3.17. *Rickettsia-Bartonella-Chlamydia* and related genera

The Rickettsiaceae are obligated intracellular parasites of both vertebrates and arthropods. *Rickettsia quintana* is the only species capable of *in vitro* growth. Included in the family Rickettsiaceae are three genera, *Rickettsia, Rochalimaea*, and *Coxiella*. The transmission of rickettsial diseases occurs by an intermediate insect vector, which may function in some cases both as the natural and primary host.

(s) (t) (u)

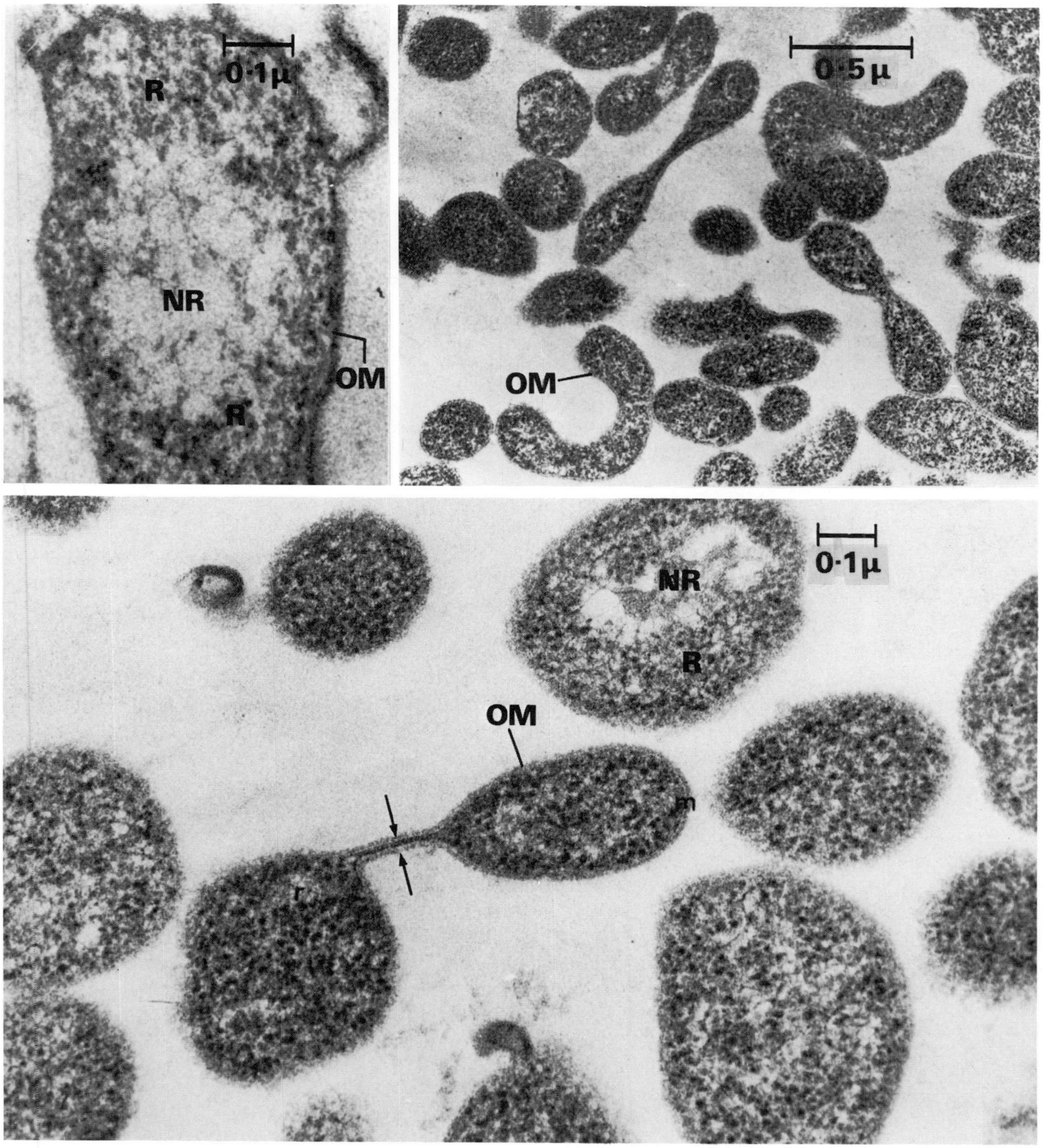

Figure 4.39. *Mycoplasma* species.
(s), (t) *M. hominis* showing the cells bounded by the unit cell membrane (OM) and varying in shape from branched, coccal cells to long filaments, depending upon the culture's physiological state, or the conditions of chemical fixation for electron microscopy. R: ribosome, NR: nuclear region.
(u) Unstained *M. meleagridis* lacks the thick ruthenium red layer (compare with (j), (k)). In addition, two daughter cells are connected by a membranous tubule (arrows), consisting of back to back unit membranes.

(v)

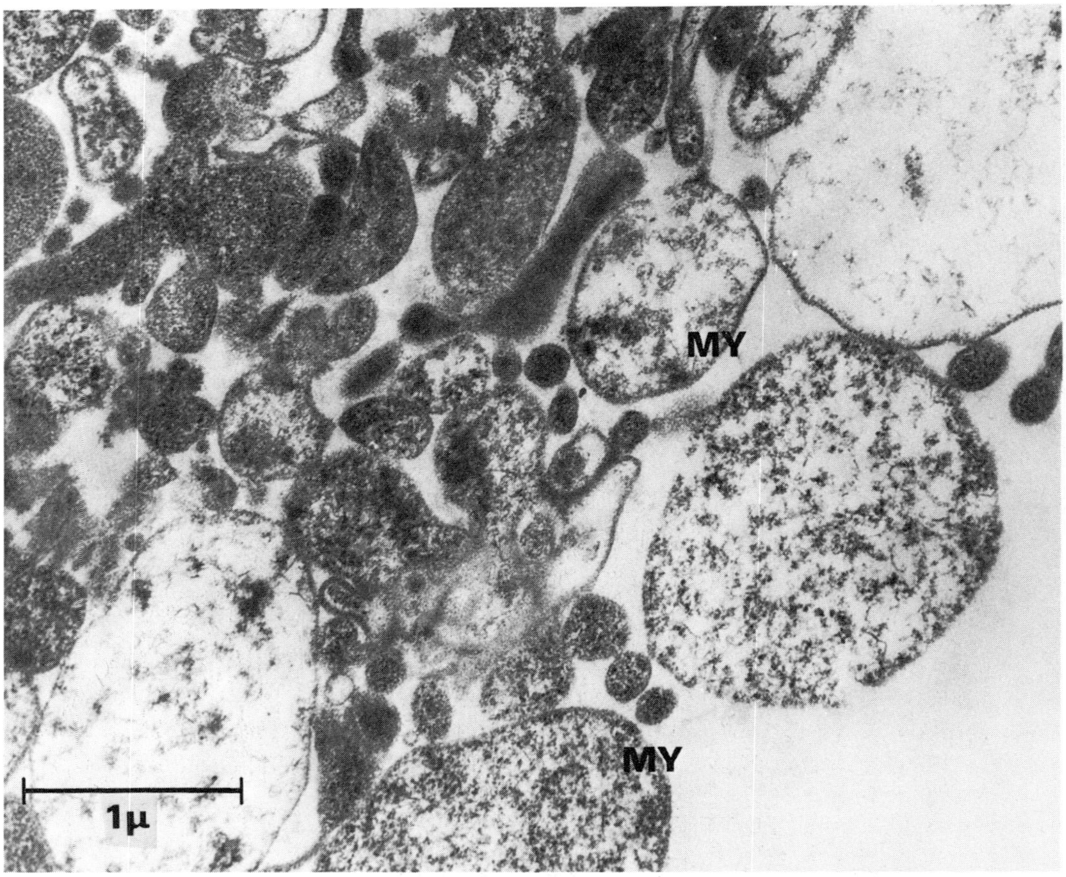

(w)

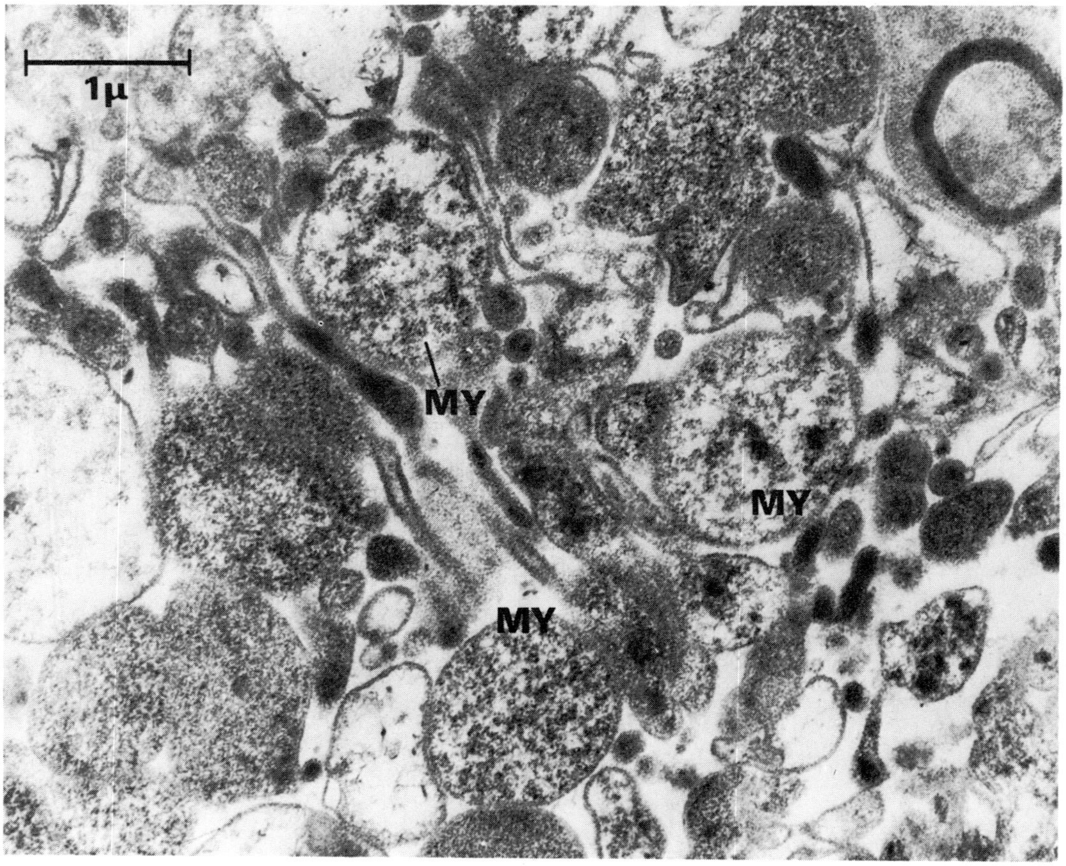

(x)
(y)
(z)

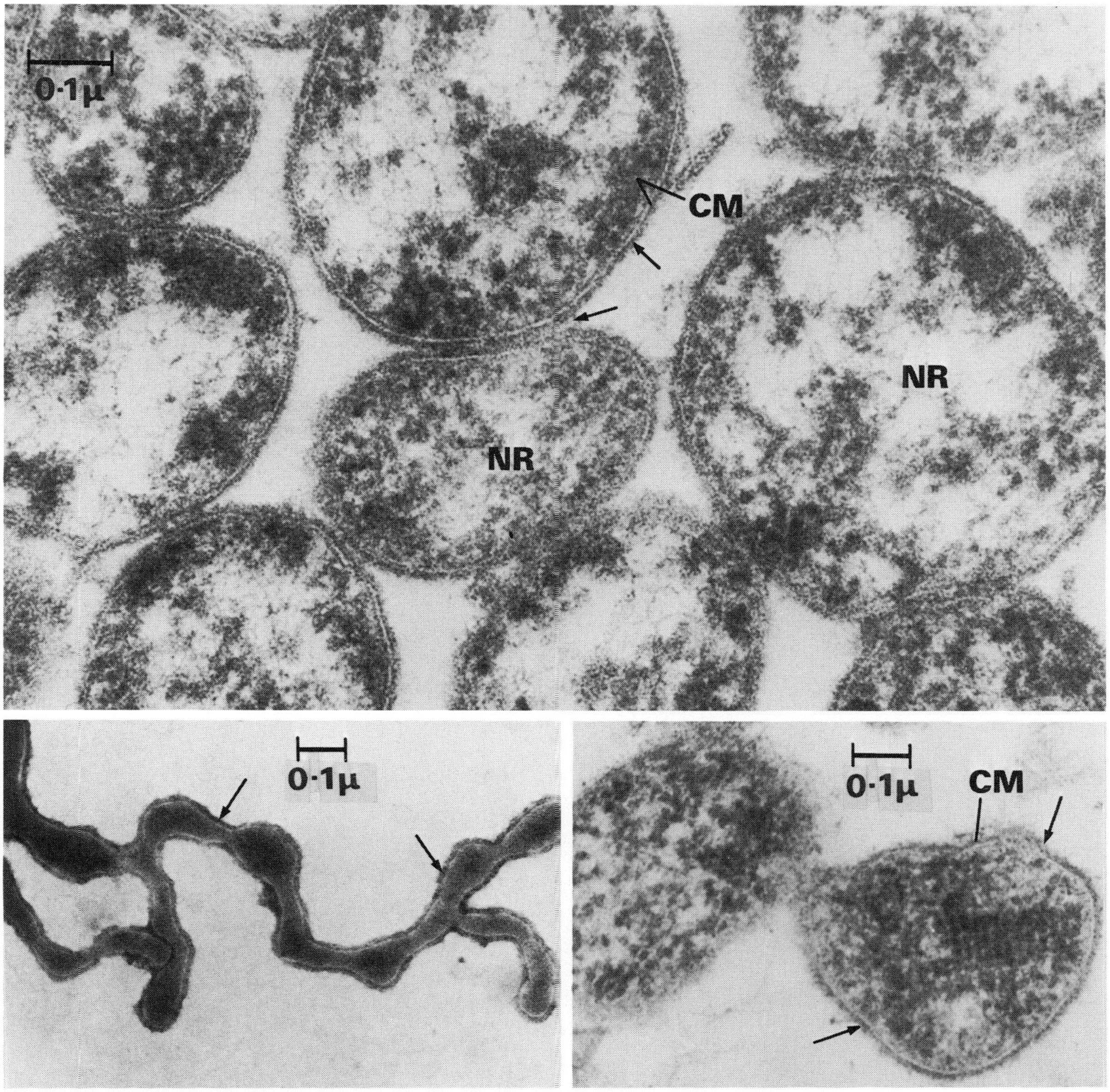

Figure 4.39. *Mycoplasma* species.
(x), (y), (z) *T-mycoplasmas* displaying morphology similar to (l), (m), (n), (s), and (t). In addition, the cell membrane (CM) is covered with a hair-like fuzz (arrows). NR: nuclear region.

Figure 4.39. *Mycoplasma* species.
(v), (w), *Mycoplasma orale* showing cells both short, coccal-shaped to long or branching filaments. My: mycoplasma.

Members of the genus *Rickettsia* are small (0·3–0·7 μm), highly pleomorphic bacilli or coccobacilli which occur singly, in pairs, chains, or in irregular clusters.[6, 158]

Electron microscopic examination of representative rickettsial species reveals them to be similar in morphology to that of other gram-negative bacteria; the cell wall is typically gram-negative, the underlying cell membrane is of unit dimension, and the cytoplasmic region is filled with electron-dense ribosomes which surround a fibrous nuclear region.

Cytologically, rickettsial infections appear to be the result of an invasion of the reticuloendothelial cell linings of the walls of small blood vessels. This invasion of the small blood vessels results in a corresponding vasculitis, with dissemination of the disease to widespread anatomic locations in the body.

The human rickettsial diseases include typhus, spotted, trench, and Q-fevers, as well as scrub typhus. With the exception of Q-fever, caused by *Coxiella* (*Rickettsia*) *burnetti*, all of the rickettsial diseases are transmitted to humans by insect vectors, which may include the louse, flea, tick, or mite.

Rickettsia prowazeki

The causative agent of epidemic typhus, *Rickettsia prowazeki* is transmitted to humans by the bite of a body louse. Members of the genus are pleomorphic, approximately 0·3–0·5 μm in diameter and sometimes reach lengths up to 4 μm (Figure 4.41(a)–(c).[6] The majority of *R. prowazeki* occur in chains in the host cell cytoplasm (Figure 4.41(c)). By negative staining (Figure 4.41(c)) *R. prowazeki* are short cocco-bacillary forms. Trilaminar membranes limit the cells and are morphologically identical to the cell wall–cell membrane observed in other gram-negative bacteria. In some cases however, the cell wall of *R. prowazeki* has been observed to be five-layered. The cytoplasm consists of electron-dense ribosomes which enclose the fibrillar nucleoid, as well as electron-dense spherical granules, 4–10 μm in diameter, which are not seen in other rickettsial species.

Staining *R. prowazeki* for varying lengths of time with uranyl acetate permits the visualization of a surface fringe consisting of regular subunit striations (Figure 4.41(b)).[158] The striations have a periodicity of 13 nm and may reach lengths up to 4 μm.

Figure 4.40. ***Spiroplasma citri.***
Representative electron micrographs of the agent of citrus "stubborn" disease. (a), (b) In log phase culture and seen by negative staining the distinct helical nature of the organisms is evident. Numerous large blebs (arrows) emerge from the cell surface, and some cells are drawn out into long filaments (double arrows). (c), (d), (e) Thin sections of glutaraldehyde-osmium fixed cells reveal the helical shaped cells to be enclosed by a thin unit membrane, similar to *Mycoplasma* species. CM: cell membrane, NR: nuclear region. (a), (b): ammonium molybdate pH 7·1.

(a) (b)

(c) (d)

(e)

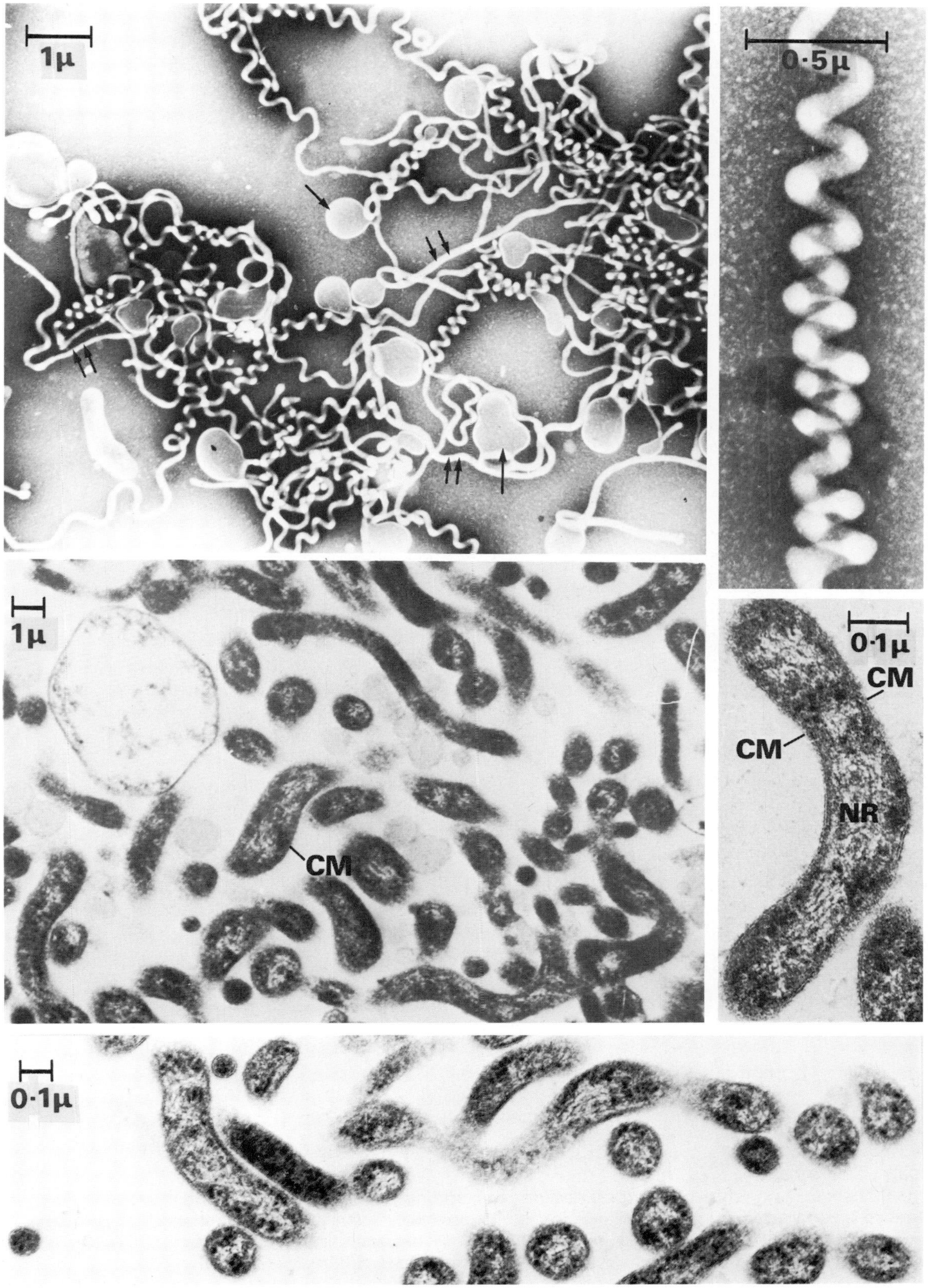

Rickettsia rickettsii–Rickettsia tsutsugamuschi

Rickettsia rickettsii, the causative organism of Rocky Mountain spotted fever (tick fever) is morphologically similar to *R. prowazeki*. However, in contrast to *R. prowazeki*, *R. rickettsii* cells are shorter, reaching lengths of only 2 μm as compared to the 4 μm maximal length reached in *R. prowazeki*. In addition, *R. rickettsii* is found in both the host cell nucleus as well as in the cell cytoplasm (Figure 4.41(d)[6] where they occur as single cells.

The rickettsiae are capable of growth in epithelial cells of the gut, hypoderm, and salivary glands of infected ticks, and in mammalian tissue cultures. The primary vector for *R. rickettsii* is the dog tick; however, in many parts of the world the organism is also found in rabbits, foxes, deer, and similar animals.

Rickettsia tsutsugamuschi, the causative agent of scrub, or mite typhus, is similar to both *R. prowazeki* and *R. rickettsii*. However, *R. tsutsugamuschi* cells are shorter, being 1·5 μm in length. *Rickettsia tsutsugamuschi* also occurs free within the host cell cytoplasm (Figure 4.41(e)).

Rickettsia sennetsu

The characteristic which distinguishes this rickettsial species from the other members of the genus is its ability to grow in membrane-lined vacuoles (Figure 4.41(g)). The other rickettsia are capable of growth within the host cytoplasm.

Morphologically, *R. sennetsu* is highly pleomorphic (Figure 4.41(f)), being approximately 0·5–1 μm in diameter. The cell wall is wrinkled, and typically gram-negative. The unit cell membrane encloses a typical procaryotic cytoplasm consisting of electron-dense ribosomes and fibrous nucleoid.

Rickettsia canada

Rickettsia canada is a member of the typhus group of rickettsia. The cells are similar in morphology and dimension (1·6 × 4 μm) to the other rickettsias. The cell wall is trilaminar and approximately 7 nm thick (Figure 4.41(h)). The external surface of the cell wall is covered with a fuzz-like, electron-dense coating. The plasma membrane is of unit dimension and construction, and encloses an electron-dense, ribosomal-filled cytoplasm. Electron-lucent vacuoles are sometimes present; a similar observation has also been made for *R. prowazeki*.

When grown in tissues of the Rocky Mountain wood tick *Dermacentor andersoni*, *R. canada* has been observed in most tick tissues, including gut (Figure 4.41(h)), hypodermal (Figure 4.41(k)–(n)), malphighian (Figure 4.41(i),

Figure 4.41. *Rickettsia* species.
(a), (b), (c) Representative rickettsia as seen after thin sectioning and negative staining. *Rickettsia prowazeki* are large pleomorphic cells; the cell wall is similar to that of other gram-negative bacteria. PTA-negative-stained cells reveal both elongate, striated forms (b) and short rod-shaped forms (c). CM: cell membrane, OM: outer cell membrane.

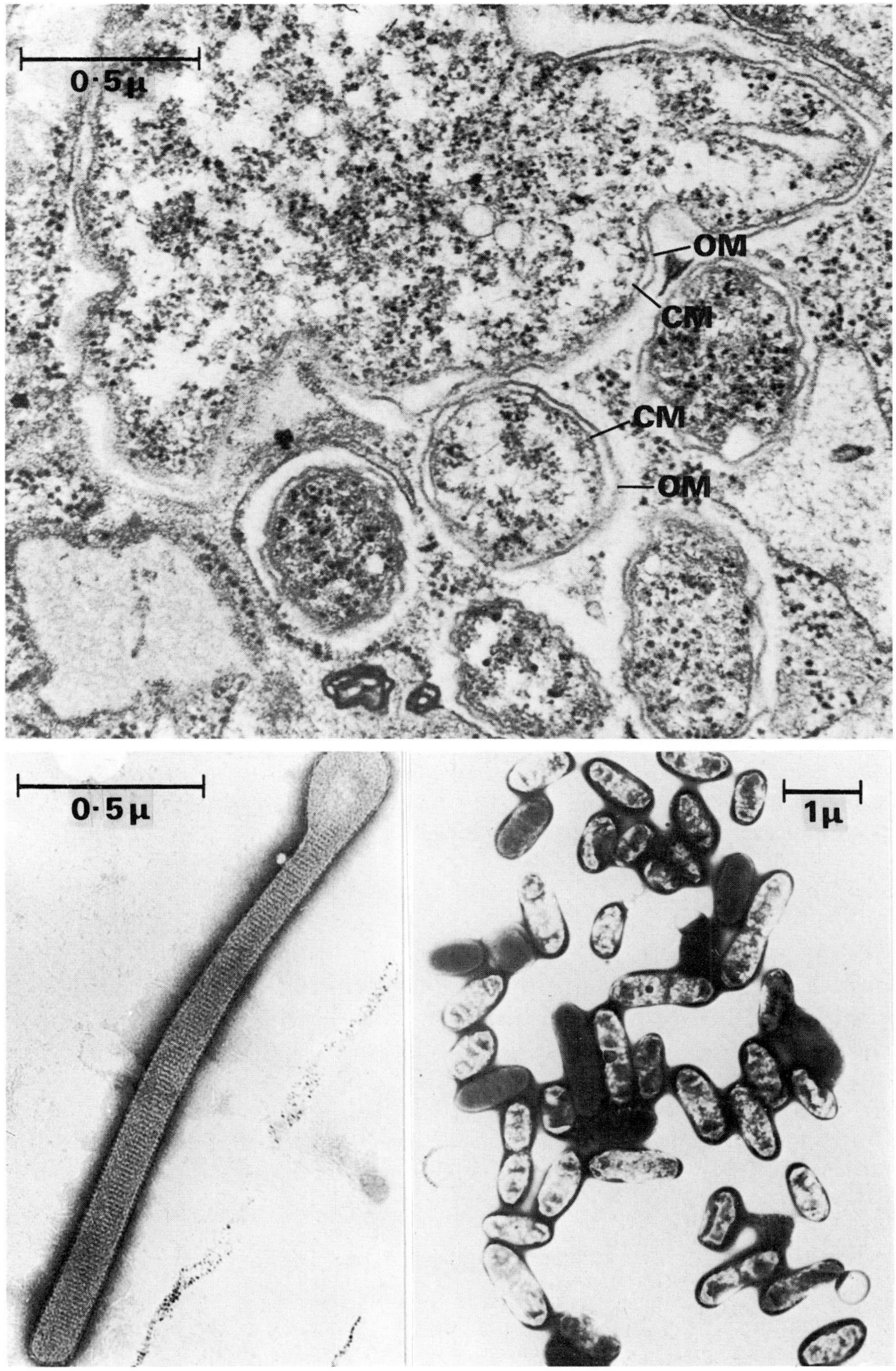

(a)

(b) (c)

and in interstitial cells of ovarian tissue (Figure 4.41(j)), and occuring as well in salivary glands and hemocytes. The electron-transparent zone surrounding several *R. canada* cells (e.g., Figure 4.41(i),(j)) may, as suggested by Brinton and Burgdorfer,[28] be an artifact of tissue preparation for electron microscopy, or represent digested host cytoplasmic matrix.

Rochalimea quintana

As already indicated, the majority of the rickettsiae are obligate intracellular parasites; however, *R. quintana*, the etiological agent of trench fever, is capable of extracellular growth in the gut of its arthropod vector, *Pediculus humanus* v. *corpuris*. In addition, *R. quintana* is capable of *in vitro* growth in cell-free media, as well as in the yolk sac of developing chick embryos. Humans appear to be the reservoir of infection, and the body louse is the only known vector. This ability to grow in the extracellular environment of the host, as well as cultivation on cell-free media, has placed *R. quintana* in the genus *Rochalimaea*.

Both louse gut and *in vitro* cultivated *R. quintana* are morphologically identical. The cells are spherical to rod-shaped, 0·2–0·5 μm wide, and up to 1·6 μm long (Figure 4.41(o)). The cell wall is gram-negative in construction and approximately 8 nm thick, with an underlying 7 nm thick cell membrane. Similar to *R. canada, R. quintana* is sometimes coated with a "fuzzy" material of moderate electron density (Figure 4.41(p)). It is not clear, however, whether this extracellular material is truly cell associated, or whether it is a product of the medium which, in some way, becomes associated with the outer cell surface.

The midgut region of infected *P. humanus* contains varying numbers of *R. quintana* in its lumen (Figure 4.41(q)), both free, and in association with the host intestinal epithelium plasma membrane.

Coxiella burnetii (Rickettsia burnetii)

The causative agent of Q-fever, *Coxiella burnetii*, exists in two serological forms, phase I and phase II. Phase I *C. burnetti* are found in nature, occurring in ticks, animals, and infected humans. Passage of *C. burnetii* through yolk sacs results in the establishment of phase II cells, a cell type most probably due to the loss of lipopolysaccharide antigenic complexes.

Both *C. burnetii* phase I and phase II cells are highly pleomorphic, varying from large "intermediate" forms 300–1,200 nm in diameter, to small "elementary bodies" 300–600 nm in diameter. Both cell types possess a cell wall similar

Figure 4.41. *Rickettsia* species.
(d) BS–C–1 tissue cell line infected with *R. rickettsia* reveals the rickettsia (arrows) distributed throughout the host cell groundplasm. The bacterial cell wall is loose-fitting. (e) In contrast, *R. tsutsugamuschi* (arrows) displays a smooth, tight-fitting cell wall when grown in BS–C–1. R: ribosome, ER: endoplasmic reticulum, OM: outer cell membrane, CM: cell membrane, SB: storage body, M: mitochondrion, NR: nuclear region.

(d)

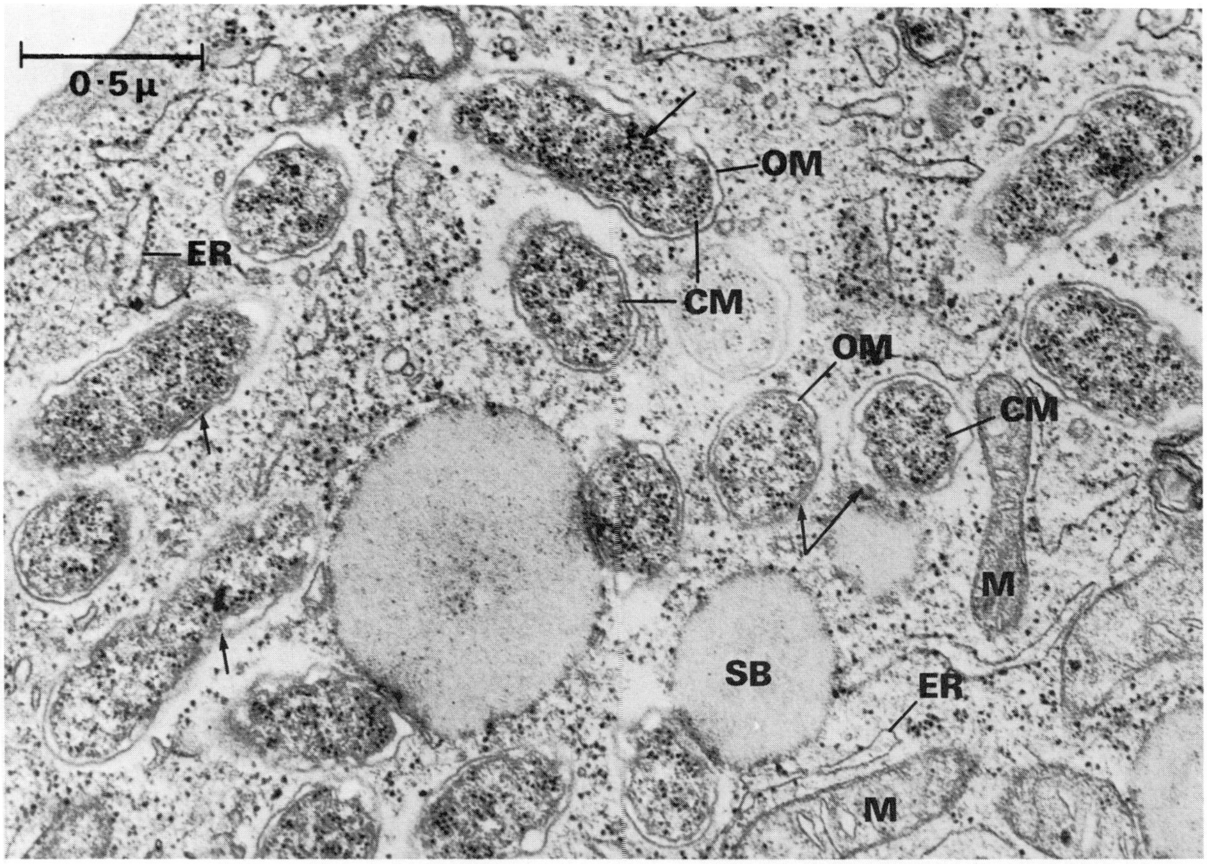

(e)

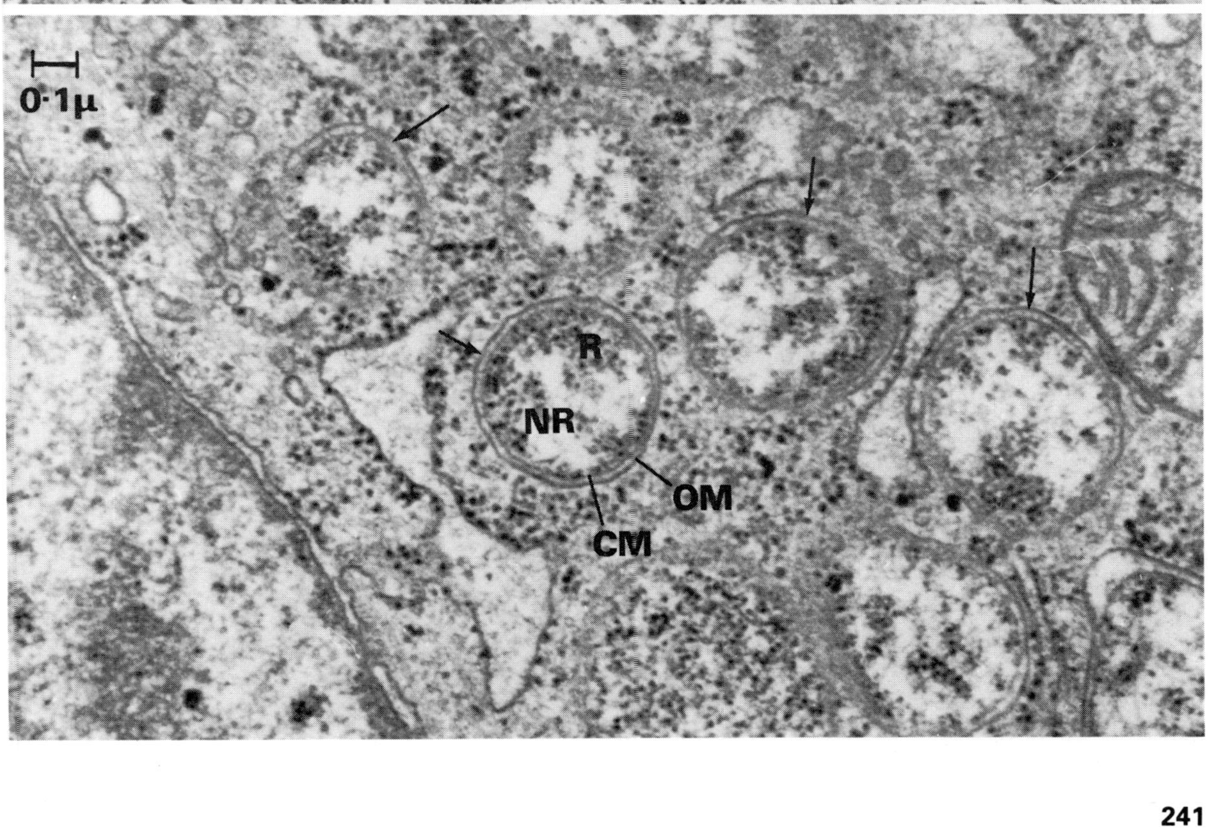

(f) (g)

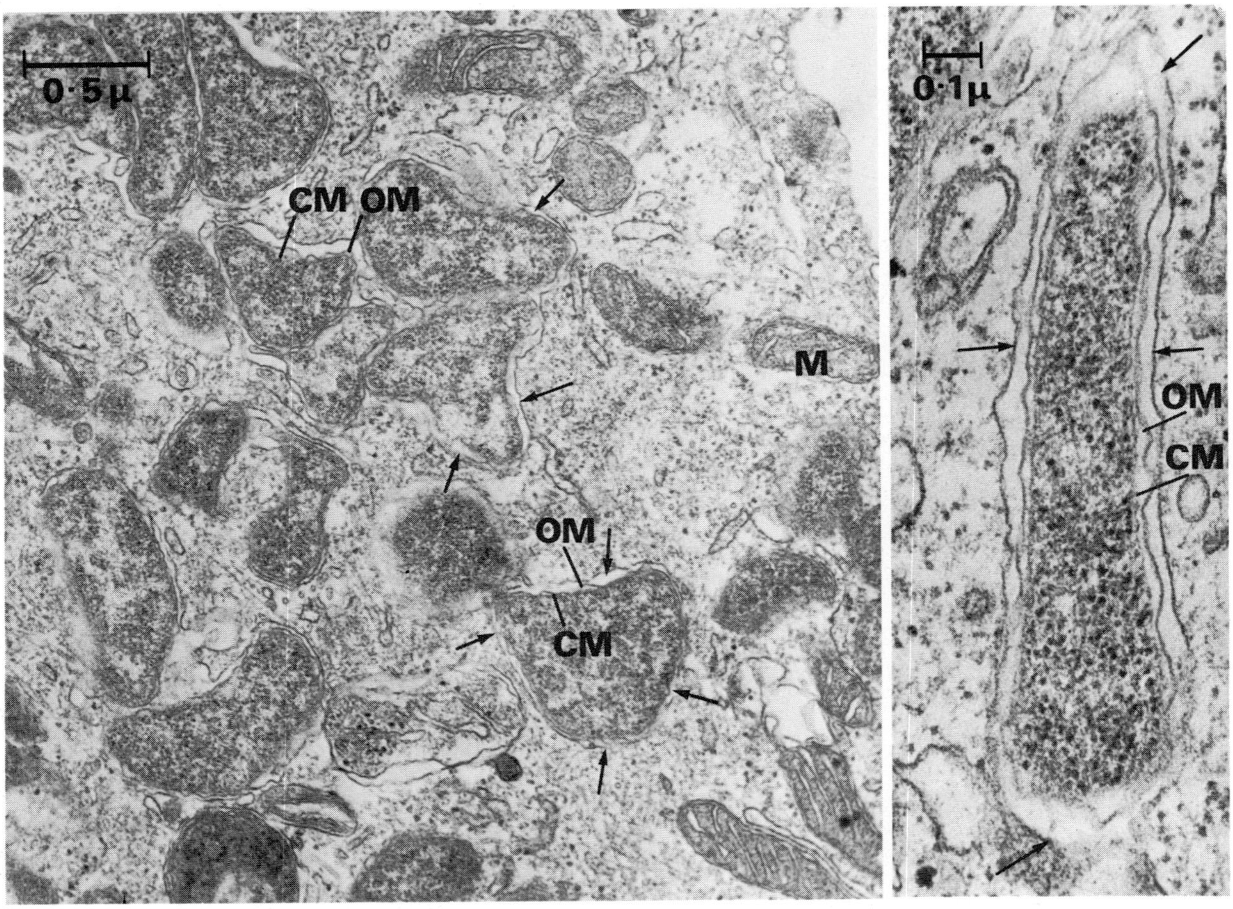

Figure 4.41. *Rickettsia* species.
(f), (g) *Rickettsia sennetsu* is morphologically similar to *R. rickettsia* but the former is enclosed in membrane limited vacuoles (arrow), clearly seen in (g). M: mitochondrion, CM: cell membrane, OM: outer cell membrane.

Figure 4.41. *Rickettsia* species.
(h), (i), (j) Infection of various tissues of the wood tick *Dermacenter andersoni* Stiles with *R. canada* 2678. At high magnification (h) the cell surface is covered with a fine hair-like fuzz (arrows). In a malpighian tubule (i) *R. canada* exists in electron-transparent spaces. The cells possess a smooth cell wall and underlying cell membrane (CM). In an interstitial cell of ovarian tissue (j) *Rickettsia canada* with another bacterium, *Wolbachia* sp. (Wob), are surrounded by an electron-transparent halo which may represent an unstained capsule, the destruction of host cell groundplasm, or an artifact of the preparative procedure for electron microscopy. ME: mesosome, OM: outer cell membrane.

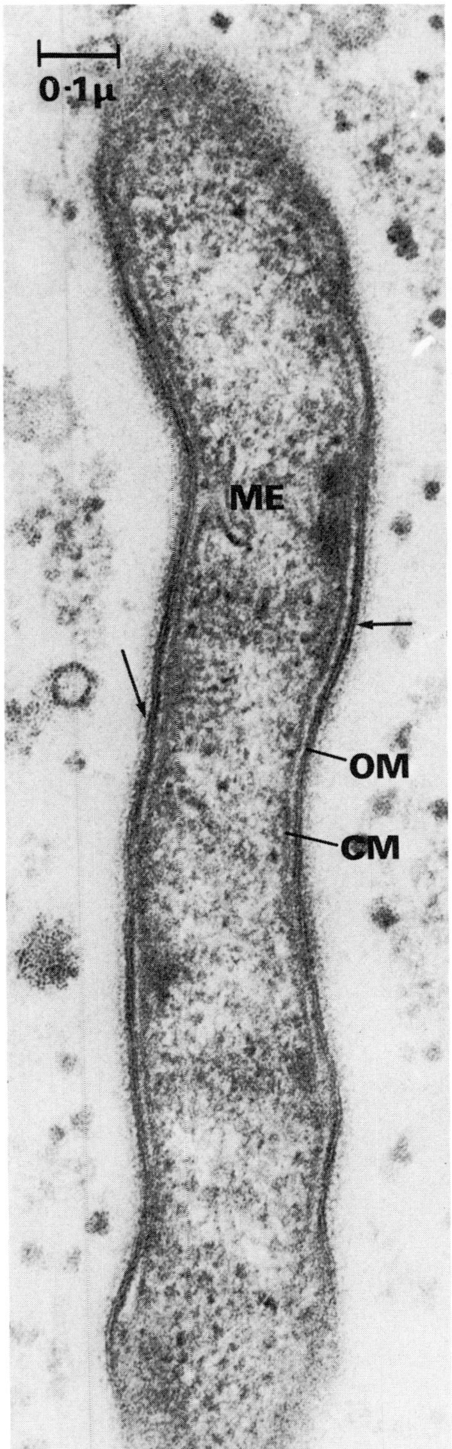

(h)

(i)

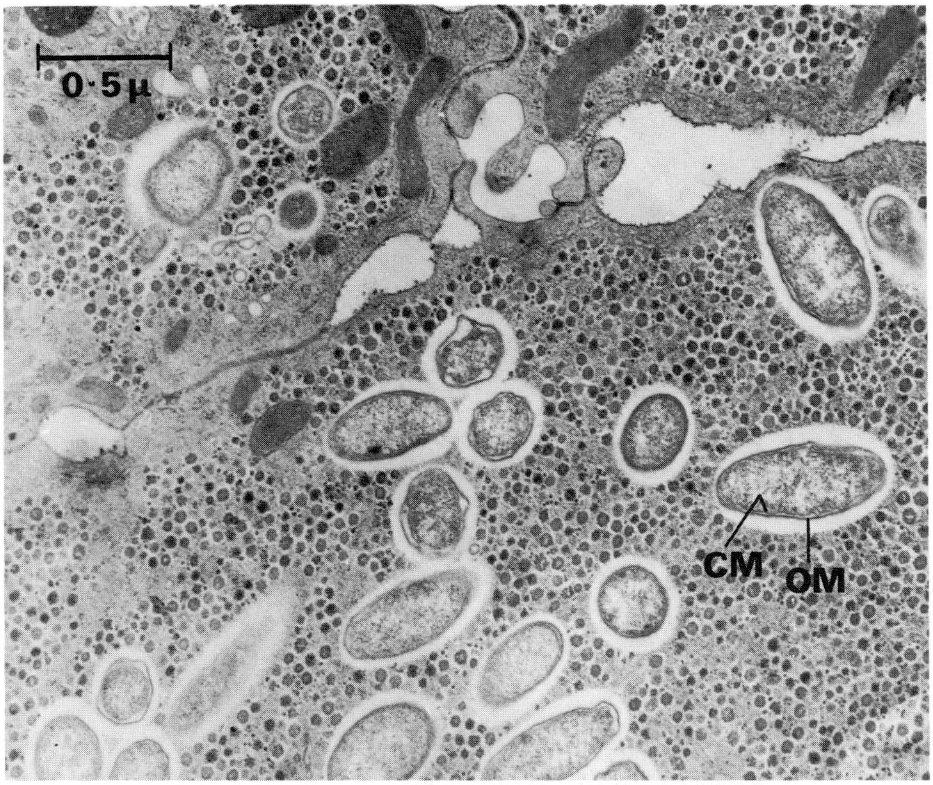

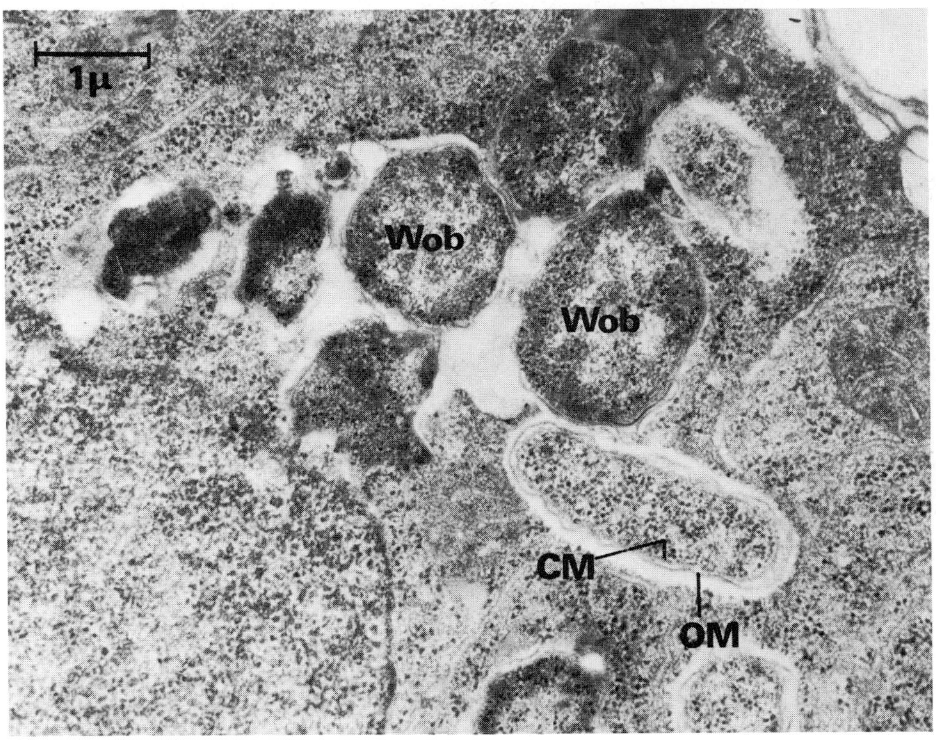

(j)

to that of other gram-negative bacteria; the outer and inner membranes of the wall are each approximately 10 nm thick (Figure 4.42(a)). The surface of the outer membrane is usually smooth, with some Phase I. *C. burnetii* containing a 20 nm thick, ruthenium red positive, fuzzy layer (Figure 4.42(b)).[32] Phase II cells are ruthenium red negative. The cytoplasmic region consists of peripherally-located ribosomes with a central nuclear region.

Of interest is the observation that *C. burnetii* phase I cells, when grown in yolk sacs and purified by differential centrifugation, are of two morphological types – small, compact rod-shaped cells and large, rounded cells (Figure 4.42(a)).[214] The nuclear region in the small cells is more dispersed. The reason for the presence of these two distinct morphological types is unknown; it is not clear, for example, whether the large cells, which are less electron dense, are viable cells.

The mechanism of host immunity to *C. burnetii* has not been clearly established; however, it has recently been found[119] that most host associated rickettsiae (i.e., *C. burnetii*) are localized *within* well-defined phagosomes at the periphery of the cell (Figure 4.42(b)). The fact that *C. burnetii* is found within phagosomes indicates that they entered these cells by phagocytic events, rather than by another mechanism(s) (i.e., penetration into cytoplasm).

When *C. burnetii* is treated with normal serum and exposed to host macrophages, two morphological forms occur: a rod-shaped form and an oval-round form (Figure 4.42(c), (d)). The oval to round cells are approximately 0·4 by 0·8 μm in diameter, have a thin cell wall and a dispersed nucleoid. The round cells occur late in the phagocytic ingestion process, and may represent degenerative cells. The rod-shaped cells are typically rickettsial in their morphology.

Bartonella

In the family Bartonellaceae, the members of the genus *Bartonella* infect both red blood cells and host tissue cells. Unlike the *Rickettsia*, *B. bacilliformis* is capable of growth on cell free media, as well as in red blood cells, and on endothelial cells of the liver, spleen, and lymph nodes. The vector is the sandfly, *Phlebotomus* sp. The bacteria are highly pleomorphic, and are usually found intracellularly.

The genus *Granhamella* is representative of the second genus in the family *Bartonellaceae*. These bacteria are similar to *Bartonella* but they are less pleomorphic and have only been observed within the red blood cells of mammals, excluding humans.

Figure 4.41. *Rickettsia* species.
(k), (l), (m), (n) *R. canada* 2678 within hypodermal tissue of wood tick. Extensive host cell groundplasm destruction is visible. Note the bacteria are surrounded by a thin basement membrane and microvilli.

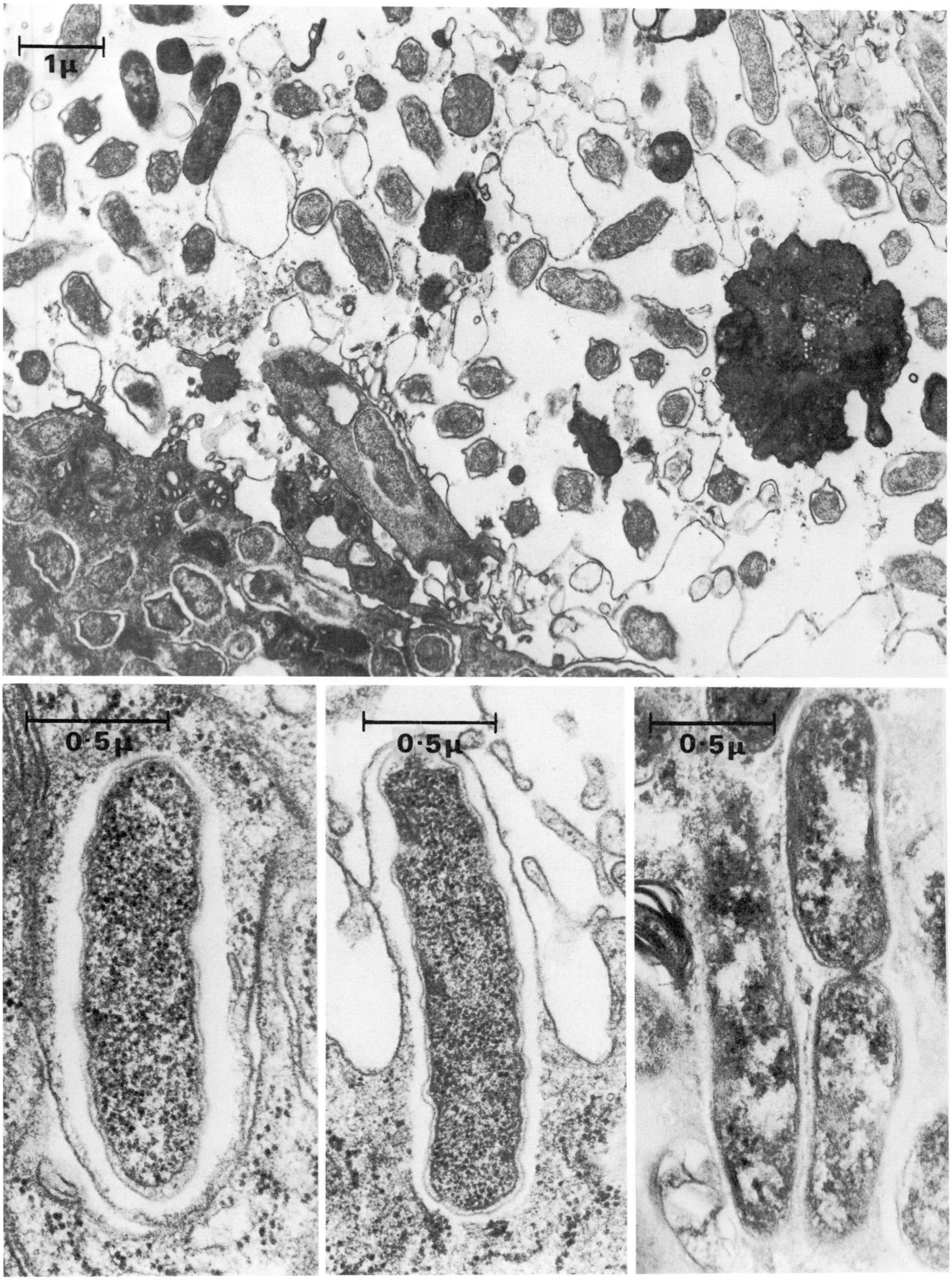

(k)

(l) (m) (n)

(o)

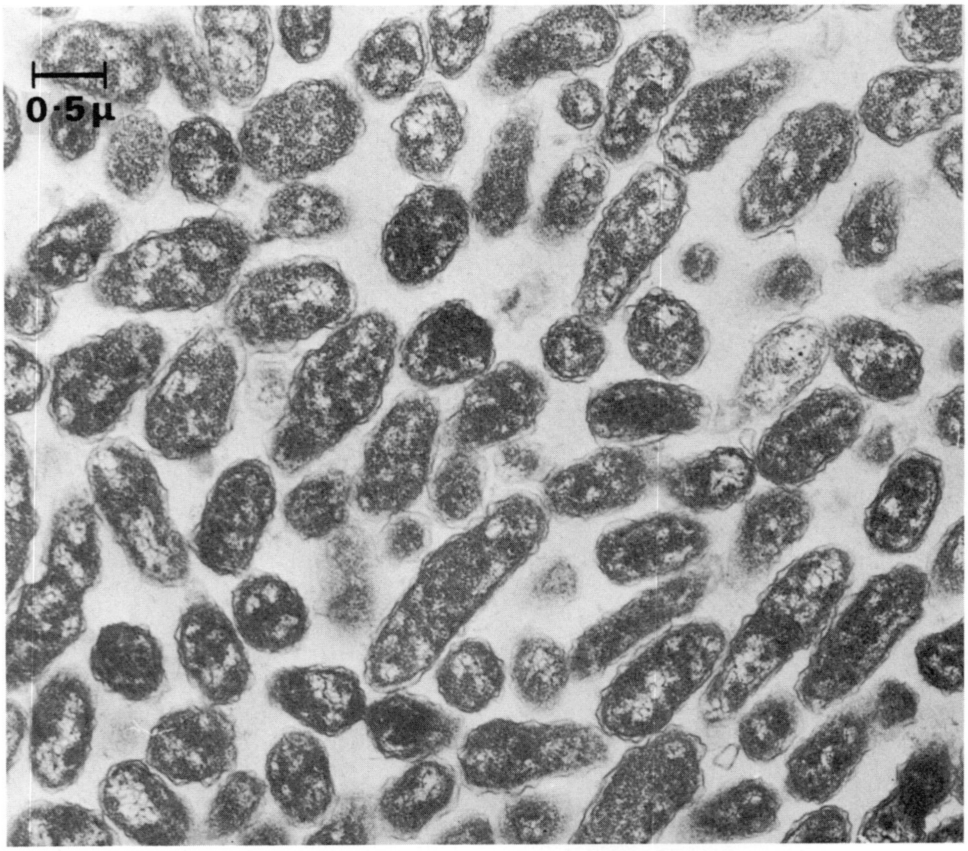

(p)

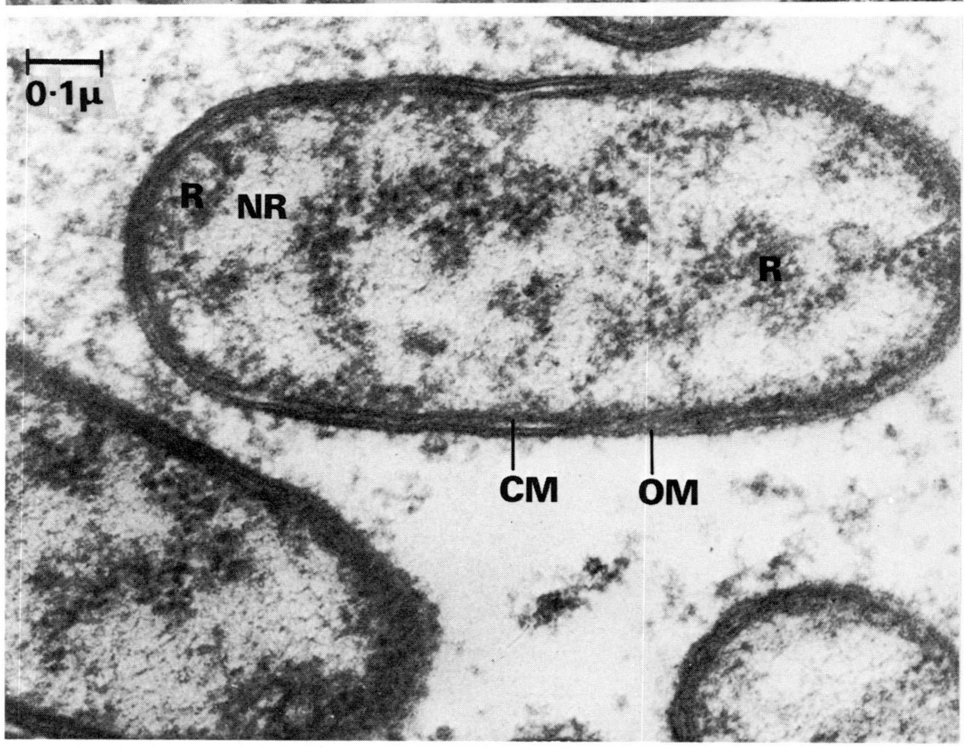

(q)

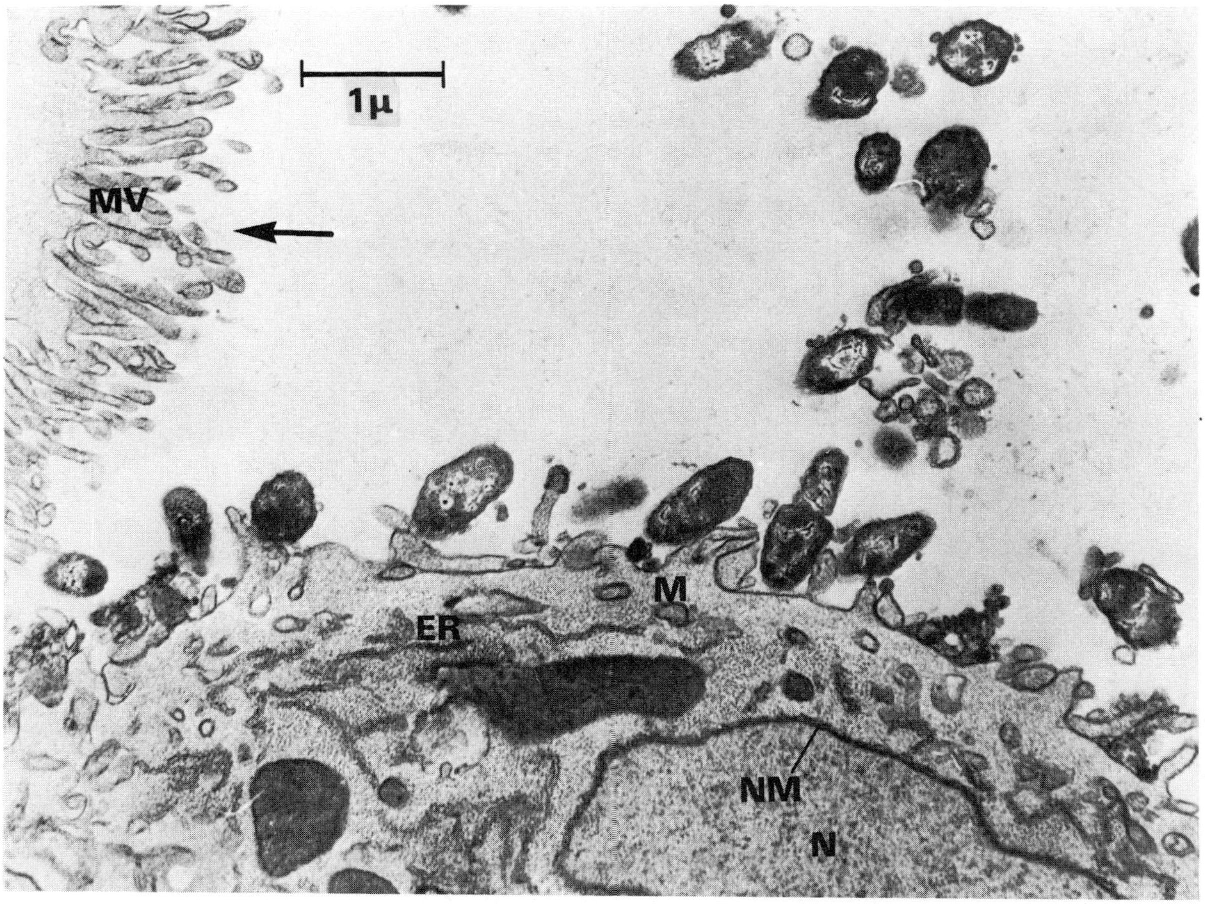

Figure 4.41. *Rickettsia* species.
(q) *R. quintana* from the intestine of an infected louse (as in (p)). At the apical portions of two surfaces of the louse intestinal epithelium, *R. quintana* is closely associated with the striated border. There is some host cell membrane destruction; note that uninfected microvilli (MV) appear normal (larger arrow). N: nucleus, NM: nuclear membrane, ER: endoplasmic reticulum.

Figure 4.41. *Rickettsia* species.
(o), (p) *Rickettsia quintana* Fuller. Cultured *in vitro* (o), the central region of a colony displays rod-shaped to slightly swollen cells. The cell wall is loose-fitting; *in situ* fixation from the intestine of an infected louse, *R. quintana* (p) possesses a smooth cell wall, electron-transparent periplasm, and dispersed nucleoid (NR) with patches of ribosomes (R). CM: cell membrane, OM: outer cell membrane.

Chlamydia

The Chlamydiae, are gram-negative, obligate intracellular parasites. In the 8th edition of *Bergey's Manual of Determinative Bacteriology* they are classified in order II Chlamydiales of the *Rickettsia*.[157]

Characteristic of this order is their mode of reproduction by constrictive binary fission as well as a complex "development cycle," in which the cells are surrounded by a dense cell wall or cloak, hence the genus epithet. This development cycle separates the Chlamydiae from the rickettsiae. Further, the *Chlamydiae* are not transmitted by insect vectors, a common feature of all *Rickettsia* species so far examined, but by direct host–host interaction.

The genus *Chlamydia* includes only two species, *C. trachomatis* and *C. psittaci*; the diseases they produce are both ecologically distinct and severe; *C. trachomatis* causes numerous oculourogenital diseases such as trachoma, urethritis, proctitis, conjunctivitis, and lymphogranuloma venereum, while *C. psittaci* has been implicated in numerous diseases of birds, cattle, sheep, and pigs. These include psittacosis, ornithosis, pneumonitis, enteritis, encephalomyelitis.

It is only in recent years that the morphology of these obligate intracellular parasites has been actively investigated. At least three morphological forms are present during their division cycle in infected tissue:[47, 120] a reticulate cell, 0·5–2 μm in diameter, an intermediate cell, 0·3–1 μm in diameter, and an elementary body, 0·26–0·35 μm in diameter. The elementary body is the infectious particle and represents the primary growth stage in the chlamydial cycle; it is housed within host cell inclusions and in several instances the host is seen to contain many elementary bodies. Reorganization of the elementary body leads to the second or developmental stage, the reticulate cell.

Entodermal explants of *C. psittaci* infected yolk sacs show a large number of the three morphological chlamydial cell types (Figure 4.43). All three types are enclosed in membrane-limited vacuoles.

The reticulate bodies (Figure 4.43(a)–(d)) are hourglass in shape, and are enclosed by two unit membranes representing the outer membrane of the cell wall and the cell membrane. The cytoplasm contains a fibrillar nuclear region with limiting electron-dense ribosomes. Cell division and cellular reorganization of the reticulate body leads to the formation of the numerous infectious elementary bodies, which can survive an extracellular environment to infect other host cells.

Numerous spherical-shaped intermediate bodies are also present in

---

Figure 4.42. *Coxiella burnetii* phase I.
(a) Two morphological cell types are present: a small, compact, electron-dense cell, and a large, rod-shaped cell with coagulated nucleoid. The larger cell probably represents a nonviable form.
(b) When grown in mouse L-cells and reacted with ruthenium red, *C. burnetti* is seen outside the host cell membrane and in close association with it. Note the dense ruthenium red positive material on the outer surface of the bacteria as well as lining the surface epithelium.

(a)

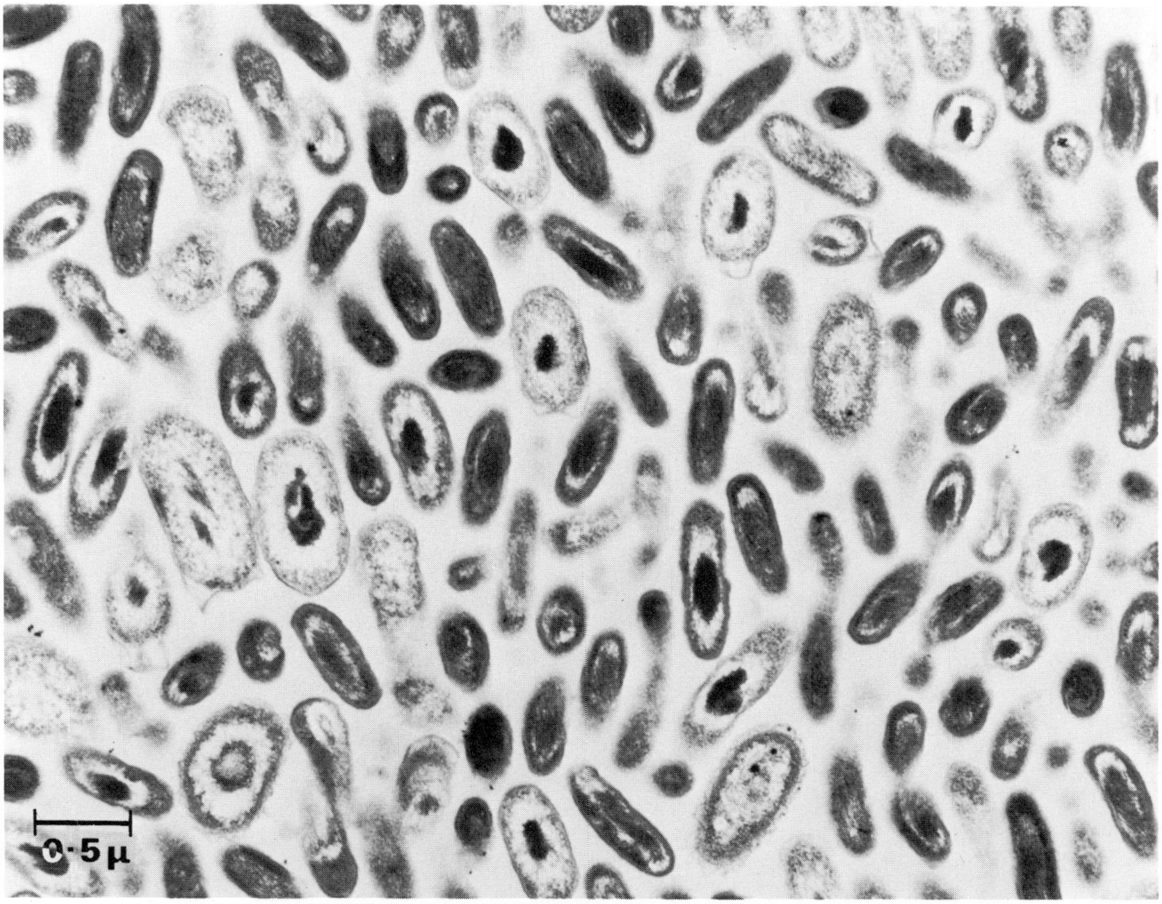

(b)

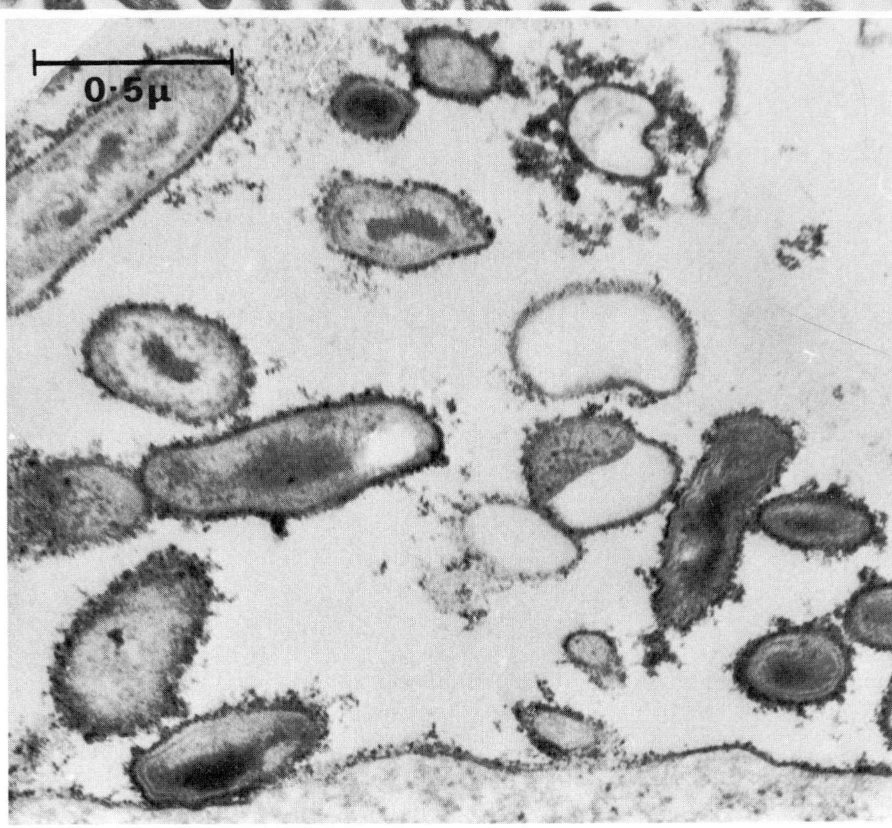

Figure 4.42. ***Coxiella burnetii*** **phase I.**
(c), (d) Numerous small, large, and dividing coxiella fill a macrophage and are present in well-defined phagosomes (PV). At high magnification (d) the bacteria have a smooth, tight fitting cell wall and central nucleoid. Some bacterial lysis is apparent.

(c)

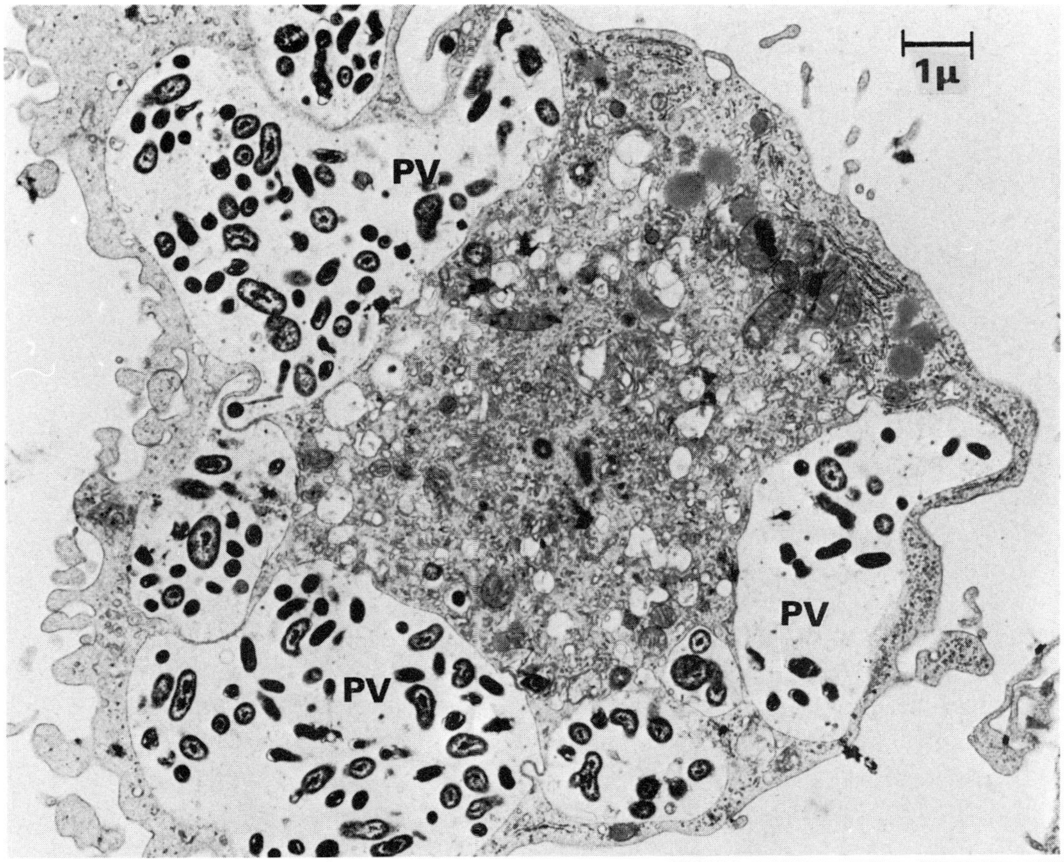

(d)

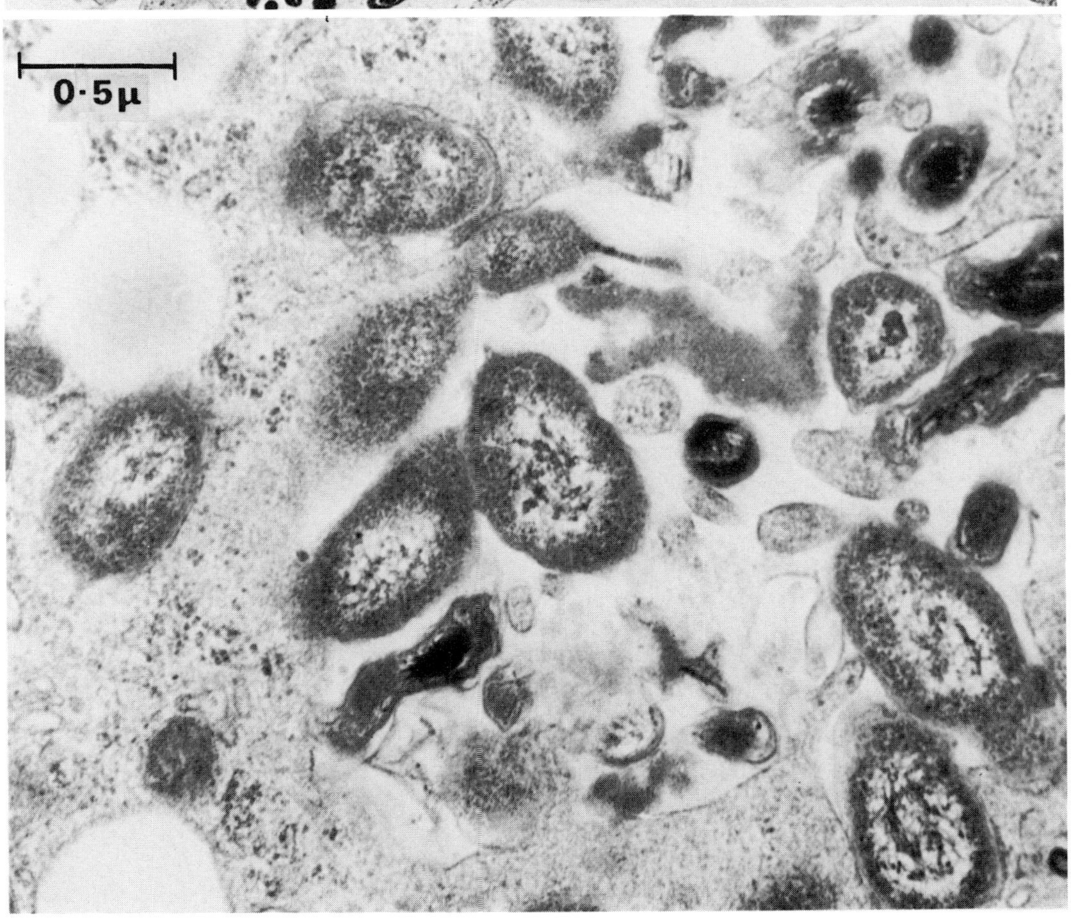

entodermal cells of infected yolk sac explants (Figure 4.43(f)). The intermediate bodies are similarly enclosed by a membrane; however, the membranes are more closely apposed to each other in the intermediate bodies than in the reticulate bodies. The electron-dense ribosomal particles are peripherally located and are separated from the condensed, fibrous nuclear region by an electron-transparent space.

The small elementary body is also enclosed by two closely apposed unit membranes (Figure 4.43(d),(e)). The cytoplasmic region of the elementary body contains condensed, electron-dense ribosomes with an eccentrically located nucleus, located in most instances adjacent to the inner membrane of the cell. This is in contrast to the reticulate and intermediate bodies, where the ribosomes and nuclear fibrils are evenly dispersed throughout the cell cytoplasm.

When elementary bodies of *C. psittaci* are fragmented, the cell wall fragments into structures consisting of regular patterns of hexagonally arranged, doughnut-shaped subunits with a periodicity of 18 nm. This pattern has been confirmed by the freeze-fracture etching technique where it appears as a rosette layer, arranged hexagonally, each rosette being 19–20 nm in diameter and composed of several subunits which surround a central zone.

The meningopneumonitis strain of *C. psittaci* cultured in L-cells has also been examined by freeze-fracture etching (Figure 4.43(g)–(i). The L-cells contain numerous smooth surfaced inclusions, the surfaces of which consist of regularly arranged particles, which occur in groups (Figure 4.43(i)). Each particle is surrounded by six other particles. Both convex and concave profiles are observed, and the reticulate bodies are surrounded by a thin layer, possibly cell wall.

4.3.18. Legionnaires bacterium (*Legionella*)

Procaryotic cells, identical in morphology (Figure 4.44) to gram-negative bacteria have been isolated from an outbreak of pneumonia from individuals attending an American Legion Convention in Philadelphia, Pennsylvannia. The epidemiological characteristics, as well as the initial isolation of the "Legionnaires' bacterium" and its possible role in pneumonia and other respiratory disease has been reviewed in previous publications.[37, 69] A recent publication[110] has described the morphological characteristics of several microorganisms isolated and cultured in yolk sac tissue from several legionnaire fatalities.

The surface of negatively stained cells (Figure 4.44(a)) of the "Legionnaire's bacterium" are highly wrinkled, again characteristic of gram-negative bacteria. Note, however, that the outermost layer of the outer membrane is

Figure 4.43. *Chlamydia psittaci.*
(a), (b), (c) Strain CAL 10. Elementary bodies (EB) are shown in (a) and reticulate bodies (RB) in (b), (c). The cells divide by constrictive division. In an infected L-cell (c), the reticulate bodies (RB) are enclosed in a limiting membrane (arrows). Pg: peptidoglycan, R: ribosome, NR: nuclear region, OM: outer cell membrane, CM: cell membrane.

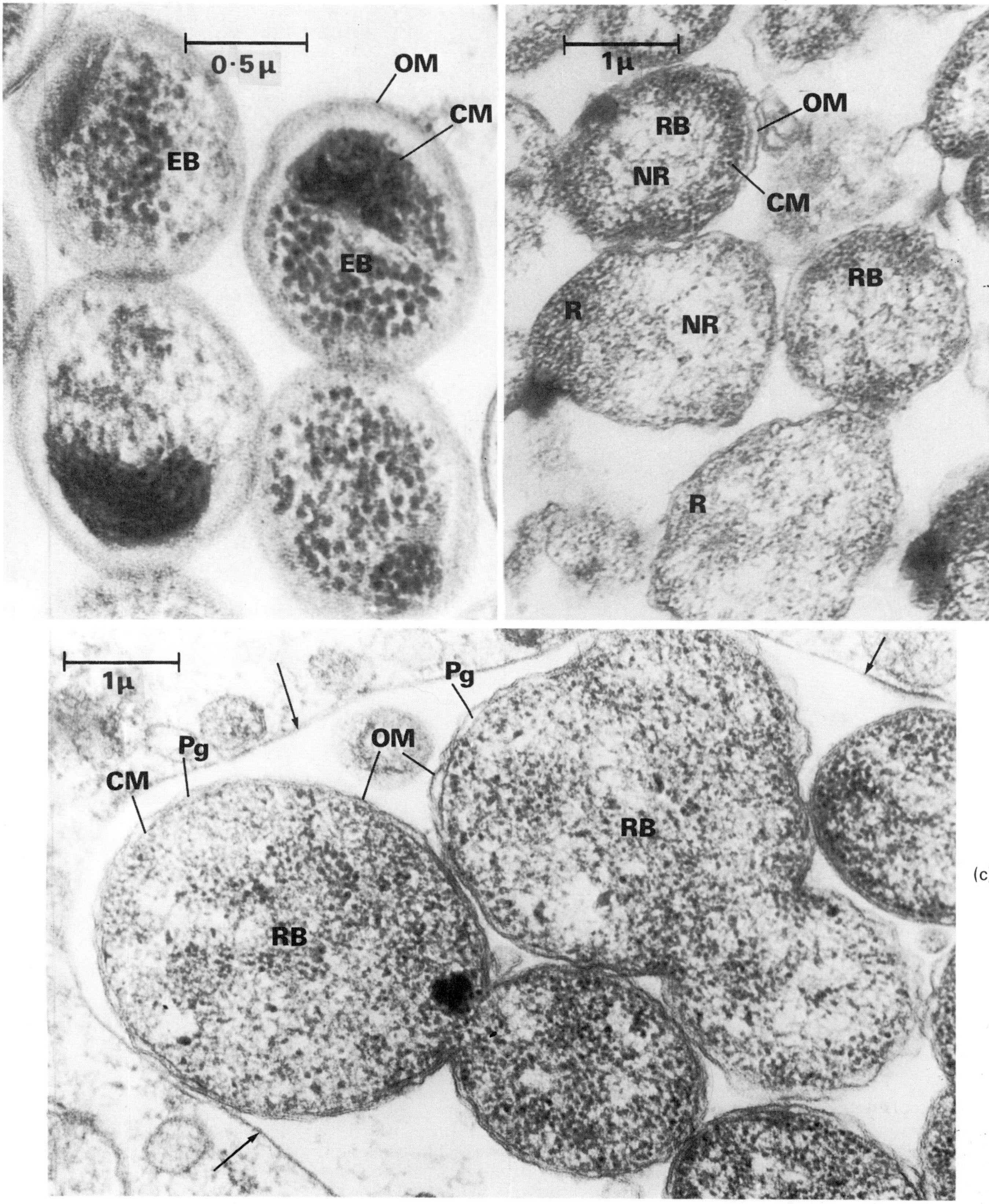
(a)
0·5μ
OM
CM
EB
EB
(b)
1μ
RB
OM
NR
CM
RB
R
NR
R
(c)
1μ
Pg
OM
Pg
CM
RB
RB

smooth, with only a few electron-dense blebs emerging from the membrane. The picture of thin sections of chemically fixed cells (Figure 4.44(b)) reveals a thin, rod-shaped cell consisting of a loose-fitting outer membrane separated from the inner cell membrane by a large electron-transparent space. One of the important in bacterial-cell interaction. These isolates have been named *Legionella pneumophila*.

Figure 4.43. *Chlamydia psittaci.*
(d), (e), (f) Strain 6BS. In an infected L-cell (d) the strain consists of numerous pleomorphic chlamydial cells similar to that seen in an infected yolk sac explant (e). Note the characteristic hourglass-shaped reticulate body (RB) in this cell section, as well as pleomorphic reticulate bodies and dividing reticulate cells. The late stage of reticulate body division is seen in the yolk sac explant section in (f); note the well-developed division planes as well as completed crosswall septa. Also present in (f) is an intermediate body (IB) and a less condensed elementary body (EB). OM: outer cell membrane, CM: cell membrane, NR: nuclear region, ER: endoplasmic reticulum.

(d)

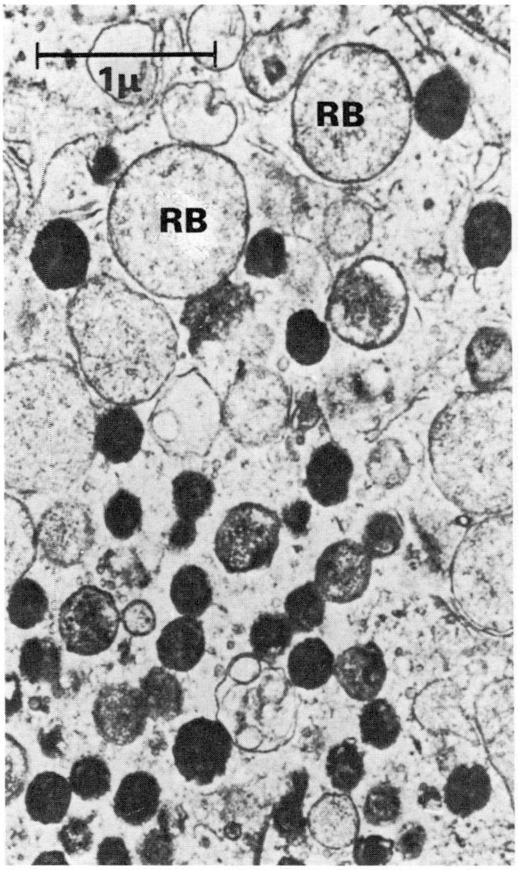

(e)

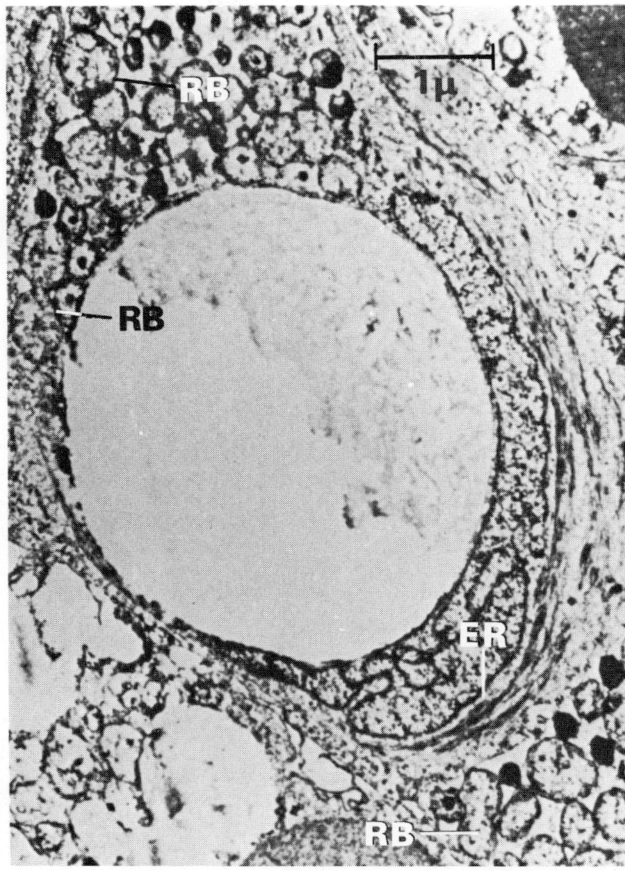

(f)

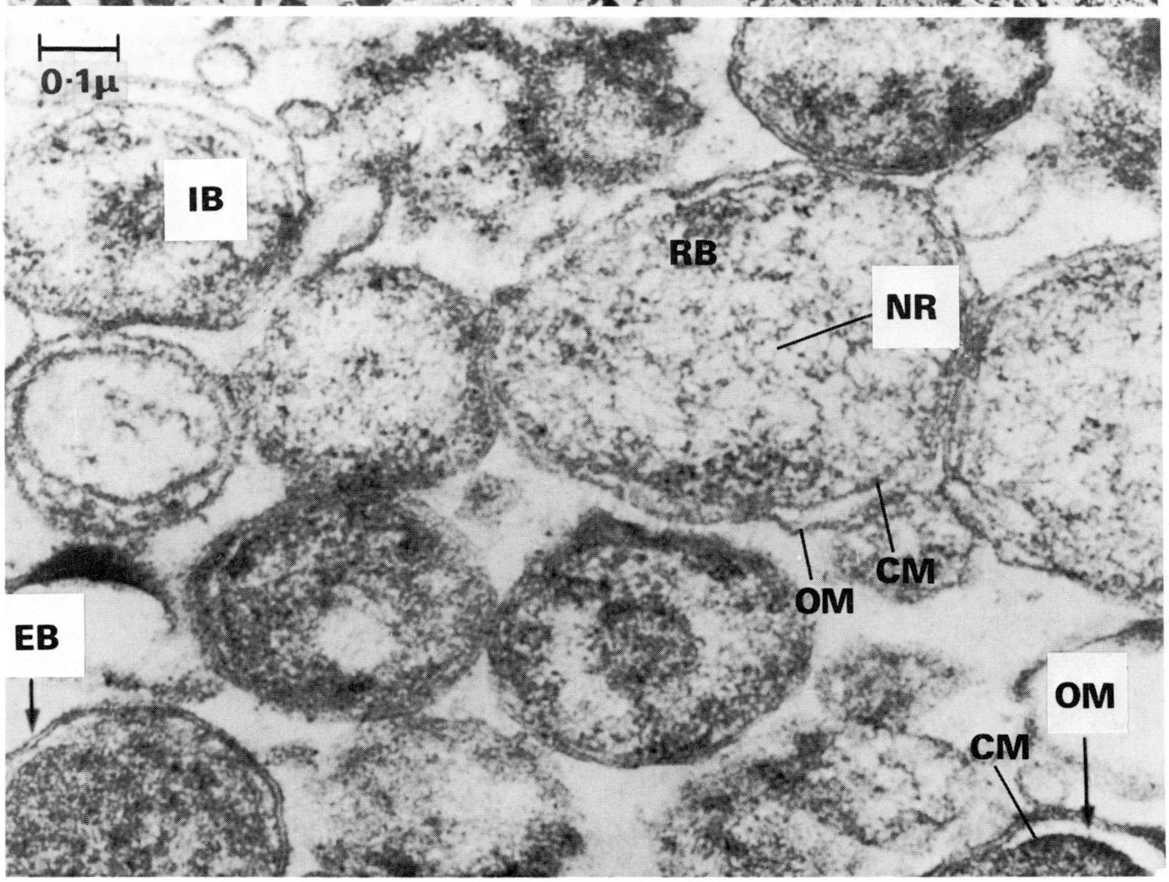

Figure 4.43. ***Chlamydia psittaci.***
(g), (h) Freeze-fracture etched meningopneumonitis strain. In (g) numerous reticulate bodies with convex and concave fracture faces are present. The large inclusion body in (h) has its outer membrane intact, the surface of which is smooth and predominantly nonparticulate; however, there are some particles in clusters of 6–30 in a definite pattern.

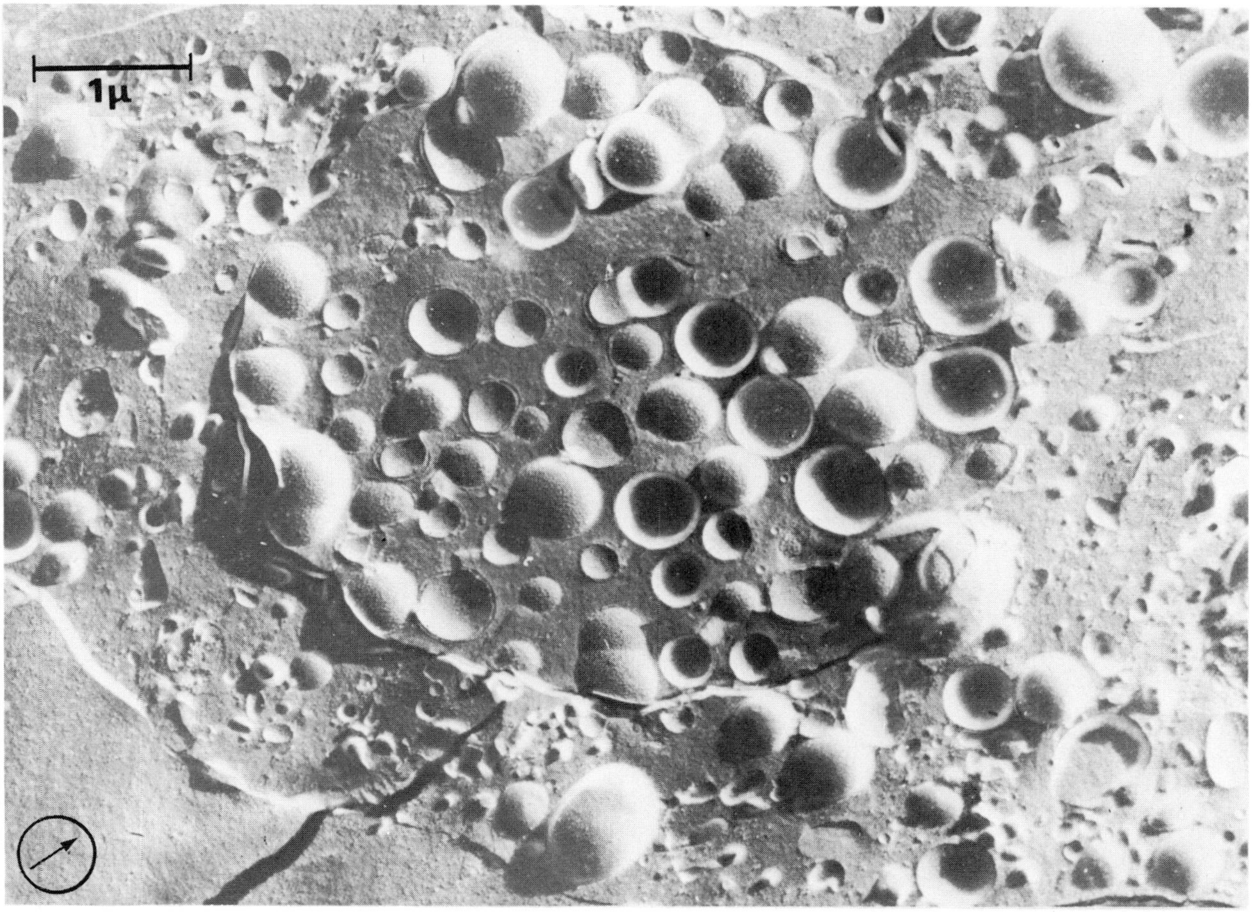

(g)

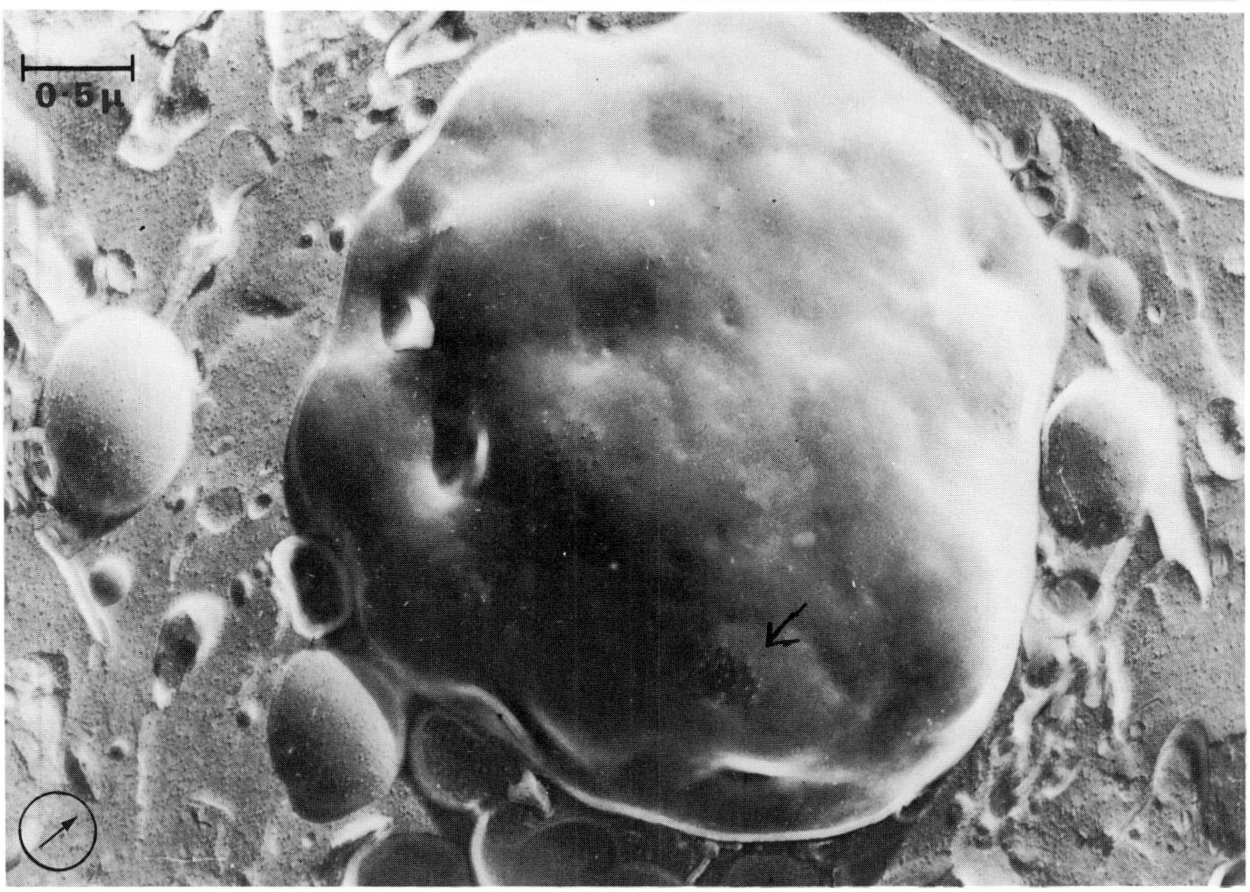

(h)

Figure 4.43. ***Chlamydia psittaci.***
(i) Meningopneumonitis. A fracture at the level of the reticulate groundplasm reveals numerous reticulate bodies, some of which are covered with particles in hexagonal array.

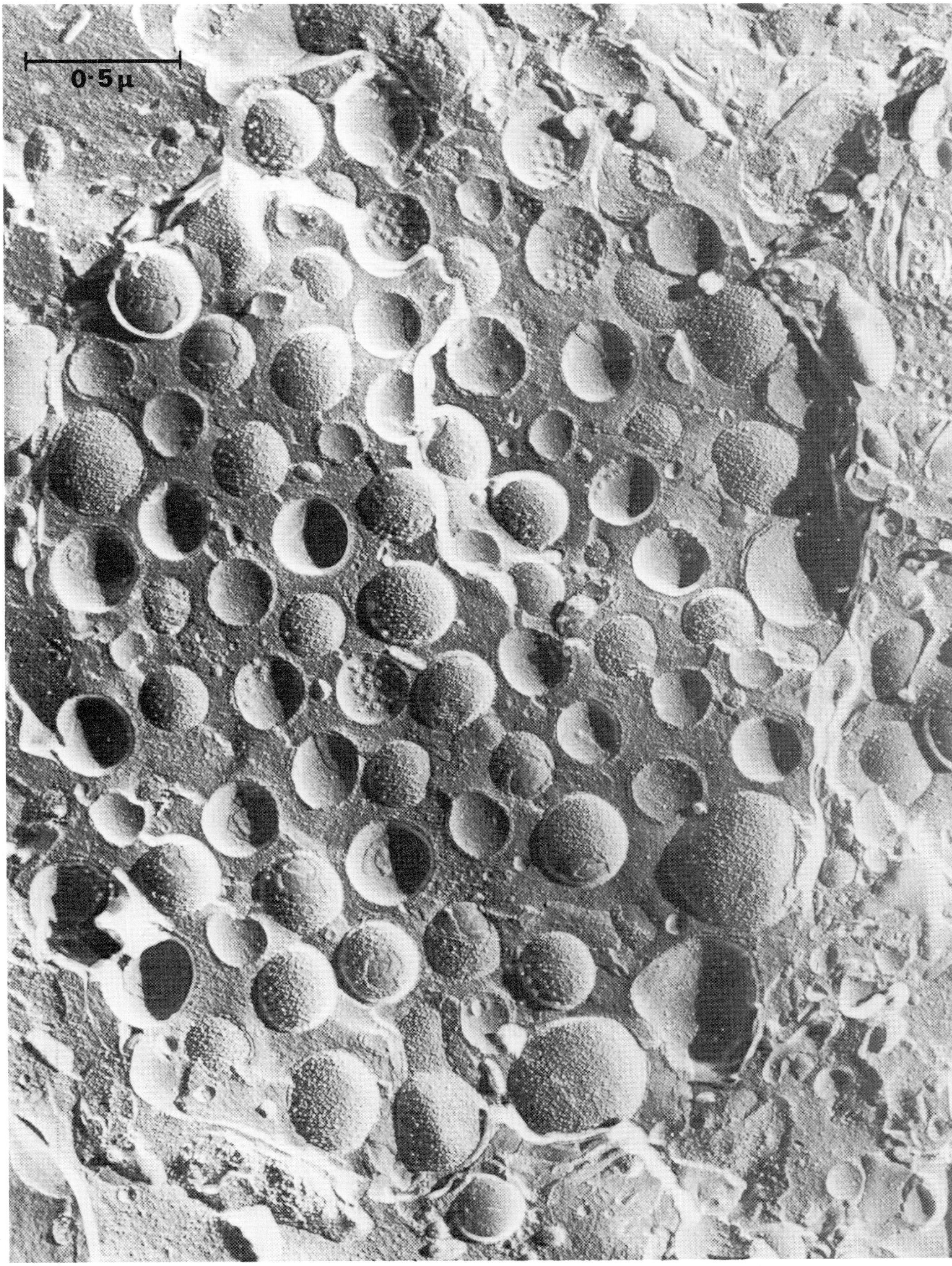

(i)

(a)

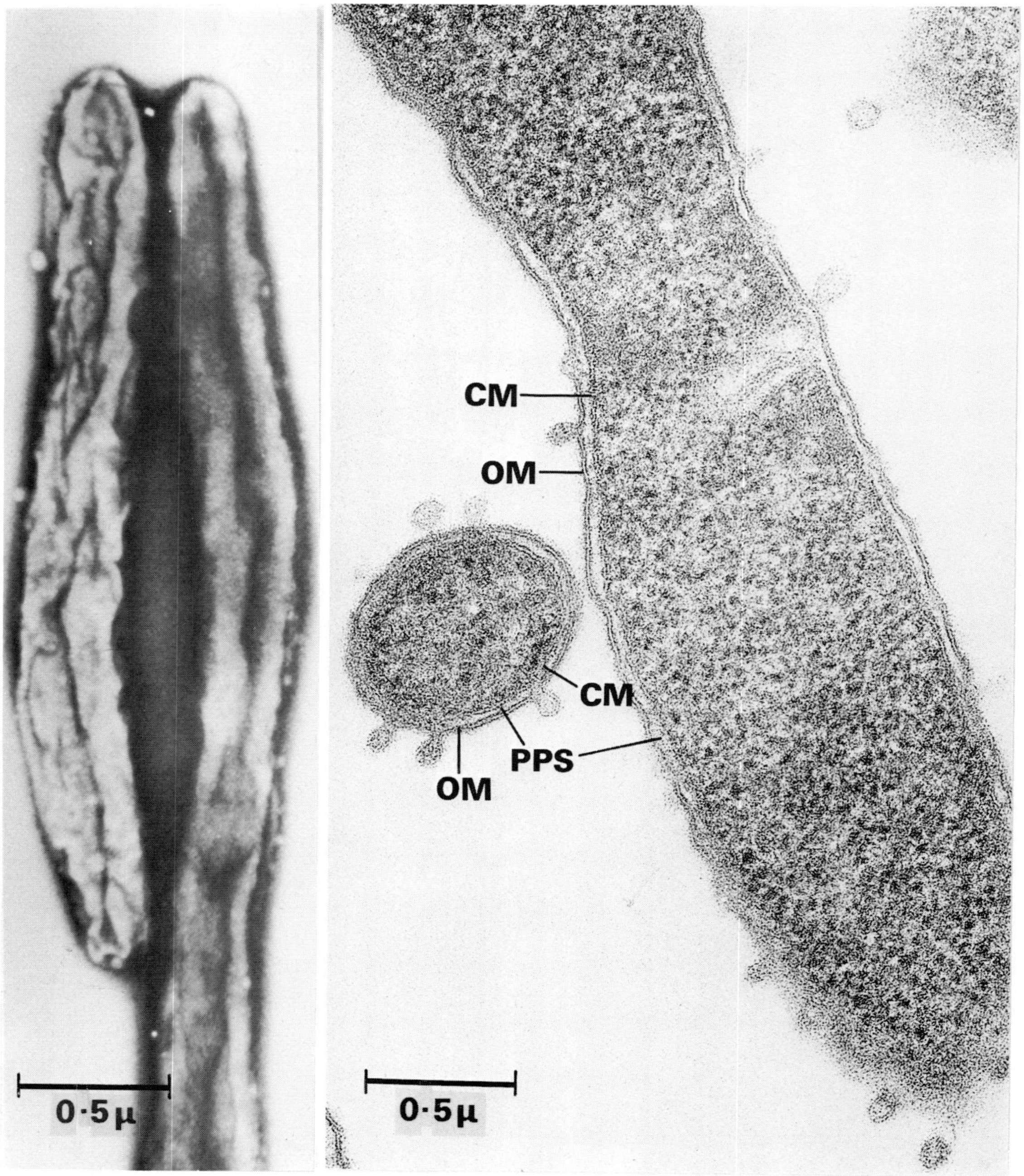

(b)

Figure 4.44. Legionnaire bacterium (*Legionella pneumophila*).
(a) Negative stain and (b) thin section of bacterial isolate from Legionnaire's disease. In (a) the cell surface is wrinkled with numerous folds or channels typical of gram-negative cells. The gram-negative cell morphology is confirmed in (b); the outer membrane (OM) is loose-fitting, enclosing a slightly dense periplasmic space (PPS). The cell membrane is distinct, and encloses an electron-dense cytoplasm (CM). Note in (b) the blebs which emerge from the surface of the outer membrane.

Table 4.1. Eucaryotic and procaryotic strains presented

Figure	Organism	Author(s)	Acknowledgement
4.1(a)	*Bacillus fastidiosus*	S. C. Holt, E. R. Leadbetter	*Bact. Rev.*, **33**, 346, 1969.
4.1(b)	Oral bacillus	L. Meyers, S. C. Holt, E. R. Leadbetter	Unpublished communication
4.2(a)	*Bacillus subtilis* + CP1	S. C. Holt, C. B. Thorne	Unpublished communication
4.2(b)	*Protaminobacter ruber*	S. C. Holt, E. R. Leadbetter	Unpublished communication
4.2(c)	Chicken chromatin	C. Woodcock	Unpublished communication
4.2(d)	Lambda DNA	G. Scangos, A. Reiner	Unpublished communication
4.3(a)–(d)	*Escherichia coli* F515–08	T. J. MacAlister *et al.*	*J. Bact.*, **130**, 318, 1977.
4.4(a),(b)	*Trichomonas vaginalis*	S. C. Holt	Unpublished communication
4.4(c)	Freeze-fracture diagram	Branton *et al.*	*Science*, **90**, 54, 1975.
4.4(d)	*Cryptococcus neoformans*	S. C. Holt, M. R. Edwards	Unpublished communication
4.4(e)	*Histomonas gallinae*	S. C. Holt	Unpublished communication
4.5(a)	Human liver cell, Whipple's disease	I. Sternlieb	Unpublished communication
4.5(b)	*Trichomonas gallinae* Rivolta	C. F. T. Mattern *et al.*	*J. Protozool.*, **14**, 320, 1967.
4.5(c)	*Porphyridium cruentum*	E. Gantt, S. F. Conti	*J. Cell Biol.*, **26**, 365, 1965.
4.6(a)	*Bacillus macroides*	R. Linder, S. C. Holt	Unpublished communication
4.6(b)	*Bacillis anthracis*	S. C. Holt	Unpublished communication
4.6(c)	*Clostridium sporogenes*	J. F. M. Hoeniger, D. E. Bradley	*Can. J. Microbiol.*, **17**, 1572, 1971.
4.6(d)	*Clostridum botulinum* type E	G. W. Brown *et al.*	*J. Bact.*, **105**, 1207, 1971.
4.6(e),(f)	*Clostridum tetani* NCTC9575	J. F. M. Hoeniger, H. D. Tauschel	*J. med. Microbiol.*, **7**, 425, 1974.
4.6(g)	*Bacillus fastidiosus*	E. R. Leadbetter, S. C. Holt	*J. gen. Microbiol.*, **52**, 299, 1968.
4.6(h)	*Clostridium botulinum* 17B	K. E. Stevenson, R. H. Vaughn	*Can. J. Microbiol.*, **18**, 1717, 1972.
4.6(i)	*Bacillus anthracis*	S. C. Holt	Unpublished communication
4.6(j)	*Bacillus fastidiosus*	see Figure 4.6(g)	
4.6(k)	*Bacillus polymyxa*	J. Murphy, L. L. Campbell	*J. Bact.*, **98**, 737, 1969.
4.6(l)	*Clostridium botulinum* type E	W. Hodgkiss, F. J. Ordal, D. C. Cann	*Can. J. Microbiol.*, **12**, 1283, 1966.
4.6(m)	*Clostridium botulinum* type E	A. J. B. Suzuki, N. Grecz	*Can. J. Microbiol.*, **18**, 1651, 1972.
4.7(a)–(c)	*Clostridium perfringens*	E. Kaiserling	*Virchows Arch. path. Anat. Physiol.*, **359**, 345, 1973.
4.8(a)	*Listeria monocytogenes*	L. Schultz, M. W. Wilder	Unpublished communication
4.8(b)	*Listeria monocytogenes* type 2	L. O. Zwillenberg	*Antonie van Leeuwenhoek J. Serol.*, **30**, 154, 1964.
4.9(a),(b)	*Actinomyces odontolyticus* ATCC17929	J. J. Duda, J. M. Slack	*J. gen. Microbiol.*, **71**, 63, 1972.
4.9(c)	*Actinomyces viscosus*	D. Birdsell	Unpublished communication
4.9(d)	*Actinomyces naeslundii*	D. Birdsell	Unpublished communication
4.9(e)	*Actinomyces israelii*	D. Birdsell	Unpublished communication
4.9(f),(g)	*Actinomyces viscosus* VP17281	A. E. Girard, B. H. Jacius	*Arch. oral Biol.*, **19**, 71, 1974.
4.9(h)	*Actinomyces viscosus* ATCC 15987 "T6"	D. Birdsell	Unpublished communication
4.9(i)	*Actinomyces naeslundii* V 16	D. Birdsell	Unpublished communication
4.9(j)	*Actinomyces viscosus* WVU745	R. A. Bulman, R. J. Stretton	*Cytobios*, **10**, 103, 1974.
4.9(k)	*Actinomyces viscosus* WVU398B	R. A. Bulman	*Cytobios*, **10**, 103, 1974.
4.10	*Nocardia asteroides* 10905	L. Bourgeois, B. L. Beaman	*J. Bact.*, **127**, 584, 1976.
4.11(a)	*Mycobacterium tuberculosis* $H_{37}Rv102$	L. Barksdale, K. S. Kim	*Bact. Rev.*, **41**, 217, 1977.
4.11(b)	*Mycobacterium tuberculosis* $H_{37}Rv102$	L. Barksdale	*Bact. Rev.*, **41**, 217, 1977.
4.11(c)	*Mycobacterium smegmatis*	S. C. Holt	Unpublished communication
4.11(d)	*Mycobacterium phlei*	S. C. Holt	Unpublished communication
4.11(e)	*Mycobacterium* sp. NQ	K. S. Kim *et al.*	*J. Bact.*, **125**, 739, 1976.
4.11(f)	*Mycobacterium* sp. NQ	K. S. Kim *et al.*	*J. Bact.*, **125**, 739, 1976.
4.11(g)	*Mycobacterium leprae*	P. D'Arcy Hart	Unpublished communication

Table 4.1. Eucaryotic and procaryotic strains presented (*continued*)

Figure	Organism	Author(s)	Acknowledgement
4.12(a),(b)	*Bifidobacterium bifidum*	J. A. Poupard *et al.*	*Bact. Rev.*, **37**, 136, 1973.
4.13(a)	*Corynebacterium diphtheriae*	See Figure 4.8(b)	
4.13(b)	*Corynebacterium diphtheriae*	See Figure 4.8(b)	
4.13(c),(d)	*Corynebacterium diphtheriae*	L. Barksdale	*Bact. Rev.*, **34**, 378.
4.13(e)	*Corynebacterium diphtheriae* var mitis. NCTC 3989	See Figure 4.22(c)	
4.13(f)	*Corynebacterium diphtheriae* C4	T. Kawata, K. Masada	*Jap. J. Microbiol.*, **16**, 515, 1972.
4.13(g),(h),(i)	*Corynebacterium minutissimum*	L. F. Montes, S. H. Black	*J. Invest. Derm.*, **48**, 342, 1967.
4.14(a),(b)	*Rothia dentocariosa*	G. D. Roth *et al.*	*J. Microscopie*, **95**, 45, 1976.
4.15	*Lactobacillus acidophilus* NCTC4504	See Figure 4.22(c)	
4.16(a)	*Staphylococcus aureus* ATCC 6538P	R. M. Cole *et al.*	*Ann. N.Y. Acad. Sci.*, **236**, 22, 1974.
4.16(b) 4.16(g)	*Staphylococcus aureus* R 3	M. Fooke-Achterrath *et al.*, B. Menge	*J. Ultrastruct. Res.*, **49**, 270, 1974.
4.16(c),(d)	*Staphylococcus aureus* R 3	D. B. Brewer	Unpublished communication
4.16(e),(f)	*Staphylococcus epidermidis*	I. W. DeVoe *et al.*	*Can. J. Microbiol.*, **19**, 525, 1973.
	Micrococcus radiodurans RI	M. J. Thornley *et al.*	*Arch. Mikrobiol.*, **51**, 267, 1965.
4.16(h)	*Micrococcus mucilaginosus* ATCC 25296	M. T. Silva *et al.*	*J. Submicrosc. Cytol.*, **9**, 53, 1977.
4.17(a)	*Sarcina maxima*	S. C. Holt, E. Canale-Parola	*J. Bact.*, **93**, 399, 1967.
4.17(b)	*Sarcina ventriculi*	S. C. Holt	*J. Bact.*, **93**, 399, 1967.
4.17(c)	*Sarcina maxima*	S. C. Holt	*J. Bact.*, **93**, 399, 1967.
4.17(d)	*Sarcina ventriculi*	S. C. Holt	Unpublished communication
4.17(e)	*Sarcina ventriculi*	See Figure 5.17(a)	
4.18(a),(b)	Group A *Streptococcus*	A. D. Glick *et al.*	*Infec. Immunity*, **4**, 772, 1971.
4.18(c)	Group D *Streptococcus*	R. G. McCandless *et al.*	*J. Bact.*, **96**, 1400, 1968.
4.18(d)	*Streptococcus faecalis* ATCC9790	M. C. Higgins and L. Daneo-Moore	*J. Cell Biol.*, **61**, 288, 1974.
4.18(e)	Group A *Streptococcus*	E. N. Fox	*Bact. Rev.*, **38**, 57, 1974.
4.18(f)	Group A *Streptococcus* type 6 543/100/15	J. Swanson *et al.*	*J. exp. Med.*, **130**, 1063, 1969.
4.18(g)	Group A *Streptococcus* 543 MA(M)	J. Swanson *et al.*	*J. exp. Med.*, **130**, 1063, 1969.
4.18(h)	Group A *Streptococcus*	See Figure 4.18(a)	
4.18(i)	*Streptococcus pyogenes* STA628	R. P. Ellen, R. J. Gibbons	*Infec. Immunity*, **5**, 826, 1972.
4.18(j)	Group A *Streptococcus*	See Figure 4.18(a)	
4.18(k)	*Streptococcus mutans* c	S. C. Holt, E. R. Leadbetter	*Can. J. Microbiol.*, **22**, 475, 1976.
4.18(l)	*Streptococcus salivarius* AHT	S. C. Holt, E. R. Leadbetter	*Can. J. Microbiol.*, **22**, 475, 1976.
4.18(m)	*Streptococcus mutans* a	S. C. Holt, E. R. Leadbetter	*Can. J. Microbiol.*, **22**, 475, 1976.
4.18(n)	*Streptococcus mutans* GS–5	M. C. Johnson *et al.*	*J. Bact.*, **118**, 304, 1974.
4.18(o)	*Streptococcus mutans*	M. C. Johnson *et al.*	*J. Bact.*, **118**, 304, 1974.
4.18(p)	*Streptococcus mutans* GS–5	M. C. Johnson *et al.*	*J. Bact.*, **118**, 304, 1974.
4.19(a)	*Diplococcus pneumoniae* type III ATCC6303	E. L. Springer, I. L. Roth	*J. gen. Microbiol.*, **74**, 21, 1973.
4.19(b)	*Diplococcus pneumoniae*	E. L. Springer, I. L. Roth	*J. gen. Microbiol.*, **74**, 21, 1973.
4.19(c)	*Diplococcus pneumoniae* type III	R. F. Baker, C. G. Looslo	*Lab. Invest.*, **15**, 716, 1966.
4.19(d)	*Diplococcus pneumoniae* type III	R. F. Baker, C. G. Looslo	*Lab. Invest.*, **15**, 716, 1966.
4.19(e)	*Diplococcus pneumoniae* type III	G. D. Cagle	*Can. J. Microbiol.*, **21**, 395, 1975.
4.20(a)	*Brucella canis* 317 (R)	G. Dubray, H. Plamnict	*Int. Symp. Brucellosis (II) Rabat, 1975. Devl. biol. Standard*, **31**, 68–91.
4.20(b)	*Brucella canis* 317 (R)	G. Dubray, H. Plamnict	
4.20(c)	*Brucella abortus* 544 (S)	G. Dubray, H. Plamnict	
4.20(d)	*Brucella abortus* 45/20(R)	G. Dubray, H. Plamnict	
4.20(e)	*Brucella melitensis* H38 (S)	G. Dubray, H. Plamnict	
4.20(f)	*Brucella abortus*-hamster kidney	Hulten	Unpublished communication

Table 4.1. Eucaryotic and procaryotic strains presented (*continued*)

Figure	Organism	Author(s)	Acknowledgement
4.21(a)	*Bordetella bronchiseptica*	D. A. Bemis, *et al.*	*J. clin. Microbiol.*, **5**, 471, 1977.
4.21(b)–(d)	*Bordetella bronchiseptica*	G. W. Richter, Y. Kress	*J. Bact.*, **94**, 1216, 1967.
4.21(e), (f)	*Bordetella bronchiseptica*	D. A. Bemis	Unpublished communication
4.22(a)	*Hemophilus vaginalis* 594 (ATCC 14018)	B. S. Criswell *et al.*	*J. Bact.*, **109**, 930, 1972.
4.22(b)	*Hemophilus vaginalis* 594 (ATCC 14018)	B. S. Criswell *et al.*	*Can. J. Microbiol.*, **17**, 865, 1971.
4.22(c)	*Hemophilus influenzae* NCTC 8476	A. Reyn *et al.*	*Can. J. Microbiol.*, **12**, 1125, 1966.
4.22(d)	*Hemophilus ducreyi* CIP 542	M. Kilian, J. Theilade	*Int. J. Syst. Bact.*, **25**, 351, 1975.
4.22(e)–(g)	*Hemophilus influenzae* type b. (HIT6).	G. V. Doern, F. L. A. Buckmire	*J. Bact.*, **127**, 523, 1976.
4.23(a)–(m)	*Bacteroides melaninogenicus*	D. Woo *et al.*	Unpublished communication
4.24(a)	*Neisseria gonorrhoeae*	J. Swanson *et al.*	*J. exp. Med.*, **134**, 886, 1971.
4.24(b)	*Neisseria gonorrhoeae* type 2	J. Swanson	*J. exp. Med.*, **136**, 1258, 1972.
4.24(c)	*Neisseria gonorrhoeae* type 2	Z. A. McGee *et al.*	*Infec. Immunity*, **14**, 266, 1976.
4.24(d)	*Neisseria gonorrhoeae* type 4	Z. A. McGee *et al.*	*Infec. Immunity*, **14**, 266, 1976.
4.24(e)	*Neisseria gonorrhoeae* type 3	See Figure 4.24(a)	
4.24(f)	*Neisseria gonorrhoeae* type 4	T. Elmros *et al.*	*J. Bact.,* **126**, 969, 1976.
4.24(g)	*Neisseria gonorrhoeae* type 4	See Figure 4.24(a)	
4.24(h)	*Neisseria gonorrhoeae* type 4	D. Danillson (ed.)	*Genital Infections and their Complications*, 1975.
4.24(i)	*Neisseria gonorrhoeae* type 4	M. E. Ward, A. A. Glynn and P. J. Watt	*Br. J. Exp. Path.*, **53**, 289–294
4.24(j)	*Neisseria haemolysans* ATCC 10379	See Figure 4.22(c)	
4.24(k)	*neisseria meningitidis* group B ATCC 13090	I. W. DeVoe, J. E. Gilchrist	*Infec. Immunity*, **10**, 872–876, 1974.
4.24(l), (m)	*Neisseria meningitidis* group B SDIC	I. W. DeVoe *et al.*	*Can. J. Microbiol.*, **19**, 1355–1359, 1973.
4.24(n)	*Neisseria gonorrhoeae*	M. E. Ward *et al.*	*Microbiology*, 188, 1975.
4.24(o)	*Neisseria gonorrhoeae*	M. E. Ward, P. J. Watt	*J. infect Dis.*, **126**, 601, 1972.
4.24(p)	*Neisseria gonorrhoeae*	See Figure 4.24(n)	
4.24(q)	*Neisseria gonorrhoeae*	See Figure 4.24(o)	
4.25(a)	*Moraxella nonliquefaciens* 3067/66	K. Bovre *et al.*	*Acta. path. microbiol. scand. B*, **78**, 765, 1970.
4.25(b), (c)	*Moraxella bovis*	A. E. Ritchie	Unpublished communication
4.26(a)	*Veillonella alcalescens* C1.	P. F. Winter, E. A. Delwiche	*Can. J. Microbiol.*, **21,** 2039, 1975.
4.26(b), (c)	*Veillonella* strain V2	H. A. Bladen, S. E. Mergenhagen	*J. Bact.*, **88**, 1482, 1964.
4.27(a)	*Leptotrichia buccalis* ATCC 14201	M. A. Listgarten, C.-H. Lai	*J. Bact.*, **123**, 747, 1975.
4.27(b)	*Leptotrichia buccalis* L–11	T. Hofstad, K. A. Selvig	*J. gen. Microbiol.*, **56**, 23, 1969.
4.27(c), (d)	Oral isolate	S. C. Holt, E. R. Leadbetter	Unpublished communication
4.28(a)–(m)	*Capnocytophaga*	S. C. Holt *et al.*	*Archs Microbiol.*, 1978.
4.29(a)	*Escherichia coli* W1655F$^+$	J. Gumpert *et al.*	*Z. Allg. Mikrobiol.*, **11**, 19, 1971.
4.29(b), (f)	*Escherichia coli* C6	J. Smarda	Unpublished communication
4.29(c), (d)	*Escherichia coli* C6	S. C. Holt	Unpublished communication
4.29(e)	*Escherichia coli* B	M. E. Bayer, C. C. Remsen	*J. Bact.*, **101**, 304, 1970.
4.30(a)	*Salmonella typhimurium*	S. C. Holt	Unpublished communication
4.30(b)	*Salmonella typhimurium* SW 1061	D. Kerridge *et al.*	*J. Mol. Biol.,* **4**, 227, 1962.
4.30(c), (d)	*Proteus vulgaris*	D. Kerridge *et al.*	*J. Mol. Biol.,* **4**, 227, 1962.
4.30(e)	*Proteus mirabilis* CA/PEN	P. H. Hofschneider, H. H. Martin	*J. gen. Microbiol.*, **51**, 23, 1968.
4.31(a), (b)	*Klebsiella pneumoniae*	E. L. Springer, I. L. Roth	*J. gen. Microbiol.*, **74**, 21, 1973.
4.31(c), (d)	*Klebsiella aerogenes*	S. C. Holt	Unpublished communication

Table 4.1. Eucaryotic and procaryotic strains presented (*continued*)

Figure	Organism	Author(s)	Acknowledgement
4.32(a), (b)	*Yersinia pestis*	P. J. Hall *et al.*	*Infec. Immunity,* **9**, 1105, 1974.
4.32(c)	*Yersinia enterocolitica* group O	A. Aleksie *et al.*	*Zentble. Parasitkde Abt I,* **234**, 513, 1976.
4.33(a), (b), (c)	*Serratia marcescens*	G. Acker	*Zentbl. Parasitkde Abt I,* **239**, 213, 1977.
4.34(a), (b)	*Alysiella filiformis*	G. E. Kaiser, M. J. Starzyk	*Can. J. Microbiol.*, **19**, 325, 1973.
4.34(c)	Oral isolate	S. C. Holt, E. R. Leadbetter	*Microb. Asp. Dental Plaque*.
4.35(a)–(c)	*Vibrio cholerae* 569B	J. R. Kennedy *et al.*	*J. Bact.*, **100**, 1393, 1969.
4.35(d)	*Vibrio eltor* NCTC 5395	See Figure 4.35(e)	
4.35(e)	*Vibrio* sp. 4716	J. M. Tweedy *et al.*	*J. gen. Microbiol.*, **51**, 235, 1968.
4.35(f)	*Vibrio parahaemolyticus* T 375	S. Ulitzur	*Archs Microbiol.*, **101**, 357, 1974.
4.35(g)	*Vibrio cholerae* NCTC 8021	See Figure 4.35(e)	
4.35(h)	*Vibrio parahaemolyticus*	E. Yabuuchi *et al.*	*Jap. J. Microbiol.*, **18**, 295, 1974.
4.35(i), (j)	*Vibrio cholerae* 3083 eltor	E. T. Nelson *et al.*	*Infec. Immunity*, **14**, 527, 1976.
4.36(a), (b)	*Vibrio fetus*	A. E. Ritchie	Unpublished communication
4.36(c)	*Spirillum serpens*	R. G. E. Murray, A. Birch-Andersen	*Can. J. Microbiol.*, **9**, 393, 1963.
4.36(d), (f)	*Vibrio fetus* (*Campylobacter*)	A. E. Ritchie *et al.*	*J. gen. Microbiol.*, **43**, 427, 1966.
4.36(e)	*Oceanospirillum atlanticum*	Krieg	Unpublished communication
4.37(a)	*Pseudomonas aeruginosa* B 10490	E. G. Hubert *et al.*	*Infect. Immunity,* **4**, 60, 1971.
4.37(b)	*Pseudomonas cepacia*	S. C. Holt	Unpublished communication
4.37(c)	*Pseudomonas aeruginosa* CaAP$^+$	C. A. Baechler, R. S. Bert	*Z. Allg. Mikrobiol.*, **14**, 267, 1974.
4.37(d)	*Pseudomonas multivorans* NCIB 9691	J. M. Tweedy *et al.*	*J. gen. Microbiol.*, **51**, 235, 1968.
4.37(e)	*Pseudomonas aeruginosa* type 1369	G. D. Cagle	*Can. J. Microbiol.*, **21**, 395, 1974.
4.37(f)	*Pseudomonas aeruginosa*	See Figure 4.37(c)	
4.37(g)	*Pseudomonas aeruginosa*	See Figure 4.37(c)	
4.38(a)	*Spirochaeta stenostrepta*	E. Canale-Parola *et al.*	*Arch. Mikrobiol.*, **63**, 385, 1968.
4.38(b)	*Treponema denticola*	R. P. Blakemore, E. Canale-Parola	*Arch. Mikrobiol.*, **89**, 273, 1973.
4.38(c)	*Leptospira interrogans*	D. Kuhn	*Bergey's Manual of Determinative Bacteriology*, 8th ed.
4.38(d)	*Borrelia anserina*	D. Kuhn	*Bergey's Manual of Determinative Bacteriology,* 8th ed.
4.38(e)	*Borrelia recurrentis*	K. Hovind-Hougen	*Acta path. microbiol. scand. B*, **82**, 799, 1974.
4.38(f)	*Treponema pallidum* Nichols	O. Jepsen *et al.*	*Acta path. microbiol. scand. B*, **74**, 241, 1968.
4.38(g)	*Treponema microdentium* FM	K. Hovind-Hougen	*Acta path. microbiol. scand. B*, **82**, 495, 1974.
4.38(h)	*Treponema pallidum* Nichols	See Figure 4.38(f)	*Acta path. microbiol. scand. B*, **74**, 241, 1968.
4.38(i)	*Treponema calligyrum*	K. Hovind-Hougen	*Acta path. microbiol. scand. B*, **82**, 495, 1974.
4.38(j)	*Borrelia hermsii*	S. C. Holt	Unpublished communication
4.38(k)	*Treponema pertenue* Gauthier	K. Hovind-Hougen *et al.*	*Acta path. microbiol. scand. B*, **84**, 101, 1976.
4.38(l)	*Leptospira pomona*	A. Birch-Andersen *et al.*	*Acta path. microbiol. scand. B*, **81**, 665, 1973.
4.38(m)	*Borrelia recurrentis*	K. Hovind-Hougen	*Acta path. microbiol. scand. B*, **82**, 799, 1974.
4.38(n)	*Spirochaeta aurantia*	S. C. Holt	*Microbiol. Rev.*, **42**, 114, 1978.
4.38(o)	Diagram	S. C. Holt	*Microbiol. Rev.*, **42**, 114, 1978.
4.38(p), (q)	*Treponema pallidum*	N. M. Ovcinnikov, V. V. Delektorskij	*Br. J. Vener. Dis.*, **51**, 10, 1975.

Table 4.1. Eucaryotic and procaryotic strains presented (*continued*)

Figure	Organism	Author(s)	Acknowledgement
4.38(r)	*Treponema pallidum*	J. Wicke *et al.*	*Arch. Derm. Res.*, **257**, 1, 1976.
4.38(s)	Biopsy, secondary syphilis	K. Hovind-Hougen	Unpublished communication.
4.38(t)	Apex, colonic epithelium rectal biopsy lupus erythematosus	E. V. Gear, Jr., W. O. Dobbins	*Gastroenterology*, **55**, 522, 1968.
4.38(u)	Epithelial cell, normal human colon	F. Minio *et al.*	*Rend. Gastro-enterol.*, **5**, 183, 1973.
4.39(a)–(c)	*Mycoplasma orale*	D. R. Anderson, M. F. Barile	*J. natn. Cancer Inst.*, **36**, 161 1966.
4.39(d), (e)	*Mycoplasma laidlawii* B	J. Maniloff	*J. Bact.*, **102**, 561, 1970.
4.39(f)	*Mycoplasma pneumoniae* FH	G. Biberfeld, P. Biberfeld	*J. Bact.*, **102**, 855, 1970.
4.39(g)–(i)	*Mycoplasma pneumoniae*	A. M. Collier, W. A. Clyde, Jr.	*Infec. Immunity*, **3**, 694, 1971.
4.39(j)	*T-mycoplasma (Ureaplasma) urealyticum* 681/74	J. Robertson, E. Smook	*J. Bact.*, **128**, 658, 1976.
4.39(k)	*Mycoplasma meleagridis* avian serotype H	F. Green III, R. P. Hanson	*J. Bact.*, **116**, 1011, 1973.
4.39(l)	*Mycoplasma arthriditis* R3	C. H. Domermuth *et al.*	*J. Bact.*, **88**, 727, 1964.
4.39(m)	*Mycoplasma pulmonis* Ash	C. H. Domermuth *et al.*	*J. Bact.*, **88**, 727, 1964.
4.39(n)	*Mycoplasma salivarium* PG20	C. H. Domermuth *et al.*	*J. Bact.*, **88**, 727, 1964.
4.39(o)–(q)	*Mycoplasma laidlawii* A	R. M. Lemcke	*J. Bact.*, **101**, 1154, 1972.
4.39(r)	*Mycoplasma orale* type 1	R. M. Lemcke	
4.39(s)	*Mycoplasma hominis*	See Figure 4.39(l)	
4.39(t)	*Mycoplasma horminis* H39	J. Maniloff	*J. Bact.*, **100**, 1402, 1969.
4.39(u)	*Mycoplasma meleagridis*	See Figure 4.39(k)	
4.39(v)	*Mycoplasma orale* serotype 1 N–1	M. Nakamura, M. Kawaguchi	*J. gen. Microbiol.*, **70**, 305, 1972.
4.39(w)	*Mycoplasma orale* serotype 1 N–1	M. Nakamura, M. Kawaguchi	*J. gen. Microbiol.*, **70**, 305, 1972.
4.39(x), (y)	*T-mycoplasma*, Pirillo	F. H. Black *et al.*	*J. Bact.*, **111**, 254, 1972.
4.39(z)	*T-mycoplasma*, Cook	F. H. Black *et al.*	*J. Bact.*, **111**, 254, 1972.
4.40	*Spiroplasma citri*	R. M. Cole *et al.*	*Ann. N.Y. Acad. Sci.*, **225**, 471, 1973.
4.41(a)	*Rickettsia prowazeki*	D. R. Anderson *et al.*	*J. Bact.*, **90**, 1387, 1965.
4.40(b)	*Rickettsia prowazeki* Breinl	E. L. Palmer *et al.*	*J. Bact.*, **118**, 1158, 1974.
4.40(c)	*Rickettsia prowazeki* Breinl	E. L. Palmer	*J. Bact.*, **118**, 1158, 1974.
4.40(d)	*Rickettsia rickettsia*	See Figure 4.41(a)	
4.40(e)	*Rickettsia tsutsugamuschi*	See Figure 4.41(a)	
4.40(f), (g)	*Rickettsia sennetsu*	See Figure 4.41(a)	
4.40(h)–(n)	*Rickettsia canada*	L. P. Brinton, W. Burgdorfer	*J. Bact.*, **105**, 1149, 1971.
4.40(o)–(q)	*Rickettsia quintana* Fuller	S. Ito, J. W. Vinson	*J. Bact.*, **89**, 481, 1965.
4.42(a)	*Coxiella burnetii* phase I Nine Mile	M. E. Wiebe *et al.*	*J. Bact.*, **110**, 368, 1972.
4.42(b)	*Coxiella burnetii* phase I	P. R. Burton *et al.*	*J. Bact.*, **122,** 316, 1975.
4.42(c)	*Coxiella burnetti* phase I	R. A. Kishimoto *et al.*	*Infec. Immunity*, **4**, 1087, 1976.
4.40(d)	*Coxiella burnetti* phase I	R. A. Kishimoto *et al.*	*Infec. Immunity*, **4**, 1087, 1976.
4.43(a)	*Chlamydia psittaci* CAI 10	A. Tamura *et al*	*J. Bact.*, **105**, 355, 1971.
4.43(b), (c)	*Chlamydia psittaci* CAI 10	A. Tamura *et al.*	*J. Bact.*, **105**, 355, 1971.
4.43(d)–(f)	*Chlamydia psittaci* 6BC	J. W. Costerton *et al.*	*Can. J. Microbiol.*, **21**, 1433, 1975.
4.43(g)–(i)	*Chlamydia psittaci*	A. Matsumoto	*J. Bact.*, **116**, 1355, 1963.
4.44(a)	*Legionella pneumophila*	J. E. McDade	Unpublished communication
4.44(b)	*Legionella pneumophila*	A. R. Flesher *et al.*	*Ann. Intern. Med.*, **90**, 4, 1979.

References to Part Four

1. Acker, G., "Feinstrukturuntersuchungen des Lipopoly-Saccharids und Polymyxin B-induzierte Veränderungen der äusseren membran (Zellwand) von *Serratia marcescens*," *Zentbl. Bakt. ParasitKde Abt. I,* **239**, 213, 1977.
2. ——, "The arrangement of lipopolysaccharides on the outer membrane of *Yersinia enterocolitica*: an electron microscopic study," *Zentbl. Bakt. ParasitKde Abt. I.*, **237**, 504, 1977.
3. Aleksic, S., Rohde, R., Muller, G., and Wohlers, B., "Examination of the envelope antigen Kl in *Yersinia enterocolitica* which was identified as fimbriae," *Zentbl. Bakt. ParasitKde Abt. I.*, **234**, 520, 1976.
4. Anderson, A. R., "Ultrastructural studies of Mycoplasmas and the L-phase of bacteria" in Hayflick (ed.), *The Mycoplasmatales and the L-phase of Bacteria*, North-Holland, Amsterdam, 1969, pp. 365–402.
5. —— and Barile, M. F., "Ultrastructure of *Mycoplasma orale* isolated from patients with leukemia," *J. natn. Cancer Inst.*, **36**, 161, 1966.
6. ——, Hopps, H. E., Barile, M. F., and Bernheim, B. C., "Comparison of the ultrastructure of several rickettsiae, ornithosis virus, and *Mycoplasma* in tissue culture," *J. Bact.*, **90**, 1387, 1965.
7. Armstrong, B. A. and Sword, C. P., "Electron microscopy of *Listeria monocytogens*–infected mouse spleen," *J. Bact.*, **91**, 1346, 1966.
8. Baechler, C. A. and Berk, R. S., "Electron microscopic observations of *Pseudomonas aeruginosa*," *Allg. Mikrobiologie.*, **14**, 267, 1974.
9. Baker, R. F. and Loosli, C. G., "The ultrastructure of encapsulated *Diplococcus pneumoniae* Type 1 before and after exposure to type specific antibody," *Lab. Invest.*, **15**, 716, 1966.
10. Barksdale, L., "*Corynebacterium diptheriae* and its relatives," *Bact. Rev.*, **34**, 378, 1970.
11. —— and Kim, K. S., "*Mycobacterium*," *Bact. Rev.*, **41**, 217, 1977.
12. Bayen, H., Frehel, C., Ryter, A., and Sebald, M., "Etude cytologique de la sporulation chez *Clostridium histolyticum*. Souche sporogene et mutants de sporulation," *Ann. Institut Pasteur.*, **113**, 163, 1967.
13. Beaman, B. L., "Structural and biochemical alterations of *Nocardia asteroides* cells walls during its growth cycle," *J. Bact.* **123**, 1235, 1975.
14. Beveridge, T. J. and Murray R. G. E., "Superficial cell-wall layers on *Spirillum* 'Ordal' and their *in vitro* reassembly," *Can. J. Microbiol.*, **22**, 567, 1976.
15. —— and ——, "Dependence of the superficial layers of *Spirillum putridiconchylium* on Ca^{++} or Sr^{++}," *Can. J. Microbiol.*, **22**, 1233, 1976.

16. Black, F. T., Birch-Andersen, A. K., and Freundt, E. A., "Morphology and ultrastructure of human *T-Mycoplasmas*," *J. Bact.* **111**, 254, 1972.
17. Bladen, H. A. and Mergenhagen, S. E., "Ultrastructure of *Veillonella* and morphological correlation of an outer membrane with particles associated with endotoxic activity," *J. Bact.*, **88**, 1482, 1964.
18. Bourgeois, L. and Beaman, B. L., "*In vitro* spheroplast and L-form induction within the pathogenic Nocardiae," *J. Bact.*, **127**, 584, 1976.
19. Bove, J. M., Saglio, P., Tully, J. G., Freundt, E. A., Lund, J., Pillot, J., and Taylor-Robinson, D., "Characterization of the mycoplasma-like organism associated with 'stubborn' disease of citrus," *Ann. N.Y. Acad. Sci.*, **225**, 462, 1973.
20. Boyd, R. F. and Hoerl, B. G., *Basic Medical Microbiology*, Little, Brown and Co., Boston, 1977.
21. Boyde, A. and Echlin, P., "Freezing and freeze-drying – A preparative technique for scanning electron microscopy," in *Proc. 6th. Ann. Scan. Electron. Microsc. Symp.*, IIT Research Institute, Chicago, 1973, p. 759.
22. Bradley, D. E. and Franklin, J. G., "Electron microscopic survey of the surface configuration of spores of the genus *Bacillus*," *J. Bact.*, **76**, 618, 1958.
23. —— and Williams, D. J., "An electron microscopic study of the spores of some species of the genus *Bacillus* using carbon replicas," *J. gen. Microbiol.*, **17**, 75, 1957.
24. Branton, D., "Fracture faces of frozen membranes," *Proc. natn. Acad. Sci., U.S.A.*, **55**, 1048, 1966.
25. ——, Bullivant, S., Gilula, N. B., Karnovsky, M. J., Moor, H., Mühlethaler, K., Northcote, D. H., Packer, L., Satir, B., Satir, P., Speth, K., Staehlin, L. A., Steere, R. L., and Weinstein, R. S., "Freeze-etching nomenclature," *Science*, **190**, 54, 1975.
26. Brewer, D. B., "Electron microscopy of phagocytosis of *Staphylococci*," *J. Path. Bact.*, **86**, 299, 1963.
27. Brieger, E. M. and Glauert, A. M., "Electron microscopy of leprosy bacillus: a study of submicroscopic structure," *Tubercle*, **37**, 195, 1956.
28. Brinton, L. P. and Burgdorfer, W., "Fine structure of *Rickettsia canada* in tissues of *Dermacentor andersoni* stiles," *J. Bact.*, **105**, 1149, 1971.
29. Brown, C. A. and Draper, P., "An electron microscope study of rat fibroblast infected with *Mycobacterium lepraemurium*," *J. Path.*, **102**, 21, 1970.
30. Buckmire, F. L. A. and Murray, R. G. E., "Substructure and *in vitro* assembly of the outer structured layer of *Spirillum serpens*," *J. Bact.*, **125**, 290, 1976.
31. Bullen, C. C., Tearle, P. V., and Willis, A. T., "Bifidobacteria in the internal tracts of infants: an *in vivo* study," *J. Med. Microbiol.*, **9**, 325, 1976.
32. Burdett, I. D. J. and Murray, R. G. E., "Electron microscope study of septum formation in *Escherichia coli* strains B and B/r during synchronous growth," *J. Bact.*, **119**, 1039, 1974.
33. Burton, P. R., Kordova, N., and Paretsky, D., "Electron microscopic

studies of the rickettsia *Coxiella burnetii*: entry, lysogonal response, and fate of rickettsia DNA in I cells," *Can. J. Microbiol.*, **17**, 143, 1971.

34. Cagle, G. D., "Fine structure and distribution of extracellular polymer surrounding selected aerobic bacteria," *Can. J. Microbiol.*, **21**, 395, 1975.
35. Canale-Parola, E., "Biology of the sugar-fermenting sarcinae," *Bact. Rev.*, **34**, 82, 1970.
36. ——, Mandel, M., and Kupfer, D. G., "The classification of sarcinae," *Arch. Mikrobiol.*, **58**, 30, 1967.
37. Chandler, F. W., Hicklin, M. D., and Blackmon, J. A., "Demonstration of the agent of Legionnaire's disease in tissue," *New Engl. J. Med.*, **297**, 1218, 1977.
38. Chapman, G. B., Hanks, J. H., and Wallace, J. H., "An electron microscope study of the disposition and fine structure of *Mycobacterium lepraemurium* in mouse spleen," *J. Bact.*, **77**, 205, 1959.
39. Cheng, K.-J. and Costerton, J. W., "Localization of alkaline phosphatase in three gram-negative rumen bacteria," *J. Bact.*, **116**, 424, 1973.
40. Cole, R. M., Chatterjee, A. N., Gilpin, R. W., and Young, F. E., "Ultrastructure of teichoic acid-deficient and other mutants of *Staphylococci*," *Ann. N.Y. Acad. Sci.*, **236**, 22, 1974.
41. ——, Tulley, J. G., Popkin, T. J., and Bove, J. M., "Morphology, ultrastructure, and bacteriophage infection of the helical Mycoplasma-like organism (*Spiroplasma citri* gen. nov. sp. Nov.) cultured from "Stubborn" disease of citrus," *J. Bact.*, **115**, 367, 1973.
42. ——, ——, ——, and ——, "Ultrastructure of the agent of citrus 'stubborn' disease," *Ann. N.Y. Acad. Sci.*, **225**, 471, 1973.
43. Collier, A. M. and Clyde, Jr., W. A., "Relationships between *Mycoplasma pneumoniae* and human respiratory epithelium. *Infect. Immunity*, **3**, 694, 1971.
44. Collier, R. J., "Diphtheria toxin: mode of action and structure," *Bact. Rev.*, **39**, 54, 1975.
45. Costerton, J. W., Geesey, G. G., and Cheng, K.-J., "How bacteria stick," *Sci. Amer.*, **238**, 86, 1978.
46. ——, Ingram, J. M., and Cheng, K.-J., "Structure and function of the cell envelope of gram-negative bacteria," *Bact. Rev.*, **38**, 87, 1974.
47. ——, Poffenroth, L., Wilt, J. C., and Kordova, N., "Ultrastructure studies of *Chlamydia psittaci* 6BC *in situ* in yolk sac explants and L cells: a comparison with gram-negative bacteria," *Can. J. Microbiol.*, **21**, 1433, 1975.
48. Criswell, B. S., Marston, J. H., Stenback, W. A., Black, S. H., and Gardner, H. L., "*Haemophilus vaginalis* 594. Gram-negative organism," *Can. J. Microbiol.*, **17**, 865, 1971.
49. ——, Stenback, W. A., Black, S. H., and Gardner, H. L., "Fine structure of *Haemophilus vaginalis*," *J. Bact.*, **109**, 930, 1972.
50. Crocker, T. T., Chen, T. H., and Meyer, K. F., "Electron microscopic study of the extracellular material of *Pasteurella pestis*," *J. Bact.*, **72**, 851, 1956.
51. Deibel, R. H. and Seeley, H. W., Jr., "Family II. Streptococcaceae fam.

nov.," in R. E. Buchanan and N. E. Gibbons (eds.), *Bergey's Manual of Determinative Bacteriology*, 8th ed., Williams and Wilkins, Baltimore, 1974, p. 490.
52. DeVoe, I. W. and Gilchrist, J. E., "Release of endotoxin in the form of cell wall blebs during *in vitro* growth of *Neisseria meningitidis*," *J. Med.,* **138**, 1156, 1973.
53. —— and ——, "Release of endotoxin in the form of cell wall blebs during *in vitro* growth of *Neisseria meningitidis*," *J. exp. Med.*, **141**, 297, 1974.
54. ——, and ——, "Ultrastructure of pili and annular structures on the cell wall surface of *Neisseria meningitidis*," *Infec. Immunity*, **10**, 872, 1974.
55. —— and ——, "Pili on meningococci from primary cultures of nasopharyngeal carriers and cerebrospinal fluid of patients with acute disease," *J. exp. Med.*, **141**, 297, 1975.
56. ——, Storm, D. W., and Gilchrist, J. E., "A study on phagocytosis of radio-labeled *Staphylococcus epidermidis* on structural events during intracellular degradation," *Can. J. Microbiol.*, **19**, 525, 1973.
57. Dipersio, J. R., Mattingly, S. J., Higgins, M. C., and Shockman, G. D., "Measurement of intracellular iodophilic polysaccharide in two cariogenic strains of *Streptococcus mutans* by cytochemical and chemical methods," *Infec. Immunity*, **10**, 597, 1974.
58. Doern, G. V. and Buckmire, F. L. A., "Ultrastructural characterization of capsulated *Haemophilus influenzae* Type B and two spontaneous nontypable mutants," *J. Bact.* , **127**, 523, 1976.
59. Dolby, J. M., Honour, P., and Valman, H. B., "Bacteriostasis of *Escherichia coli* by milk. I. Colonization of breast-fed infants by milk-resistant organisms," *J. Hyg.*, **78**, 85, 1977.
60. Domermuth, C. H., Nielsen, M. H., Freundt, E. A., and Birch-Anderson, A., "Ultrastructure of *Mycoplasma* species," *J. Bact.*, **88**, 727, 1964.
61. Drews, G., "Elektronenmikroskopische Untersuchungen an *Mycobacterium phlei*," *Arch. Mikrobiol.*, **35**, 53, 1960.
62. Dubray, G., "Etude ultrastructure des bacteries de colonies lisses (s) et rugueuses (R) du genre 'Brucella'," *Ann. Institut Pasteur*, **123**, 171, 1972.
63. Duda, J. J. and Slack, J. M., "Ultrastructural studies on the genus *Actinomyces*," *J. gen. Microbiol.*, **71**, 63, 1972.
64. Echlin, P. and Moulton, R., "The preparation, coating and examination of frozen biological materials in the SEM," in *Proc. 6th. Ann. Scan Electron Microsc. Symp.*, IIT Research Institute, Chicago, 1973, p. 325.
65. Edwards, M. R. and Stevens, R. W., "Fine structure of *Listeria monocytogenes*," J. Bact., **86**, 414, 1963.
66. Ellen, R. P. and Gibbons, R. J., "M protein associated adherence of *Streptococcus pyogenes* to epithelial surfaces: prerequisite for virulence," *Infec. Immunity*, **5**, 826, 1972.
67. Fooke-Achterrath, M., Lickfeld, K. G., Reusch, Jr., V. M., Aebi, U., Tschope, U., and Menge, B., "Close-to life preservation of *Staphylococcus aureus* mesosomes for transmission E.M.," *J. Ultrastruct. Res.*, **49**, 270, 1974.

68. Fox, E. N., "M proteins of Group A Streptococci," *Bact. Rev.*, **38**, 57, 1974.
69. Fraser, D. W., Tsai, T. R., Orenstein, W., Parkin, W. E., Beechan, H. J., Sharrar, R. G., Harris, J., Mallison, G. F., Martin, S. M., McDade, J. E., Shepard, C. C., Brachman, P. S., and the field investigation team, "Legionnaire's disease. Description of an epidemic of pneumonia," *New Engl. J. Med.*, **297**, 1189, 1977.
70. Freeman, V. J., "Studies on the virulence of bacteriophage-infected strains of *Corynebacterium diphtheriae*," *J. Bact.*, **61**, 675, 1951.
71. Freundt, E. A., "Morphology and ultrastructure of the mycoplasmas," in J. T. Sharp (ed.), *The Role of Mycoplasmas and L-forms of Bacteria in Disease*, Thomas, Springfield, Illinois, 1970, pp. 29–81.
72. Ghosh, B. K. and Murray, R. G. E., "Fine structure of *Listeria monocytogenes* in relation to protoplast formation," *J. Bact.*, **93**, 411, 1967.
73. Gibbons, R. J. and van Houte, J., "Bacterial adherence in oral microbial ecology," *A. Rev. Microbiol.*, **29**, 19, 1975.
74. Giesbrecht, P., Wecke, J., and Reinicke, B., "On the morphogenesis of the cell wall of staphylococci," *Int. Rev. Cytol.*, **44**, 225, 1976.
75. Gilchrist, J. E. and DeVoe, I. W., "Ultrastructure of *Staphylococcus epidermidis* after freeze-etching and thin sectioning," *Can. J. Microbiol.*, **19**, 294, 1973.
76. Girard, A. E. and Jacius, B. H., "Ultrastructure of *Actinomyces viscosus* and *Actinomyces naeslundii*," *Arch. oral Biol.*, **19**, 71, 1974.
77. Glauert, A. M. and Phillips, R., "The preparation of thin-sections," in D. Kay (ed.), *Techniques for Electron Microscopy,* Blackwell, Oxford, 1967.
78. Glick, A. D., Getnick, R. A., and Cole, R. M., "Electron microscopy of group A *Streptococci* after phagocytosis by human monocytes," *Infec. Immunity*, **4**, 772, 1971.
79. Goodman, J. R., Moore, R. E., and Baker, R. F., "Electron microscopic study of phagocytosis of *Staphylococci* by human leukocytes. II. Virulent and non-virulent *Staphylococci*," *J. Bact.*, **72**, 736, 1956.
80. Green, F., III. and Hanson, R. P., "Ultrastructure and capsule of *Mycoplasma meleagridis*," *J. Bact.*, **116**, 1011, 1973.
81. Grimble, A. and Armitage, L. R. G., "Surface structure of the *Gonococcus*," *Br. J. vener. Dis.*, **50**, 354, 1974.
82. Habeeb, A. F. S. A. and Hiramoto, R., "Reactions of proteins with glutaraldehyde," *Arch. Biochem. Biophys.*, **126**, 16, 1968.
83. Hard, G. C., "Electron microscopic examination of *Corynebacterium ovis*," *J. Bact.*, **97**, 1480, 1969.
84. ———, "Examination by electron microscopy of the interaction between peritoneal phagocytes and *Corynebacterium ovis*," *J. Med. Microbiol.*, **5**, 483, 1972.
85. Haschemeyer, R. H. and Myers, R. J., "Negative staining," in M. A. Hayat (ed.), *Principles and Techniques of Electron microscopy*, *Biological Applications*, vol. 2, Van Nostrand Reinhold, New York, 1972, chapter 3.
86. Henrichsen, J., "Bacterial surface translocation: a survey and a classification," *Bact. Rev.*, **36**, 478, 1972.

87. Hewitt, J. H. and Rigby, J., "Effect of various milk products on numbers of *Eschericia coli* and *Bifidobacterium* in the stools of newborn infants," *J. Hyg.*, **77**, 129, 1976.
88. Higerd, T. B., Baechler, C. A., and Berk, R. S., "*In vitro* and *in vivo* characterization of pyocin," *J. Bact.*, **93**, 1976, 1967.
89. Higgins, M. L. and Daneo-Moore, L., "Factors influencing the frequency of mesosomes observed in fixed and unfixed cells of *Streptococcus faecalis*," *J. Cell Biol.*, **61**, 288, 1974.
90. —— and Shockman, G. D., "Early changes in the ultrastructure of *Streptococcus faecalis* after amino acid starvation," *J. Bact.*, **103**, 244, 1970.
91. Hodgkiss, W., Ordal, Z. J., and Cann, D. C., "The comparative morphology of the spores of *Clostridium botulinum* Type E and the spores of the 'OS mutant'," *Can. J. Microbiol.*, **12**, 1283, 1966.
92. Hoeniger, J. F. M. and Bradley, D. E., "Structural changes in cells of *Clostridium sporogenes*," *Can. J. Microbiol.*, **15**, 1061, 1969.
93. ——, Stuart, P. F., and Holt, S. C., "Cytology of spore formation in *Clostridium perfringens*," *J. Bact.*, **96**, 1818, 1968.
94. —— and Tauschel, H. D., "Sequence of structural changes in cultures of *Clostridium tetani* grown on a solid medium," *J. Med. Microbiol.*, **7**, 425, 1974.
95. Hoffman, H. P., Geftic, S. G., Gelzer, J., Heymann, H., and Adair, F. W., "Ultrastructural alterations associated with the growth of resistant *Pseudomonas aeruginosa* in the presence of benzalkonium chloride," *J. Bact.*, **113**, 409, 1973.
96. Hofstad, T. and Selvig, K. A., "Ultrastructure of *Leptotrichia buccalis*," *J. gen. Microbiol.*, **56**, 23, 1969.
97. Holt, S. C., "Anatomy and chemistry of spirochetes," *Microbiol. Rev.*, **42**, 114, 1978.
98. —— and Canale-Parola, E., "Fine structure of *Sarcina maxima* and *Sarcina ventriculi*," *J. Bact.*, **93**, 399, 1967.
99. —— and Leadbetter, E. R., "Comparative ultrastructure of selected aerobic spore-forming bacteria: a freeze-etching study," *Bact. Rev.*, **33**, 346, 1969.
100. —— and ——, "Comparative ultrastructure of selected oral *Streptococci*: thin-sectioning and freeze-etching studies," *Can. J. Microbiol.*, **22**, 375, 1976.
101. ——, ——, and Socransky, S. S., "*Capnocytophaga*: New genus of oral gram-negative bacteria. II. Morphology and ultrastructure," *Archs Microbiol.*, **122**, 17, 1979.
102. Horne, R. W., "Negative staining methods," in D. Kay (ed.), *Techniques for Electron Microscopy*, 2nd ed., Blackwell, Oxford, 1967, p. 328.
103. Hovind-Hougen, K., Birch-Andersen, A., and Skovgaard Jensen, H. J., "Ultrastructure of cells of *Treponema pertenue* obtained from experimentally infected hamsters," *Acta path. microbiol. scand. B.*, **84**, 101, 1976.
104. Imaeda, T. and Convit, J., "Electron microscope study of *Mycobacterium leprae* and its environment in a vesicular leprous lesion," *J. Bact.*, **83**, 43, 1961.

105. Irving, J. T., Newman, M. G., Socransky, S. S., and Healey, J. D., "Histological changes in experimental periodontal diseases in rats monoinfected with a gram-negative organism," *Arch. oral Biol.*, **20**, 219, 1975.
106. Johnson, M. C., Bozzola, J. J., and Schechmeister, I. L., "Morphological study of *Streptococcus mutans* and two extracellular polysaccharide mutants," *J. Bact.*, **118**, 304, 1974.
107. Kaiser, G. E. and Starzyk, M. J., "Ultrastructure and cell division of an oral bacterium resembling *Alysiella filiformis*," *Can. J. Microbiol.*, **19**, 325, 1973.
108. Karnovsky, M. J., "Simple methods for staining with lead at high pH in electron microscopy," *J. biophys. biochem. Cytol.*, **11**, 729, 1961.
109. ———, "A formaldehyde-glutaraldehyde fixative of high osmolarity for use in electron microscopy," *J. Cell Biol.*, **27**, 137A, 1965.
110. Katz, S. M. and Nash, P., "Legionnaires' disease: Structural characteristics of the organism," *Science*, **199**, 896, 1978.
111. Kawata, T. and Inoue, T., "Electron microscopic observations of a remarkable body in aged *Corynebacterium diphtheriae*," *J. Bact.*, **89**, 1613, 1965.
112. ——— and Masuda, K., "Extracellular crystalline lattice material of *Corynebacterium diphtheriae* revealed by electron microscopy," *Jap J. Microbiol.*, **16**, 515, 1972.
113. Kellenberger, E., Ryter, A., and Séchaud, J., "Electron microscope study of DNA-containing plasms. II. Vegetative and mature phage DNA as compared with normal bacterial nucleoids in different physiological states," *J. biophys. biochem. Cytol.*, **4**, 671, 1958.
114. Kelstrup, J., "The incidence of *Bacteroides melaninogenicus* in human gingival sulci, and its prevalence in the oral cavity at different ages," *Periodontics*, **4**, 14, 1966.
115. Kennedy, J. R. and Richardson, S. H., "Fine structure of *Vibrio cholerae* during toxin production," *J. Bact.*, **100**, 1393, 1969.
116. Kilian, M. and Theilade, J., "Cell wall ultrastructure of strains of *Haemophilus ducreyi* and *Haemophilus piscium*," *Int. J. Syst. Bact.*, **25**, 351, 1975.
117. Kim, K. S. and Barksdale, L., "Crystalline inclusions of bacterium 22 M," *J. Bact.*, **98**, 1390, 1969.
118. ———, Salton, M. R. J., and Barksdale, L., "Ultrastructure of superficial mycosidic integuments of *Mycobacterium* sp.," *J. Bact.*, **125**, 739, 1976.
119. Kishimoto, R. A., Veltri, B. J., Canonico, P. G., Shirey, F. G., and Walker, J. S., "E.M. study on the interaction between normal guinea pig peritoneal macrophages and *Coxiella burnetii*," *Infec. Immunity*, **4**, 1087, 1976.
120. Lai, G. C., Listgarten, M., and Rosan, B., "Serology of *Streptococcus sanguis*: Localization of antigens with unlabelled antisera," *Infec. Immunity*, **8**, 475, 1973.
121. Lambe, D. W., Jr., "Determination of *Bacteroides melaninogenicus* serogroups by fluorescent antibody staining," *Appl. Microbiol.*, **28**, 561, 1974.

122. Leadbetter, E. R. and Holt, S. C., "The fine structure of *Bacillus fastidiosus*," *J. gen. Microbiol*, **52**, 299, 1968.
123. ——, ——, and Socransky, S. S., "*Capnocytophaga*: New genus of gramnegative gliding bacteria. I. General chacteristics, taxonomic considerations and significance," *Archs Microbiol.*, **122**, 9, 1979.
124. Lemcke, R. M., "Osmolar concentration and fixation of *Mycoplasmas*," *J. Bact.*, **110**, 1154, 1972.
125. Lewin, R. A., "Rod-shaped particles in *Saprospira*," *Nature, Lond.*, **198**, 103, 1963.
126. Listgarten, M. A. and Lai, C. H., "Unusual cell wall ultrastructure of *Leptotrichia buccalis*," *J. Bact.*, **123**, 747, 1975.
127. Luft, J. H., "Permanganate – a new fixative for electron microscopy," *J. biophys. biochem. Cytol.*, **2**, 799, 1956.
128. ——, "Improvements in epoxy resin embedding methods," *J. biophys. biochem. Cytol.*, **9**, 409, 1961.
129. ——, "Fine structure of capillary and endo-capillary layer as revealed by ruthenium red," *Fed. Proc.*, **25**, 1773, 1966.
130. ——, "Ruthenium red and violet. I. Chemistry, purification, methods of use for electron microscopy and mechanism of action," *Anat. Rev.*, **171**, 347, 1971.
131. MacDonald, J. B. and Gibbons, R. J., "The relationship of indigenous bacteria to periodontal disease," *J. dent. Res.*, **41**, suppl., 320, 1962.
132. ——, ——, and Socransky, S. S., "Bacterial mechanisms in periodontal disease – in the metabolism of oral tissue," *Ann. N.Y. Acad. Sci.*, **85**, 467, 1960.
133. Maniloff, J., "Electron microscopy of small cells: *Mycoplasma hominis*," *J. Bact.*, **100**, 1402, 1969.
134. ——, "Ultrastructure of *Mycoplasma laidlawii* during culture development," *J. Bact.*, **102**, 561, 1970.
135. Margaritis, L. H., Elgsaeter, A., and Branton, D., "Rotary replication for freeze-etching," *J. Cell Biol.*, **72**, 47, 1977.
136. McCandless, R. G., Cohen, M., Kalmanson, G. M., and Guze, L. B., "Cores, microbial organelles possibly specific to Group D *Streptococci*," *J. Bact.*, **96**, 1400, 1968.
137. McDade, J. E., Shepard, C. C., Fraser, D. W., Tsai, T. R., Redus, M. A., Dowdle, W. R., and the laboratory investigation team, "Legionnaires' disease. Isolation of a bacterium and demonstration of its role in other respiratory disease," *New Engl. J. Med.*, **297**, 1197, 1977.
138. McGee, Z. A., Gross J., Dourmashkin, R., and Taylor-Robinson, D., "Nonpilar surface appendages of colony Type 1 and colony Type 4 *Gonococci*," *Infec. Immunity*, **144**, 266, 1976.
139. Minio, F., Tonietti, G., and Torsoli, A., "Spontaneous spirochete infestation in the colonic mucosa of healthy men," *Rend. Gastroenterol.*, **5**, 183, 1973.
140. Montes, L. F. and Black, S. H., "The fine structure of *Corynebacterium minutissimum*," *Separatum experientia*, **23**, 345, 1967.
141. —— and ——, "The fine structure of diphtheroids of *Erythrasma*," *J. Invest. Derm.*, **48**, 342, 1967.

142. ——, McBride, M., Johnson, W., Owens, D., and Knox, J., "Ultrastructural study of the host-bacterium relationship in *Erythrasma*," *J. Bact.*, **90**, 1489, 1965.
143. Moor, H. and Mühlethaler, K., "Fine structure in frozen-etched yeast cells," *J. Cell Biol.*, **17**, 609, 1963.
144. Muller, R., "Zur Stellung der Krankheitserreger im Natursystem," *Münch. med. Wschr.*, **58**, 2246, 1911.
145. Murphy, J. A. and Campbell, L., "Surface features of *Bacillus polymyxa* spores as revealed by scanning electron microscopy," *J. Bact.*, **98**, 737, 1969.
146. Murray, E. D. G., *The Meningococcus, Special Report Series*, no. 124, Medical Research Council, London, 1929.
147. Nakamura, M. and Kawaguchi, M., "Ultrastructure of *Mycoplasma orale* serotype 1 in agar growth," *J. gen. Microbiol.*, **70**, 305, 1972.
148. Nelson, E. T., Clements, J. D., and Finkelstein, R. A., "*Vibrio cholerae* adherence and colonization in experimental cholera: E.M. studies," *Infec. Immunity*, **14**, 527, 1976.
149. Newman, M. G. and Socransky, S. S., "Predominant cultivable microbiota in periodontosis," *J. Periodont. Res.*, **12**, 120, 1977.
150. ——, ——, Savitt, E. D., Propas, D. A., and Crawford, A., "Studies of the microbiology of periodontosis," *J. Periodont. Res.*, **47**, 373, 1976.
151. Ovcinnikov, N. M. and Delektorskij, V. V., "Morphology of *Treponema pallidum*," *Bull. Wld. Hlth. Org.*, **35**, 223, 1966.
152. —— and ——, "Further study of ultrathin sections of *Treponema pallidum* under the electron microscope," *Br. J. vener. Dis.*, **44**, 1, 1968.
153. —— and ——, "Further studies of the morphology of *Treponema pallidum* under the electron microscope," *Br. J. vener. Dis.*, **45**, 87, 1969.
154. —— and ——, "*Treponema pertenue* under the electron microscope," *Br. J. vener. Dis.*, **46**, 349, 1970.
155. —— and ——, "Electron microscopy of phagocytosis in syphilis and yaws," *Br. J. vener. Dis.*, **48**, 227, 1972.
156. —— and ——, "*Treponema pallidum* in nerve fibers," *Br. J. vener. Dis.*, **51**, 10, 1975.
157. Page, L. A., "Order II. Chlamydiales Stone and Page 1971, 334," in R. E. Buchanan and N. E. Gibbons (eds.), *Bergey's Manual of Determinative Bacteriology*, 8th ed., Williams and Wilkins, Baltimore, 1974.
158. Palmer, E. L., Mallavia, L. P., Tzianbos, T., and Obijeski, J. F., "E.M. of the cell wall of *Rickettsia prowazekii*," *J. Bact.*, **118**, 1158, 1974.
159. Pawley, J. B., Hayes, T. L., and Nowell, J. A., "Microdissection," in M. A. Hayat (ed.), *Principles and Techniques of Scanning Electron Microscopy. Biological Applications* vol. 3, Van Nostrand Reinhold, New York 1975, p. 17.
160. Pessac, B. and Defendi, V., "Cell aggregation: Role of acid mucopolysaccharides," *Science*, **175**, 898, 1972.
161. Petitprez, A., Roos, P., and Tacquet, A. "Aspect ultrastructural de *Mycobacterium phlei*," *J. Microsc. (Fr.)*, **6**, 229, 1967.
162. Poupard, J. A., Husain, I., and Norris, R. F., "Biology of *Bifidobacteria*," *Bact. Rev.*, **37**, 136, 1973.

163. Powell, D. A., Hu, P. C., Wilson, M., Collier, A. M., and Baseman, J. B., "Attachment of *Mycoplasma pneumoniae* to respiratory epithelium," *Infec. Immunity*, **13**, 956, 1976.
164. Punsalang, A. P. and Sawyer, W. D., "Role of pili in the virulence of *Neisseria gonorrhoeae*," *Infec. Immunity*, **8**, 255, 1973.
165. Reun, A., Birch-Andersen, A. K., and Larage, S. P., "An e.m. study of thin sections of *Haemophilus vaginalis* (Gardener and Dukes) and some possibly related species," *Can. J. Microbiol.*, **12**, 1125, 1966.
166. Reynolds, E. S., "The use of lead citrate at high pH^+ as an electron opaque stain in electron microscopy," *J. Cell Biol.*, **17**, 208, 1963.
167. Richter, G. W. and Kress, Y., "Electron microscopy of a strain of *Bordetella bronchiseptica*," *J. Bact.*, **94**, 1216, 1967.
168. Ritchie, A. E., Keeler, R. F., and Bryner, J. H., "Anatomical features of *Vibrio fetus*: electron microscope survey," *J. gen. Microbiol.*, **43**, 427, 1966.
169. Robertson, J. and Smook, E., "Cytochemical evidence of extramembranous carbohydrates on *Ureaplasma urealyticum* (T-strain Mycoplasma)," *J. Bact.*, **128**, 658, 1976.
170. Rosenberg, M., Bartl, P., and Lesko, J., "Water-soluble methacrylate as an embedding medium for the preparation of ultrathin sections," *J. Ultrastruct. Res.*, **4**, 298, 1960.
171. Roth, G. D. and Flanagan, V., "The pathogenicity of *Rothia dentocariosa* inoculated into mice," *J. dent. Res.*, **48**, 957, 1969.
172. ———, Garrison, R. G., and Boyd, K. S., "Ultrastructure of *Rothia dentocariosa*," *J. gen. Microbiol.*, **95**, 45, 1976.
173. Ryter, A. and Pillot, J., "Structure des spirochetes. II. Etude du genre *Cristispira* au microscope optique et au microscope électronique," *Ann. Institut Pasteur, Paris*, **109**, 552, 1965.
174. Sabatini, D. D., Bensch, K., and Barrnett, R. J., "Cytochemistry and electron microscopy. The preservation of cellular ultrastructure and enzymatic activity by aldehyde fixation," *J. Cell Biol.*, **17**, 19, 1963.
175. Saglio, P., Lafleche, D., Bonissol, C., and Bove, J. M., "Isdement, culture, et observation au microscope electronique des structures de type mycoplasm associés à la maladie du Stubborn des argumes et leur comparaison avec les structures observées dans le cas de la maladie du greening des argumes," *Physiolog. Veg.*, **9**, 569, 1971.
176. Sarkany, I., Taplin, D., and Blank, H., "Organisms causing *Erythrasma*," *Lancet*, **7250**, 304, 1962.
177. Savage, D. C., "Microbial ecology of the gastrointestinal tract," *A. Rev. Microbiol.*, **31**, 107, 1977.
178. Sherwin, R. P. and Wilkins, J., "The ultrastructure of *Hemophilus influenzae*," in S. H. W. Sell and D. T. Karzon (eds.), *Hemophilis influenzae*, Vanderbilt University Press, Nashville, Tennesee, 1973.
179. Shively, J., "Inclusion bodies of prokaryotes," *A. Rev. Microbiol.*, **28**, 167, 1974.
180. Simons, H., "Saprophytische Oscillarien des Menschen und der Tiere," *Zentbl. Parasitkde Abt. I*, 88, 501, 1922.

181. Singer, S. J., "Preparation of an electron-dense antibody conjugate," *Nature, Lond.*, **183**, 1523, 1959.
182. ——— and Schick, A. F., "The properties of specific stains for electron microscopy prepared by the conjugation of antibody molecules with ferritin," *J. biophys. biochem. Cytol.*, **9**, 519, 1961.
183. Smith, A. L., *Microbiology and Pathology*, 11th ed., Mosby, Saint Louis, 1976.
184. Springer, E. L. and Roth, I. L., "The ultrastructure of the capsules of *Diplococcus pneumoniae* and *Klebsiella pneumoniae* stained with ruthenium red," *J. gen. Microbiol.*, **74**, 21, 1973.
185. Spurr, A. R., "A low-viscosity epoxy resin embedding medium for electron microscopy," *J. Ultrastruct. Res.*, **26**, 31, 1969.
186. Stanier, R. Y., Adelberg, E. A., and Ingraham, J. L., *The Microbial World*, 4th ed., Prentice-Hall, Engelwood Cliffs, New York, 1976, p. 672.
187. Steed, P. D. M., "Simonsiellaceae fam. nov. with characterization of *Simonsiella crassa* and *Alysiella filiformis*," *J. gen. Microbiol.*, **29**, 615, 1963.
188. Suganuma, A., "Some observations on the fine structure of *Staphylococcus aureus*," *J. infect. Dis.*, **111**, 8, 1962.
189. ———, "Some features of membranous structures in *Staphylococcus aureus*," *J. infect. Dis.*, **113**, 179, 1963.
190. ———, "Fine structure of *Staphylococcus aureus*," *Ann. N.Y. Acad. Sci.*, **128**, 26, 1965.
191. ———, "Studies on the fine structure of *Staphylococcus aureus*," *J. Electron Microsc., Jap.*, **15**, 257, 1966.
192. ———, "Electron microscopic studies of staphylococci by serial sections." *J. Electron Microsc., Jap.*, **17**, 315, 1968.
193. Suzuki, J. B. and Grecz, N., "Electron microscopy of leukocyte interaction with spores of *Clostridium botulinum* Type A," *Can. J. Microbiol.*, **18**, 1651, 1972.
194. Swanson, J., "Studies on gonococcus infection. II. Freeze-fracture, freeze-etch studies on gonococci," *J. exp. Med.*, **136**, 1258, 1972.
195. ———, Hsu, K. C. and Gotschlich, C., "Electron microscopic studies on streptococci I. M. antigen," *J. exp. Med.*, **130**, 1063, 1969.
196. ———, Kraus, S. J. and Gotschlich, E. G., "Studies on gonococcus infection," *J. exp. Med.*, **134**, 886, 1971.
197. ——— and McCarty, M., "Electron microscopic studies on opaque colony variants of group A streptococci," *J. Bact.*, **100**, 505, 1969.
198. Takagi, A., Nakamura, K., and Ueda, M., "Electron microscope studies of the intracytoplasmic membrane system in *Clostridium tetani* and *Clostridium botulinum*," *Jap. J. Microbiol.*, **9**, 131, 1965.
199. Tamura, A., Matsumoto, A., Manire, G. P., and Higashi, N., "Electron microscopic observations on the structure of the envelopes of mature elementary bodies and development of reticulate forms of *Chlamydia psittaci*," *J. Bact.*, **105**, 355, 1971.
200. Thornley, M. J., Horne, R. W., and Glauert, A. M., "The fine structure of *Micrococcus radiodurans*," *Arch. Mikrobiol.*, **51**, 267, 1965.

201. Tissier, M. H., "La reaction chromophile d'Escheriche et le *Bacterium coli*," *C.r. Séanc. Acad. Sci., Paris*, **51**, 943, 1889.
202. Tomasz, A., Jamieson, J. D., and Ottolenghi, E., "The fine structure of *Diplococcus pneumoniae*," *J. Cell Biol.*, **22**, 453, 1964.
203. Tweedy, J., Park, R. W. A., and Hodgkiss, W., "Evidence for the presence of fimbriae (pili) on *Vibrio* species," *J. gen. Microbiol.*, **51**, 235, 1968.
204. Ultizur, S., "Induction of swarming in *Vibrio parahaemolyticus*," *Archs. Microbiol.*, **101**, 357, 1974.
205. Ushijima, T., "Fine structure of strictly anaerobic, non-spore-forming, gram-negative bacilli. I. *Bacteroides insolitus*," *Jap. J. Microbiol.*, **11**, 275, 1967.
206. ———, "Fine structure of strictly anaerobic, non-spore-forming, gram-negative bacilli. II. Mesosomes and the cell division of a strain of *Bacteriodes convexus*," *Jap. J. Microbiol.*, **13**, 51, 1969.
207. ———, "Morphology and chemistry of the bacterial cell wall. I. The location of mucopeptide in the cell wall of *Bacteriodes convexus* and its chemical composition," *Jap. J. Microbiol.*, **14**, 15, 1970.
208. Van Houte, J. and Jansen, H. M., "The iodophilic polysaccharide synthesized by *Streptococcus salivarius*," *Caries Res.*, **2**, 47, 1968.
209. Vysotskii, V. V., "*Brucella* ultrastructure. I. The ultrastructure of S-cells," *J. Microbiol. Epidem. Immunobiol.*, **42**, 3, 1965.
210. ———, Jurdina, D. S., and Ustrovskaya, N. W., "Ultrastructure of *Brucella*. II. Ultrastructure of R-cells," *J. Microbiol. Epidem. Immunobiol.*, **44**, 19, 1967.
211. Ward, M. E., Glynn, A. A., and Watt, P., "The fate of gonococci in polymorphonuclear leucocytes: An electron microscopic study of the natural disease," *Br. J. exp. Path.*, **53**, 289, 1972.
212. ——— and Watt, P. J., "Adherence of *Neisseria gonorrhoeae* to urethral mucosal cells: an electron microscopic study of human gonorrhoeae," *J. infect. Dis.*, **126**, 601, 1976.
213. Whitcomb, R. F. and Williamson, D. C., "Helical wall-free prokaryotes in insects: multiplication and pathogenicity," *Ann. N.Y. Acad. Sci.*, **266**, 260, 1975.
214. Wiebe, M. E., Burton, P. R., and Shankel, D. M., "Isolation and characterization of two cell types of *Coxiella burnetii* Phase I," *J. Bact.*, **110**, 368, 1972.
215. Williams, R. C. and Wyckoff, R. W. G., "Application of metallic shadow-casting to microscopy," *J. appl. Physiol.*, **17**, 23, 1946.
216. Williamson, D. L., "Unusual fibrils from the Spirochete-like sex ratio organism," *J. Bact.*, **117**, 904, 1974.
217. Winter, P. F. and Delwiche, E. A., "Cell wall composition and incorporation of radio-labelled compounds by *Veillonella alcalescens*," *Can. J. Microbiol.*, **21**, 2030, 1975.
218. Yabuuchi, E., Watani, T. M., and Takeda, Y., "Flagellar morphology of *Vibrio parahaemolyticus*," *Jap. J. Microbiol.*, **18**, 295, 1974.
219. Yamamoto, T., Nishiura, M., Harada, N., and Imaeda, T., "Electron

microscopy of *Mycobacterium lepraemurium* in ultra-thin sections of murine leprosy lesions," *Int. J. Lepr.*, **26**, 111, 1958.

220. Joklik, W. K. and Willett, H. P. (eds.), *Zinsser Microbiology*, 16th ed., Appleton-Century-Crofts, New York, 1976.

Part Five

Protozoa

5. Protozoa

Jean-Marie Jadin and Jean-Baptiste Jadin

5.1. Introduction

The Protozoa are unicellular organisms living in fresh water and oceans, in soil, on plants, and in the body of other organisms.

They are generally microscopic in size, and although they consist of but a single cell this always poses many questions about the fundamental structure of the cell, the cell division, the sexuality, the endocytic process, for example. Answers to these questions can be of great help in understanding the many problems of multicellular organisms.

The Protozoa include some of the greatest powerful enemies of man, e.g., the amoeba with *Amoeba histolytica*, the trypanosomes with *Trypanosoma cruzi* and *Trypanosoma rhodesiense*, plasmodium with *Plasmodium falciparum*. All these Protozoa kill people in tropical and subtropical countries; many species of Protozoa attack and kill cattle depriving man of food rich in proteins.

Until about twenty years ago, knowledge of the morphology of Protozoa was based on preparations viewed by light microscopy. Laboratory workers were familiar with the morphology of the Protozoa as seen in dry, fixed blood films stained by cytological methods or seen by phase contrast with living material.

During the last two decades special new techniques, particularly that of conventional and scanning electron microscopy and that of ultracentrifugation and biochemistry, have increased our information about cells of Protozoa.

The development of these new techniques has placed morphological and biochemical tools in the hands of the cell biologists. In protozoology, structures with dimensions in nanometres have been described, and new concepts of cell structure have arisen.

Methods used routinely for electron microscopy often involve fixation of living material by treatment with chemical agents like glutaraldehyde and/or osmium tetroxide. This method retains life-like preservation at anatomical and molecular levels but it is not yet certain that micrographs obtained in this way reliably represent the living tissue.

Freezing-etching or scanning electron microscopy have obvious advantages over chemical fixation for the preservation of biological materials when tissues have been subjected to suitable treatments to avoid the direct action of chemical fixatives or the formation of ice-crystal. The possibility must be considered that stimulation and alterations caused by the application of chemical fixatives might induce movement of extracellular electrolytes, sodium, and chloride into cells, giving an artificial picture of this organization.[114, 184]

The final electron microscopical picture of a protozoan or any other kind of cell is affected by each step of the processing procedure. The nature of the final image can be affected by fixation, dehydration, infiltration, and polymerization of the final embedding medium. Each of these steps may interact with the preceding and/or subsequent steps. With biochemical processing, some artifacts are obtained unintentionally.

Therefore, it is not surprising that in protozoological papers many artifacts have been published by some authors. In this contribution we aim to avoid the inclusion of information involving artifacts.

Consideration of so vast a subject in a single chapter may mean that some of the interesting references have been omitted.

5.2. Classification

The classification of the Protozoa is constantly being revised and improved. The system adopted in this chapter is a simplified one: only the biological groups that contain human parasites are listed and only the most representative families of human pathogenic protozoa are included.[45]

PHYLUM PROTOZOA
Subphylum I – SARCOMASTIGOPHORA
Superclass I – SARCODINA
Class RHIZOPODEA
Subclass LOBOSA
Order AMOEBIDA
Genera: *Entamoeba, Endamoeba, Endolimax, Iodamoeba, Hartmonella, Naegleria*
Superclass II – MASTIGOPHORA
Class ZOOMASTIGOPHOREA
Order KINETOPLASTIDA
Suborder 1 BODONINA
Genera: *Bodo, Cryptobia* (nonhuman parasites)
Suborder 2 TRYPANOSOMATINA (important human parasites)
Genera: *Trypanosoma, Leishmania*
Order RETARTOMONADIDA
Genera: *Retartomonas, Chilomastix* (unimportant parasites)
Order DIPLOMONADIDA
Genera: *Giardia, Hexamita* (unimportant parasites)
Order TRICHOMONADIDA
Genera: *Trichomonas*
Superclass III – OPALINATA
Order OPALINIDA
Genus: *Opalina* (parasites of amphiabians)

Subphylum II SPOROZOA
Class 1 TELOSPOREA
Subclass 1 GREGARINIA
Subclass 2 COCCIDIA
Order EUCOCCIDIA
Suborder 1 EIMERIINA
Genera: *Eimeria, Isospora, Lankesterella*, etc.
Suborder 2 HAEMOSPORINA
Genera: *Plasmodium, Haemoproteus, Hepatocystis, Leucocytozoon*, etc.

Subphylum III – CNIDOSPORA *(nonhuman parasites)*

Subphylum IV – CILIOPHORA
Subclass HOLOTRICHIA
Order TRICHOSTOMATIDA
Genera: *Balantidium* (unimportant human parasite)

5.3. Class Rhizopodea

An enormous variety of amoebae exist naturally either free-living in water, soil, or plants, or living as commensals or parasites in the digestive tract of vertebrates. All the amoebae are grouped in the superclass Sarcodina of the subphylum Sarcomastigophora. They float or move slowly in fresh or salt water or in the viscera of vertebrates, locomotion being accomplished by pseudopodia and occasionally by flagella (*Naegleria*). All the parasitic forms are naked amoebae and they are included with many nonparasitic ones, like the well-known *Amoeba proteus* in the order Amoebida, of the subclass Lobosa in the class Rhizopodea. They move with lobose pseudopodia. A few genera are only obligate parasites of their host's gut (*Entamoeba* (a constant source of confusion with *Endamoeba*, a parasitic amoeba of insects), *Endolimax, Iodamoeba, Dientamoeba*) or their host's brain (*Naegleria, Acanthamoeba*). Only few species are known to be harmful parasites: *E. histolytica* of man and other mammals, *E. invadens* of reptiles, *N. fowleri* and *A. culbertsoni* of man.[52] These amoebae sometimes invade the tissue of their hosts. No vectors are known for any parasitic amoebae although trophozoites or cysts may be carried by insects (e.g., cockroaches, flies) or by dirty hands from fecal matter to food.

Only two parasitic genera of amoebae will be considered: genus *Entamoeba* and genus *Naegleria*.

5.4. Genus *Entamoeba*

Amoebae belonging to this genus are distinguished by their nuclear structure. The Feulgen-positive heterochromatin is located in a peripheral ring, which appears granular in fixed material, lining the nuclear membrane and in a small approximately central karyosome. During mitrosis, peripheral granules do not move to polar masses, and the nuclear membrane remains intact. No temporary flagella are formed. Type species: *E. histolytica*

5.4.1. External appearance and life cycle

Observed with a phase contrast microscope, trophozoites of *E. histolytica* move with a flowing movement sometimes called limax-type movement. The entire cell, elongated, flows smoothly in one forward direction, its anterior end acting as a broad but single pseudopodium. Trophozoites of *E. histolytica* found in the host's gut usually measure 10–20 μm in diameter. Sometimes, for reasons unknown, trophozoites invade the mucosa and the submucosa of the intestine and may spread via the blood vessels to other tissues (e.g., liver, lung, spleen, kidney). In these new environments they become larger in size (20–40 μm).

Only in the gut lumen can the trophozoites transform into cysts. They round up and secrete the cyst wall. Cysts of *E. histolytica* have four nuclei and chromatoid bodies. They cannot develop further in the original host but are excreted with the feces, here representing the transmission stage. The cyst can survive outside for some time, waiting to be swallowed by another susceptible mammal host. Then it passes unharmed through the gastric juice to the small intestine where an amoeba emerges from the cyst with four nuclei. Each nucleus divides once more, followed by cytoplasmic division, giving eight, small uninucleate amoebae. These reach the large intestine where they grow to the full size. Subsequently, the amoebae multiply by binary fission in both the large intestine and in the tissues if an invasion occurs.

5.4.2. Ultrastructure

Cell membrane

The cell membrane is a unit membrane 7 μm in thickness, composed of a continuous trilaminar plasma membrane of about 100 nm and a prominent external filamentous coat. In this coat numerous thin filaments radiate away from the plasma membrane for a distance of about 1,500–2,000 nm. The appearance of the thin filaments and their diameter vary from one preparation to another and they are also highly sensitive to small variations in the technical procedure. In some instances, the filaments appear contracted at intervals into small spheres (influence of the fixative procedure ?). With osmium fixative, the filaments are about 40–80 nm, but after staining with uranyl acetate they are about 120 nm with a light core of 60–80.

Amoeba cell membrane has been used extensively as experimental material for a great number of important studies on endocytosis (pinocytosis, phagocytosis) and cell surface structure.[54, 197] Many of the Carlsberg Laboratory's papers of Copenhagen were devoted to these subjects.[27, 28, 29, 30, 31, 32, 33, 34, 35, 38, 47]

The chemical composition of the amoeba cell coat has been thoroughly studied.[22]

Nucleus

The nuclear membrane is composed of an electron-lucent layer sandwiched between two electron-dense layers of about 7 nm. The nuclear envelope thickness averages from 16 to 37 nm, following the difference in thickness of the middle lucent layer. Nucleopores of about 70 nm in outside diameter are present.

Nuclear matrix. The nucleoprotein of *E. histolytica* is arranged in a peripheral ring lining the nuclear membrane, and in a small karyosome.

Nuclear division. According to their mode of division, the amoebae fall into three main groups as illustrated in Figure 5.1(a),(b),(c) using Feulgen reaction.[172]

In the first group, the example used is *Entamoeba histolytica*. The nucleoli or karyosomes, a term commonly utilized in protozoology, migrate centrally during prophase and break up prior to metaphase. Meanwhile the peripheral granules do not give rise to polar masses during mitosis. The nucleoli slowly reform during late telophase and early interphase. The nucleolar membrane is always intact (Figure 5.1(a)).

The second group includes, for example, *Naegleria fowleri*. Nuclear division begins with a swelling of the nucleus and an elongation of the nucleolus. Later the nucleolus divides into two equal halves known as polar masses (prophase). After the formation of polar masses, the position of the equatorial plate is visible, occupied by a solid mass of Feulgen-positive chromatin. Spindle fibers connecting the polar masses can be seen (metaphase).

On Feulgen preparations counterstained with light green or in iron-alum-hematoxylin preparations of mitotic *Naegleria*, it is easy to see achromatic caps situated between the ends of the polar masses and the nuclear membrane. These caps are called polar caps.

The band of chromatic material divides into two and each half moves towards its pole (anaphase). The polar caps disappear. The nuclear membrane persists during all the steps of the nuclear division. At the beginning, it becomes elongated and finally it constricts giving rise to two daughter nuclei (telophase) (Figure 5.1(b)).

In the third group, *Hartmanella castellani* is the example used. At the early stage, the Feulgen-positive granules of the nucleolus move to the center giving the appearance of fragmentation of the nucleolus. The nucleolus gradually disappears, probably giving rise to the spindle which is globular with a loop of Feulgen-positive material in the center (prophase).

The chromatic material assumes the shape of a solid band at the equatorial plate stage. The chromosomes are not distinguishable. The chromatic band divides in two (metaphase). The globular spindle gradually becomes elongated while the nuclear membrane disappears. When the mixtures of chromosomes reach the two ends of the poles, they seem to be connected by a thread-like structure (anaphase).

After the chromatic material has reached the two poles, a new nuclear membrane begins to grow up and surrounds the chromatic materials forming two daughter nuclei (telophase) (Figure 5.1(c)).

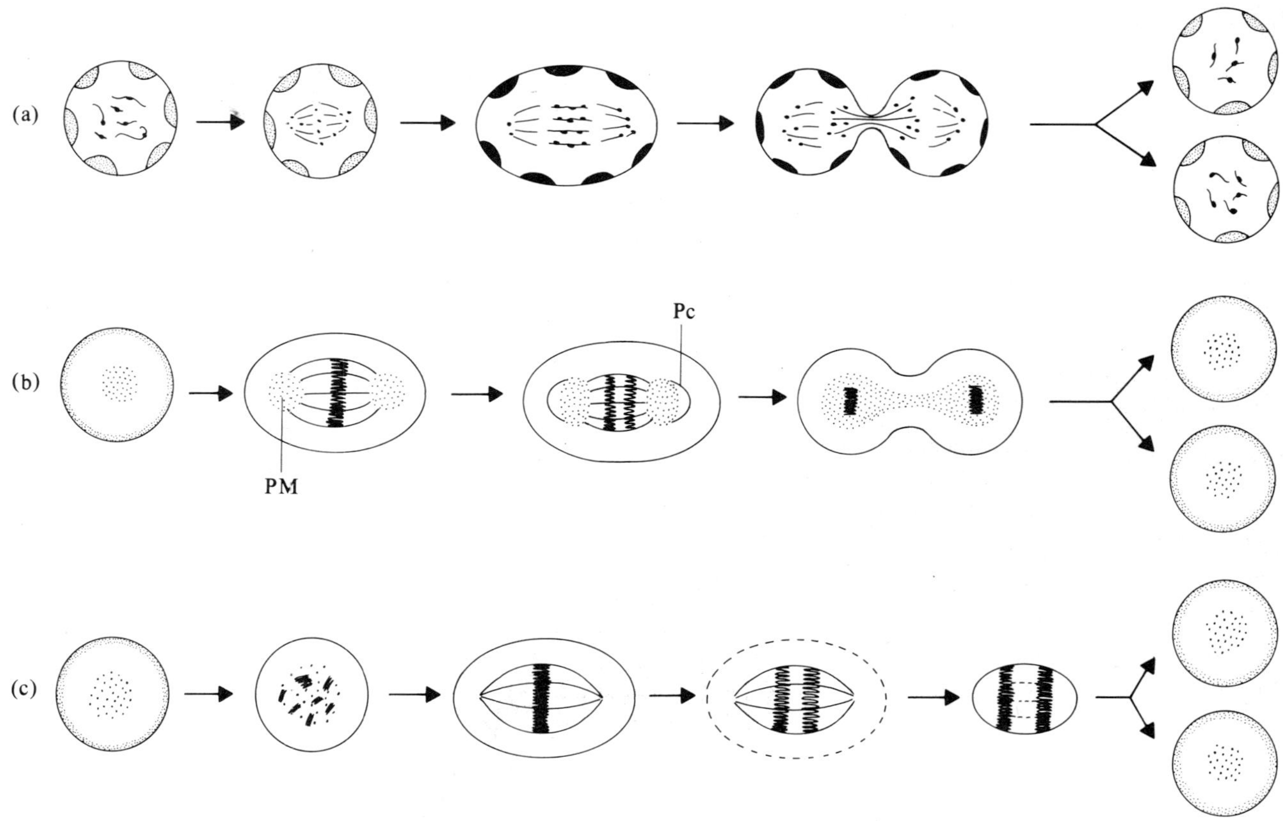

Figure 5.1. Mode of division of amoebae.
(a) *Entamoeba histolytica.*
(b) *Naegleria fowleri*. Pc: polar cap. PM: polar mass.
(c) *Hartmanella castellani.*

Mitochondria

Most of the mitochondria of *E. histolytica* are 1–2 μm spheroids. Two types of mitochondria are known:

1. A rod-shaped mitochondrion with a dense matrix and tubular cristae of variable width.
2. A more spherical mitochondrion with lighter matrix and narrower tubular cristae which are of more uniform width.

The amoeba can develop one or the other type of mitochondrion following an aerobic or anaerobic environment. It was demonstrated that *Pelomyxa palustris*, a nonpathogenic amoeba which feeds on only nonmotile algae and diatoms, has a respiratory metabolism in direct relation with the presence of endosymbionts.[53]

Correlations between electron microscopy studies and biochemical data are needed for *E. histolytica*.

Glycogen bodies

The glycogen bodies are spheroidal and usually measure 5–20 μm. They vary in size, form, and number, presumably following differences in physiological, nutritional, and environmental conditions.

Vacuolar apparatus

Like all the Protozoa, the amoebae include in their cytoplasm many cellular organelles characterized by their content of acid hydrolases; these organelles are defined by de Duve[55] as lysosomes. Many cytochemical studies on the localization of acid phosphatase and esterase have shown the presence of digestive enzymes in lysosomes and in phagosomes of amoebae.[27, 28, 29, 30, 31, 32, 33, 34, 35, 36, 37, 38]

Golgi apparatus is present but reduced in size, and sometimes numerous. Following nutritional and physiological conditions, an amoeba contains several hundred Golgi apparatuses. Each is composed of curved, centrally flattened cisternae with expanded ends. The number of cisternae varies from one body to another, but it is usually six to eight, and their diameter is about 1 μm. The Golgi apparatus is often closely associated with some elements of the endoplasmic reticulum. They are involved in the formation of the amoeba cell surface, particularly in the production of the polysaccharide-rich filamentous material for the cell coat. Some Golgi cisternae may be stained by the Gomori method for acid phosphatase, thus they are also involved in the packaging of hydrolases into small vesicles to form primary lysosomes.

Endocytosis is initated after an interaction with the material to be ingested with the cell coat. Pseudopodia are formed at the sides of the touched areas of the cell coat and close round the material which is finally ingested in a phagosome. The prey is a ciliate, the death of which, due to dehydration of the vacuole with resulting anoxia, takes place within 10 to 15 minutes after ingestion.

Many drugs or local conditions (e.g., pH, concentration of solutions) may affect the endocytic process. This has been extensively studied in the Laboratory of Carlsberg in Copenhagen.[33, 37, 76]

Changes in charge and hydration of surface material, correlated with calcium-dependent changes in membrane structure, modify the surface charge pattern of the cell and induce the formation of pinocytic channels. The pseudopodia bearing the channels are composed of hyaline cytoplasm containing many or few organelles. During channel formation, the plasmalemma moves rapidly and cytoplasmic streaming is minimal. The pinocytic channels divide up into vacuoles and vesicles. In the Carlsberg Laboratory,[31] in studies using Alcian blue, it was shown that channel subdivision occurs two hours after endocytosis. The label material (Alcian blue) could be identified in fewer and larger vacuoles up to 72 hours later.

Some pathogenic amoebae, *E. histolytica*, and many strains of free-living amoebae contain intact microorganisms (bacteria, viruses) within cytoplasmic vacuoles. The relationship of these intracellular organisms (endosymbionts?) to the host amoeba may vary from one type of amoeba to another. In fact the nature of that relationship is unknown and the electron microscope is not particularly valuable here.

Contractile vacuole

The contractile vacuole consists of a central vacuole membrane surrounded on its cytoplasmic surface by a thick, close layer of smooth vesicles and tubules, and by a layer of closely packed mitochondria. These vesicles and tubules vary in size from 300 to 2,000 nm, and may come from different sources including pinocytosis, phagocytosis, release of vesicles from blebs on the outer membrane of the nuclear envelope, or from Golgi bodies. The contractile vacuole plays a role in osmoregulation, pH regulation, water balance, and in the recycling of the plasmalemma.

Ribosomes

The ribosomes appear to be sparsely distributed in the cytoplasm.

Chromatoid bodies

Chromatoid bodies are specific for cysts of *E. histolytica*. In fresh preparations of cysts, the chromatoid bodies appear glass-like and stain intensely with hematoxylin. They disappear when the cyst ages, because at this stage the ribonucleoprotein becomes dispersed throughout the cytoplasm. They contain ribonucleoproteins arranged as ribosome-like particles in a regular array.

Chromatoid bodies function as a reservoir, their substance being used by the amoeba just before it escapes from the cyst.

5.5. Genus *Naegleria*

Carter[198] first isolated an amoeboflagellate from a fatal human case of primary meningoencephalitis in Australia. Up to July 1976, about 95 other fatal human cases of primary amoebic meningoencephalitis had been reported. The amoeboflagellates responsible for meningoencephalitis belong to only two species, *Naegleria fowleri* (Figure 5.2) and *Acanthamoeba culbertsoni* (strain Lilly). *N. fowleri* is cosmopolite and frequently observed. *A. culbertsoni* is also cosmopolite, but not so frequently seen. These pathogenic amoebae cause fatal cerebral meningoencephalitis in man and in experimental animals.[198, 199, 200]

5.5.1. Primary amoebic meningoencephalitis

N. fowleri and *A. culbertsoni* are free-living in water.[96, 150] The amoebae enter through the nasal passages[115, 192] where they penetrate and destroy the olfactory mucosal epithelium, and then invade the brain via the cribriform plate and the olfactory nerve. The disease is fatal in a few days.

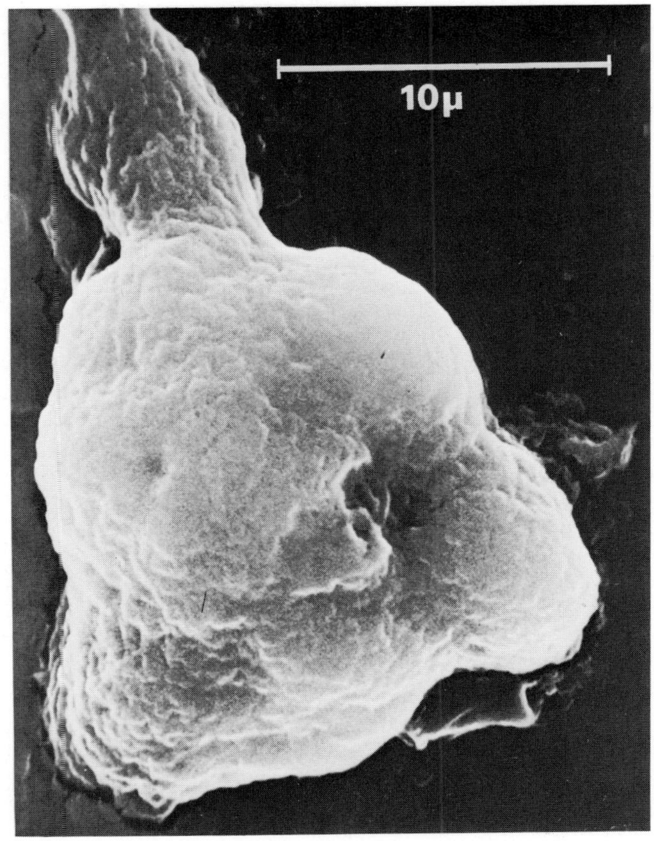

Figure 5.2. ***Naegleria gruberi.***
Scanning electron micrograph. (× 4,400)

Anatomopathology of the lesions has been performed by Martinez[115] and by Visvesvara and Callaway.[192]

5.5.2. Other pathological roles of *Naegleria*

Naegleria, like other strains of amoebae, may carry in their cytoplasm some pathogenic microorganisms.

Schuster[152, 153] has described the presence of virus-like particles in the non-pathogenic *N. gruberi*. Such studies of pathogenic amoebae are greatly needed.

Jadin[79] demonstrated that pathogenic mycobacteria (*Mycobacterium leprae*) can grow in some parasitophorous vacuoles (phagosomes, lysosomes ?) of *Naegleria.*

5.5.3. Cysts

Identification of free-living amoebae is difficult when no distinctive features exist among the different genera and species. In such instances cyst structures are very useful in the indentification process (Figure 5.3).

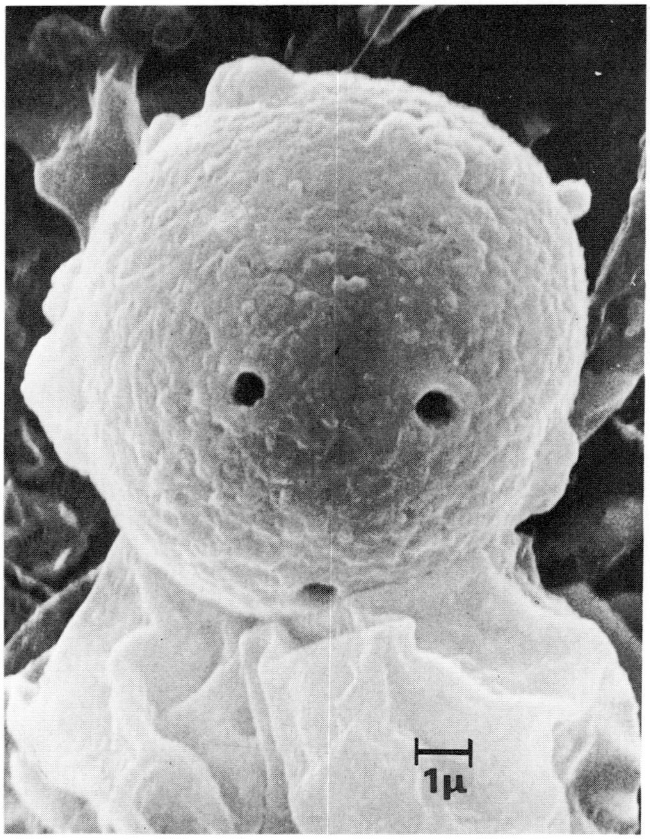

Figure 5.3 ***Naegleria fowleri*** **cyst.**
Scanning electron micrograph showing the pores of the cyst. (× 8,000)

The more characteristic features are the shape of the cyst, the appearance of the wall, and the number and morphology of the pores. Schuster[151] has published a comparative study on the ultrastructure of cysts of *N. gruberi*, *N. fowleri*, and *N. jadini*.

With scanning electron microscopy it is easy to observe that all cysts of any strain of amoeboflagellate have at least one pore. The morphological characteristics of these pores are of a great interest in systematics.[97]

5.5.4. Flagellate form

When a cyst is put into distilled water it transforms immediately into flagellate form with two flagella (Figure 5.4); the ameoboflagellate moves rapidly and transforms into the trophozoite with an amoeboid form.

5.6. Class Zoomastigophorea

The Zoomastigophorea are flagellates. They have one or more flagella used for locomotion. In some species the flagellum is anterior in position and pulls the

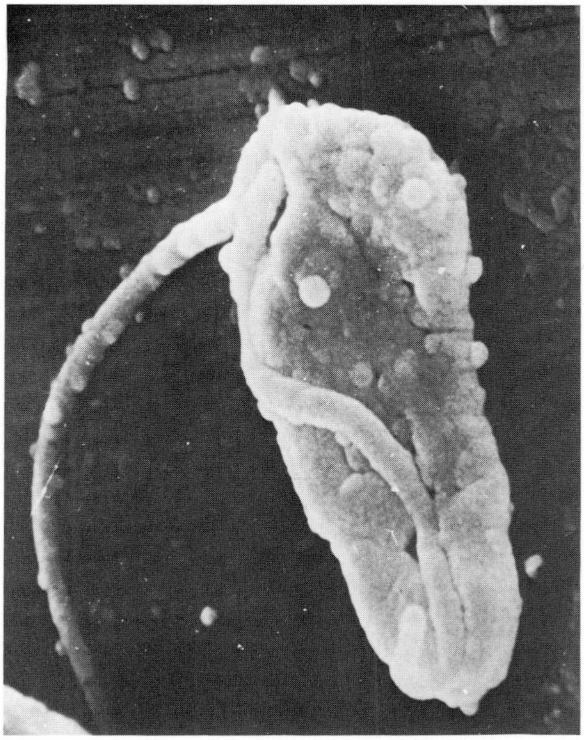

Figure 5.4. ***Naegleria fowleri***
Scanning electron micrograph of a flagellate form. (× 3,000)

protozoan; in other species it is posterior and pushes it. Asexual reproduction occurs by binary fission, but in some groups the reproduction is sexual. Protozoa belonging to the class Zoomastigophorea are solitary or colonial and their nutrition may be saprozoic, holozoic, or holophitic.

Zoomastigophorea include many parasitic Protozoa distributed in several orders. The order Dinoflagellida includes free-swimming Protozoa, a few of which are parasitic for other Protozoa or for numerous invertebrates. For instance, *Coccodinium* parasitizes other dinoflagellates; *Blastodinium* lives in the gut of *Cyclops*, a crustacean. Dinoflagellida present two grooves, one longitudinal and one transverse with a flagellum in each groove. Their cells contain chromatophores.

The order Rhizomastigida includes Protozoa which possess amoeboid bodies often producing pseudopodia, and one to four flagella. The best-known example is *Histomonas meleagridis* which has an intermediate host, a caecal nematode worm (*Heterakis gallinae*), and a definitive host, chickens or pheasants.

The order Kinetoplastida includes suborders Bodoninia and Trypanosomatina. Protozoa belonging to the Bodonina are free-living and have two or more flagella. Most of them live in the gut of vertebrates or invertebrates and usually are not pathogenic. The suborder Trypanosomatina includes many human parasites (e.g., *Trypanosoma, Leishmania*) and will be carefully examined.

The order Retartamonadida comprises harmless flagellates, e.g., *Chilomastix mesnili* which lives saprophytically in the cecum and colon of man, primates, and pig.

The order Diplomonadida includes flagellates with a bilaterally symmetrical body, two nuclei, and eight flagella. The two main genera, *Giardia* and *Hexamita*, are easily distinguished by the fact that *Hexamita* does not encyst, is narrower, lacks an interior sucker, and has two pairs of flagella emerging at the extreme front end. *Hexamita* species are not pathogenic while *Giardia lamblia* is more or less pathogenic for humans.

Order Trichomonadida includes almost all Protozoa parasitic for vertebrates and invertebrates. They do not produce cysts. *Trichomonas vaginalis* is perhaps the best known species of this order. The most diagnostic feature of *Trichomonas* is the axostyle, a hyaline rod passing lengthwise through the cell and often projecting from the posterior end. These flagellates are also characterized by the presence of several anterior flagella, a costa, a pelta, and an undulating membrane (silver staining area). *T. vaginalis* may induce a leukorrheic discharge. The most important ultrastructural characteristic is the presence of hydrogenosomes and the absence of recognizable mitochondria. Hydrogenosomes are the site of the anerobic formation of H_2 but also of the reduction of O_2 in the presence of certain substrates.[25, 109, 110, 111, 117, 118]

The presence of hydrogenosomes in trichomonad flagellates is of chemotherapeutic importance.[111, 118]

5.7 Genera *Trypanosoma* and *Leishmania*

Both of these parasites play an important role in human pathology. At some stage of their life cycle they have an elongated form with a single nucleus, a single flagellum, and a kinetoplast. The end where the flagellum becomes free is conventionally called the anterior. The kinetoplast is always situated near the origin of the flagellum.

5.7.1. Terminology

Hoare and Wallace[75] suggested a new terminology for developmental stages of trypanosomatid flagellates. The new terminology is not connected with the names of taxa such as genera, but it is descriptive, uniform in construction, and universally acceptable. The arrangement of the flagellum serves as a basis for the classification. In this new terminology, the term *mastigote* (from Greek μαστιξ = whip) is combined with an appropriate prefix showing the starting point of the flagellum, its course, and its point of emergence.

We agree with the terminology of Hoare and Wallace[75], but we prefer the term "micromastigote" in place of "amastigote". On this basis, we use the following new terms: micromastigote, promastigote, opisthomastigote, epimastigote, trypomastigote, and choanomastigote (Figure 5.5).[80, 81]

Micromastigote, for former leishmanial stage, or for amastigote,[75] is represented by rounded forms with a small flagellum easily discernible by electron

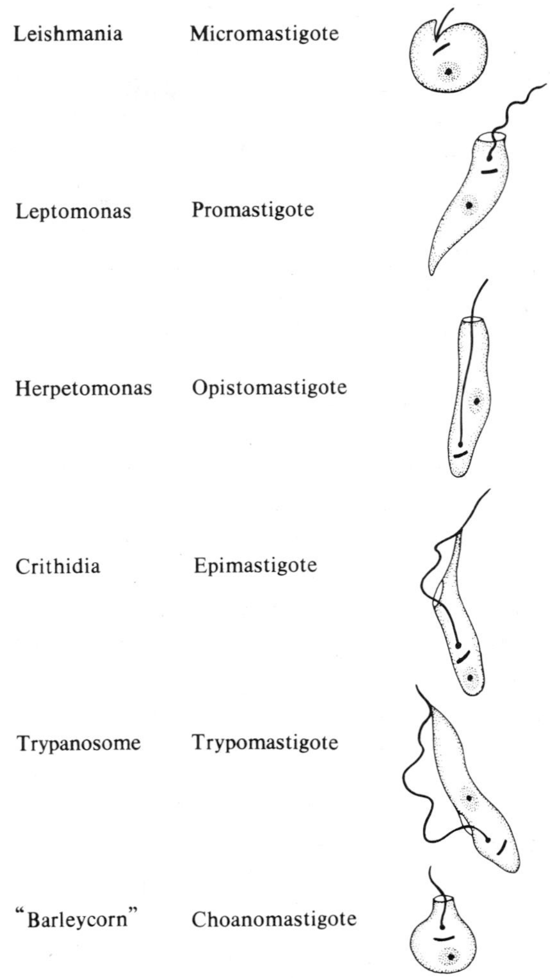

Figure 5.5. Old and new terminology of trypanosomatid flagellates.

microscopy (Figure 5.6). Thus, we think it is necessary to use the appropriate prefix micro- and not the prefix a-, which is quite appropriate in optical microscopy but inappropriate in electron microscopy, and may lead to further confusion. In this cell, only the kinetoplast and the nucleus are clearly apparent in optical microscopy, but in electron microscopy all the normal components of the Trypanosomidae are easily observed, e.g., periplast, blepharoplast, flagellar pocket, endoplasmic reticulum, nuclear membrane, mitochondria, ribosomes. Such forms are seen in the genus *Leishmania* and some species of the genus *Trypanosoma.*

Promastigote for the former leptomonad stage is represented by an elongated form with the starting point of the flagellum at the anterior end of the cell. The flagellum rapidly becomes free. The kinetoplast lies between the flagellum and the nucleus. Promastigote forms can be seen in the genus *Leptomonas* and some culture forms of the genus *Trypanosoma.*

Opisthomastigote for the former trypanomorphic form in which the starting

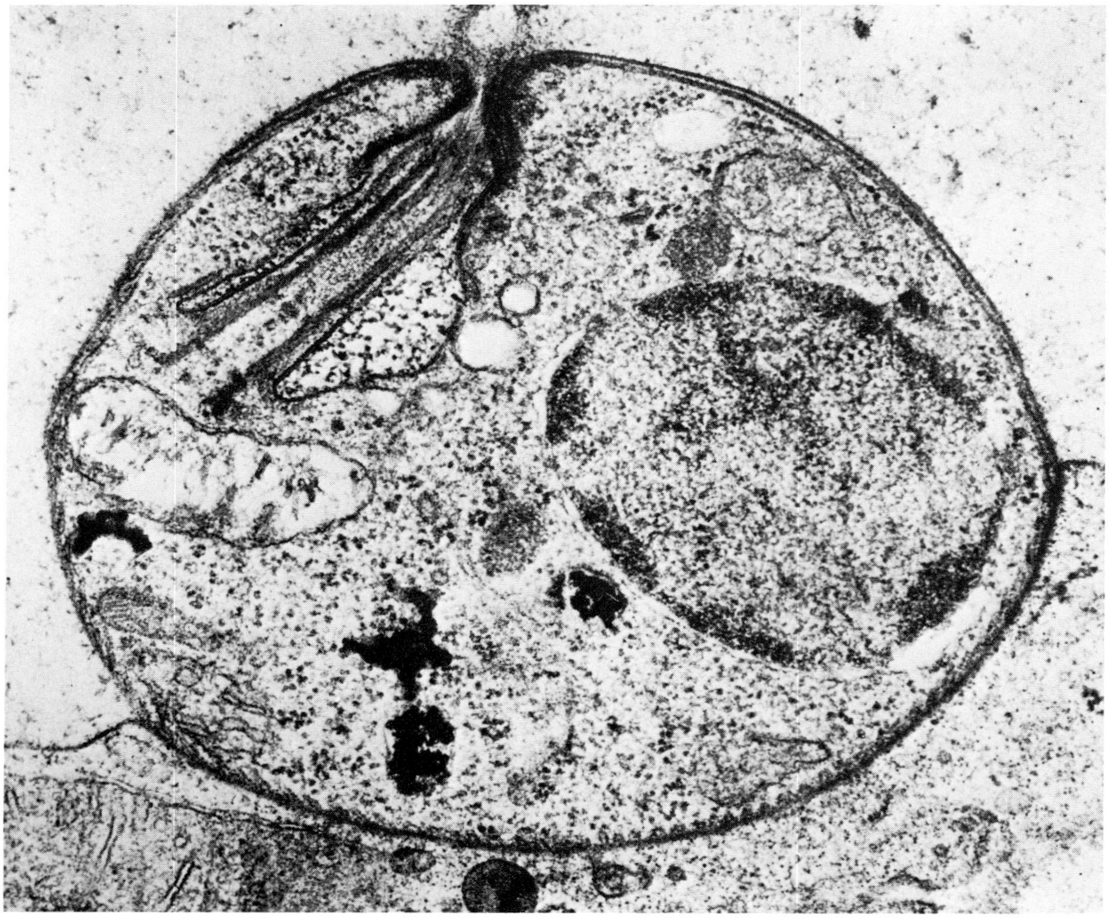

Figure 5.6. Micromastigote of *Leishmania mexicana*
Section of a micromastigote inside a parasitophorous vacuole of a spleen cell of a hamster. The flagellum is visible inside the flagellar pocket near the kinetoplast. (× 53,100)

point of the flagellum is at the posterior end of the cell. The flagellum passes through the cell and emerges at the level of the anterior end. Thus, the flagellar pocket is elongated. The kinetoplast is in a posterior situation. The nucleus has a central localization. These forms can be observed in the genus *Herpetomonas*.

Epimastigote for the former crithidia form, in which the starting point of the flagellum is at one side of the cell, from where it continues along the cell forming the so-called undulating membrane and eventually runs free from the anterior end. The nucleus lies near the kinetoplast at the posterior end. Epimastigote forms can be observed in the genera *Blastocrithidia*, *Trypanosoma*, and *Crithidia*.

Trypomastigote for the former trypanosome form in which the flagellum is lateral and there is an undulating membrane. The nucleus is centrally located and thus more anterior than the kinetoplast. Genus *Trypanosoma* (Figure 5.7) illustrates these forms.

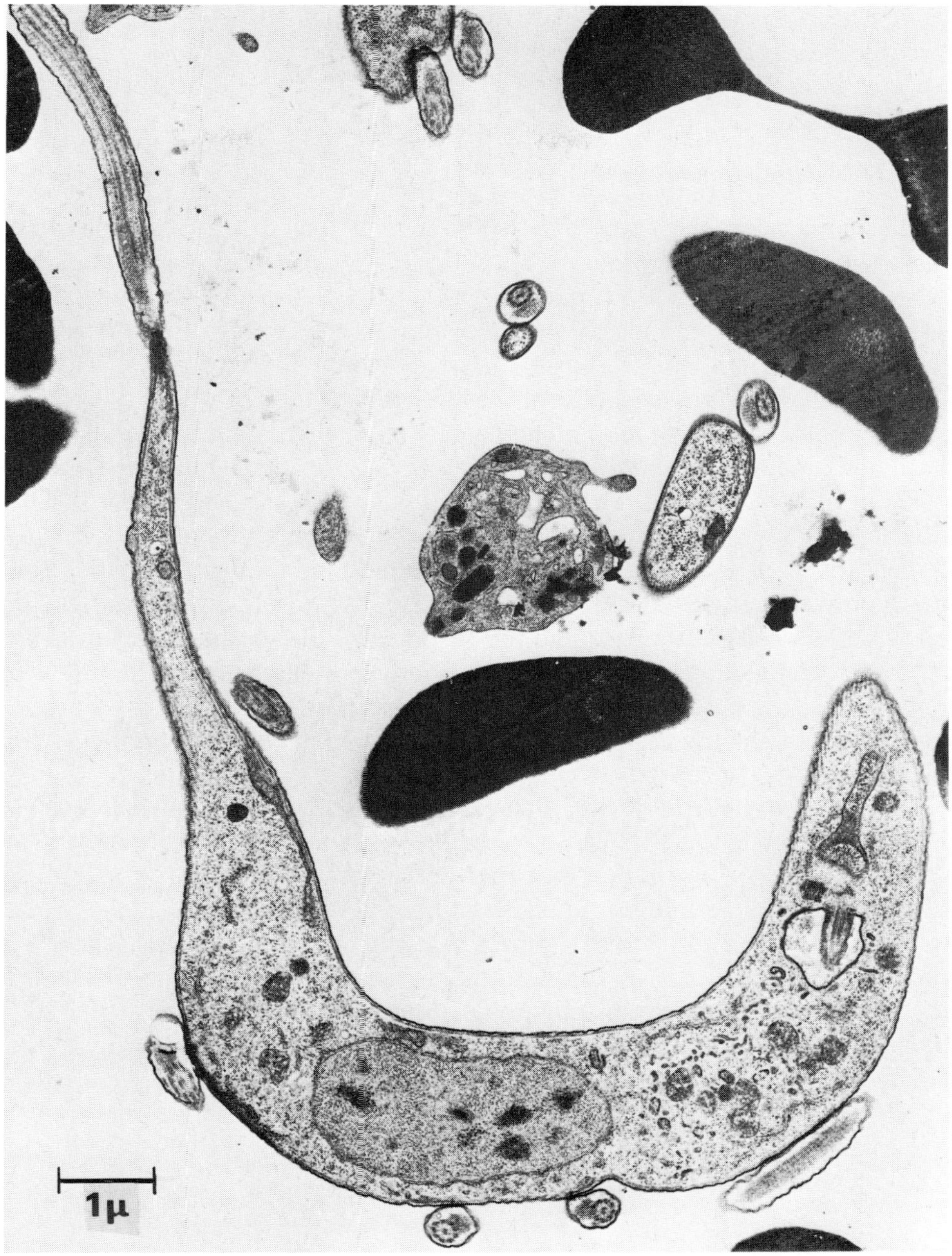

Figure 5.7. Trypomastigote of *Trypanosoma brucei.*
Section. (× 13,000)

Choanomastigote is the new term used for small and large promastigote. The flagellar pocket is well developed. These forms can be seen in the genera *Strigomonas* and *Crithidia*.

5.7.2. External appearance

Until two decades ago knowledge of the morphology of trypanosomes was based on studies by light microscopy.[80] In this section morphological data will be reduced and confined to characters which are interesting for study and comprehension of general medical protozoology. More exhaustive data are available elsewhere.[80]

The cell of blood-stream forms of mammalian trypanosomes (trypomastigote) (Figure 5.8) is curved and lanceolate in shape and elongated like a flattened blade. The extremity which is directed forwards during locomotion is customarily described as the anterior end, the other is the posterior end. The flagellum starts at the posterior end from where it emerges and runs forwards along the cell membrane, forming the so-called undulating membrane. Formerly, it was thought that the flagellum was a part of the cell membrane, but electron microscopical investigations have shown that desmosomes exist as structural connections between flagellum and cell membrane. When the flagellum reaches the anterior end of the cell, a free-portion is either present or absent, depending on the species. The base of the flagellum is known as the blepharoplast.

The flagellar pocket is a permanent invagination of the cell wall in the area where the flagellum starts. Close to the blepharoplast lies a characteristic and dark staining structure called the kinetoplast, an integral part of the mitochondrial system, rich in bundles of fibrils containing DNA.

The nucleus is situated in a different part of the body, in the center or in the anterior or posterior half of the cell. The respective positions of the nucleus, the kinetoplast, the starting point of the flagellum, and the degree of development of the undulating membrane are of great taxonomical importance (Figures 5.5, 5.8).

Other structures of the cytoplasmic contents are the Golgi apparatus, the endoplasmic reticulum, the lysosomes, the phagosomes, the ribosomes, and lipid vacuoles.

5.7.3. Ultrastructure

The cell membrane

This section looks at the periplast, the flagellar pocket, and the filopodium like-processes.

The periplast. The cell membrane of the Trypanosomidae is characterized by a structure called the periplast (Figure 5.8). The periplast includes a cell wall of 8–10 nm and an internal layer of microtubular fibers, also called subpellicular microtubules. According to several authors[50, 139, 149] the architecture of the cell wall corresponds to the general model described by Robertson.[134]

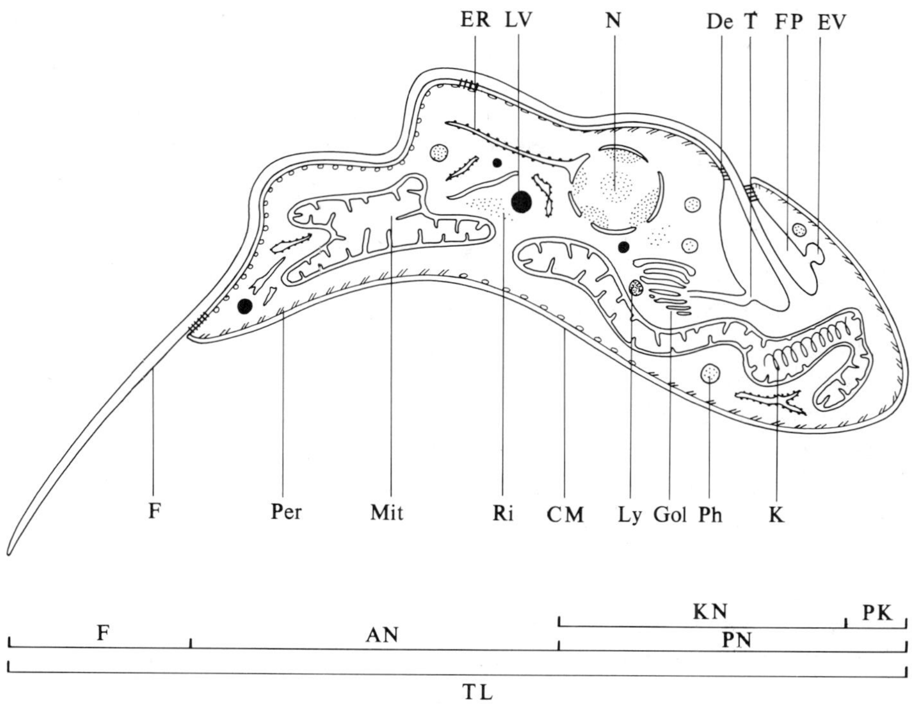

Figure 5.8. Diagrammatic representation of a trypomastigote.
F: flagellum, Per: Periplast, Mit: mitochondrion, ER: endoplasmic reticulum, Ri: ribosomes, CM: cell membrane, LV: lipid vacuole, N: nucleus, Ly: lysosome, Gol: Golgi apparatus, De: desmosome, FP: flagellar pocket, T: tubule, Ph: phagosome, EV: endocytic vacuole, K: Kinetoplast.
F: length of the free flagellum, TL: total length, AN: length between the base of the free flagellum and the nucleus, PN: length between the posterior end and the nucleus, KN: length between the kinetoplast and the nucleus, PK: length between the posterior end and the kinetoplast.

The cell wall becomes deformed only in the internal area of the flagellar pocket during the process of endocytosis.

The external surface of every cell is normally covered by a cellular layer called the extraneous coat or glyocalyx by several authors, or cell coat by Rambourg and Leblond.[131] The cell coat has a very variable structure. It includes sugars, lipids, amino acids, glycoproteins, mucopolysaccharides, and other components, and occasionally it is impregnated with calcium phosphate and carbonate. The components of the glyocalyx form a web-like network.

Such a cell coat is found in trypanosomes. It is compact, 12–15 nm thick, and overlies the entire plasma membrane but not the flagellar pocket. Some fixative procedures for electron microscopy can damage the trypanosome cell coat. The best one to use is ruthenium red in osmium, followed by rinsing in a physiological buffer with ruthenium red and dehydrating with alcohol/water solutions containing ruthenium red, to avoid the solubilization of the stain during washing and the first subsequent steps of dehydration. Using exclusively the procedure described we can conclude that all the pathogenic and non-pathogenic Kinetoplastida have a cell coat. However, the thickness of the cell coat can change during the cycle of the trypanosome and it is also evident that the thickness differs from one species of trypanosome to another, investigated under the same conditions.

The cell coat proteins are frequently renewed. Vickerman observed a relationship between the modifications of the biochemical components of the cell coat and the cyclical variations of the antigenicity of the trypanosomes.[187, 188]

Leptomonas ctenocephali, a nonpathogenic trypanosome living in the intestinal tract of fleas is covered with a thin glyocalyx like a spider's web.[85]

All the salivarian trypanosomes pathogenic for mammals are covered with a compact coat. This coat can be visualized by freeze-etching[155] and contains a carbohydrate layer next to the underlying plasma membrane[202] that can be removed by proteolytic enzymes.[190, 202]

The cell coat of *Leptomonas ctenocephali* is not discarded after many years of culture.[85] In contrast, the normally compact cell coat of *Trypanosoma brucei* and *T. congolense* is greatly reduced within 24 hours of placing the trypanosomes in a culture tube or in a vector. It is subsequently regained when the epimastigote trypanosomes transform into metacyclic forms.[189, 190, 191]

The cell coat contains the variant antigens demonstrated by agglutination reactions, by immuno-electrophoresis, and by immuno-electron microscopy.[56, 107, 187]

Loss of variant antigenic character is correlated with loss of the coat. The surface coat and variant antigen are glycoproteins. Peroxidase or ferritin-conjugated antibody to a specific variant do not bind to the coat of heterologous trypanosomes or to homologous trypanosomes from which the coat has been removed, but only to the surface of homologous trypanosomes whose cell coat is intact. We did not observe a cell coat in the area of the flagellar pocket.

The cell wall is located 8–10 nm from the layer of microtubular fibers. The microtubules are regularly spaced, every 20 to 30 nm (Figure 5.11) and run from one end of the cell to the other. Their structure has been thoroughly studied.[15, 80] On perpendicular section of their axis, the microtubular fibers have a diameter varying from 15 to 23 nm. The partition wall is about 5 to 6 nm.

The great variations in fixative procedures in electron microscopy must partly be held responsible for the variable size of diameter of microtubular fibers mentioned in the literature.[50, 80, 135, 149]

Fuge[60] studied the ultrastructure of the microtubules of the periplast of *T. brucei* and *T. gambiense*. He showed clearly that the microtubules are composed of protofilaments. The latter are formed by globular subunits arranged like the beads of a rosary, measuring 4·5–5 nm. The periplast's microtubules are anastomosed every 15 nm. An important contribution to the ultrastructure of the microtubules[14] shows that they run parallel with each other, and that the general disposition is that of a helix.

They originate near the area of insertion of the flagellum and stretch as far as the apical area of the cell. The terminal end is free.

The physiological functions of the subpellicular microtubules are not known. It is possible that they play the role of myonemes and are responsible for contraction of the trypanosome cell.

These microtubular fibers exist also in the Bodonidae[127] and the Euglenidae.

Flagellar pocket. This is a permanent invagination of the cell wall like the finger of a glove in the area where the flagellum originates. Thus, a cylindrical space is clearly defined. The center is limited by the cell wall of the flagellum and the bottom is the bottom of the flagellar pocket. The lateral sides are formed by the cell wall. The apex of the flagellar pocket is formed by junctions of the cell wall of the flagellum with the cell wall of the lateral side of the flagellar pocket. On good electron micrographs junctional structures like desmosomes are clearly visible at the apex of the flagellar pocket.[22]

The lumen of the flagellar pocket is always in communication with the outside, with the pericellular fluid.[80]

Many terms have been used to describe the structure, e.g., blepharoplast, parabasal grain, reservoir, gullet, cinetovacuole. We feel that these terms have been used wrongly as synonyms by some authors to describe the flagellar pocket.

The term blepharoplast indicates the proximal portion of the flagellum. The term parabasal grain actually refers to an artifact. It was formerly used to describe a concentration of staining drugs in the flagellar pocket, giving, under the optical microscope, the illusion of an organelle distinct from the kinetoplast. The term reservoir implies exclusively an alimentary function. Gullet (or goulot, in French) is probably better, but it is used specifically to describe a distinct formation in *Euglena*. In order to avoid confusion in the future, it seems preferable to reserve the term gullet for exclusive use in *Euglena*. Flagellar pocket is certainly the most suitable term, since it is purely descriptive and avoids theoretical considerations.

At the bottom of the flagellar pocket the cell wall is deprived of subpellicular microtubules and can be more easily distorted. Some tubules of the endoplasmic reticulum emerge in the lumen of the flagellar pocket where they discharge their enzymatic contents.[51, 80, 82, 83, 89, 90, 91, 95]

Sometimes a diverticular formation of the flagellar pocket is visible[49, 88] but the structure is not always present and the physiological role is still unknown.

It was actually demonstrated[80] that the flagellar pocket plays an important role in the process of endocytosis. The prey in the pericellular liquid is engulfed in the flagellar pocket where it is processed by the proteolytic enzymes coming from the specialized tubules. At the bottom of the flagellar pocket the cell wall invaginates, surrounding the digestible material, and finally giving rise to a phagosome.[21, 23, 24, 26, 80]

The filopodium-like processes. After *in vitro* examination of living blood-stream forms of *T. brucei*, Wright and Lumsden[201] and Wright *et al.*[203] have observed that the trypomastigotes of *T. brucei* possess long filaments arising from both the posterior end and the tip of the flagellum. When the mobility of the trypanosomes was lowered by suspension in methyl cellulose, they produced filopodium-like processes which were easily studied and photographed. Not all trypanosomes produce filopodia; they appear *in vitro* at varying periods after isolation from the host. The formation of the filopodium-like process is not related to the death of the animal but is activated by *in vitro* incubation of trypanosomes in physiological salt solution (Figure 5.9).

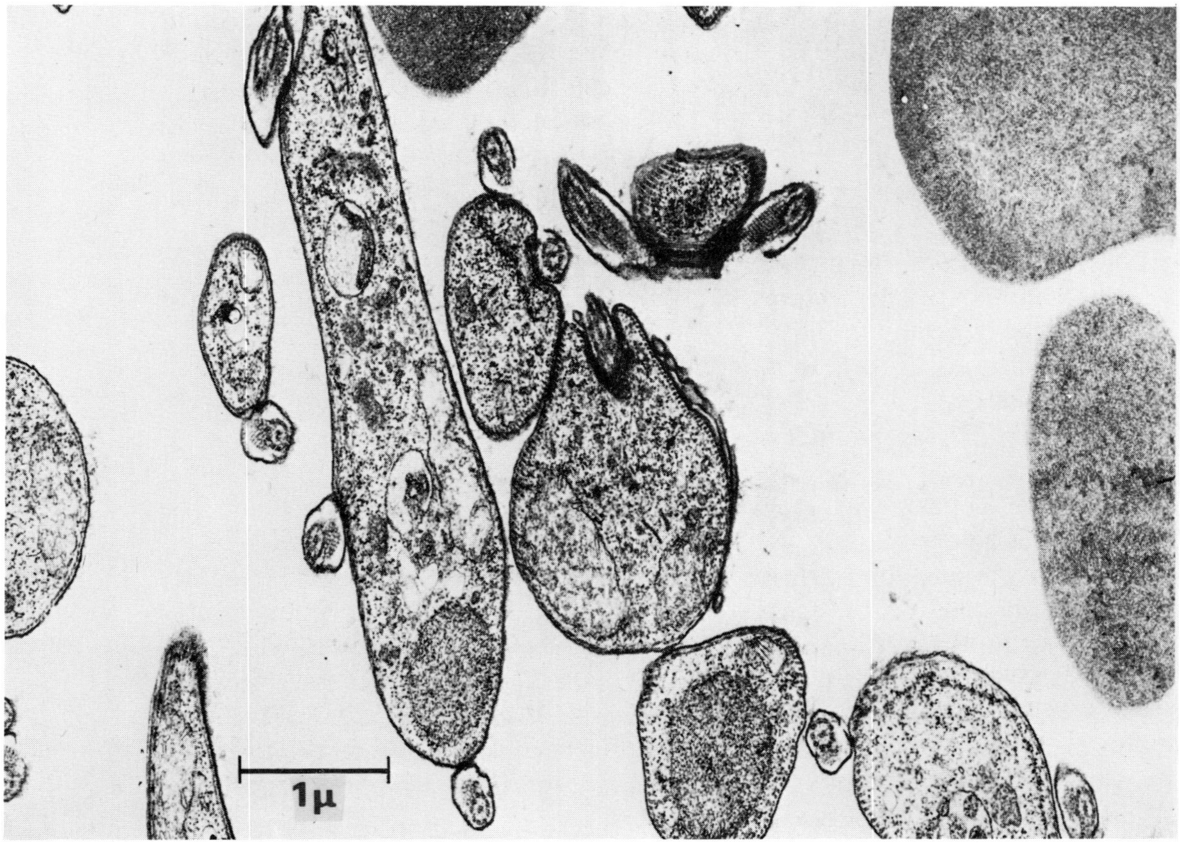

Figure 5.9. ***Trypansoma brucei* filopodia.**
Section of trypomastigotes from blood of mouse. One trypomastigote, in the center, shows the filopodium-like processes. (× 20,000)

Nucleus

A large part of the cell of a trypanosome is occupied by the nucleus. According to the species or the developmental stage of the trypanosome, the position of the nucleus is variable, being found in the anterior, median, or posterior third of the cell. It measures about 1·5 μm long and a little over 1 μm wide. Only electron microscopy shows the structural details.

The nucleus is limited by a double, clearly defined membrane, the space between these two membranes being approximately 20–30 nm. The external side of the external membrane is covered by some ribosomes. The nuclear membrane is perforated by large pores with a diameter of 80–100 nm. The space between the double nuclear membrane is in continuity with the endoplasmic reticulum.

The nucleus has a matrix consisting of granular and fibrillar material. The nuclear DNA differs from the kinetoplastic DNA in that it holds in check the metabolic activity and the morphological features of the cell. The DNA is combined with structural proteins to form the chromatin. In the center of the nucleus there are one or two oval, regularly shaped, dense bodies about 220 nm in shorter diameter, the nucleoli. In a number of cells dense or less dense material can be seen, presumably the chromosomes. In optical microscopy, Feulgen's reaction shows dense bodies which are too small to be accurately counted. Nevertheless, some authors have tried to determine the chromosome numbers (Table 5.1).

Table 5.1. Chromosome number in some species of trypanosome[80]

T. brucei	6–8
T. congolense	6–7
T. cruzi	6–10
T. dimorphon	10
T. equinum	6
T. gambiense	6
T. rhodesiense	5–6

A particularly memorable work of Schaudinn,[141] who inspired all the protozoologists of his time, consisted of a description of male and female trypanosomes of *T. noctuae* from the owl *Athene noctuae*. It seems that Schaudinn confused the morphology of the trypanosomes with that of the host's spermatozoa![70] Such an attempt to prove the existence of sexuality in the trypanosomes was also made by Vanderplank[181] with *T. rhodesiense* and *T. congolense*, and by Fiennes[59] with *T. congolense*.

Mitochondrial system

Trypanosomidae like Bodonidae have special mitochondria. Each trypanosome has a single mitochondrion with a very large portion located directly behind the blepharoplast. This large portion is called the kinetoplast.[20, 40, 49, 80, 108, 119, 128, 130, 173, 196] The kinetoplast contains and synthesizes DNA, seen in the form of a granule, easily stained by Giemsa's staining for optical microscopy, and by many osmiophilic stains for electron microscopy (Figures 5.10 and 5.11).
Kinetoplast. The kinetoplast is surrounded by a double membrane like the mitochondrion; infoldings of the inner membrane are not present at the side nearest the blepharoplast, but are clearly seen at the other sides. The kinetoplastic DNA differs in density from the nuclear DNA and also, in some species, from the mitochondrial DNA bounded by the internal membrane of the mitochondrion. Following Riou and Pautrizel[132] for *T. gambiense*, the density of the nuclear DNA is 1·707 and of the mitochondrial DNA 1·701 and 1·690.
Mitochondria. As indicated above, the kinetoplast is a hypertrophied portion of the mitochondrial system. In fact, the double membrane of the kinetoplast at the side opposite to the blepharoplast shows infoldings of the inner membrane like cristae. The lateral side of the kinetoplast is extended by mitochondria containing cristae, resembling more the mitochondria of mammals and differing from those of ciliates, amoebae, or *Toxoplasma*.

In the various morphologically differentiated cells of the Trypanosomidae (Figures 5.10, 5.11) the size and the structure of the kinetoplast differs accordingly.[80] For example, the extracellular promastigote of *Leishmania mexicana*

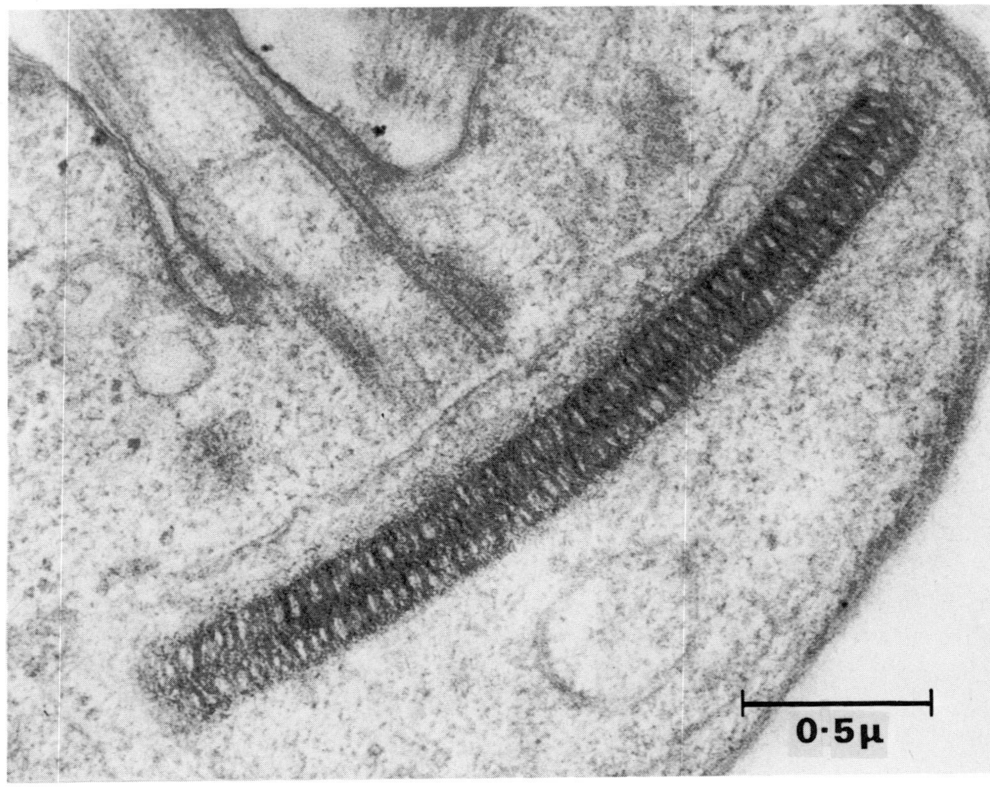

Figure 5.10. DNA in *Trypanosoma cruzi.*
Section of a trypomastigote of *T. cruzi* showing the two layers of minicircles of DNA.[80] (× 50,000)

has a large kinetoplast with many cristae while the intracellular micromastigote has a small one with few cristae.[50, 80]. It is probable that this change in size and structure is connected with a fundamental difference of response of the Trypanosomidae surrounded by different pericellular media. In *L. mexicana* the transformation of micromastigote into promastigote was studied by Creemers and Jadin[50] and in *L. donovani* by Rudzinska *et al.*[135]

T. brucei is a polymorphic trypanosome.[173, 185, 186] In the slender multiplicative blood-stream forms, the mitochondrion is reduced in size to a single peripheral canal extending laterally from the kinetoplast. Mitochondrial cristae become sparse or absent. These trypanosomes break down glucose to pyruvate which they cannot degrade further, and they oxidize their $NADH_2$ following a glycerophosphate oxidase cycle insensitive to cyanide.[19] Transformation of slender forms into stumpy forms occurs with a swelling of the mitochondrial area and the development of tubular cristae. These stumpy forms can oxidize the 2-oxoglutarate.

The function of the kinetoplastic DNA is completely unknown.

After treatment with curative doses of certain dyes like pyronin or acridine (acriflavine, ethidium bromide), a variable percentage of trypanosomes take the staining properties with Giemsa's stain of the kinetoplast. In fact, these drugs

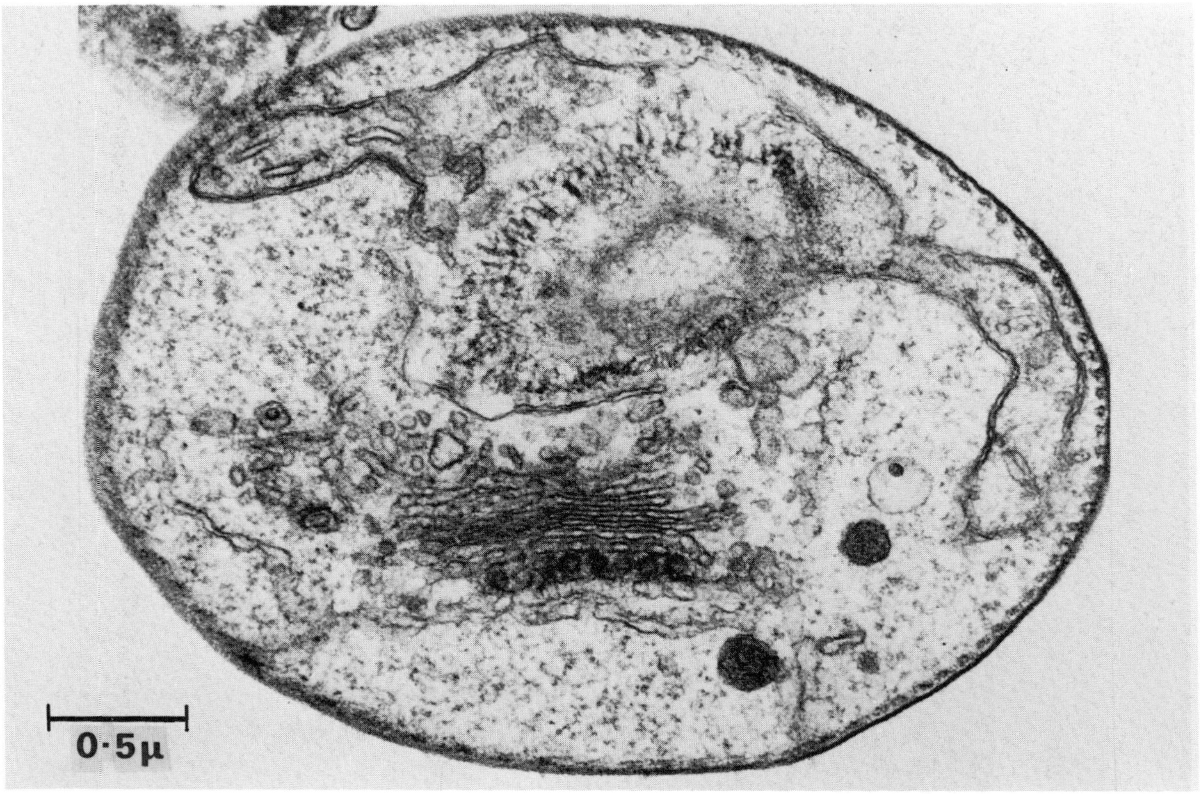

Figure 5.11. ***Leishmania tropica*** **kinetoplast.**
Section of a promastigote showing the kinetoplast ending with two mitochondria and containing a ring of minicircles of DNA. Golgi apparatus is well developed.[80] (× 36,500)

alter and reduce the kinetoplastic DNA structure so that the kinetoplast is no longer stainable with basic dyes, but the mitochondrial structure still remains. Trager and Rudzinska[177] proposed that these treated and modified trypanosomes should be considered as dyskinetoplastic and not as akinetoplastic.

Dyskinetoplastic trypanosomes (trypomastigote) can live and multiply in the blood stream of the host, a mammal, but they cannot undergo transformation to promastigote or epimastigote (insect vector forms) which require normal mitochondrial respiration.[175]

Some species (*T. equinum, T. equiperdum*), which lack a developmental phase in an insect vector, are normally dyskinetoplastic, the percentage varying from 70 to 90, depending on the strain.

Vacuolar apparatus

The vacuolar apparatus of Kinetoplastida is very active. For example, acid phosphatase, which is synthesized by the Golgi apparatus, is excreted through tubular structures with other enzymes into the flagellar pocket where it is stored in large quantities (Figures 5.12, 5.13). If the trypanosome is an intracellular

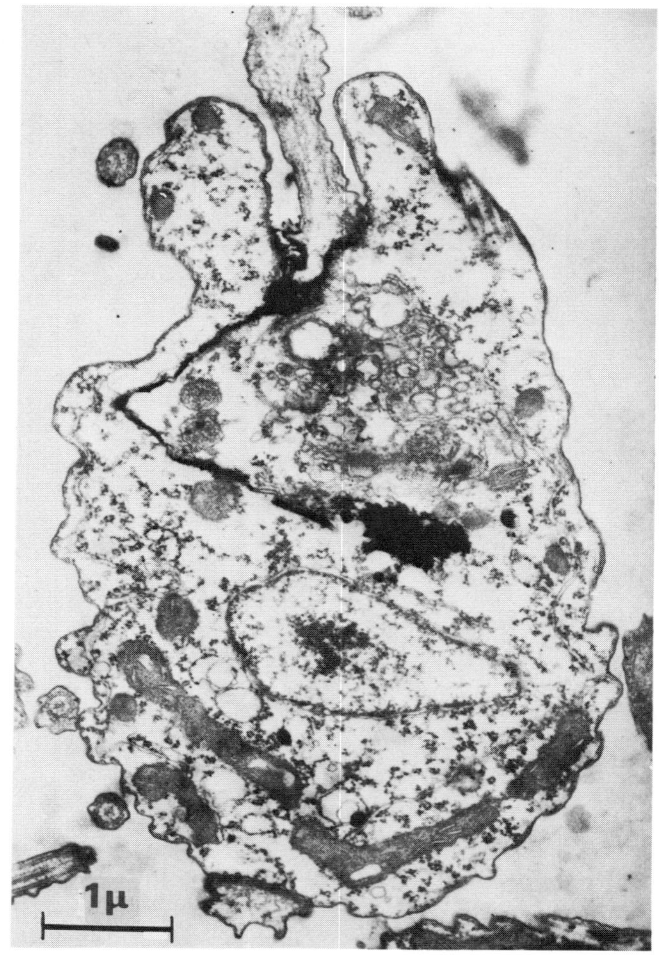

Figures 5.12. Acid phosphatase in *Trypansoma cruzi.*
Section of a trypomastigote after the Gomori's reaction. Acid phosphatase seems to be synthesized by the Golgi apparatus and excreted through tubular structure into the lumen of the flagellar pocket.[51] (× 30,000)

species, e.g., *T. cruzi*, the enzymatic contents of the flagellar pocket are excreted inside the parasitophorous vacuole.[11] The problem is the same with the intra-cellular toxoplasms.[98, 99]

Lipid vacuoles

Osmiophilic lipid vacuoles are present in trypanosomes. They correspond to the type II granules of Molloy and Ormerod.[116]

Unidentified cytoplasmic structures

Some of the structures found in trypanosomes are difficult to identify. Among these are the volutin granules, the cytosome, the bipolar body of *Strigomonas oncopelti*, and the appareil élaborateur singulier.

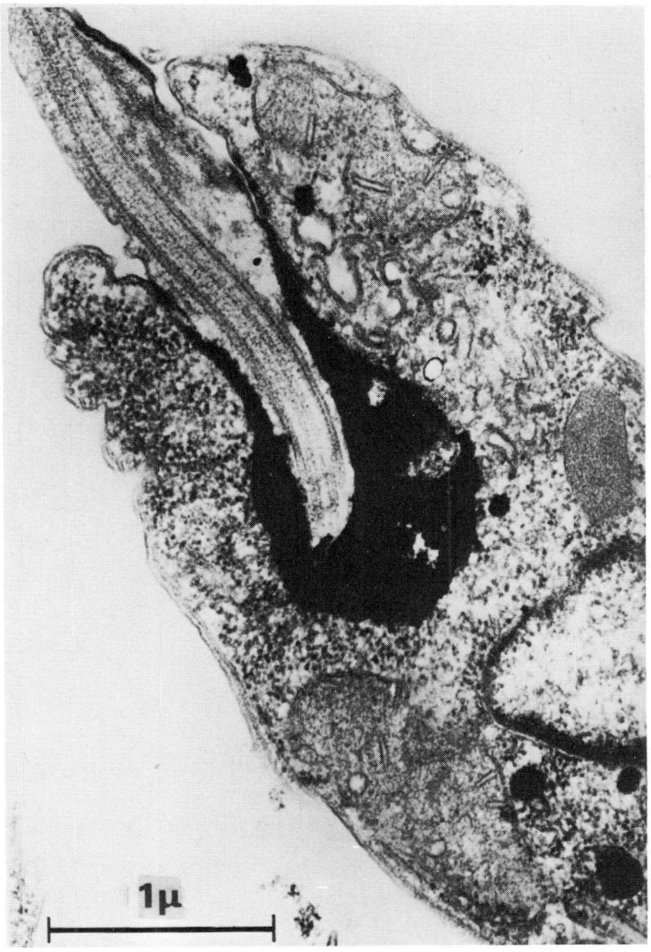

Figure 5.13. Acid phosphatase in *Trypanosoma cruzi.*
Section of a trypomastigote after the Gomori's reaction showing the presence of acid phosphatase in the flagellar pocket. (× 30,000)

Swellengrebel[176] described the so-called volutin granules, which where variously described by other authors, e.g., as pentose nucleic acid granules,[180] as methylene blue stainable granules, and as type I granules.[17, 116] In optical microscopy, they represent some of the cytoplasmic vesicles (from 0·5 μm in diameter) but in electron microscopy it seems impossible to identify them.[80] In our opinion, they may correspond to vacuoles of the vacuolar system.

In 1960, Steinert and Novikoff[174] described in *T. mega* a distinct flagellar pocket formation: the cytostome. Their pictures show a deep invagination of the cell membrane unrelated to the flagellar pocket described in texts. The so-called cytostome appeared to be a channel ending deep in the cell from the anterior part. It seemed to be a permanent structure. Some pinocytic vacuoles arose from the cytostome. In fact, such a description of a cytostome seems to be the result of confusion with a tangential section of a flagellar pocket (Figure 5.14). No cytostome is visible by scanning electron microscopy. It may be useful to use the name cytostome for the diverticular formation sometimes

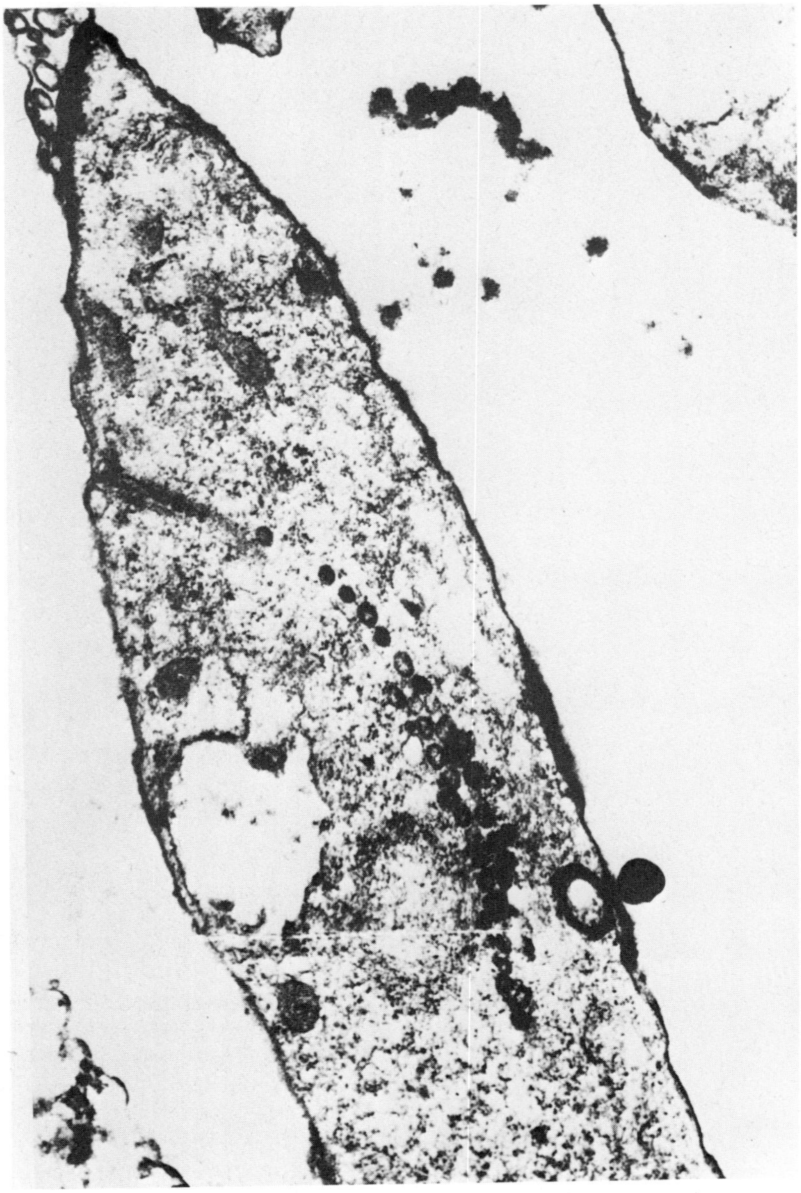

Figure 5.14. *Trypanosoma brucei* flagellar pocket.
Tangential section of a trypomastigote of *T. brucei* showing the lateral side of the flagellar pocket. The unusual section may induce confusion with a cytostome-like body, but with serial section it is easy to identify the flagellar pocket. (× 38,600)

observed at the bottom of the flagellar pocket although we disagree with this since it seems to give a specific function to tubular structures which have other functions (transport of excreted enzymes such as acid phosphatase).

The cytoplasm of *Strigomonas oncopelti* includes bipolar bodies which could be saprophytic mycoplasma.[77, 80, 120, 129]

Boisson *et al.*[18] related the presence of a so-called appareil élaborateur singulier in *T. congolense.* These structures seem to belong to the Golgi apparatus and the endoplasmic reticulum.

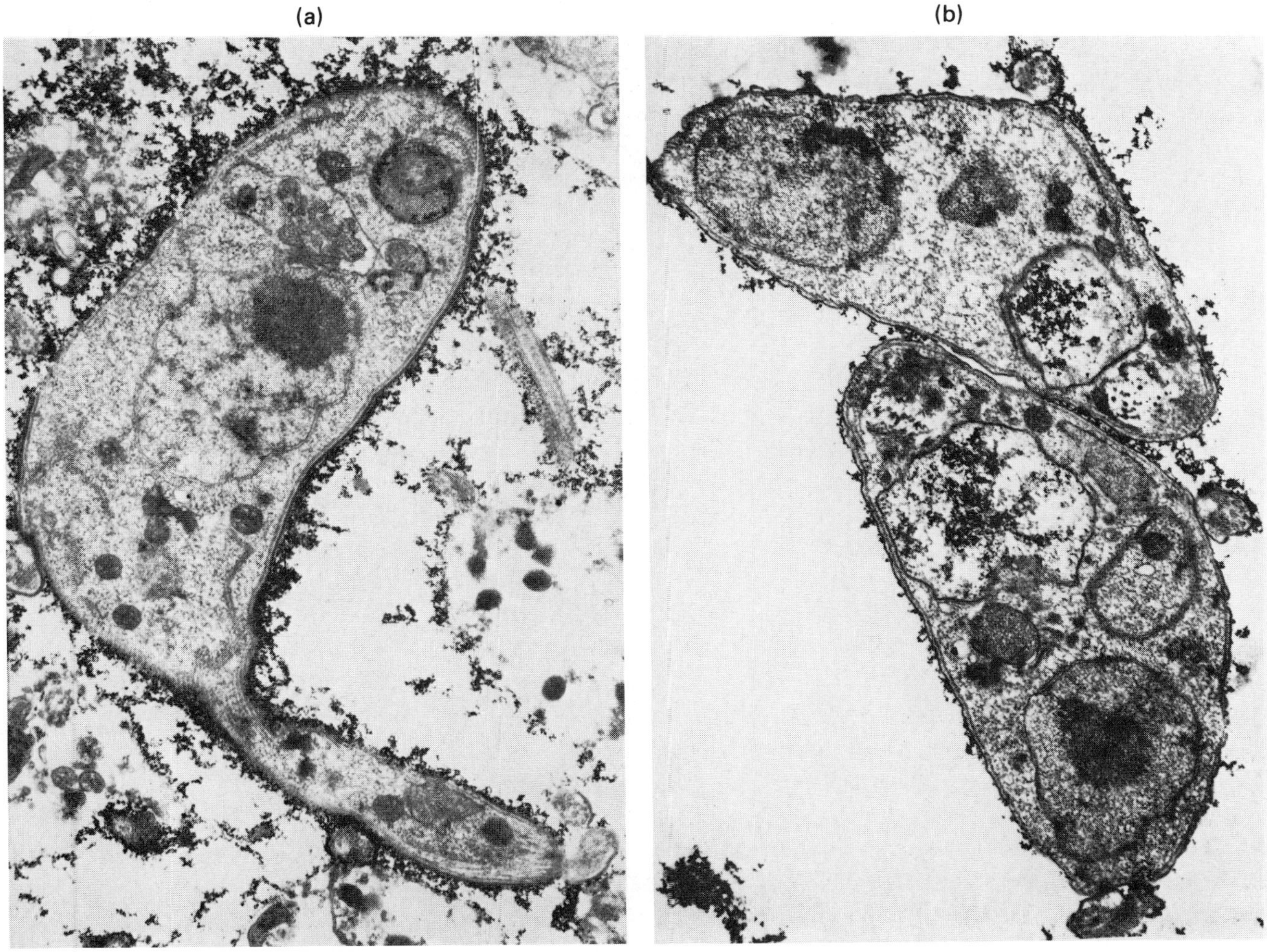

Figure 5.15. Uptake of thorotrast in *Trypanosome brucei.*
(a) Section of a trypomastigote showing accumulation of endocytotic marker (thorotrast) in the flagellar pocket after a few minutes of contact.[80] (× 20,000)
(b) Section of a trypomastigote. After five minutes of contact, the thorotrast particles are concentrated via the flagellar pocket in big phagosomes.[80] (× 20,000)

5.7.4. Uptake and digestion of nutrients

In all forms of trypanosomes (e.g., epimastigote, trypomastigote), the flagellar pocket is the site of uptake of nutrients. It is an easily observed phenomenon. When endocytotic markers (e.g., thorotrast, ferritin, peroxidase) are added to a culture medium or injected into a mammal infected with trypanosomes, the exogenous marker appears rapidly in the lumen of the flagellar pocket and may not be removed by washing (Figure 5.15(a)). At the bottom of the flagellar pocket, some large vesicles invaginating from the cell wall (endocytotic vesicles) take up the marker into the cell, in the space of a few minutes (Figure 5.15(b)).[80]

Capture of the endocytotic marker requires energy and cannot be done at 0 °C or in the presence of the classical endocytotic inhibitor. In fact, the flagellar pocket is a place of transit for digestive enzymes[51, 80, 82, 83, 90, 94, 105, 154] and during the uptake of nutrients all the alimentary prey are in contact with digestive enzymes (external digestive process) inside the flagellar pocket and thus only predigested material is engulfed in the endocytotic vesicles, which eventually become phagosomes. The small phagosomes fuse together to form one, two, or three large phagosomes, which fuse with lysosomes and then the internal digestive process occurs.

5.7.5. The rosette forms

In the usual culture medium, some strains of trypanosomes grow mainly on the surface. After a few days, they are so numerous that they form little clusters or rosette forms visible to the naked eye.[45, 106] These clusters are formed by cell death. Cytoplasmic organelles are liberated after rupture of the cell wall and accumulate in the flagellar pocket of surrounded cells (Figure 5.16).[41, 80, 83, 85, 106]

5.8. Class Telosporea

The class Telosporea is one of the three classes that form the subphylum Sporozoa. All the Protozoa in this subphylum are parasites. There are two subclasses: Gregarinia and Coccidia. The gregarines are all parasites of the gut and hemocoel of invertebrates. Subclass Coccidia is subdivided in two orders: Protococcidia, parasites of marine annelids, and Eucoccidia. This second order includes three suborders: Adelina (nonhuman parasites), Eimeriina, and Haemosporina. The latter two suborders concern important human parasites.

The suborder Eimeriina includes parasites living usually in the cells of the intestinal epithelium of vertebrates, and sometimes in deeper tissues of invertebrates and vertebrates. The oocysts of most species of Eimeriina enter and leave the host via the alimentary canal. Both schizogony and sexual development (fertilization and sporogony, too) occur in various cells of the vertebrate host. The thick-walled oocyst can survive for long periods outside the host and is infectious for other potential hosts.

Eimeriina includes many genera such as *Lankesterella, Isospora* (to which belong the toxoplasms), *Eimeria*, *Hoarella*, *Pythonella*, *Aggregata*.

The suborder Haemosporina includes the malaria parasites and will be considered in sections 5.10 and 5.11.

5.9. Suborder Eimeriina

The fine structure of the Coccidia is similar to the fine structure of the protozoan or metazoan cell with certain typical ultrastructural characteristics. These typical fine structural features are predominantly in the motile stages (sporozoites, merozoites) with an apical complex (conoid, polar ring, rhoptries, and micronemes), with the subpellicular complex and the micropore.

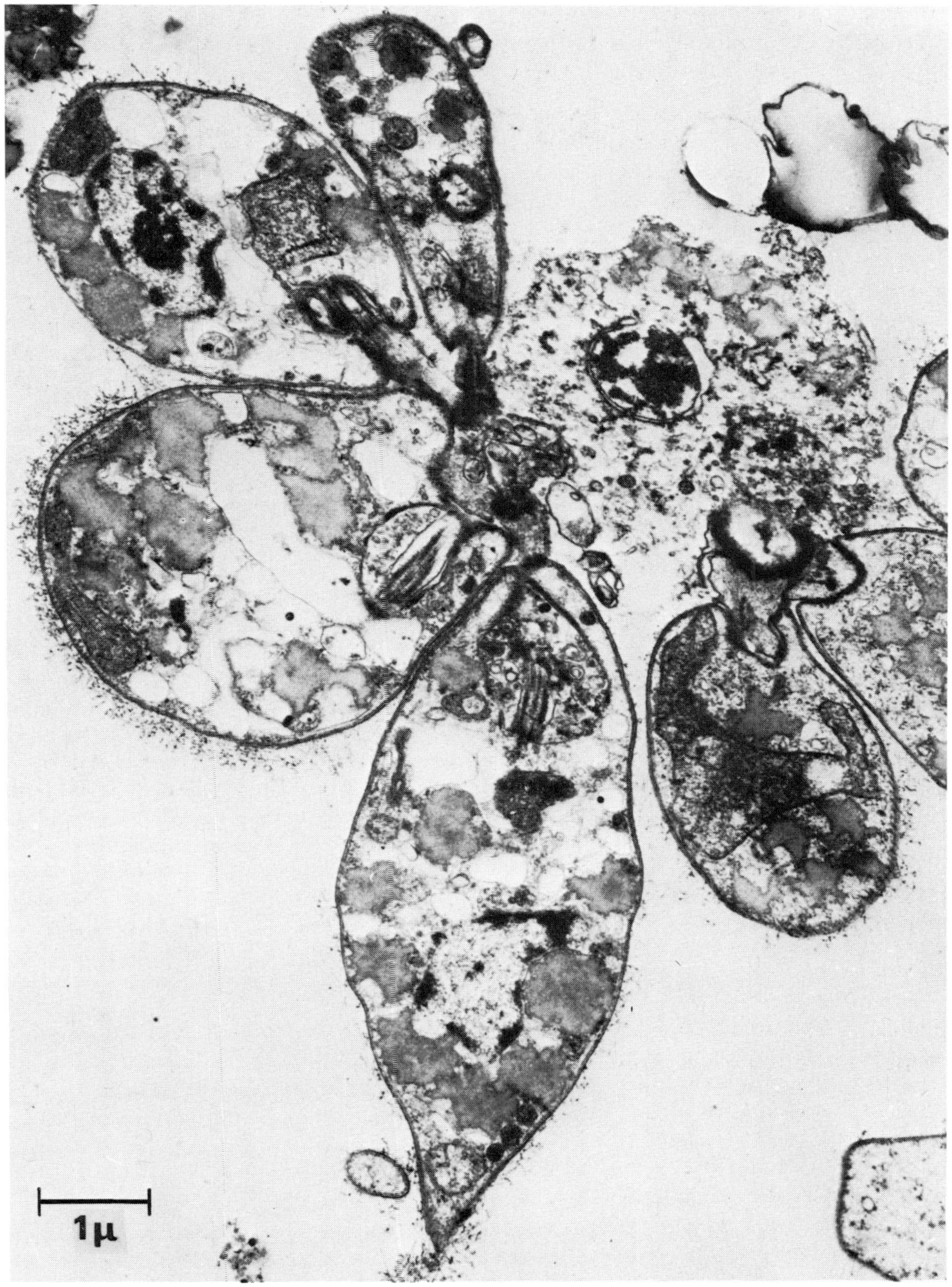

Figure 5.16. *Leishmania ctenocephali* rosette form.
Section of rosette form or cluster of *L. ctenocephali* showing a case of cannibalism with the trypanosomes.[85] (× 15,700)

The nature of the division process (endodyogeny, endopolygeny, internal schizogony) and of the asexual reproduction (schizogony) are characteristic of the Coccidia (Figure 5.7).[44, 69, 92, 93, 124, 156, 167, 170, 183, 193]

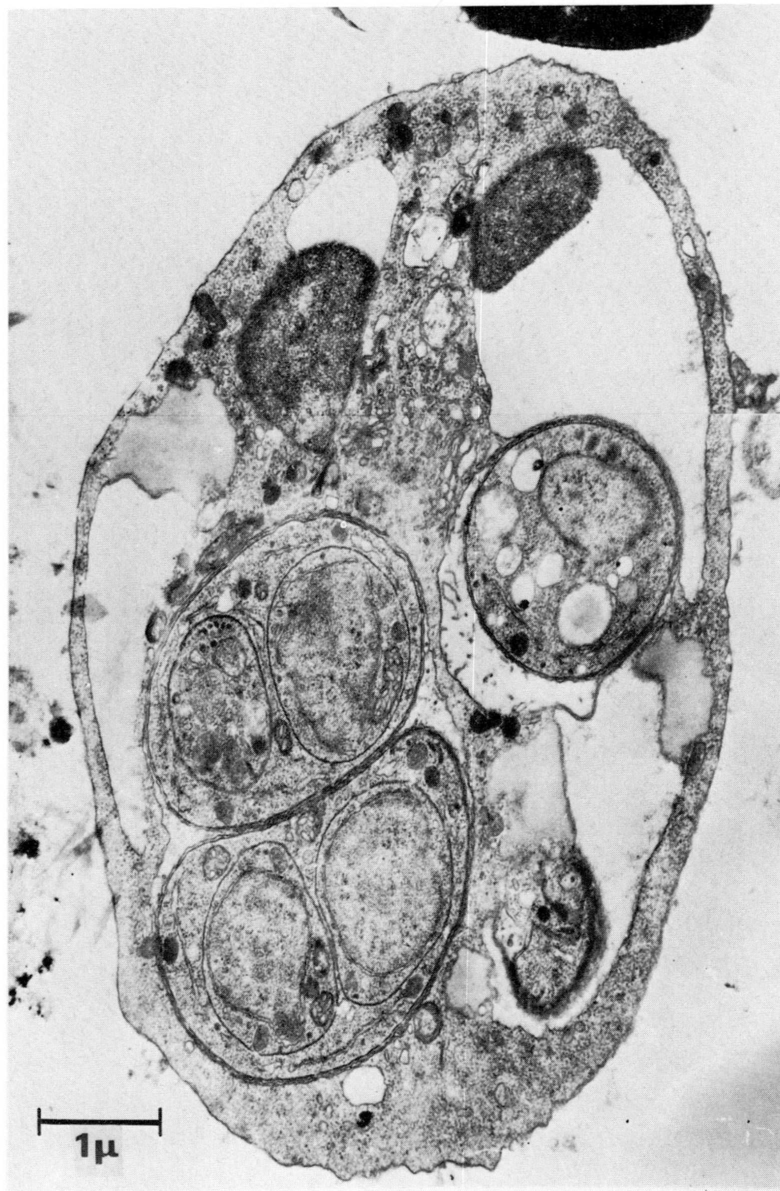

Figure 5.17. Endodyogeny in *Isospora gondii*.
Section of mouse macrophage containing three *Isospora gondii* toxoplasms, two of which are dividing by endodyogeny.[84] (× 16,000)

5.9.1. General fine structural characteristics

Cell wall

The cell wall of the sporozoites and of the merozoites consists of an outer unit membrane and an inner layer. The inner layer is composed of two unit membranes closely attached to one another. An intermediate osmiophobic space of

150–200 Å lies between the outer unit membrane and the inner layer. The cell wall measures about 400 Å with some variations according to the species.[133, 142, 145, 182, 194] The structure of the cell wall is modified in the area of the conoid end of the micropore.

Polar ring

At the anterior end of the cell, the polar ring is formed by the inner membranous layer of the cell wall. In the nondividing cell, one polar ring is present. The function of the polar ring is still unknown (perhaps a supporting structure with a function related to locomotion?).

Microtubules

Microtubules or subpellicular fibrils originate at the polar ring and extend posteriorly as far as the nucleus.[133] The number of microtubules of Coccidia varies according to the species. Microtubules are composed of an osmiophilic central core about 100 Å in diameter and a cortex about 60 Å wide. By negative staining, the microtubules appear to be composed of 8–13 subunits consisting of beaded filaments running parallel to the long axis of the microtubules.

The number of microtubules may differ also according to the developmental stage of the coccidia.

Conoid

A conoid is a truncated hollow cone of spirally arranged fibrillar structures from 260 to 300 Å in diameter. Usually the size of a conoid is 0·08–4 μm long and 0·15 μm to 0·4 μm wide at the anterior end. There are 6 to 8 spiral fibrillar structures. Two preconoidal rings are located at the anterior side of the hollow cone in sporozoites and merozoites of many species.

At the apex of the conoid there is the orifice of the rhoptries.

The conoid is an organelle which can be protruded and retracted and probably plays a function in penetration of host cells.[84, 87, 121, 122, 195]

Rhoptries or paired organelles

In the anterior portion of the sporozoites of *Isospora* and many *Eimeria*, and in merozoites of *Babesia* and many *Eimeria*, there is a pair of electron-dense, club-shaped structures. These paired organelles, called rhoptries by Senaud,[158] are elongated or tortuous in shape. The anterior portion of the rhoptries extends into the conoid area. The posterior portion, about 200–400 μm in diameter, lies in the cytoplasm without morphological connexions with any other cytoplasmic organelle. After osmic fixation, the rhoptries appears as a homogeneous osmiophilic organelle sometimes with alveolar portions, giving a vacuolar, gland-like appearance. The function of the rhoptries is probably to secrete proteolytic enzymes.[142, 148]

Micronemes or toxonemes

The micronemes are small osmiophilic and cord-like structures in the anterior portion of sporozoites of many Sporozoa. Micronemes may vary considerably in length from 600 to 900 μm. Their structures appear completely dense and osmiophilic.

These structures were first described as toxonemes in *Toxoplasma* and later as Cytoplasmatränge in *Eimeria*, as sarconemes in *Sarcocystis*, as lankesterellonemes in *Lankesterella*, as convoluted tubules in *Plasmodium*. Micronemes is the most frequently used term.[142]

The function of the micronemes is unknown but a suitable hypothesis is that rhoptries and micronemes belong to the vacuolar apparatus.

Micropore

This is an organelle on the cell wall, present in almost every stage of the life cycle of Sporozoa. The micropore is an invagination of the outer unit membrane and the inner membranous layer, meanwhile the outer membrane continues through the invagination without interruption. A thickened cylindrical structure is formed by the inner membranous complex and is surrounded by the outer membrane. Thus, a transverse section of a micropore shows two concentric rings. In size the micropore is about 200 μm wide at the surface and about 350 μm deep.

The micropore may function as a feeding organelle.[7, 8, 63, 157, 166, 168, 169]

5.9.2. Infective stages

The motile infective stages, capable of penetrating host cells, are the *sporozoites* and *merozoites*. The initial stage is the sporozoite which penetrates the host cell and after schizogony forms the merozoites.

Sporozoites

Relatively little is known of the ultrastructure of sporozoites (Figure 5.18).

All the sporozoites of *Eimeria* species are morphologically quite the same; the morphological differences are mainly in their shape and size.

They have 24–25 microtubules. *Eimeria* species have a so-called refractile body or paranuclear body, which is osmiophilic and homogeneous.[74, 142] The refractile body disappears usually after the sporozoites have entered the host cells. The function of the refractile body is unknown. Sporozoites of *Eimeria* have a single micropore.

There is a close affinity between the sporozoites of *Eimeria* species and their counterparts of the genera *Lankesterella*, *Aggregata*, *Eucoccidium*, *Coelotropha*.[125, 126, 142, 143, 144, 146, 147, 158, 159, 160, 171] The refractile body was not observed in the sporozoites of *Aggregata*, *Eucoccidium*, *Coelotropha*, *Myrioporides*, and *Angeiocystis*.

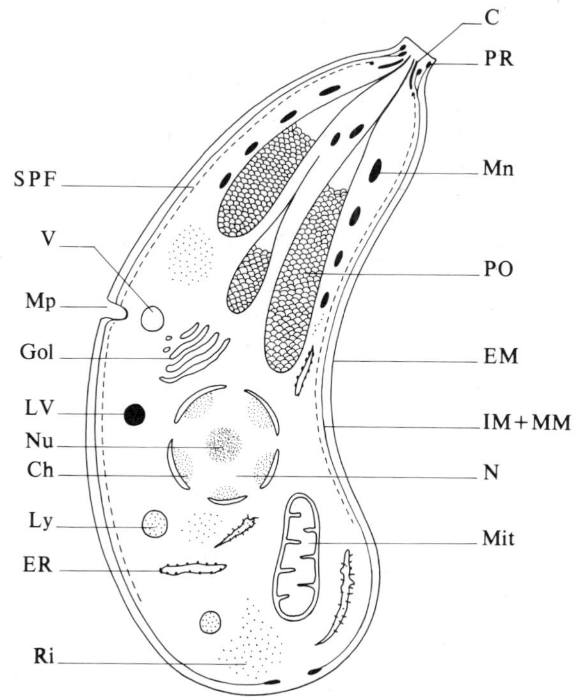

Figure 5.18. Diagrammatic representation of a sporozoite.
C: conoid, PR: polar ring, SPF: sub-pellicular fibrilles, V: vacuole, Mp: micropore, LV: lipidic vacuole, Nu: nucleolus, Gol: Golgi apparatus, Ch: chromatin, Ly: lysosome, ER: endoplasmic reticulum, Ri: ribosomes, Mit: mitochondrion, N: nucleus, IM + MM: inner and median membranous layer, EM: external membrane, PO: paired organelles, Mn: microneme or rhoptries or toxoneme.

Merozoites

The merozoites derive from asexual reproduction (schizogony). They have been extensively studied by means of electron microscopy.[13, 39, 42, 43, 100, 101, 140, 142, 161, 162, 163, 164, 165]

They are distinguished from the sporozoites by their absence of inclusions like the refractile bodies.

Merozoites of *Eimeria* are fusiform and their anterior region contains numerous micronemes. They have a single micropore. The nucleus is located at the posterior one-third of the merozoite.

In some species peculiar structures resembling centrioles were described near the nucleus. Some ovoid bodies containing presumably amylopectin are concentrated in the central area of the merozoite.

It is difficult to recognize the different genera of Coccidiomorpha by the structure of their merozoites.

The crescent-shaped merozoite of *Isospora gondii*, the toxoplasm, has been more extensively investigated by electron microscopy than the merozoite of any other Sporozoa. The toxoplasm has a pointed anterior end and a rounded posterior pole. Usually, there are 22 subpellicular microtubules and 4 to 8 rhoptries. The mitochondria of toxoplasms have a sponge-like structure.[78, 84, 86, 87] Biochemical studies on this subject would be of value.

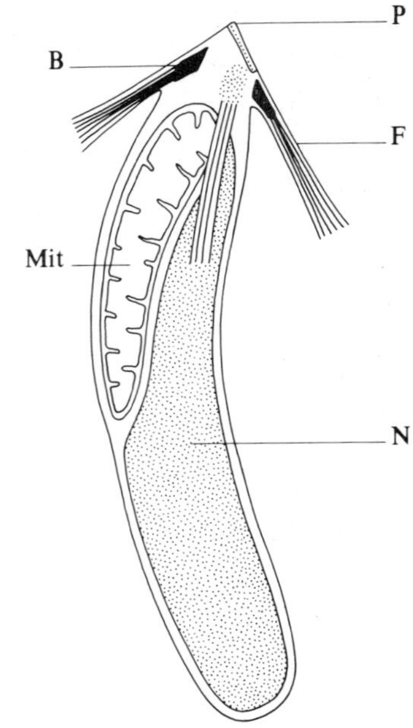

Figure 5.19. Diagrammatic representation of a microgamete.
P: perforatorium, F: flagellum, B: base of the flagellum, Mit: mitochondrion, N: nucleus.

5.9.3. Sexual reproductive cells

Coccidia have a sexual phase in their life cycle with differentiated gametes.

Microgamonts

Microgamonts are parent cells of the microgametes. They vary from 10 to 14 μm in size. The number of developing nuclei of future microgametes may be a few or many thousands according to the species. Microgamonts possess a pair of centrioles. The cytoplasm contains numerous membrane-enclosed compartments like small to large vesicles, and also amylopectin-like granules.

Microgametes

Coccidia contain three morphological types of microgametes:

(1) Elongated microgametes; they are slightly curved, e.g., *Eimeria*, *Isospora*
(2) Cup-shaped microgametes, e.g., *Eucoccidium*, *Coelotropha*.
(3) Filiform microgametes, e.g., *Haemosporina*.

The microgametes of *Eimeria* are from 4 to 7 μm in length. They consist mainly of nuclear substance bounded by a unit membrane derived from the unit membrane of the microgamont (Figure 5.19).[71, 72, 73]

The conoid of the microgamont shows a reinforced aspect which has been given the name perforatorium.[101]

It is surprising to find that the microgamete is flagellated, but flagella are probably present as phylogenetic remnants of the Mastigophora. Three flagella occur in microgametes of *Eimeria*, and two in microgametes of *Isospora gondii*.

Macrogametes

Macrogametes are located in a narrow, electron-pale, parasitophorous vacuole.

The cell wall of the macrogamete is derived from the pellicle of the merozoite stage. Numerous invaginations are observed at the surface of *Eimeria* macrogametes and they are associated with pinocytotic invaginations. Micropores are present in large number.

Close to the surface of the macrogamete but within the parasitophorous vacuole, there are groups of tubular formations. Very little is known about the development of these tubules. They may be associated with transport of materials between the macrogamete and its host cell.

The macrogametes of all Coccidia, except those of Haemosporidia, contain many granules consisting of carbohydrate reserve.[142]

The mitochondria have an internal tubular structure; they are more numerous in the peripheral regions. The Golgi complex is represented only by small lamellae and vesicles. The endoplasmic reticulum is well developed mainly in young macrogametes.

Very little is known about the vacuolar apparatus of macrogametes; by cytochemical methods acid phosphatase was found in association with membrane-closed compartments along vesicles and canaliculi of the endoplasmic reticulum.

The nucleus is spheroidal in young macrogametes and becomes irregular in shape in older macrogametes.

5.10. Suborder Haemosporina

The Haemosporina includes many different genera of parasites, for example, *Plasmodium*, *Haemoproteus*, *Hepatocystis*, *Leucocytozoon*.

Genus *Plasmodium* concerns *malarial parasites* which parasitize reptiles, birds, and mammals. The other genera are not considered to be human parasites.

Members of the family Plasmodiidae are obligate intracellular parasites for almost all of their life cycle and have two hosts: an invertebrate and a vertebrate. They undergo sexual reproduction in mosquitoes (Cubicidae) and asexual reproduction or schizogony in vertebrates. Schizogony occurs in certain fixed tissue cells and also in erythrocytes of vertebrates. The gametocytes form in the erythrocytes.

The following genera are usually accepted:

1. In reptiles: *Sanramoeba*, *Carinamoeba*.
2. In birds: *Haemamoeba*, *Huffia*.
3. In lower mammals and in lemurs: *Vinckeia*.
4. In primates and in man: *Plasmodium*, *Laverania*.

5.11. Genus *Plasmodium*

Four species of *Plasmodium*, *P. falciparum*, *P. vivax*, *P. ovale*, and *P. malariae*, have man as their main vertebrate host. After ingestion of red cells infected with micro- and macrogametes by a mosquito, the gametes become mature inside the midgut of the insect. The microgametes (male gametes) transform by extraflagellation.[61, 62]

Microgametes must join macrogametes to form zygotes. These develop and become motile ookinetes (or wandering vermicules) which penetrate through the intestinal wall of the mosquito and reach the outer layer of the midgut wall. There they become oocysts, each producing several thousand sporozoites, through an intermediary form, the sporoblast. The sporozoites, which are liberated wherever the body cavity extends, are attracted probably by chemotaxis towards the salivary glands. They penetrate the secretory gland cells to reach, finally, the salivary gland ducts.

During the process of feeding, the female mosquito injects into the subcutaneous tissues a small portion of salivary fluid containing infectious sporozoites. These sporozoites enter the circulation and are carried with the blood cells. Here the next phase of the life cycle of *Plasmodium* starts: primary exoerythrocytic schizogony. Each sporozoite enters a host tissue cell, liver being concerned for human *Plasmodium*. Inside the liver cell the sporozoite forms (by schizogony) numerous daughter cells called merozoites or schizonts. These cells break down the host cells, and may invade new host tissue cells and multiply inside by secondary exoerythrocytic schizogony, or may enter erythrocytes to multiply by erythrocytic schizogny. The preerythrocytic stage or tissular phase with exoerythrocytic schizogony is essential to allow this protozoan to undergo the necessary adaptations for the modified conditions of life from a poikilothermic insect to a warm blooded vertebrate.

The life history of human *Plasmodium* presents some fundamental differences. *P. falciparum* undergoes only one generation of preerythrocytic schizogony. After one tissular phase, the erythrocytic cycle continues alone, and if the erythrocytic parasites are destroyed by chemotherapy, there can be no relapse. But *P. vivax*, *P. ovale*, and *P. malariae* may develop secondary exoerythrocytic schizonts which can give rise in the future to further generations of blood-form parasites.

5.11.1. Erythrocytic stages

These are the most familiar forms from their appearance on stained preparations under the light microscope.

An intraerythrocytic plasmodium has a signet-ring form at the beginning of its development, after which it increases considerably in size until it occupies a large portion of the red cell. When fully grown, this stage, the *trophozoite*, undergoes nuclear division followed by cytoplasmic division, in which process of sporogony new merozoites are formed. The number of new merozoites may vary from 8 to 24, following the host reaction, the species, and the age of infection.

Merozoites

The ultrastructure of merozoites has been extensively studied.[1, 3, 4, 5, 9, 16, 46, 48, 65, 66, 67, 103, 123, 137] They are ovoid organisms 0·9 to 1·5 μm in length. The cell wall is composed of three layers: a thin outer membrane, a thick interrupted inner membrane, and a row of microtubules.[2]

The nucleus is rounded and centrally located. The outer nuclear membrane is often limited with ribosomes. A large quantity of compact chromatin material (hererochromatin) lies at the periphery of the nucleus.

Like the Coccidia, the anterior end of the plasmodium merozoite ends with a conoid, but the plasmodium conoid is not as sophisticated as the conoid of *Eimeria*. The erythrocytic merozoite invades the host cell by pressing the conoid against the host cell membrane. It was suggested by Ludvik[112] that the conoid has a screw-like function.

Rhoptries (paired organelles) are located close to the anterior end of the erythrocytic merozoites. Micronemes seem to be associated with the rhoptries. A micropore occupies a position about midway between the anterior and posterior ends of the merozoite. A crescent-shaped mitochondrion with irregular folds is located posterior to the nucleus.

Trophozoites

Trophozoites of *Plasmodium* are in fact merozoites after entering the host cells.[6, 8, 9, 10, 102, 103, 104, 178, 179]

Firstly, a merozoite contacts the red blood cell with the anterior end orientated against the blood cell membrane, forming a depression of that membrane. The depression may become a cavity which will be occupied by the parasite. When that kind of penetration becomes complete, the trophozoite is surrounded by a membrane originating from that of the host cell (Figure 5.20).[93]

One to five pigment granules (hemozoin) are enclosed in smooth-membraned vesicles. Merozoites contain lysosomes and Golgi apparatus of little size.

The ingestion of host cell cytoplasm by the trophozoites of *Plasmodium* occurs not through the plasma membranes but through an invagination of these membranes followed by the formation of a food vacuole. The feeding endocytotic process follows two patterns. In the first group concerning *P. falciparum* and *P. lophurae* (a nonhuman parasite), digestion occurs within the food vacuoles where hemozoin, the residue of hemoglobin digestion, accumulates. In the second group concerning human parasites *P. ovale* and *P. vivax* and nonhuman parasites *P. berghei*, *P. vinckei*, *P. gonderi*, and *P. coatneyi*, the same phenomenon occurs with an additional selective step: the food vacuole is pinched off in small vesicles in which the digestion occurs; in the food vacuole of such trophozoites we never observed pigment granules containing hemozoin. The pigment is always located within male vesicles in the close vicinity of food vacuoles. Plasmodium of the second group are thus characterized by their secondary internal microendocytosis process.

After digestion of hemoglobin, the nondigestible material remains in the

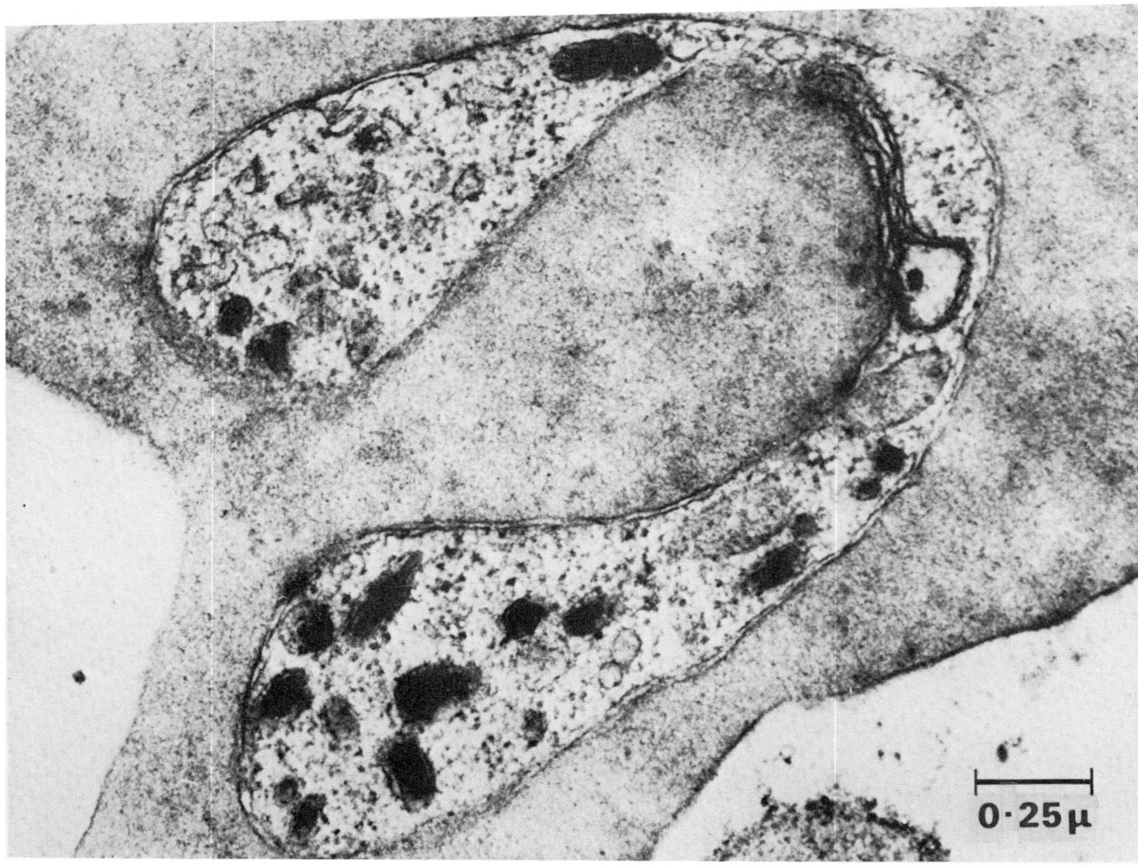

Figure 5.20. *Plasmodium vinckei* trophozoite.
Section of a trophozoite inside a red cell showing hemozoin granules and a multilaminated membraned body.[95] (× 60,000)

cell in a telolysosome called the multilaminated membraned body with a myelin-like pattern.[93, 136, 138]

The trophozoite contains one nucleus. The chromatin material is finely granular and filamentous. If a large quantity of euchromatin is present, this suggests that the trophozoite is undergoing preparation for the next stage: the schizont, in which nuclear division occurs.

Schizonts

Schizonts are larger than either the merozoite or the trophozoite. The two principal events during schizogony are nuclear division and differentiation of the cytoplasmic organelles.

For electron microscopic studies it is necessary to arrest or to slow down the nuclear division of the parasite. This can be done by administration of pyrimethanine. After treatment with the drug, many nuclei of the schizonts appear in metaphase or in anaphase. The most noticeable changes are the appearance of spindle fibers within the nucleus (for more details see Aikawa and Beaudoin[4]). During the nuclear division, the nuclear membrane does not disappear.

At the same time, the mitochondrion increases in size, becomes irregular and finally undergoes fission.

The process of merozoite formation progresses. The cell membrane of young merozoites begins to protrude in the space formed between the parasite and the host-cell membrane. With the progression of merozoite budding many cytoplasmic organelles migrate into the developing merozoites from the mother schizont. These organelles are nucleus mitochondria, rhoptries, micronemes, endoplasmic reticulum, ribosomes, lysosomes.

Further, when the new merozoites develop and grow, the size of the original schizont decreases, and only a large residual body remains.

Exoerythrocytic stages

Use of tissue culture systems provides a great number of exoerythrocytic forms for electron microscopical and immunological studies.

Exoerythrocytic merozoites possess rhoptries and micronemes. Their feeding process is different from that of the trophozoites, because the exoerythrocytic merozoites cannot form pseudopodia to surround prey. They may feed through the micropore or by simple diffusion of nutrients through the host cell.

Exoerythrocytic stages of *Plasmodium* do not contain malaria pigment like hemozoin. When schizonts are liberated from the host's cell after schizogony, all the telolysosomes and related structures (multilaminated membraned bodies, hemozoin) are expelled.

5.11.2. Sexual forms

A gametocyte is a large parasite the same size as a schizont, possessing a disproportionately large single nucleus and scattered malarial pigment.

The more interesting characteristic of a gametocyte, identifiable by electron microscopy is the three-layered cell wall. This three-layered structure is particularly pronounced among the gametocytes of the reptilian and ovian plasmodium but it is not so apparent in mammalian *Plasmodium* species.

The microgametocyte can be differentiated from the macrogametocyte by the staining properties of the cytoplasm. The macrogametocytes have a greater number of ribosomes than the microgametocytes. Osmiophilic bodies are also more frequent in the macrogametocytes.

Nothing is known about the feeding of the gametocytes; probably ingestion of nutrients takes place in the cytostome.

Little information is available about microgametogenesis. Exflagellation of microgametocytes has been described by Garnham in *P. berghei*[66] and *P. cynomolgi*[66] and by Aikawa in *P. cathemerium*[1] and *P. gallinaceum*[1]; all are nonhuman plasmodium.

5.11.3. Mosquito stages

There are four stages in the mosquito: ookinete, oocyst, sporoblast, and sporozoite.

Ookinetes

The ookinete is surrounded by a pellicular complex similar to that of the erythrocytic and exoerythrocytic merozoites.[57, 64] The anterior end of the ookinete is cone-shaped and several electron-dense bodies similar to micronemes have been observed with little rhoptries.

The ookinete penetrates the midgut of the mosquito by pushing the anterior end between the microvilli of the epithelial cells of the midgut. Probably proteolytic substances are secreted by the anterior end and make possible the entry of the ookinete. Further information is needed about the next step of the life cycle: migration and maturation of the ookinete inside the midgut.

Oocysts

Oocysts are spherical or ovoid structures of about 30–40 μm diameter. They are surrounded by a thick electron-dense capsule of 1 μm thickness. Oocysts contain abundant ribosomes, endoplasmic reticulum, mitochondria, lysosomes, several nuclei, and several spherical vacuoles.[58]

The cell-wall often invaginates deeply into the oocyst cytoplasm. No micropore has been observed.

During the process of sporozoite formation, the central large region of the oocyst differentiates into sporoblasts which are extremely basophilic during the process of sporozoite budding.

Sporoblast

The fine structure of the sporoblast is similar in many respects to that of erythrocytic and exoerythrocytic schizonts, but rhoptries are very small.

Sporozoites

Sporozoites are needle-shaped cells measuring 0·8 μm in diameter and 11 μm in length. They are bounded by two membranes which originate from the membranes of the sporoblast.

Sporozoites contain the same organelles as merozoites. The rhoptries are extremely long and extend to the midportion of the sporozoite. Micronemes are present in the anterior and in the midregion.

References to Part Five

1. Aikawa, M., "The fine structure of the erythrocytic stages of three avian malarial parasites, *Plasmodium fallax, P. lophurae* and *P. cathemerium*," *Am. J. trop. Med. Hyg.*, **15**, 449, 1966.
2. ———, "Ultrastructure of the pellicular complex of *Plasmodium fallax*," *J. Cell Biol.*, **35**, 103, 1967.
3. ——— and Antonovych, T., "Electron microscopic observations of *Plasmodium berghei* and the Kupffer cell in the liver of rats," *J. Parasit.*, **50**, 620, 1964.
4. ——— and Beaudoin, R. L., "Studies on nuclear division of a malarial parasite under pyrimethanime treatment," *J. Cell Biol.*, **39**, 749, 1968.
5. ——— and Cook, R. T., "Ribosomes of the malarial parasite, *Plasmodium knowlesi.* II. Ultrastructural features," *Comp. Biochem. Physiol.*, **39B**, 913, 1971.
6. ———, ———, Sakoda, J. J., and Sprinz, H., "Fine structure of the erythrocytic stages of *Plasmodium knowlesi*: A comparison between intracellular and free forms," *Z. Zellforsch. mikrosk. Anat.*, **100**, 271, 1969.
7. ———, Hepler, P. K., Huff, C. B., and Sprinz, H., "The feeding mechanism of avian malarian parasites," *J. Cell Biol.*, **28**, 355, 1966.
8. ———, Huff, C. G., and Sprinz, H., "Comparative feeding mechanism of avian and primate malarial parasites," *Milit. Med.*, **131**, 969, 1966.
9. ———, ———, and ———, "Fine structure of the asexual stages of *Plasmodium elongatum*," *J. Cell Biol.*, **34**, 229, 1967.
10. ———, ———, and ———, "Comparative fine structure study of the gametocytes of avian, reptilian and mammalian malarial parasites," *J. Ultrastruct. Res.*, **26**, 316, 1969.
11. Alexander, J. and Vickerman, K., "Fusion of host cell secondary lysosomes with the parasitophorous vacuoles of *Leishmania mexicana*-infected macrophages," *J. Protozool.*, **22**, 502, 1975.
12. Allen, H. J. and Winzler, R. J., "The chemistry of amoeba surface," in L. K. Jeon (ed.), *The Biology of Amoeba*, Academic Press, New York, 1973, p. 451.
13. Andreassen, J. and Behnke, O., "Fine structure of merozoites of a rat coccidian *Eimeria miyarii* with a comparison of the fine structure of other sporozoa," *J. Parasit.*, **54**, 150, 1968.
14. Angelopoulos, E., "Pellicular microtubules in the family Trypanosomatidae," *J. Protozool.*, **17**, 39, 1970.

15. Babudieri, B. and Tomassini, N., "Fine structura dei Tripanosomi," *Parassitologia*, **4**, 89, 1962.
16. Baker, J. R., "Fine structure of *Plasmodium* in relation to locomotion," *Proc. IInd Intern. Congr. Parasitol.*, Washington, D.C., 1970, p. 16.
17. Bird, R. G., Molloy, J. O., and Ormerod, W. E., "Granules and tubes in the cytoplasm of the sleeping sickness trypanosome: an electron microscope study," *Trans. R. Soc. trop. Med. Hyg.*, **60**, 753, 1966.
18. Boisson, M. E., Boisson, C., and Mattei, X., "Présence d'un appareil élaborateur singulier chez *Trypanosoma congolense*," *C. r. Séanc. Soc. Biol.*, **161**, 1772, 1967.
19. Bowman, I. B. R., "Intermediary metabolism of pathogenic flagellates," in *Trypanosomiasis and Leishmaniasis with Special Reference to Chaga's Disease*, Ciba Foundation Symposium 20 (New Series), Elsevier Excerpta Medica, North-Holland, Amsterdam, 1974, p. 255.
20. Bresslau, E. and Scremin, L., "Die Kerne der Trypanosomen und ihr Verhalten zur Nuclealreaktion," *Arch. Protistenk.*, **48**, 509, 1924.
21. Brooker, B. E., "Lysosome-like bodies in trypanosomatid flagellates," *Progress in Protozoology*, London, 1965, p. 138.
22. ———, "Desmosomes and hemidesmosomes in the flagellates *Crithidia fasciculata*," *Z. Zellforsch. mikrosk. Anat.*, **105**, 155, 1970.
23. ——— and Vickerman, K., "Acid phosphatase in trypanosomes," *Trans. R. Soc. trop. Med. Hyg.*, **58**, 293, 1964.
24. Brown, K. N., Amstrong, J. A., and Valentine, R. C., "The ingestion of protein molecules by blood forms of *Trypanosoma rhodesiense*," *Expl. Cell Res.*, **39**, 129, 1965.
25. Brugerolle, G. and Metenier, G., "Localisation intracellulaire et caractérisation de deux types de malate déshydrogénase chez *Trichomonas vaginalis*," *J. Protozool.*, **20**, 320, 1973.
26. Byram, J. and Seed, J. R., "Biochemical and cytochemical studies of the acid phosphatase of *Trypanosoma gambiense*," *J. Protozool.*, **13**, suppl. 18, 1966.
27. Chapman-Andresen, C., "Pinocytosis of inorganic salts by *Amoeba proteus* (*Chaos diffluens*)," *C. r. Trav. Lab. Carlsberg*, **31**, 77, 1958.
28. ———, "Studies in pinocytosis in amoeba," *C. r. Trav. Lab. Carlsberg*, **33**, 73, 1962.
29. ———, "Surface renewal in *Amoeba proteus*," *J. Protozool.*, **11**, suppl., 11, 1964.
30. ———, "The induction of pinocytosis in *Amoeba*," *Arch. Biol., Liège*, **76**, 189, 1965.
31. ———, "Studies on endocytosis in *Amoeba*. The distribution of pinocytically ingested dyes in relation to food vacuoles in *Chaos chaos*. I. Light microscopic observations," *C. r. Trav. Lab. Carlsberg*, **36**, 161, 1967.
32. ———, "The effect of starvation on the vacuolar digestive system of *Chaos chaos*," *Progress in Protozoology*, Leningrad, 1969, p. 23.
33. ———, "Endocytic process," in K. W. Jeon (ed.), *Biology of Amoeba*, Academic Press, New York, 1973, p. 319.
34. ——— and Holter, H., "Studies on the ingestion of ^{14}C glucose by

pinocytosis in the *amoeba Chaos chaos*," *Expl. Cell Res.*, suppl. 3, 52, 1955.

35. —— and ——, "Differential uptake of proteins and glucose by pinocytosis in *Amoeba proteus*," *C. r. Trav. Lab. Carlsberg*, **34**, 211, 1964.
36. —— and ——, "The uptake of fluorescent albumin by pinocytosis in *Amoeba proteus*," *J. Biophys. Cytol.*, **8**, 288, 1967.
37. —— and Nilsson, R. J., "Studies on endocytosis in amoeba: the distribution of pinocytically ingested dyes in relation to food vacuoles in *Chaos chaos*. II. Electron microscopic observations using alcian blue," *C. r. Trav. Lab. Carlsberg*, **36**, 189, 1967.
38. —— and Prescott, D. M., "Studies in pinocytosis in the *Amoeba proteus*," *C. r. Trav. Lab. Carlsberg*, **30**, 57, 1956.
39. Cheissin, E. M. and Snigirevskaya, E. S., "Nouvelle contribution à l'étude infrastructurale des mérozoites d'*Eimeria intestinalis* (Sporozoa, Eimeridea)," *Protistologica*, **1**, 123, 1965.
40. Clark, T. B. and Wallace, F. G., "A comparative study of kinetoplast ultrastructure in the Trypanosomatidae," *J. Protozool.*, **7**, 115, 1960.
41. Colas-Belcour, J. and Vervent, G., "De l'obtention de culture en voile sur milieu liquide avec différentes souches de *Leishmania tropica*," *C. r. Séanc. Acad. Sci., Paris*, **236**, 532, 1953.
42. Colley, F. C., "Fine structure of sporozoites of *Eimeria nieschulzi*," *J. Protozool.*, **14**, 217, 1967.
43. ——, "Fine structure of schizonts and merozoites of *E. nieschulzi*," *J. Protozool.*, **15**, 374, 1968.
44. —— and Zaman, V., "Observations on the endogenous stages of *Toxoplasma gondii* in the cat ileum," *Southeast Asian J. trop. Med. Pub. Hlth.*, **1**, 465, 1970.
45. Committee on taxonomy and taxonomic problems of the society of protozoologists. "A revised classification of the Phylum Protozoa," *J. Protozool.*, **11**, 1, 1964.
46. Cook, R. T., Rock, R. C., Aikawa, M., and Fournier, M. J., "Ribosomes of the malarial parasite, *Plasmodium knowlesi*. I. Isolation, activity and sedimentation velocity," *Comp. Biochem. Physiol.*, **39B**, 897, 1971.
47. Cooper, B. A., "Quantitative studies of pinocytosis induced in *Amoeba proteus* by simple cations," *C. r. Trav. Lab. Carlsberg*, **36**, 385, 1968.
48. Cox, F. E. G. and Vickerman, K., "Pinocytosis in *Plasmodium vinckei*," *Ann. trop. Med. Parasit.*, **60**, 293, 1966.
49. Creemers, J. and Jadin, J. M., "Ultrastructure et biologie de *Trypanosoma rotatorium*," *Acta Zool. Pathol. Antwerp*, **41**, 119, 1966.
50. —— and ——, "Etude de l'ultrastructure et de la biologie de *Leishmania mexicana*. I. Les modifications qui surviennent lors de la transformation *leishmania-leptomonas*," *Bull. Soc. Pathol. exot.*, **60**, 53, 1967.
51. —— and ——, "Phosphatase acide chez les Trypanosomidae," in *Résumé des Communications, 7e Congr. Int. Micr. Electr., Grenoble*, Favard, Paris, 1970, p. 391.
52. Culbertson, C. G., "The pathogenicity of soil amebas," *A. Rev. Microbiol.*, **25**, 231, 1971.

53. Daniels, E. D., "Ultrastructure," in K. W. Jeon (ed.), *The Biology of Amoeba*, Academic Press, New York, 1973, p. 125.
54. Daniels, E. W. and Breyer, E. P., "Ultrastructure of the giant amoeba *Pelomyxa palustris*," *J. Protozool.*, **14**, 167, 1967.
55. de Duve, C., "The lysosome in retrospect," in J. T. Dingle and H. B. Fell (eds.), *Lysosomes in Biology and Pathology*, North-Holland, Amsterdam, 1969, p. 3.
56. de Raadt, P., "Immunity and antigenic variation. Clinical observations suggestive of immune phenomena in African trypanosomiasis," in *Trypanosomiasis and Leishmaniasis with Special Reference to Chaga's Disease*, Ciba Foundation Symposium, 20 (New Series), Elsevier-Excerpta Medica, Amsterdam, 1974, p. 199.
57. Desser, S. S. and El-Nahal, H. M. S., "Electron microscope studies on motile stages of malaria parasites. IV. The ookinete of *Plasmodium berghei yoelii* and its transformation into the early oocyst," *Trans. R. Soc. trop. Med. Hyg.*, **63**, 194, 1969.
58. Duncan, D., Eades, J., Julian, S. R. and Micks, D., "Electron microscope observations on malaria oocysts (*Plasmodium cathemerium*)," *J. Protozool.*, **7**, 18, 1960.
59. Fiennes, R. N., "A sexual reproduction cycle of *Trypanosoma congolense*," *Nature, Lond.*, **156**, 390, 1945.
60. Fuge, H., "Zum Feinbau des Periplasten und der Geissel in *Trypanosoma brucei* und *Trypanosoma gambiense*," *Z. Zellforsch. Mikrosk. Anat.*, **89**, 201, 1968.
61. Garnham, P. C. C., "The ultrastructure of Haemosporidia and allied protozoa with special reference to the motile stages," *Proc. 1st Int. Conf. Protozool.*, Prague, 427, 1963.
62. ———, "Locomotion in the parasitic protozoa," *Biol. Rev.*, **41**, 561, 1966.
63. ———, "The structure and function of the cytostome (micropyle) in the Sporozoa, in *Progress in Protozoology, IIIrd Int. Congr. on Protozoology*, Leningrad, 1969, p. 8.
64. ———, Bird, R. G., and Baker, J. R., "Electron microscope studies of motile stages of malaria parasites. III. The ookinetes of *Haemamoeba* and *Plasmodium*," *Trans. R. Soc. trop. Med. Hyg.*, **56**, 116, 1962.
65. ———, ———, and ———, "Electron microscope studies of motile stages of malaria parasites. IV. The fine structure of the sporozoite of four species of *Plasmodium*," *Trans. R. Soc. trop. Med. Hyg.*, **57**, 27, 1963.
66. ———, ———, and ———, "Electron microscope studies of motile stages of malaria parasites. V. Exflagellation of *Plasmodium*, *Hepatocystis* and *Leucocytozoon*," *Trans. R. Soc. trop. Med. Hyg.*, **61**, 58, 1967.
67. ———, ———, ———, and Bray, R. S., "Electron microscope studies of motile stages of malaria parasites. II. The fine structure of the sporozoites of *Laverania* (= *Plasmodium*) *falciparum*," *Trans. R. Soc. trop. Med. Hyg.*, **55**, 98, 1961.
68. ———, ———, ———, and Killick-Kendrick, R., "Electron microscope studies on the motile stages of malarial parasites. VII. The fine structure of the merozoites of exoerythrocytic schizonts of *Plasmodium berghei*

yoelii," *Trans. R. Soc. trop. Med. Hyg.*, **63**, 328, 1969.
69. Gavin, M. A., Wanko, T., and Jacobs, L., "Electron microscope studies of reproducing and interkinetic *Toxoplasma*," *J. Protozool.*, **9**, 222, 1962.
70. Grassé, P. P., *Traité de Zoologie*, Tome I, *Protozoaires*, Masson, Paris, 1952.
71. Hammond, D. M., Scholtyseck, E., and Chobotar, B., "Fine structures associated with nutrition of the intracellular parasite *Eimeria auburnensis*," *J. Protozool.*, **14**, 678, 1967.
72. ———, ———, and ———, "Fine structural study of the microgametogenesis of *Eimeria auburnensis*," *Z. ParasitKde*, **33**, 65, 1969.
73. ———, ———, and Miner, M. L., "The fine structure of microgametocytes of *Eimeria perforans, E. stiedae, E. bovis, and E. auburnensis*," *J. Protozool.*, **53**, 235, 1967.
74. ———, Speer, C. A., and Roberts, W., "Occurrence of refractile bodies in merzoites of *Eimeria* species," *J. Parasit.*, **56**, 189, 1970.
75. Hoare, C. A. and Wallace, F. G., "Developmental stages of trypanosomatid flagellates: a new terminology," *Nature, Lond.*, **212**, 1385, 1966.
76. Holter, H. and Marshall, J. M., "Studies on pinocytosis in the amoeba *Chaos chaos*," *C.r. Lab. Carlsberg*, **29**, 7, 1954.
77. Horne, R. B. and Newton, B. A., "Intracellular structures in *Strigomonas oncopelti.* II. Fine structure of the kinetoplast–blepharoplast complex," *Expl. Cell Res.*, **15**, 103, 1958.
78. Jacobs, L., "*Toxoplasma* and toxoplasmosis," *Adv. Parasitol.*, **5**, 1, 1967.
79. Jadin, J. B., "Amibes Limax vecteurs possibles des Mycobactéries et de *Mycobacterium leprae*," *Acta Leprologica*, **59**, 57, 1975.
80. Jadin, J. M., "Cytologie et cytophysiologie des Trypanosomidae," *Acta Zool. Pathol., Antwerp*, **53**, 1, 1971.
81. ———, "Etude comparative des Leishmanies," in *Abstract Colloque Int. Ecologie des Leishmanioses, Montpellier, août 1974.*
82. ——— and Creemers, J., "L' ultrastructure des formes en rosaces de *Leishmania tropica*," *Annis Soc. belge Méd. trop.*, **46**, 349, 1966.
83. ——— and ———, "Etude de l'ultrastructure et de la biologie *Leishmania mexicana*. II. Les formes en rosaces," *Bull. Soc. Pathol. exot.*, **60**, 58, 1967.
84. ——— and ———, "Ultrastructure et biologie des Toxoplasmes. I. Formes libres et formes intracellulaires," *Annls Soc. belge Méd. trop.*, **47**, 385, 1967.
85. ——— and ———, "Etude de la pinocytose dans les formes en rosaces de *Leptomonas ctenocephali*," *Acta trop.*, **25**, 363, 1968.
86. ——— and ———, "Etude de l' ultrastructure et de la biologie de *Toxoplasma gondii*," *Louvain Médical*, **87**, 361, 1968.
87. ——— and ———, "Ultrastructure et biologie des Toxoplasmes. III. Observations de Toxoplasmes intraérythrocytaires chez un Mammifère," *Acta trop.*, **25**, 267, 1968.
88. ——— and ———, "L'ultrastructure des formes de culture *d'Endotrypanum schaudinni*," *Protistologica*, **5**, 383, 1969.
89. ——— and ———, "Le processus d'endocytose chez les Trypanosomidae,"

in O. Otto (ed.), *Abstract 2nd International Congress of Parasitology, Washington*, 1. 1970, p. 305.

90. ——— and ———, "Etude en microscopie électronique des localisations des hydrolases acides chez les Trypanosomidae," *Archs Int. Physiol. Biochim*., **78**, 17, 1970.
91. ——— and ———, "The role of excretion of acid phosphatase in *T. brucei* and *T. Cruzi*," *Trans. Soc. trop. Med. Hyg*., **66**, 8, 1972.
92. ———, ———, and Heremans, H., "Ultrastructure et biologie des Toxoplasmes. II. Les modes de reproduction," *Annls Soc. belge Méd. trop*., **47**, 391, 1967.
93. ———, ———, and Giroud, P., "Les stades initiaux du mode de division binaire de *Toxoplasma gondii*, *C. r. Séanc. Acad. Sci., Paris*, **268**, 193, 1969.
94. ———, ———, and Jacques, P., "Etude de la pinocytose chez les Trypanosomidae," *Prog. Protozool., Leningrad*, **64**, 1969.
95. ———, ———, and Le Ray, D., "Ultrastructure et biologie de *Plasmodium vinckei*. I. L'absorption et la digestion de l'hémoglobine chez les trophozoïtes," *Annls Soc. belge Méd. trop*., **48**, 497, 1968.
96. ———, ———, and Van der Schueren, B., "Contribution à l'étude de l'endocytose chez les formes amibiennes des *Naegleria*," *J. Protozool*., **18**, suppl., 49, 1971.
97. ———, Esbach, H. L., Verheyen, F., and Willaert, E., "Etudes comparatives des kystes de *Naegleria* et d'*Acanthamoeba*," *Ann. Soc. belge Méd. trop*., **54**, 35, 1974.
98. Jones, T. C. and Hirsch, J. G., "The interaction between *Toxoplasma gondii* and mammalian cells. II. The absence of lysosomal fusion with phagocytic vacuoles containing living parasites," *J. Expl. Med.*, **136**, 1173, 1972.
99. ———, Yeh, S., and Hirsch, J. G., "The interaction between *Toxoplasma gondii* and mammalian cells. I. Mechanism of entry and intracellular fate of the parasite," *J. Expl. Med.*, **1**, 1157, 1972.
100. Kheysin, E. M., "Electron microscopic study of microgametes in *Eimeria intestinalis* (Sporozoa, Coccidia)," *Zool. J*., **43**, 647, 1964.
101. ——— and Snigirevskaya, E. S., "Some new data on the fine structure of the merozoites of *Eimeria intestinalis* (Sporozoa, Eimeriidea)," *Protistologica*, **1**, 121, 1965.
102. Ladda, R., Aikawa, M., and Sprinz, H., "Penetration of erythrocytes by merozoites of mammalian and avian malarial parasites," *J. Parasit.*, **55**, 633, 1969.
103. ———, Arnold, J., and Martin, D., "Electron microscopy of *Plasmodium faciparum*. I: The structure of trophozoites in erythrocytes of human volunteers," *Trans. R. Soc. trop. Med. Hyg*., **60**, 369, 1966.
104. ———, ———, ———, and Luehrs, F., "Electron microscopy on *Plasmodium berghei*. On the migration of trophozoites from infected erythrocytes in the rats," *Trans. R. Soc. trop. Med. Hyg*., **59**, 420, 1965.
105. Langreth, S. G. and Balber, A. E., "Protein uptake and digestion in bloodstream and culture forms of *Trypanosoma brucei*," *J. Protozool*, **22**, 40, 1975.

106. Laveran, A. and Pettit, A., "Sur les cultures de *Leishmania donovani* en milieu liquide," *Bull. Soc. Pathol. exot.*, **3**, 216, 1910.
107. Le Ray, D., "Structure antigénique de *Trypanosoma brucei brucei* (Protozoa, Kinetoplastida). Analyse immunoélectrophorétique et étude comparative," *Thesis*, Louvain, 1975.
108. Lewis, D. H., "Ultrastructural study of promastigotes of *Leishmania* from reptiles," *J. Protozool.*, **22**, 344, 1975.
109. Lindmark, D. G. and Müller, M., "Hydrogenosomes as cytoplasmic organelle of the anaerobic flagellate *Tritrichomonas foetus*, and its role in pyruvate metabolism," *J. biol. Chem.*, **248**, 7724, 1973.
110. —— and ——, "Biochemical cytology of Trichomonad flagellates. II. Subcellular distribution of oxidoreductases and hydrolases in *Monocercomonas* sp.," *J. Protozool.*, **21**, 374, 1974.
111. ——, ——, and Shio, H., "Hydrogenosomes in *Trichomonas vaginalis*," *J. Parasit.*, **61**, 552, 1975.
112. Ludvik, J., "Morphology of *Toxoplasma gondii* in electron microscope," *Věst. Čsl. zool. Spol.*, **22**, 130, 1958.
113. Lycke, E., Norby, R., and Remington, J., "Penetration-enhancing factor extracted from *Toxoplasma gondii* which increases its virulence for mice," *J. Bact.*, **96**, 785, 1968.
114. Malhotra, S. K., "Freezing-substitution and freeze-drying in electron microscopy," in McGee-Russell (ed.), *Cell Structure and its Interpretation*, Edward Arnold London, 1968, p. 11.
115. Martinez, A. J., Duma, R. J., Nelson, E. C., and Moretta, F., "Experimental *Naegleria* meningoencephalitis in mice. Penetration of the olfactory mucosal epithelium by *Naegleria* and pathogenic changes produced: a light and electron microscope study," *Lab. Invest.*, **29**, 121, 1973.
116. Molloy, J. O. and Ormerod, W. E., "Two types of cytoplasmic granules in *Trypanosoma rhodesiense*," *Exptl Parasitol.*, **17**, 57, 1965.
117. Müller, M., "Biochemical cytology of trichomonad flagellates. I. Subcellular localization of hydrolases, dehydrogenases, and catalase in *Trichomonas foetus*," *J. Cell Biol.*, **57**, 463, 1973.
118. —— and Lindmark, D. G., "Uptake of metronidazole and its effect on viability in trochomands and *Entmoeba invadens* under anaerobic and aerobic conditions," *Antimicrobial Agents and Chemotherapy*, **9**, 696, 1976.
119. Newton, B. A., "Fine structure of the kinetoplast in a trypanosomid flagellate," *Verh. IV int. Kongr. Elektronenmikr.*, **2**, 515, 1958.
120. —— and Horne, B. W., "Intracellular structure in *Strigomonas oncopelti*," *Expl. Cell Res.*, **13**, 563, 1957.
121. Norby, R., Lindholm, L., and Lycke, E., "Lysosomes of *Toxoplasma gondii* and their possible relation to the host-cell penetration of toxoplasma parasites," *J. Bact.*, **96**, 916, 1968.
122. —— and Lycke, E., "Factors enhancing the host-cell penetration of *Toxoplasma gondii*," *J. Bact.*, **93**, 53, 1967.
123. Peters, W., "Penetrations of a mouse erythrocyte by a merozoite of *Plasmodium berghei* as revealed by electron microscopy," *Trans. R. Soc. trop. Med. Hyg.*, **57**, 269, 1963.

124. Piekarski, G., Pelster, B., and Witte, H. M., "Endopolygenie bei *Toxoplasma gondii*," *Zentbl. Parasitkde Abt. I*, **36**, 122, 1971.
125. Porchet-Henneré, E., "Etude de premiers stades de développement de la coccidie *Coelotropha durchoni*," *Z. Zellforsch. mikrosk. Anat.*, **80**, 556, 1967.
126. —— and Vivier, E., "Ultrastructure comparée des germes infectieux (sporozoites, mérozoites, schizozoites, endozoites, etc. . . .) chez les Sporozoaires," *Annls Biol.*, **10**, 77, 1971.
127. Pyne, C. K., "L'ultrastructure de l'appareil basal des flagelles chez *Cryptobia helicis* (Flagellé, Bodonidae)," *C.r. Séanc. Acad. Sci., Paris*, **250**, 1912, 1960.
128. ——, "Etudes sur la structure inframicroscopique du cinétoplaste chez *Leishmania tropica*," *C.r. Séanc. Acad. Sci., Paris*, **251**, 2776, 1960.
129. ——, "Etude de la structure inframicroscopique de *Strigomonas oncopelti*," *Abstracts I, Int. Conf. Prot.*, 396, 1961.
130. ——, "L'ultrastructure des flagellés des familles Trypanosomidae et Bodoniadae," *Ann. Sci. Natur. Zool. Biol. Anim.*, **9**, 345, 1967.
131. Rambourg, A. and Leblond, C. P., "Electron microscope observations on the carbohydrate rich cell coat present at the surface of cells in the rat," *J. Cell Biol.*, **32**, 27, 1967.
132. Riou, G. and Pautrizel, R., "Nuclear and kinetoplastic DNA from trypanosomes," *J. Protozool.*, **16**, 509, 1969.
133. Roberts, W. L. and Hammond, D. M., 'Ultrastructural and cytological studies of sporozoites of four *Eimeria* species," *J. Protozool.*, **17**, 76, 1970.
134. Robertson, J. D., "Unit membranes. A review with new studies of experimental alterations and a new subunit structure in synaptic membranes," in M. Locke (ed.), *Cellular Membranes in Development, 22nd Symposium for the Study of Development and Growth*, Academic Press, New York, 1964, vol. I.
135. Rudzinska, M. A., d'Alessandro, P. A., and Trager, W., "The fine structure of *Leishmania donovani* and the role of the kinetoplast in the *Leishmania-leptomonad* transformation," *J. Protozool.*, **11**, 166, 1964.
136. —— and Trager, W., "Intracellular phagotrophy by malaria parasites: an E.M. study of *Plasmodium lophurae*," *J. Protozool.*, **4**, 190, 1957.
137. —— and ——, "The fine structure of trophozoites and gametocytes in *Plasmodium coatneyi*," *J. Protozool.*, **15**, 73, 1969.
138. ——, ——, and Bray, R. S., "Pinocytotic uptake and the digestion of hemoglobin in malaria parasites," *J. Protozool.*, **12**, 563, 1965.
139. —— and Vickerman, K., "The fine structure," in D. Weinman and M. Ristio (eds.), *Infectious Blood Diseases of Man and Animal Diseases caused by Protist*, Academic Press, New York, 1968, p. 217.
140. Ryley, J. F., "Ultrastructural studies on the sporozoite of *Eimeria tenella*," *Parasitology*, **59**, 67, 1969.
141. Schaudinn, F., "Generation und Wirtswechsel bei Trypanosoma und Spirochaeta," *Arch. Kais. Gesundheitsamt.*, **20**, 287, 1904.
142. Scholtyseck, E., "Ultrastructure" in Datus M. Hammond, and P. Long

(eds.), *The Coccidia (Eimeria, Isospora and Toxoplasmea and related genera)*, University Park Press, Baltimore and Butterworth, London, 1973, pp. 81–144.
143. ——, Friedhoff, K., and Piekarski, G., "Über mikromorphologische Übereinstimmungen bei Entwicklungsstadien von Coccidien, Toxoplasmen und Piroplasmen," *Zentbl. Parasitkde Abt. I*, **35**, 119, 1970.
144. —— and Hammond, D. M., "Electron microscope studies of macrogametes and fertilization in *Eimeria bovis*," *Z. ParasitKde*, **34**, 310, 1970.
145. —— and Piekarski, G., "Elektronenmikroskopische Untersuchung an Merozoiten von Eimeren (*Eimeria perforans* und *Eimeria stiedae*) und *Toxoplasma gondii* zur systematischen Stellung von *T. gondii*," *Z. ParasitKde*, **26**, 91, 1965.
146. Schrevel, J., "L'ultrastructure de la région antérieure de la grégarine *Selenidium* et son intérêt pour l'étude de la nutrition chez les sporozoaires," *J. Microsc.*, **7**, 391, 1968.
147. ——, "Les polysaccharides de réserve chez les Sporozoaires," *Annls Biol.*, **10**, 31, 1971.
148. —— and Vivier, E., "Etude de l'ultrastructure et du rôle de la région antérieure (mucron et épimérite) des Grégarines parasites d'Annélides polychètes," *Protistologica*, **2**, 17, 1966.
149. Schulz, H. and Mac Clure, E., "Elektronenmikroskopische Untersuchung des *Trypanosoma cruzi* mit besonderer Berücksichtigung der Periplasten und des Blepharoplasten," *Z. Zellforsch. mikrosk. Anat.,* **55**, 389, 1961.
150. Schuster, F. L., "An electron microscope study of the amoebo-flagellate, *Naegleria gruberi* (Schardinger). II. The cyst stage," *J. Protozool.*, **10**, 313, 1963.
151. ——, "Ultrastructure of cysts of *Naegleria* spp: a comparative study," *J. Protozool.*, **22**, 352, 1975.
152. —— and Dunnebach, T. H., "Formation of bodies associated with virus-like particles in the amoebo-flagellate *Naegleria gruberi*," *J. Ultrastruct. Res.*, **36**, 659, 1971.
153. —— and ——, "Transmission of virus-like particles in the EGS strain of *Naegleria gruberi*," *J. Protozool.*, **20**, suppl., 520, 1973.
154. Seed, J. R., Byram, J., and Gam, A., "Characterization and localization of phosphatase activity of *Trypanosoma gambiense*," *J. Protozool.*, **14**, 117, 1967.
155. ——, Gnau, J. M., Krier, J. P., and Pfister, R. M., "*Trypanosoma congolense*: fine structure study by the carbon replica and freeze-etch technique," *Expl Parasitol.*, **31**, 399, 1972.
156. Senaud, J., "Ultrastructure comparée des endodyozoites de *Sarcocystis tenella* et de *Toxoplasma gondii*," *2nd International Conference of Protozoology*, Excerpta Medica Foundation, Amsterdam, 1965, p. 183.
157. ——, "L'ultrastructure du micropyle des Toxoplasmidae," *C.r. Séanc. Acad. Sci., Paris,* **262**, 119, 1966.
158. ——, "Contribution à l'étude des Sarcosporidies et des Toxoplasmes," *Protistologica*, **3**, 167, 1967.
159. ——, "Ultrastructure des formations kystiques de *Besnoitia jellisoni*

Protozoaire, Toxoplasmea, parasite de la souris (*Mus musculus*)," *Protistologica*, **5**, 413, 1970.

160. ———, "Le cycle de développement des toxoplasmes *Toxoplasma gondii*," *Bull. Institut Pasteur*, **70**, 3, 1972.
161. ——— and Cerna, Z., "Etude en microscopie électronique des mérozoites et de la mérogonie chez *Eimeria pragensis*. Coccidie parasite de l'intestin de la souris (*Mus musculus*)," *Annls Soc. Biol. Besse-en-Chandesse*, **3**, 221, 1968.
162. ——— and ———, "Ultrastructure comparée des endodyocyte des toxoplasmes et des mérozoites de quelques *Eimeria*," *Prog Protozool., Leningrad*, **72**, 1969.
163. ——— and ———, "Etude ultrastructurale des mérozoites et de la schizogonie des coccidies (Eimeriina). *Eimeria magna* de l'intestin des lapins et *E. tenella* des coecums des poulets," *J. Protozool.*, **16**, 155, 1969.
164. ——— and ———, "La microgamétogénès chez *Eimeria pragensis, Sporozoa, Telosporea, Coccidia, Eimeriina*, parasite de l'intestin de la souris: étude au microscope électronique," *Protistologica*, **6**, 5, 1970.
165. ——— and de Puytorac, P., "Observations complémentaries sur l'ultrastructure de la "spore" de *Sarcocystis tenella* Rail (Sarcosporidie du mouton)," *C.r. Séanc. Soc. Biol., Paris*, **156**, 1630, 1962.
166. Sheffield, H. G., "The function of the micropyle in the cyst organisms of *Besnoitia jellisoni*," *J. Parasit.*, **53**, 888, 1967.
167. ———, "Schizogony in *Toxoplasma gondii*: An electron microscope study," *Proc. helminth. Soc.Wash.*, **37**, 237 (1970).
168. ——— and Hammond, D. M., "Fine structure of first-generation merozoites of *Eimeria bovis*," *J. Parasit.*, **52**, 595, 1966.
169. ——— and ———, "Electron microscope observations on the development of first-generation merozoites of *Eimeria bovis*," *J. Parasit.*, **53**, 831, 1967.
170. ——— and Melton, M. L., "The fine structure and reproduction of *Toxoplasma gondii*," *J. Parasit.*, **54**, 209, 1968.
171. ——— and ———, "*Toxoplasma gondii*: the oocyst, sporozoite and infection of cultured cells," *Science*, **167**, 892, 1970.
172. Singh, B. N., *Pathogenic and Non-Pathogenic Amoebae*, Macmillan, London, 1975.
173. Steiger, R. F., "On the ultrastructure of *Trypanosoma* (*Trypanozoon*) *brucei* in the course of its life cycle and some related aspects," *Acta Tropica*, **30**, 64, 1973.
174. Steinert, M. and Novikoff, A. B., "The existence of a cytostome and the occurrence of pinocytosis in the trypanosome, *Trypanosoma mega*," *J. biophys. biochem. Cytol.*, **8**, 563, 1960.
175. Stuart, K. D., "Evidence for the retention of kinetoplast DNA in an acriflavine-induced dyskinetoplastic strain of *Trypanosoma brucei* which replicates the altered central element of the kinetoplast," *J. Cell. Biol.*, **49**, 189, 1971.
176. Swellengrebel, N. H., "La volutine chez les trypanosomes," *C.r. Séanc. Soc. Biol., Paris* **64**, 38, 1908.

177. Trager, W. and Rudzinska, M. A., "The riboflavin requirement and the effects of acriflavin on the fine structure of the kinetoplast of *Leishmania tarentolae*," *J. Protozool.*, **11**, 133, 1964.
178. Vandenberg, J. and Rhodin, J., "Differentiation of nuclear and cytoplasmic fine structure during sporogonic development of *Plasmodium berghei*," *J. Cell Biol.*, **32**, C7, 1967.
179. ——, ——, and Yoeli, M., "Electron microscopic and histochemical studies of sporozoite formation in *Plasmodium berghei*," *J. Protozool.*, **14**, 82, 1967.
180. Van den Berghe, L., "Sur la présence d'acide pentosenucléique sous forme de grains de 'volutine' dans les Trypanosomes," *Acta biol. belg.*, **2**, 464, 1942.
181. Vanderplank, F. L., "Identification of trypanosomes by chromosomes," *Nature , Lond.*, 3896, 19, 1944.
182. Van Der Zypen, E. and Piekarski, G., "Zur Ultrastruktur des Cystenwand von *Toxoplasma gondii* im Gehirn der Weissen Maus," *Z. ParasitKde*, **28**, 45, 1966.
183. —— and ——, 'Die Endodyogenie bei *Toxoplasma gondii*. Eine morphologische Analyse," *Z. ParasitKde*, **29**, 15, 1967.
184. Van Harreveld, A., Crowell, J., and Malhorta, S. K., "A study of extracellular space in central nervous tissue by freeze-substitution," *J. Cell. Biol.*, **25**, 117, 1965.
185. Vickerman, K., "Polymorphism and mitochondrial activity in sleeping sickness trypanosomes," *Nature, Lond.*, **208**, 762, 1965.
186. ——, "Electron microscopy of tsetse salivary gland stages in the life-cycle of *Trypanosoma rhodesiense*," *Trans. R. Soc. trop. Med. Hyg.*, **60**, 8, 1966.
187. ——, "Fine structure and function in the life cycle of the African trypanosomes," *Prog Protozool., Leningrad*, **46**, 1969.
188. ——, "On the surface coat and flagellar adhesion in trypanosomes," *J. Cell Sci.*, **5**, 163, 1969.
189. ——, "The fine structure of *Trypanosoma congolense* in its bloodstream phase," *J. Protozool.*, **16**, 54, 1969.
190. ——, "The host-parasite interface of parasitic protozoa," *Symp. Br. Soc. Parasit.*, **10**, 71, 1972.
191. ——, "The ultrastructure of pathogenic flagellates," in *Trypanosomiasis and Leishmaniasis with Special Reference to Chaga's Disease, Ciba Foundation Symposium 20* (*New Series*), Elsevier Excerpta Medica, North-Holland, Amsterdam, 1974, p. 171.
192. Visvesvara, G. S., Callaway, C. S., and Healy, G. R., "Light and electron microscopic studies on the mechanism of pathogenesis of *Naegleria fowleri* in mouse brain and tissue culture," *J. Protozool.*, **20**, suppl., 498, 1973.
193. Vivier, E., "Observations nouvelles sur la reproduction asexuée de *Toxoplasma gondii* et considérations sur la notion d' endogénèse," *C.r. Séanc. Acad. Sci., Paris*, **271**, 2123, 1970.
194. —— and Petitprez, A., "Le complexe membranaire superficiel et son

évolution lors de l'élaboration des individus-fils chez *Toxoplasma gondii*," *J. Cell Biol.,* **43**, 329, 1969.

195. —— and ——, "Données ultrastructurales complémentaires morphologiques et cytochimiques sur *Toxoplasma gondii*," *Protistologica*, **8**, 199, 1972.

196. Wery, M. and De Groodt-Lasseel, M., "Ultrastructure de *Trypanosoma cruzi* en culture sur milieu semisynthétique," *Annls Soc. belge Méd. trop.*, **46**, 337, 1966.

197. Wigg, D., Bovee, E. C., and Jahn, T. L., "The evacuation mechanism of the water expulsion vesicle ('Contractile vacuole') of *Amoeba proteus*," *J. Protozool.*, **14**, 104, 1967.

198. Willaert, E., "Primary amoebic meningoencephalitis. A selected bibliography and tabular survey of cases," *Annls. Soc. belge Méd. trop.*, **54**, 205, 1974.

199. ——, "Etude immunotaxonomique des genres *Naegleria* et *Acanthamoeba* (Protozoa, Amoeba)," *Thesis*, Lille, 1975.

200. —— and le Ray, D., "Caractéres morphologiques, biologiques et immunochimiques de *Naegleria jadini* sp. nov. (Amoebida, Vahlkampfiidae)," *Protistologoca*, **9**, 417, 1973.

201. Wright, K. A. and Lumsden, W. H. R., "The fine structure of *Trypanosoma* (*Trypanozoon*) *brucei*," *Trans. R. Soc. trop. Med. Hyg.*, **63**, 13, 1964.

202. —— and Hales, H., "Cytochemistry of the pellicle of *Trypanosoma* (*Trypanozoon*) *brucei*," *J. Parasit.*, **56**, 671, 1970.

203. ——, Lumsden, W. H. R., and Hales, H., "The formation of filopodium-like processes by *Trypanosoma* (*Trypanozoon*) *brucei*," *J. Cell Sci.*, **6**, 285, 1970.

Part Six

Fungi

6. Fungi

Johannes Müller

6.1. Introduction

Electron microscopic studies have contributed a remarkable amount to the understanding of the nature of mycotic diseases in humans. This field of research, however, needs special knowledge of the causative agent, the fungus, which is a plant of lower organization possessing a well-developed cell wall and the capacity for sexual and nonsexual reproduction. Human pathogenic fungi belong to various systematic taxa. Transmission as well as scanning electron microscopy have proved to be useful tools for morphological investigations with fungi *in vitro* and with mycotic tissue foci in humans; some questions could only be solved with experimental infections of laboratory animals. Freeze-etching techniques have rendered understanding of membrane-bound processes, important in mycotic infections. In recent years ultrahistochemistry and immuno-electron miscroscopy have allowed access to a broad means for studying physiological processes relevant in host–fungus interrelationship. Last, but not least, the mechanisms of action of antimycotic drugs on fungi have aroused the interest of electron microscopists.

New questions of physiological, immunological, biochemical, and biophysical nature have arisen which can be solved by complex studies only, using ultrahistochemistry, immuno-electron miscroscopy, and ultrastructural autoradiography and this must be regarded as a strong challenge for the future research activities in medical mycology.

6.2. The ultrastructure of fungi cultured *in vitro*

Precise information on ultrastructure is necessary for comparing the pathogenic fungi with nonpathogenic organisms, and for evaluating the ultrastructural aspects of fungi situated in mycotic lesions of the host. By analogy with bacterial infections no clearly defined constituent(s) can be pointed out in fungi to which pathogenic potency can be ascribed or restricted. Although the exploration of fungal ultrastructure is expected to be a necessary and useful instrument in understanding the character of mycoses, it must be emphasized that all morphological details in the final analysis have to be related to their physiological significance concerning the host–fungus relationships. Considering this high aim it must be stated that the present knowledge of fungal ultrastructure is not yet sufficient in all aspects.

The medical mycologist is in the advantageous situation of being able to utilize the broad spectrum of excellent information on the ultrastructure of fungi compiled by botanical mycologists.[4, 15, 22, 30, 114, 148, 192, 330] This must be the basis of all special considerations on ultrastructural aspects in pathogenic fungi which themselves are not the subject of the above mentioned botanical literature.

6.2.1. Keratinophilic dermatophytes

The *in vitro* aspects of the keratinophilic fungi have been thoroughly investigated. The superficial characteristics of the *Microsporum–Nannizzia* group are accessible with SEM studies[117, 128, 274, 310] and in parallel by TEM.[62] A series of TEM publications provides information on ultrastructural cytology.[120, 174, 175, 319, 321] Ascosporogenesis has been described for *Nannizzia gypsea*.[122] The normally pigmented species *Microsporum cookei* has been compared with an albino mutant strain for their cell wall and mitochondrial characteristics,[130] and the cell wall features of *M. gypseum* were compared with those of *Trichophyton* by means of freeze-etching techniques.[111] The *Trichophyton-Arthroderma-Keratinomyces* group has been examined under the auspices of morphological cytology,[127, 129, 131, 173, 176, 179, 257, 269, 302] while freeze-etching studies rendered information on the structure of inner membrane surfaces.[109, 110, 112] Similar morphological structures are displayed in experiments on *Epidermophyton floccosum*.[154, 213, 320]

Beyond mere morphological descriptions, studies have been made applying biochemical methods in combination with electron microscopy.[107, 213] In this way, three layers are made evident in the composition of the cell wall of *T. mentagrophytes*: an external one with a microfibrillary structure, poor in polysaccharides; a middle layer with a periodic texture and a moderate polysaccharide content; and an internal layer rich in polysaccharides which is not observed in old cells.[237] Finally, synoptical reviews with regard to cytochemical and histochemical studies revealed the aspect of a differentiated ultrastructural and functional organization of the dermatophytic entity. Ultrastructural and biochemical differentiations of the hyphae are related to various biological activities in different sites of the coenocytic mycelium. A characteristic difference between this coenocytic organization and the differentiation of tissue cells in higher organisms is the organizational flexibility and reversibility of the fungal organisms, especially in relation to its pathogenic potency.[172, 177, 178]

6.2.2. Causative agents of deep mycoses

Most fungi of this group are dimorphic: the fungus appears yeast-like *in vivo*, but converts into the mycelial form when growing in culture at temperatures lower than the temperature of warm-blooded organisms. Studies on *Histoplasma capsulatum* concerned the TEM of the yeast phase[78] and of protoplasts,[40] as well as TEM[96, 97] and SEM[93] of the yeast-mycelium conversion showing aberrant transitional cells. Moderate interest has been shown in cell wall architecture.[258, 259] A strange phenomenon was described with the selfparasitism of the fungus.[157] The morphology of two varieties in pigmentation has been compared by SEM[94] with special regard to the conspicuous architecture of the *Histoplasma* conidia. Ultrastructural data on the ascospores of the perfect form *Emmonsiella capsulata* have been contrasted with those of *Ajellomyces dermatitidis.*[101] The content of several cell wall constituents has been determined quantitatively with *Histoplasma farciminosum*, and the fibrillar texture of the different fractions was demonstrated with negative staining.[260]

Ajellomyces (*Blastomyces*) *dermatitidis* has been intensely examined for its cell wall structures with special consideration of budding processes,[36, 37] intracellular membrane systems,[74] and ribosomes and polyribosomes.[90] Likewise, clear information is available on the characteristics of microconidia,[95] on cleistothecia in the perfect form,[98, 101] on the yeast-mycelium conversion,[96] as well as on selfparasitism.[157] Sufficient basic knowledge has likewise been obtained for *Paracoccidioides brasiliensis* concerning its cell wall structure and budding features,[36, 37] its yeast phase morphology,[41, 42, 43] and the mycelium-yeast phase conversion.[91] Ultrahistochemical studies concern the proof of glycogen and enzymes in glycolysis and the citric acid cycle in the yeast and mycelial forms of this fungus.[139, 140]

Two distinct morphological cycles are known for *Coccidioides immitis*: a mycelium-arthroaleuriospore cycle is observed in saprophytic environments. Airborne spores initiate the self-perpetuating spherule-endospore cycle in infected hosts. The morphology of both cycles and of the conversion stages is well analyzed.[33, 80, 215, 275, 276]

In compiling the ultrastructural morphology of *Sporothrix schenkii*, special interest was shown in the analysis of the outer cell wall constituents which seem to prove the existence of a capsular-like layer.[92, 156, 158] The conidiogenesis was studied in comparison with other fungal genera.[55]

The morphology of *Emmonsia crescens*, the causative agent of the adiaspiromycosis in animals and man, was studied with TEM[147] and with freeze-etching techniques.[121] SEM investigations revealed the outstanding multilocular germination of the huge adiaspores during the conversion into the mycelium phase.[115, 116, 118, 119] Little information exists on the causative agents of chromomycosis: *Fonsecaea pedrosoi* has been examined by freeze-etching methods comparing the fungus with *Trichophyton.*[113] The conidial ontogeny was demonstrated for *Phialophora* species with SEM.[57, 217]

Cryptococcus neoformans possesses a well-developed capsule, *in vitro* as well as *in vivo*, which has found special interest concerning the pathomechanisms in cryptococcosis by the analogy of infections caused by encapsulated bacteria. Therefore, the texture especially of the cell wall was demonstrated in several TEM studies[7, 48, 60, 76, 77, 153, 299] besides other ultrastructural features. Freeze-etching experiments rendered concepts on the physiological processes of the cell wall formation and on the dependence of the aspects of the plasmalemma from constituents of culture media.[282, 283] Capsules could also be proved to be present on a hyphae-forming strain.[164]

6.2.3. Opportunistic pathogenic fungi

The most detailed and extensive information on the ultrastructure of a pathogenic fungus has been obtained with *Candida albicans*.[323, 325, 326] Comparative examinations of several *Candida* species suggests ultrastructural differences[5, 8] which are plausible if consideration is given to the fact that the genus *Candida* is an artificial taxon partially comprising unrelated species. Authors often emphasize the difficulties of preparation of intracellular details due to the resistance of the *Candida albicans* cell wall to fixation procedures.[27, 244, 251] The intracytoplasmic membrane systems,[132, 133, 187] the mitochondria and nuclei,[132, 162, 133, 293] the appearance of glycogen,[252, 253] and the morphology in chlamydospores[296] have been well studied. A still better approach to the construction of the membrane systems was achieved by freeze-etching examinations.[108, 246, 277] A detailed series of studies contributed to the knowledge of the dependence of the ultrastructure on defined nourishing hydrocarbon substrates;[123, 218, 220, 222, 223, 224, 225] these publications concern several *Candida* species. Since the diphasic *Candida* fungi are pathogenic in the yeast phase as well as in the mycelium phase, the germination processes during the yeast–mycelium conversion were subject to analysis with blastospores[47, 264, 270, 272, 290, 292] and with chlamydospores.[50, 291] Cryoultrathin section techniques reveal better membrane structures and details which cannot be proved by conventional methods.[294, 296] A series of SEM examinations has described surface phenomena of *Candida albicans*.[137, 219, 287, 288, 330]

Special interest was dedicated to the cell wall architecture regarding the fact that this fungus constituent plays a preeminent role among the various host–fungus relationships. In order to optimize such analysis special preparation methods have been developed.[68, 235] Eight different cell wall layers can be distinguished by using appropiate fixation and staining methods.[235, 239, 240, 241, 254] The cloning of *Candida albicans* populations from cultures or from clinical specimens led to cell lines with genetically stable richness or deficiency in the external electron-dense cell wall layer which is suspected to have a pathogenetic effect on infected hosts.[239, 240] This external layer is regarded as the site of antigenic properties recorded by agglutination reactions. The exploration of immunobiological properties and phenomena is possible by immuno-electron microscopy using ferritin-labeled antibodies and was started in 1973 with mycological subjects. The first step was the adaptation of these methods to fungi and the proof of specificity with indirect immunoferritin tests,[210] demonstrating that cultured *Candida albicans* cells bind antibodies from human sera to the external electron-dense cell wall layer, and rendering topographic details of agglutination sites. The direct immunoferritin test, using ferritin-labeled polyvalent[309] or monospecific antisera against polysaccharide (mannane) antigens, allows the investigation of the distribution pattern of the superficially located antigens in yeast cells,[211] still more impressive when examined by SEM techniques (Figure 6.1(a),(b)).[286] The application of such immuno-electron microscopic methods not only serve to increase the basic knowledge of fungal surface antigenicity which has become important as a means of yeast systematics but the exploration of superficial antigenic structures on the molecular level is also a

(a)

(b)

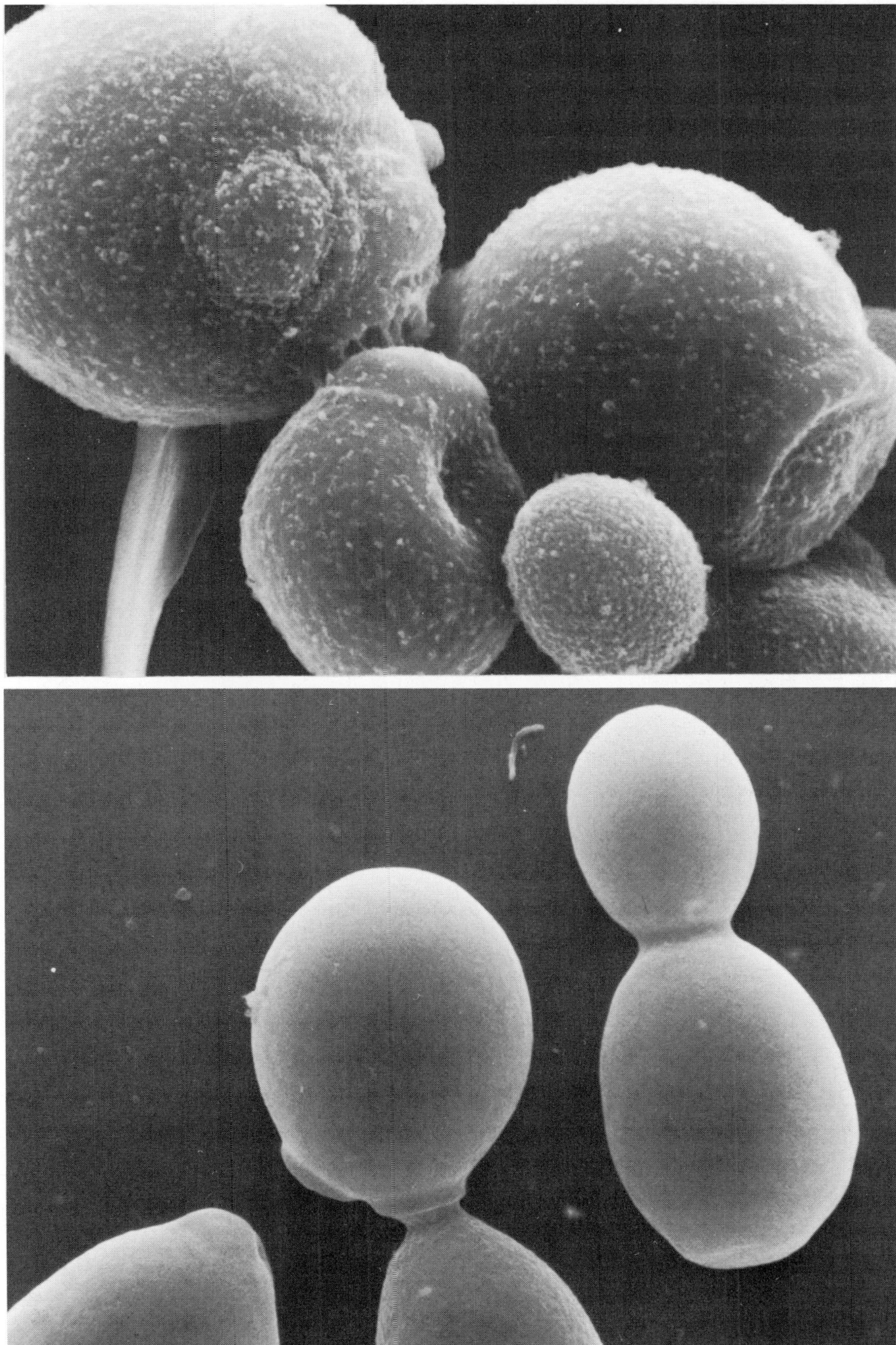

Figure 6.1. *Candida albicans* blastospores (J. Tokunaga).

(a) Blastospores treated with ferritin-labeled polyvalent anti-*Candida albicans* serum (direct immunoferritin test). The rough surface demonstrates that the external cell wall layer consists of readily accessible antigenic material. (× 12,000)

(b) Nontreated control cells. The cell surface appears smooth. (× 10,000)

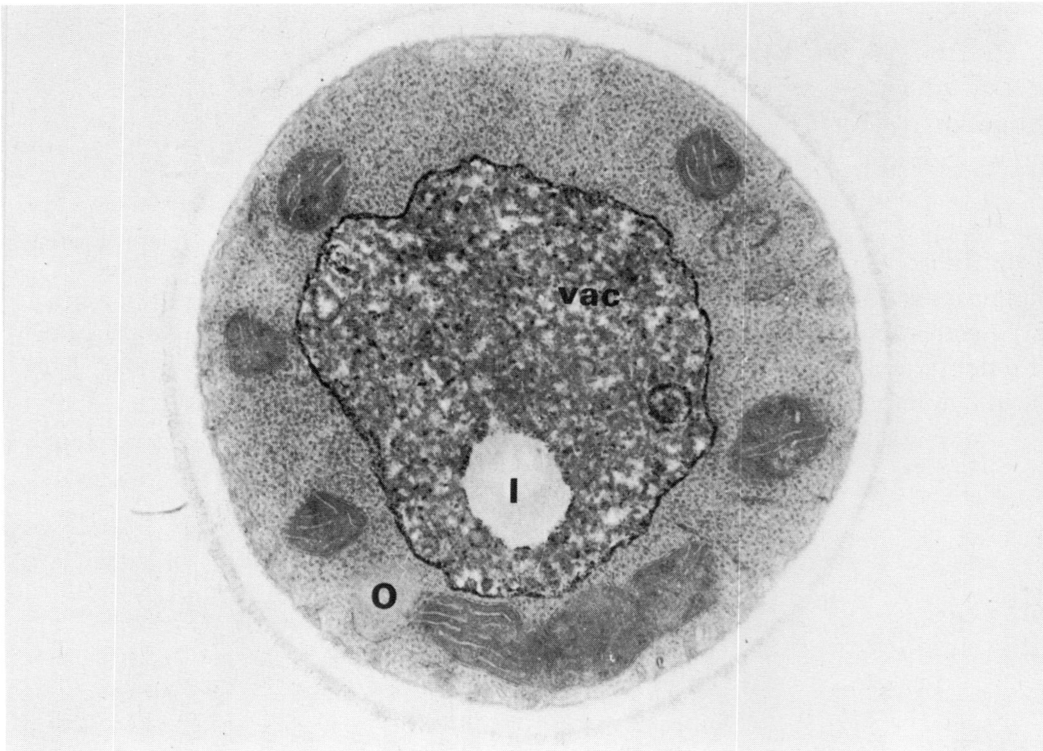

Figure 6.2. Alkaline phosphatase in *Candida albicans* (M. Borgers).
Ultrahistochemical proof of alkaline phosphatase shown by the strong deposition of precipitate on the limiting membrane of the vacuole (vac). Staining of the amorphous or vesicular material inside the vacuole is less intense. The vacuole generally is a center of high metabolic activity. The bulk of hydrolase activity is concentrated within this organelle. o: ovoid body, l: lipid. (× 32,800)

fundamental requirement for the understanding of host–fungus relationships in the infectious process.

The development of ultrahistochemistry has offered possibilities for proving that enzymes are related to distinct ultrastructures. Studies of this type have likewise been performed mainly with *Candida*. Catalase activity has been proved to correspond with the composition of its nourishing medium.[82, 83, 221, 227] This distribution of most specific and nonspecific phosphatases and peroxidases is different from that found in mammalian cells; oxidative and peroxidative activities are demonstrated only in mitochondria (Figure 6.2); specific marker enzymes for cell organelles are not found.[28, 65, 66, 308] Dehydrogenases and reductases are localized in both mitochondria and cytoplasm.[106, 184] The phospholipase A secreted by *C. albicans* could play a part in the invasion of host tissues in lesions of candidosis, while lysophospholipase at the same cell sites may protect the cell from high concentrations of phospholipase secreted in the process of attacking animal tissues.[248] The intracellular distribution and morphology of glycogen particles[328] and their behavior to amylase exposition has been studied with regard to the pathways of carbohydrate metabolism.[252, 253] A similar process is the gradual digestion of the cell wall by enzymes such as glycosidases, proteases, and chitinases in order to examine morphology and reactivity of the remaining constituents.[46, 146, 231] Such techniques lead to the formation of protoplasts lacking cell walls; nevertheless, in digestion experiments with *C. albicans* some constituents of the cell wall appeared to be resistant in that complete digestion was not obtained.

Pityrosporum orbiculare, a lipophilic fungus responsible for Pityriasis versicolor, has been studied in serial sections in order to demonstrate the three-dimensional construction of the single cell.[144, 145] Striking features are the regular, electron-dense speckling of the cell wall sections and the spiral arrangement of indentations demonstrated by freeze-etching on the ex-fracture face of the plasma membrane.[32]

Some ultrastructural and cytochemical information relevant to the problems discussed here has been obtained with nonpathogenic yeast fungi. In these respects some early studies and review articles must be cited concerning morphology and developmental stages of cell organelles with TEM[14, 30, 114, 169, 170, 183, 193, 194, 196, 212, 285] and with freeze-etching techniques,[191] concerning the formation of mycelia and spores,[56, 167, 303] as well as the relationships between ultrastructure and systematic classification.[141] Broad attention is given to the peculiarities of budding processes[3, 19] and to the exploration of the cell wall constitution.[126, 195, 255, 256] In these respects studies on protoplast formation and reconstitution of the cell wall earn high interest[20, 138, 149, 181, 182, 230, 231] and promise to give further information on this important cell organelle for studies on pathogenicity, for which protoplasts have not yet been used within the scope of electron microscopy.

The most important opportunistic pathogenic species among the filamentous fungi are to be found in the genus *Aspergillus*. Most interest was addressed to the electron microscopy of morphology, ultrastructure, and germination processes of the conidiospores with *A. niger*, *A. fumigatus*, and *A. oryzae*: TEM studies[25, 35, 280, 300, 301] were completed by cryoscanning investigations[289, 295, 297, 330] on the fine morphology of the conidiospore chains with high systematic significance. The cultivation of *A. parsiticus* on manganese-deficient medium demonstrated morphogenetic alterations which resemble a hyphae-yeast-like dimorphism.[89] In protoplasts of *A. nidulans*, obtained by enzymatic digestion, two types of cells were found, differing in content of ribosomes and size of vacuoles; the formation of the new cell wall was described.[100]

Zygomycetes are rarely found as causative agents, mainly infecting blood vessels and the brain. These fungi form polynuclear thalli and represent a lower level of phylogenic development within the fungi. The ultrastructure of the vegetative as well as of the reproductive fungal elements is excellently reviewed.[22] In studies on the biosynthesis of the *Mucor* cell wall chitosomes have been detected, and their role as the cytoplasmic container and conveyer of chitin synthetase, forming chitin microfibrils for constructing the cell wall, has been resolved;[31] this new ultrastructural element seems to be the site of chitin synthesis in other fungal taxa too. Ultrastructural and electron cytochemical investigations have been performed with *Entomophthora*,[99] freeze-etching studies on mycelia, yeast forms and sporangiospores under aerobic and anaerobic conditions with *Mucor* species,[278, 279, 281] and examination of septal structures with *Basidiobolus*.[104]

The only taxon of Algae with human pathogenic significance, the genus *Fissuricella* (former genus *Prototheca*), has been demonstrated as cuboidal cell clusters resulting from incomplete separation of daughter cells after multiple cleavages.[232]

Remarkable information on fungal viruses has been gathered in the last two decades. Such viruses contain double-stranded RNA, and they can be made evident by means of infectivity tests, by various serological techniques, and by electron microscopy.[6, 124, 161, 197] The latter method is handicapped by the difficulty of differentiating virus-like particles against the background of host cell components. Human pathogenic fungi, such as *Histoplasma capsulatum*, *Sporothrix schenckii*, *Candida albicans*, and *Torulopsis glabrata* have been screened also for the presence of virus-like particles by electron microscopy.[2] Possible implications of viruses with pathogenetic effects of their fungal hosts in human mycoses (as proved in plant pathology) remain to be seen.

6.3. Ultrastructural features of host–fungus relationships in mycoses

Host–fungus relationships can be discussed under the first two different aspects below, but regarding newer concepts of infectious diseases the third aspect must be added.

1. How can virulence factors of the fungus against the host be demonstrated electron microscopically?
2. What ultrastructural information is available on defence mechanisms of the host against the fungus?
3. In which way does the host himself contribute to pathogenetic processes directed against his own tissues?

Some of these aspects have been reviewed earlier according to the knowledge available at that time.[17, 18, 38]

6.3.1. Evidence of fungal effects on the host

Dermatomycoses

Dermatomycoses caused by keratinophilic fungi is a field in which pathomechanisms have been explored systematically. Since these microörganisms colonize merely the epidermis it has been suggested that the fungi utilize the keratin as a nourishing substrate and that they possess keratinolytic enzymes. Furthermore, mechanical pressure and nonkeratinolytic enzymes contribute likewise to the aspects of the dermatomycosis.

The infection of human hair *in vitro* by *Trichophyton mentagrophytes*, *T. ajelloi*, and *T. rubrum*[21] is maintained by eroding and perforating fungal elements digesting the cuticle as well as the cortex keratin; differences in the keratinolytic potency between the three fungus species are obvious. The aspects of the invasion and destruction of the human hair *in vitro* by *Microsporum gypseum* has been explored by SEM.[150]

A comparative study using SEM and TEM was performed with *Trichophyton rubrum* in its interaction with human skin[268] and with various, nonkeratinous cell cultures.[269] SEM studies show the focal penetration of *Trichophyton rubrum* hyphae into the nail plate[70] and in other sites of the epidermis.[16] The keratinolytic activity of *Epidermophyton floccosum* in human epidermis seems to be low;[261] a surprising result of this publication was the proof of colonization of the cutis, so far unknown as an infection site for keratinophilic fungi. The hyphae of *Microsporum gypseum* naturally infecting human hairs differ conspicuously in size and shape from saprophytically grown fungi;[143] most of the hyphae follow the long axis of the hair, only very few are aligned concentrically around the cuticle and transversely through the cortex. Keratinolysis was not reliably proved, as it was found in artificially infected hairs. Obviously, parasitizing fungi utilize soluble proteins and glycogen available in differentiating host cells for the fungal nutrition. The controversal opinions on keratinolysis in naturally infected hair were discussed, including observations with experimental infections of guinea pigs by *Trichophyton mentagrophytes*, which explain the difficulties of proving keratinolysis *in vivo*.[233, 234, 236] Likewise keratinolysis was proved in the stratum corneum of guinea pigs infected with *Trichophyton mentagrophytes*[238] (Figure 6.3) as in hairs infected *in vitro* with *Trichophyton mentagrophytes* and *Microsporum gypseum* from germ-free guinea pigs.[81]

Piedra hortai[54] forms septate hyphae, asci, and free ascospores which penetrate into the cuticle of the hair shaft; the nodules of black piedra are thus formed by the broken up strands of the cuticle together with the fungal elements. The fungus seems to be unable to penetrate further into the shaft.

Dermatomycoses caused by nonkeratinophilic fungi have more rarely been the subject of ultrastructural studies. The aspects of *Pityrosporum* in tinea versicolor were examined by TEM[142] and SEM.[69, 263] *Candida* interactions with the human skin have been described,[185, 186, 265] but information on the factors of aggression is still lacking.

Endomycoses

Mechanical forces may play a role in endomycotic infections if the initial invasion is based on filamentous elements.[67] This is the case with Aspergilli and Zygomycetes, both undocumented electron microscopically. EM studies with *Candida* demonstrate the pathogenetic significance of the pseudohyphae;[10, 51, 52, 53, 199, 202, 266, 271, 322] the most important aspect is the capacity for invading blood vessels in the course of tissue invasion.[199] Since most of the fungi filamentous in culture pass over to a more or less spherical shape while infecting the host, the importance of the pathogenetic role of fungal filaments might have been overestimated.

Tissue areas infected by fungi are clearly pathologically changed to a large extent. This concerns the autochthonous tissues of the respective organs as well as the host defence cells which have infiltrated the infected focus. The damage

Figure 6.3. *Trichophyton mentagrophytes* (D. Poulain).
Keratinolytic activity effectuated by the fungus in the hair cortex of a guinea pig. The zone of digestion is indicated by arrows. F: fungus, Co: cortex, Cu: cuticula, R: residues of keratin digestion. (× 37,000)

which can be observed in host cells is rather uniform and does not clearly depend on the host cell type nor on the fungus species: nuclei, mitochondria, and endoplasmic reticulum are swollen, the total cell is edematous and contains lipid droplets, fibrous structures, electron-optically empty vacuoles, and an increased amount of glycogen particles.

In analogy to pathomechanisms in bacterial infections many efforts have been made to demonstrate special toxins with complex effects responsible for pathological phenomena in mycoses. Although a certain number of pharmacologically active toxins have been found in pathogenic fungi, it seems rather doubtful that such substances alone may support the mycotic process. The toxicity of these mycotoxins is about 10^5 times lower than the toxicity of gram-negative bacteria. The ultrastructural manifestation of a toxin action can be assumed when host cells closely located to fungi are damaged in the immediate vicinity of the fungus, and when the intensity of such damage is clearly decreasing with increasing distance from the fungus. A cytolytic effect of fungi to host cells as well as the formation of exudates, as assumed from light microscopical observations, could not be made evident by electron microscopy, or remains doubtful in such an interpretation.[75, 171, 312, 314] A phenomenon, however, which can be interpreted within that meaning is demonstrated with *Candida albicans* in Figure 6.4. The fungus cell located in a mouse kidney is surrounded by a granulocyte but the phagosome is not completely closed; the cytoplasmic area around the phagosomal membrane is within a certain, rather narrow zone, free of ribosomes and other cell organelles;[199, 200, 201, 202] similar observations are reported with *Candida* in oral candidiasis[188] and *Histoplasma* in mouse spleen.[75] The formation of electron microscopically empty halos around fungus cells can likewise be explained as a giving way of the host cell to immediate fungal toxic effects (*Candida albicans*,[73, 199, 202, 203, 312] *Aspergillus fumigatus*,[314] *Sporothrix schenckii*[165]). On the other hand, such halos might also be caused artificially by unequal shrinking due to preparation procedures,[199] or they may reflect the presence of a capsule-like cell wall layer: this interpretation is supported by comparing the above mentioned fungal infections with electron micrographs of *Cryptococcus neoformans* which indisputably possesses a capsule.[17, 59, 105] In immuno-electronmicroscopy the direct immunoferritin test using ferritin-labeled monospecific antifungal antisera has been developed as an important tool for exploring the release of fungal antigenic constituents into the host milieu;[180, 198, 209] toxic effects caused by liberated fungal substances as such could not have been made evident. The possibility of demonstrating known fungal toxins *in vivo* by ferritin-labeled antitoxic sera has not yet been applied experimentally.

Otherwise, the pathological alterations of fungus-infected tissues could be explained by the activity of the normal enzymatic (mainly hydrolytic) endowment of the fungi. Ultrastructural evidence of defined fungal enzymes acting against the host is scanty. A pioneer study demonstrates that phospholipase A is secreted by *Candida albicans*, especially at the tip of hyphae penetrating chick chorioallantoic epithelial cell (Figure 6.5); phospholipase A activity is likewise observed with blastospores[248, 249] where it is promoting budding processes controlled by lysophospholipase. Ultrastructural enzymology with fungi *in vitro* and *in vivo* remains a wide field open to numerous studies essentially to

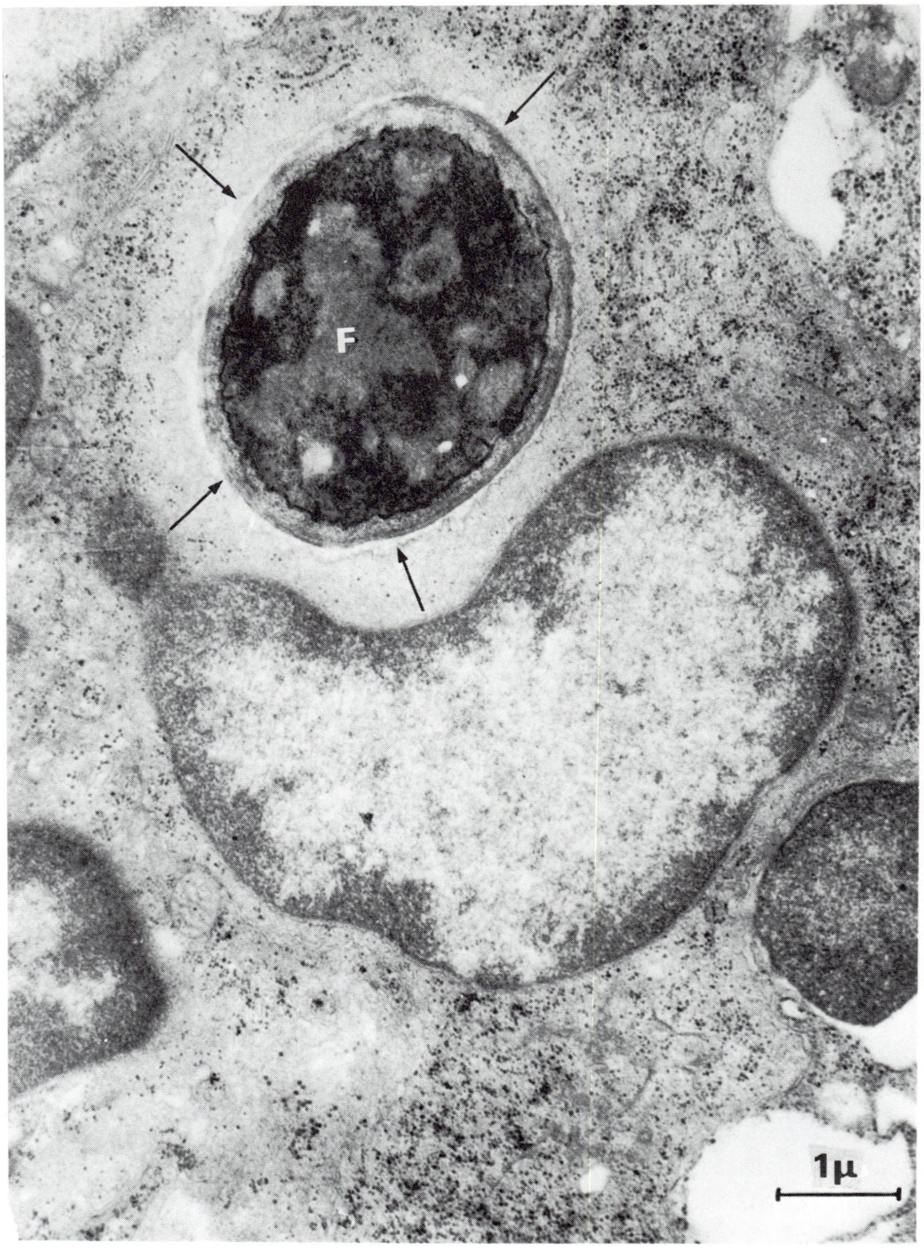

Figure 6.4. Granulocyte surrounding *Candida albicans* blastospore (J. Müller). Organelle-free cytoplasmic zone (arrows) of a granulocyte surrounding a singly located blastospore (F). The empty zone indicates a toxic effect caused by the fungus to the host cell. Experimental *Candida albicans* infection of mouse kidney. (× 16,000)

be performed on the pathogenetic role of fungal enzyme patterns.[28, 29, 125] If, however, the normal enzyme endowment of a fungus must be made responsible for its pathogenicity, the question is raised again as to what are the differences between pathogenic and nonpathogenic fungus species. In summary, it seems surprising that the causative principles of fungal pathogenic effects on the host cannot be demonstrated so far electron microscopically even in classical

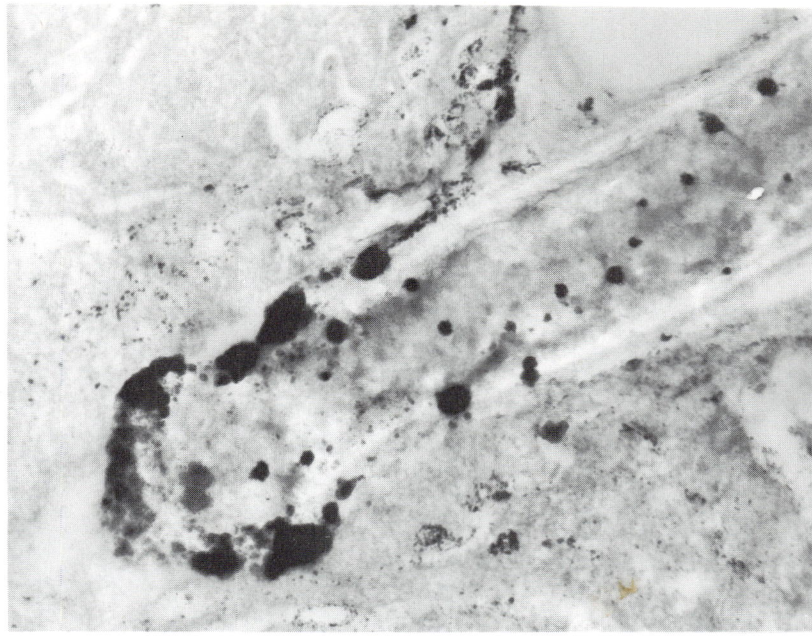

Figure 6.5. Penetrating hypha of *Candida albicans* (D. Pugh).
The host cell, epithelial cell of chick chorioallantoic membrane, has been broken down by the advancing hypha and scattered phospholipase activity is visible in the chick cytoplasm and in the space between the fungus and the invaded cell. The highest activity is found at the tip of the hypha. (× 30,000)

mycoses caused by obligate human pathogenic fungi such as histoplasmosis,[39, 75] blastomycosis,[58] paracoccidioidomycosis,[44, 61, 84, 85, 86] and chromomycosis.[305] All information available, however, indicates that the pathological changes observed in the infected host tissue are of a complex nature which will be discussed further in section 6.3.3.

6.3.2. Evidence of host effects on fungi

Fungal elements invading the host face the defence barrier of phagocytosing cells. This phenomenon has been studied *in vitro* with *Candida albicans* by incubating the fungus cells with human or animal cells with phagocytosing capacity.[11, 12, 23, 73, 216, 294] It can easily be demonstrated that such host cells readily engulf blastospores (Figure 6.6), and even filaments can be attacked by phagocytes[294] (Figure 6.7). Electron micrographs in combination with kinematographic studies make it evident that chemotaxis plays a role in phagocytosis (Figure 6.8).[294] Opsonization favors the uptake of fungal cells, but there seems to be no difference between cell wall coatings by specific antibodies or by unspecific serum proteins (Figure 6.9).[73, 202] Immuno-electron microscopic studies reveal that specific antibodies coating the *Candida* cells are removed from the cell wall surface in the course of the engulfment by a phagocyte.[152, 180]

Phagocytosis of fungi is effected by different host cell types. All types of granulocytes as well as macrophages, histiocytes, and monocytes are most frequently involved, and endothelial and epithelial cells also show this capacity.

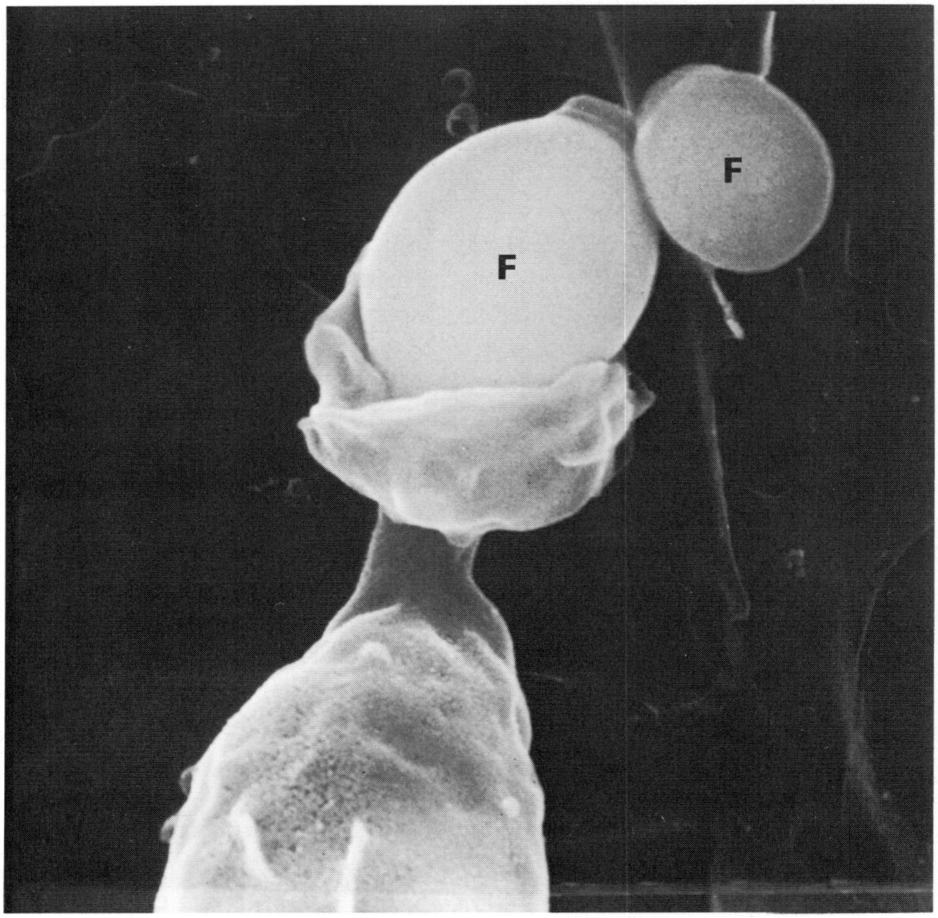

Figure 6.6. Phagocytosis of yeast fungi (J. Tokunaga).
A polymorphonuclear leukocyte engulfing a nonopsonized *Candida albicans* cell (F) by a cup-like pseudopod. *In vitro* study. (× 10,000)

The complete cycle of phagocytosis has been demonstrated including the evidence of phagolysosomes in fusion with the phagosomes.[294] (Figure 6.10) Phagocytosis itself does not necessarily imply fungal death; intracellular killing depends on the presence of hydrolytic enzymes; a main candidacidal factor has been proved to be the myeloperoxidase-halide-hydrogen peroxidase system.[11] Ultrahistochemical information on the decomposition of fungi in phagosomes, however, is not available. *Candida* cells located in phagocytes which do not possess fungicidal activity can escape from the host cell by germinating and penetrating the host cell membrane. This capacity of the genus *Candida* seems to be a virulence factor within a primary infection which is not controlled by immune defence mechanisms; in contrast, immune host organisms show a high fungicidal activity in phagocytosis.[152]

Observations on phagocytosis of fungi in the course of human or animal infections are rare.[152] Obviously phagocytosis occurs mainly very early during the infection process in vessels or body cavities; here it seems to display a

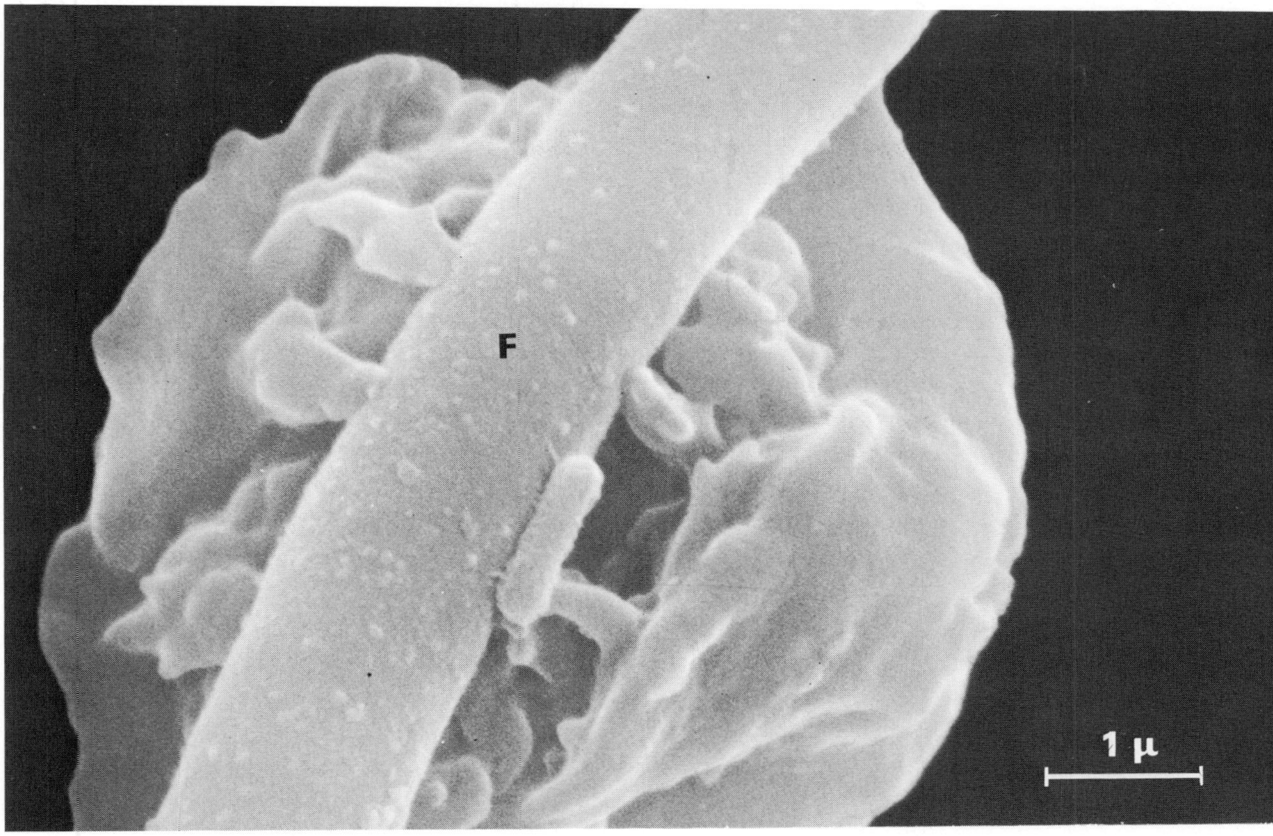

Figure 6.7. Phagocytosis of *Candida albicans* filament (J. Tokunaga).
A polymorphonuclear leukocyte interacting with a filament (F). Several pseudopodia are attaching to the fungus. (× 20,000)

marked effect in controlling the infection, since systemic mycoses, at least those caused by opportunistic fungi, are rare events compared with the frequency of fungemic situations. A most striking result of EM studies is the rarity of phagocytosed fungal elements in mycotic abscesses and granulomas.

Although some fungi are engulfed by phagocytes, the process of phagocytosis mostly is not complete, that means a real phagosome is not formed as it is with *in vitro* experiments.[168, 198, 199, 200, 201] More often fungus cells are surrounded by several host defence cells, even of different cell types.[199, 267] This seems to be the most common cellular reaction of the host against fungi in a mycotic focus, disregarding the fungus species causing the mycosis. This situation may lead to a giant cell formation with the fungus cell located in a phagosome-like organelle (for example, Paracoccidioides[44, 61, 84, 85]). Nevertheless, it is obvious that the nonphagocytosed fungal cells are undergoing a rapid destruction, and especially in a mycotic process with a strong inflammatory reaction mycoses are based on a rather low number of living fungal elements.[17, 198, 202] Even the immunized host who is able to eliminate pathogenic fungi rapidly in early stages of a mycotic infection does not display a higher phagocytic activity in mycotic tissue foci.[152]

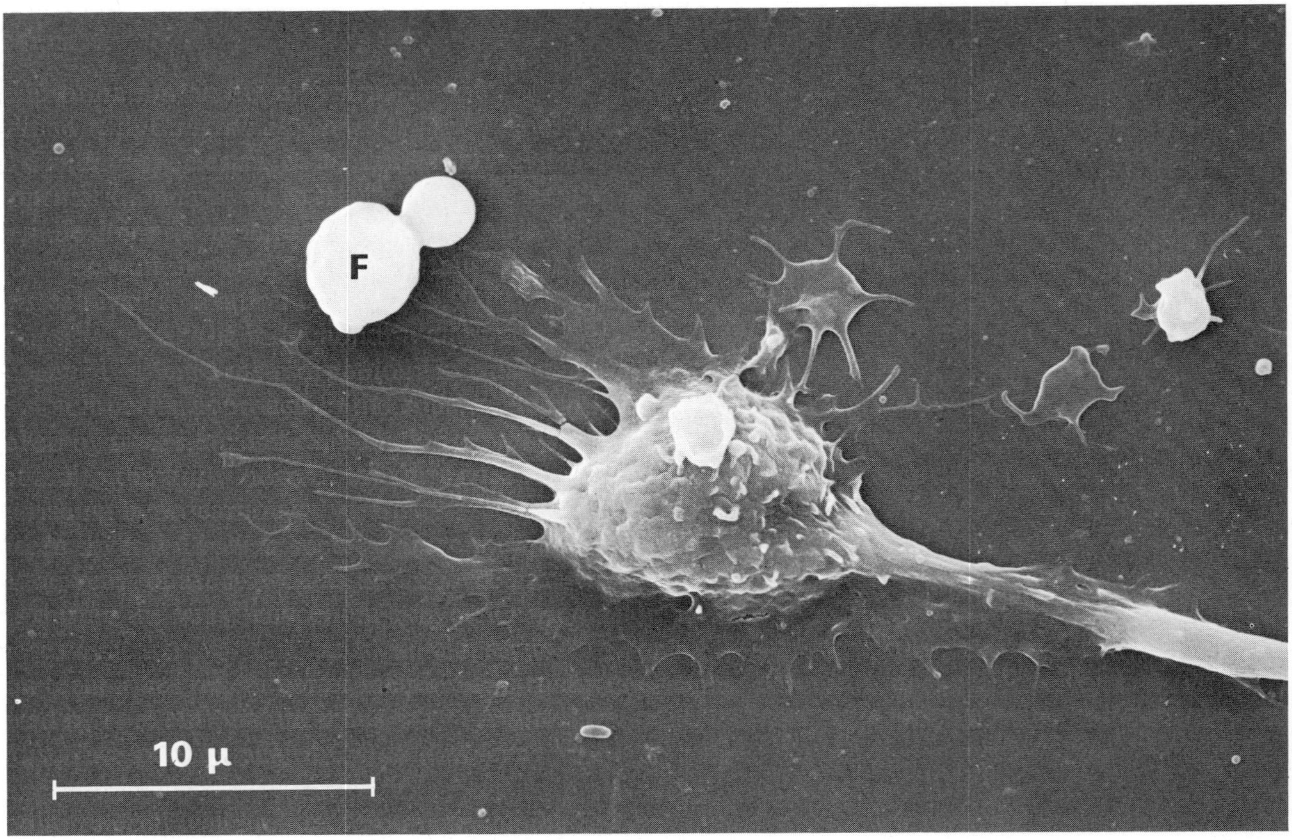

Figure 6.8. Chemotaxis in phagocytosis (J. Tokunaga).
Polymorphonuclear leukocyte approaching *Candida albicans* cells (F) in the course of phagocytosis *in vitro*. The numerous pseudopodia indicate the chemotactic attraction. (× 4,200)

The fact that phagocytosis is rare in mycotic foci, although the destruction of single fungus cells is obvious, poses the major question for the mechanism of that fungus destruction. Humoral principles such as lysozyme and the complement system do not seem to possess a remarkable fungicidal effect.[198] A more detailed analysis offers the interpretation that the cells of host defence can release lysosomal enzymes into the extracellular space in order to act against the nonphagocytosed fungi.[203] An aspect of a possible role of such free enzymes as a factor of pathogenicity is discussed in section 6.3.3.

EM studies have revealed a lot of information on morphological differences between fungi cultured *in vitro* and those found finally as parasites in the host. This touches the problems of dimorphism in its widest sense.[58] Generally, it can be said that the living conditions for the fungus in the host are not optimal in respect of physical and nutritional requirements, and that this situation is aggravated by host defence mechanisms.[51, 52, 53]

Consequently, morphological features can be observed which differ from cultured fungi and which are more or less characteristic for mycoses in general. The single fungus element is larger when located in the host, indicating that the

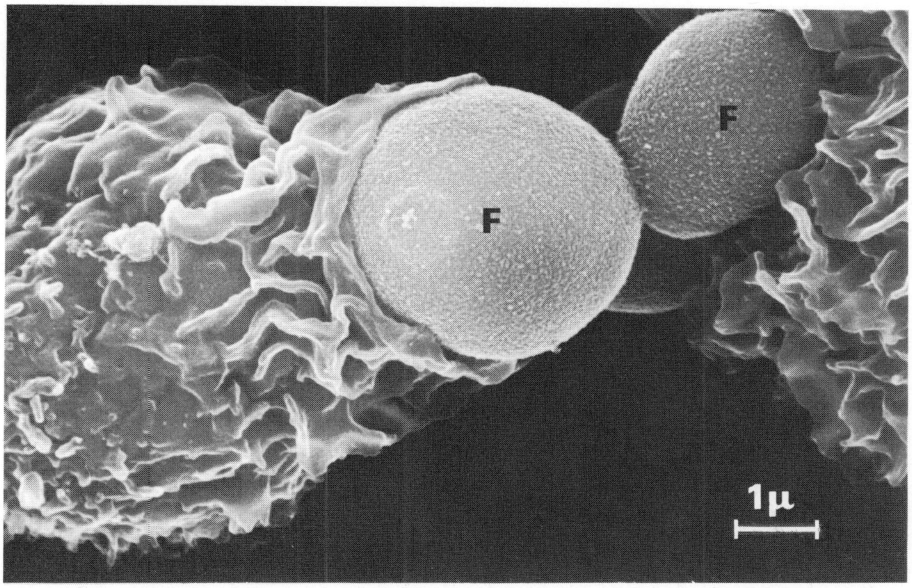

Figure 6.9. Phagocytosis of yeast fungi *in vitro* (T. Arai).
Engulfment of unspecifically opsonized *Candida albicans* cells (F) by rabbit alveolar macrophages. (× 10,000)

fungus cannot perform cell division in the least possible generation time.[284] Fungus filaments are not so straightened as in culture, giving the impression that they are searching in the host for the locus minoris resistentiae.[58] This is especially conspicuous with keratinophilic fungi colonizing the epidermis. Morphological characteristics which differ most notably from cultural features are found with fungi in immunized hosts.[151] Since the fungi are rapidly eliminated from these hosts EM investigations studying the conditions of immunity are still rare due to technical difficulties in demonstrating the fungi. The few fungal elements in immunized hosts seem to be blown up into grotesque shapes, sometimes making it difficult to identify these subjects as being fungi. The internal structures often show an enrichment of glycogen particles, of lipid droplets or vacuoles, if they are not already completely empty leaving only the resistant cell wall for the final host–fungus conflict,[105] which is especially impressive with the encapsulated cell wall of *Cryptococcus neoformans*.[49, 59] Generally, the host attack effective in nonimmune and immune hosts leads often to a better demonstrability of fungal elements than with cultured cells (which is most remarkable with *Candida albicans*[52, 199, 201, 202, 312, 313, 315, 317]).

Ultrastructural indications of interfering cell-mediated immunity are the appearance of activated lymphocytes and activated macrophages in the mycotic focus. Only two pioneer studies exist which discuss these situations on the background of recent advances in cellular immunology, the one case of dermatomycoses caused by keratinophilic fungi,[234, 238] the other case with a model of deep-seated mycoses caused by *Candida albicans*.[151] Experimental infections of animals with a defined status of immunity are essential hereto. It is surprising to learn that the basic mechanisms of cell-mediated immunity characterizing

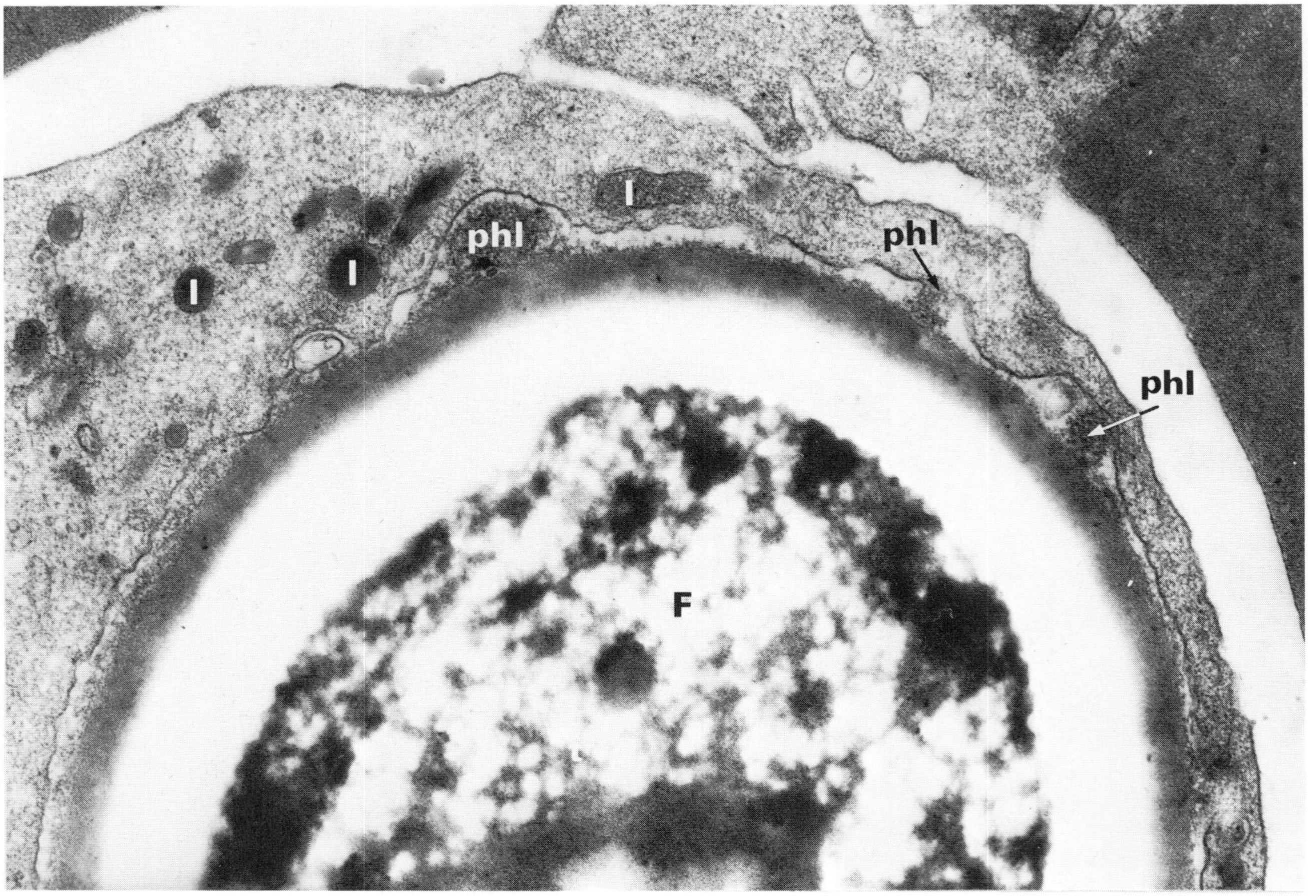

Figure 6.10. Phagocytosis of *Candida albicans in vitro* (J. Tokunaga).
The fungus cell (F) is incorporated in a phagosome of a polymorphonuclear leukocyte. Phagolysosomes (phl) have been fused with the phagosome, the transferred lysosomal material is in contact with the fungal cell wall surface. l: lysosomes located in the cytoplasm of the host cell. (× 40,000)

mycotic processes seem to be rather similar in the two types of mycoses examined which are so different in various clinical aspects. For dermatomycosis it has become clear that activated lymphocytes support the main part of the host reaction to the fungus, thus probably contributing lymphocytotoxic effects to the aspects of the disease in addition to the damage effected by the fungus itself (Figure 6.11). The macrophages seem to be of minor significance which can well explain the chronicity of dermatomycoses in humans. In contrast, the deep-seated *Candida* mycosis is well controlled by cellular immunity; activated macrophages containing numerous electron-dense lysosomes infiltrate the mycotic focus, discharging their lysosomal enzyme endowment for the rapid and effective decomposition of the fungal parasite (Figure 6.12). Activated lymphocytes seem to play a minor role in this situation. Plasmocytes are rare in mycotic lesions.

Concerning humoral immunity, it has become evident with immuno–electron microscopic studies that even high concentrations of circulating antibodies

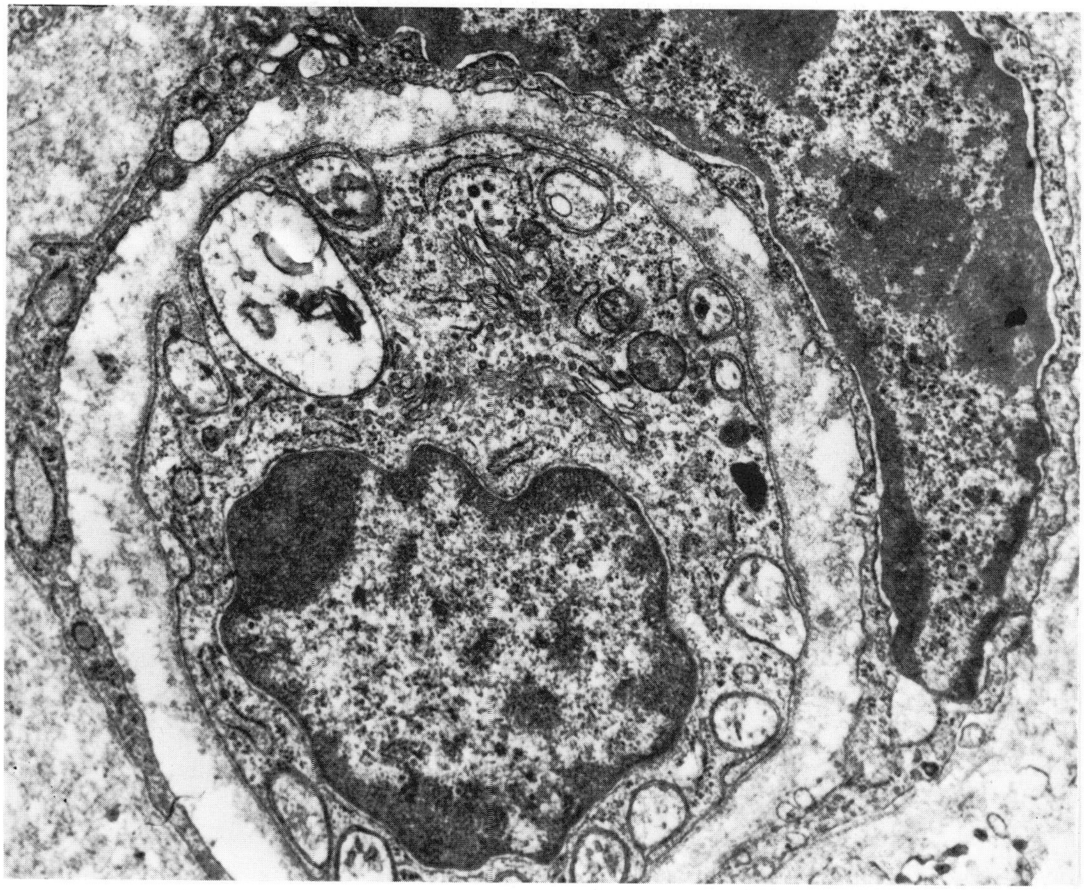

Figure 6.11. Activated lymphocyte (D. Poulain).
Host cell showing the characteristics of an activated lymphocyte located in a dermal capillary near the epidermis. Such cells are suspected of being able to effectuate a lyphocytotoxic action as a host contribution to a dermatomycosis. (× 11,000)

reacting with fungal cell constituents are not able to inhibit fungal actions to the host nor do they lead to fungal death mediated by complement or lysozyme.[198, 204, 208, 209]

Otherwise it is obvious that fungi exposed to the effects of the host are stimulated to form antigenic material in abundance on the outer cell wall surface which readily reacts with circulating antibodies.[180, 198, 208] These antigen–antibody complexes are able to fix complement[180] and (to a minor extent, if present in the focus) lysozyme. It has been made evident by immuno–electron microscopic investigations that such antigen–antibody (–complement) complexes are liberated in considerable amounts into the infected area (Figure 6.13).[180, 209] At the present state of knowledge it must be taken into account that these vagabondizing complexes could have a cytotoxic effect mediated by complement on host tissues. This would be another pathomechanism of fungi with contribution by the host.

Fungi maintaining chronic disease sometimes show conspicuous alterations on their outer surface in the form of thick depositions mostly with star-like

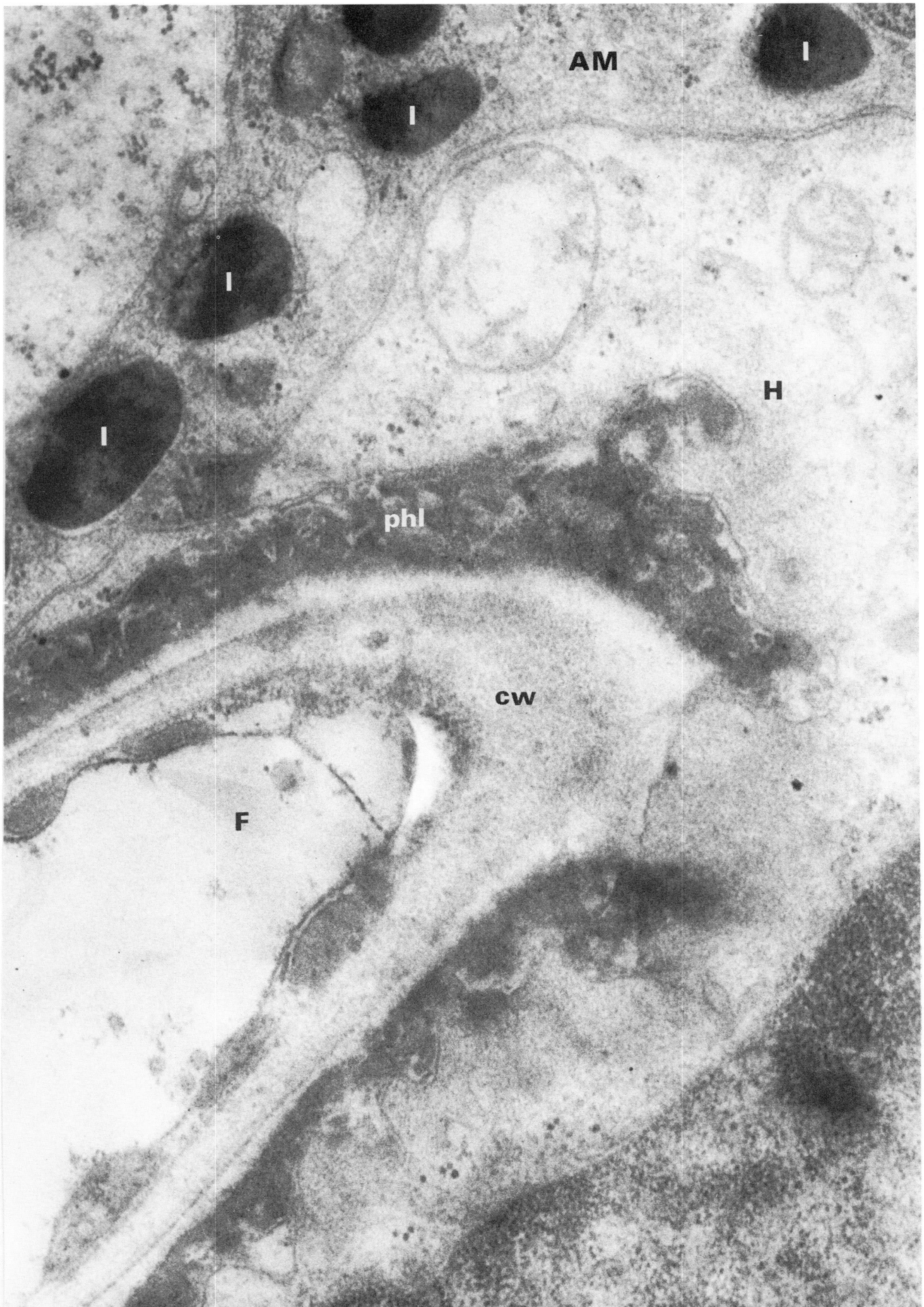
AM
I
I
I
I
I
H
phl
cw
F

Figure 6.12. Destruction of fungal filament in immunized host (E. S. Kuttin).
Decomposition of the filament is effected by enzymes released by activated macrophages. Experimental infection of immunized rabbits (kidney) with *Candida albicans.* F: fungal cell, cw: fungal cell wall, phl: phagolysosomal material, l: macrophagic lysosomes, AM: activated macrophage, H: histiocyte. (× 26,000)

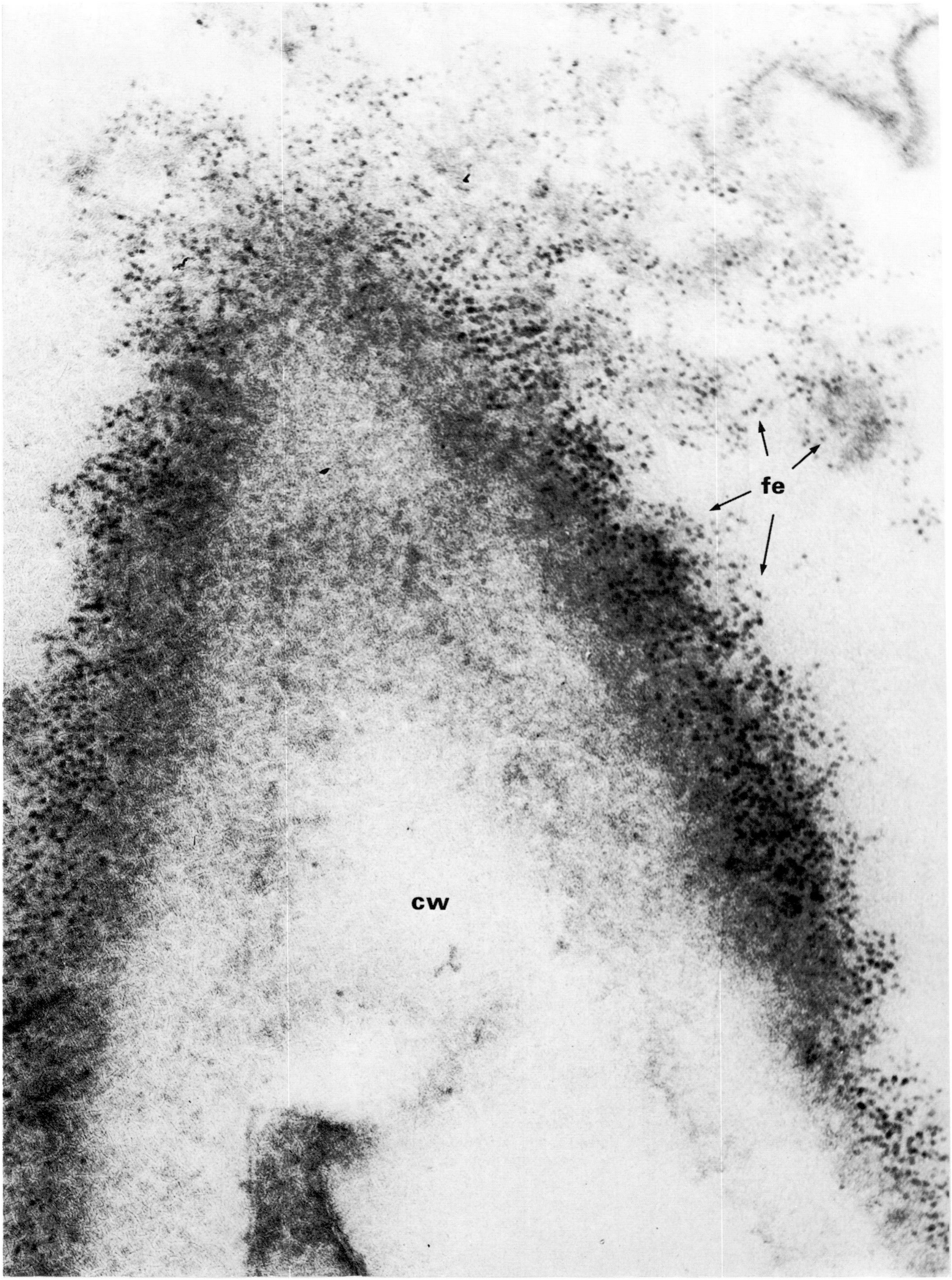
fe
cw

Figure 6.13. Demonstration of antigen–antibody complexes in *C. albicans* (H. Melching-Müller).

Candida albicans in vaginal scrapings incubated with ferritin-labeled anti-human IgA. The pattern of the ferritin molecules (fe: dark points) indicates that antigen-antibody complexes are liberated from the fungal cell wall (cw) in large amounts into the infected area. (× 160,000)

indentations. Such phenomena have been described by light microscopic investigations as "asteroid bodies;" EM studies on this subject are rare.[24] The ultrastructure of asteroid bodies was first studied with *Sporothrix schenckii*;[163, 165] these authors discussed the possible nature of the asteroid bodies. Since they could induce the asteroid body formation by incubating the fungi in an anti-*Sporothrix* serum they assumed that the asteroid body might be related to immune reactions of the host. Later immuno-electron microscopic studies with *Candida albicans* demonstrated that the asteroid bodies of these fungi are formed by complexes of fungal cell wall antigens and host antibodies representing an antigen–antibody precipitate *in vivo*, not released in the form of free antigen–antibody complexes into the host tissue[207] (Figure 6.14). Otherwise the asteroid body formation could not be induced with *Candida* by incubation in specific antisera.[210] Further studies revealed that antibody-coated fungus cells are quite common in mycoses with chronic character, and that the asteroid bodies, evident in light microscopic observations, are merely the endpoint in a scale of variable dimensions of antigen–antibody layers.[180, 205, 206, 208] Checking previous publications shows that several authors have found such electron-dense layers on fungal cells *in vivo* without discussing a relationship to the asteroid body phenomenon.[34, 52, 73, 186, 188, 271, 273, 313, 315, 322] An ultrastructural study on asteroid body formation with *Aspergillus* in human liver supports the request for ultrahistochemical examinations besides immuno-electron microscopic methods, in order to answer the question whether enzyme proteins could build up asteroid body substance also[71] (Figure 6.15). Up to now the question of whether the asteroid body formation protects or attacks the fungus or the host, respectively, must remain unanswered.

6.3.3. Evidence for pathogenetic concepts in mycoses

Reviewing the available ultrastructural information on morphological features and, as far as possible to date, on physiological processes in mycoses, it seems imperative to outline conceptions of pathomechanisms effective in mycotic diseases. Such ideas are hypothetical in character, and their legitimation is based on the resulting possibility of pioneering further productive experimental pathways in the electron microscopy of fungal diseases.

The acute, primary mycotic infection (of a nonimmunized host) seems to be based:

On the part of the fungus:

1. On mechanical power as far as filamentous fungi are concerned.
2. To a minor extent on toxic constitutents which may damage locally the host tissue.

Figure 6.14. Asteroid bodies (J. Müller).
Candida albicans cells incubated with ferritin-labeled anti-human IgA in the urine of a *Candida* nephritis patient. The electron-transparent part of the cell wall is covered with an electron-opaque layer consisting of fungal cell wall antigens and host antibodies. icw: internal cell wall layer, cpm: cytoplasmic membrane, fe: ferritin molecules marking the sites of antibodies. (× 150,000)

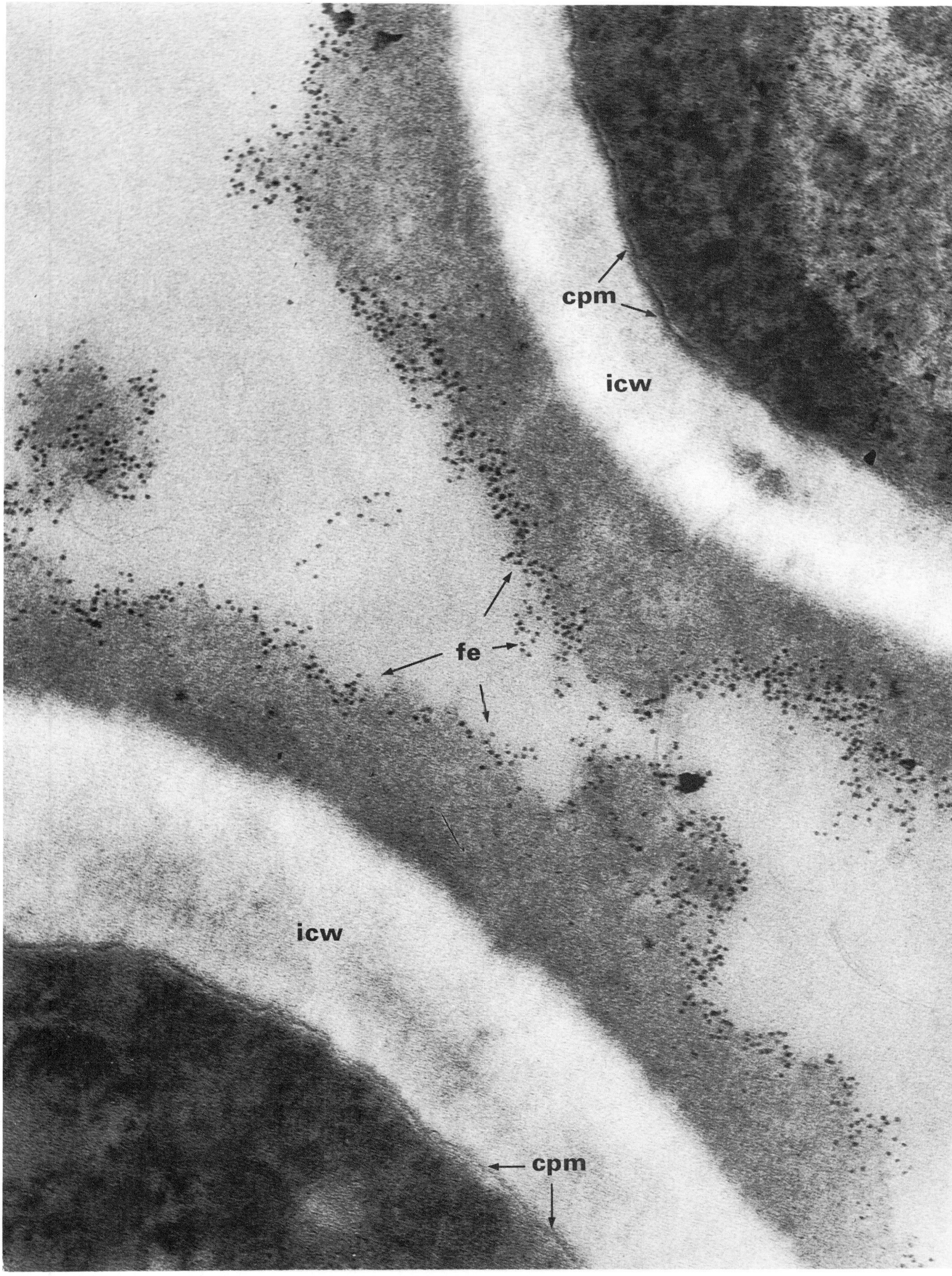
cpm
icw
fe
icw
cpm

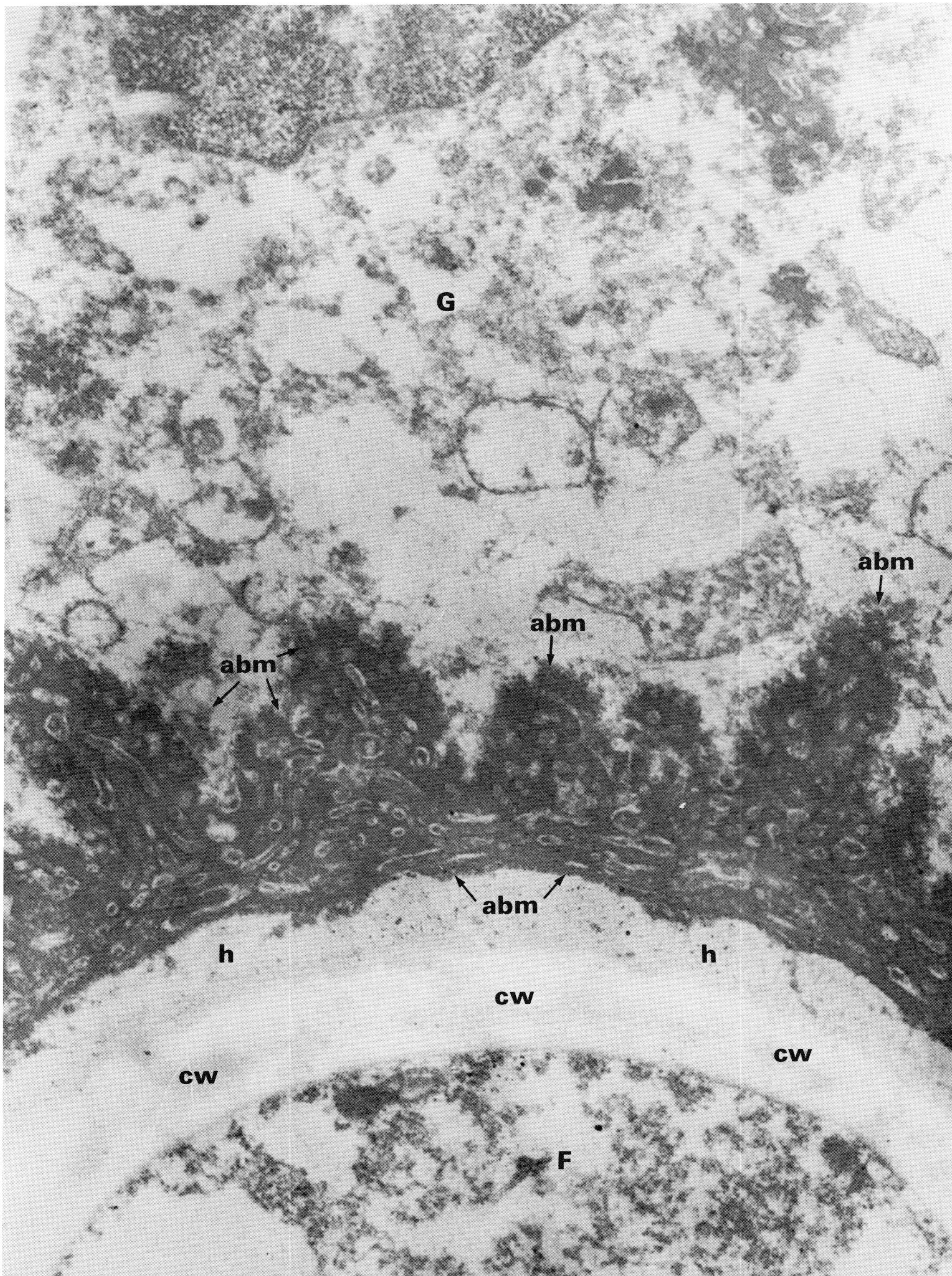
G
abm
abm
abm
abm
h
h
cw
cw
cw
F

3. On the endowment with hydrolytic enzymes.
4. On rapid propagation under the ecological conditions of the infection site in order to counteract the host defence.

On the part of the host:
1. On the release of hydrolytic enzymes (and other compounds?) from the cells of the host defence system into the host tissue due to the fact that the large plant cells do not undergo an effective phagocytosis which alone could protect the host tissues from being destroyed by its own enzymatic activities.

In this sense the acute, primary mycotic infection can be interpreted as a special form of a foreign body reaction with the peculiarity that the "foreign body" is selfproducible and that action of the fungus and reaction of the host thus become permanent.

In contrast, the well-immunized host is able to control a mycotic infection by the formation of activated macrophages which rapidly wall-in the fungus elements, destroying them by enzymes of lysosomal origin especially adapted to the decomposition of fungus constitutents.

The interpretation of the chronic mycotic disease is difficult. The most important results of electron microscopic studies are the following observations.

1. The fungi are stimulated to form antigenic cell wall material in excess.
2. Circulating antibodies react with cell wall antigens to a large extent without being able to neutralize the antigenic substance.
3. Antigen–antibody complexes are released into the infection area; at least partially complement factors are fixed and possibly lytic properties against host cell membranes are activated (pathomechanism I).
4. Activated lymphocytes infiltrate the infection site; lymphocytotoxic activities must be regarded as another possibility of damaging the host tissue (pathomechanism II).
5. The role of mast cells in fungal infections with the possibility of releasing suddenly large amounts of histamine, serotonin, etc., with the symptoms of delayed hypersensitivity has not yet been examined by EM studies. Clinical features of mycotic disease lead to the assumption that mast cells likewise may contribute to such syndromes.

Electron microscopy always struggles with problems of sampling. Each interpretation of an ultrastructural situation, even if correct in all details, is valid for the examined site at a distinct moment only. A mycotic disease in its spatial and chronological entirety must always be considered as a permanently changing mixture of the above mentioned basic concepts with altering importance of the one or the other side.

Figure 6.15. Asteroid bodies in *Aspergillus* (H. Drexler).
Human liver containing few fungal elements (probably *Aspergillus*) covered with electron-dense vesicular material. F: fungal cell. cw:fungal cell wall, h: halo surrounding the fungal cell wall, abm: vesicular asteroid body material, G: neutrophil granulocyte showing signs of damage. (× 150,000)

6.4. Action of drugs on fine structure of fungi

Since the main target of medical mycology is the successful treatment of mycotic diseases, it is of great interest to investigate the ultrastructural changes under drug influence. The mechanisms of action of the antimycotics are well known. This offers the possibility of analyzing the drug-induced changes against the background of fungal metabolism or function elements.

6.4.1. Griseofulvin

Studies on the ultrastructural changes caused by griseofulvin in keratinophilic fungi are few. The hyphal cells of *Trichophyton rubrum* and *T. mentagrophytes* become pleomorphic, round, and swollen, especially the apices. The most obvious intracytoplasmic change is the accumulation of lipid droplets. The cell organelles are no longer identifiable, while the cytoplasmic membrane persists; the outer layer of the cell wall becomes granular.[26, 298] In *Microsporum canis* similar changes are observed; additionally, the cytoplasmic vacuolization is conspicuous. The macroconidia remain unchanged, thus proving to be more resistant to griseofulvin than the vegetative mycelium, although with SEM the disappearance of the riblike surface structures on the macroconidia can be shown.[62]

6.4.2. Polyene antimycotics

These drugs form complexes with membrane sterol. Thus, they attack the cytoplasmic membrane which loses its semipermeability; this is followed by a rapid release of important soluble cell constituents.

Amphotericin B

The primary attack of the drug on the cytoplasmic membrane could not be observed in early studies with *Candida albicans* exposed *in vitro* to fungicidal doses of amphotericin B and filipin. Similarly, the destruction of other cell organelles was not observed.[87, 155] The application of amphotericin B and hamycin on the yeast phase of *Histoplasma capsulatum* and *Blastomyces dermatitidis in vitro* leads to irregularities of the cytoplasmic membrane and the cell wall, obviously in connection with lomasomes. This damage results in rapid intracellular degeneration, plasmolysis, and ultimate death of the cell.[159] An *in vivo* assay allows the study of the effect of amphotericin B on the mycotic

Figure 6.16. Effect of amphotericin B on *Candida albicans* (J. Müller).
Infected mouse under treatment. *Candida* blastospore (F) engulfed by an eosinophilic granulocyte (G). The fungal cytoplasmic membrane (cpm) is destroyed, fungus cell organelles have disappeared. The fungus cell wall (cw) is persisting, but clearly stratified. Prevailing stratification in the innermost cell wall layers indicates that the fungus decomposition is mainly caused by autolysis following the cell death due to amphotericin B than by host cell actions. The amphotericin B effect is restricted to the destruction of the cytoplasmic membrane. (× 20,000)

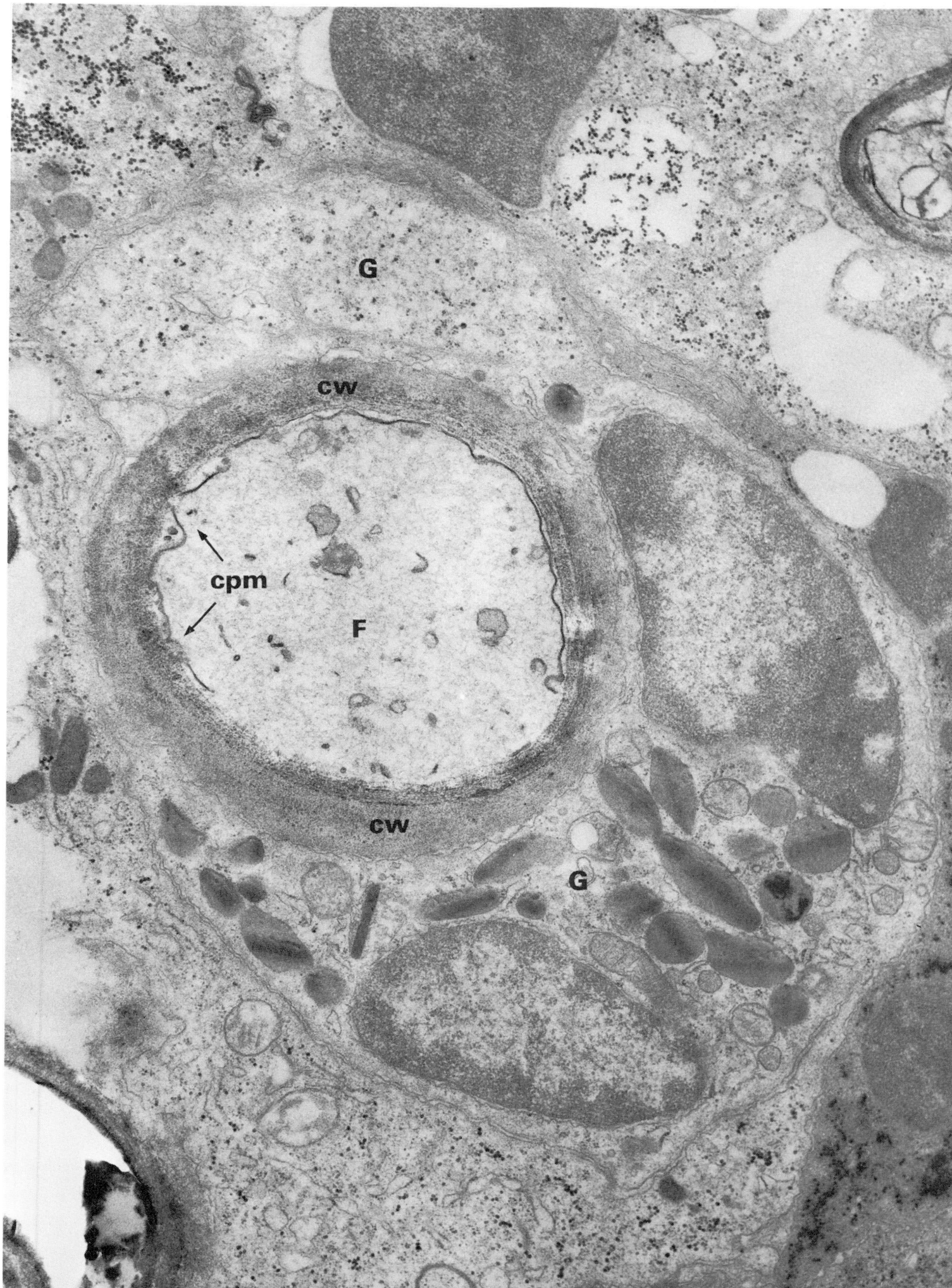
G
cw
cpm
F
cw
G

nephritis of *Candida albicans*-infected mice. The fungi are hindered from forming fungal abscesses. The drug-induced destruction of the cytoplasmic membrane is recognizable simultaneously with the damage of other fungus organelles. The cell wall is the most stable part of the fungus;[200] its decomposition leads to a stratification due to the deaggregation of the cell wall substances (Figure 6.16).[199]

Nystatin

Nystatin-induced alterations in fungi resemble those caused by amphotericin B.[102, 155] SEM studies prove that nystatin effects deep depressions on the fungal surface which must be combined with an increasing flexibility of the normally very rigid cell wall.[9, 79, 250] This effect is specific for nystatin among other polyens and cannot be explained sufficiently by the well-known action of the polyen compounds.

6.4.3. Imidazole antimycotics

The mechanism of action of the imidazole compounds is complex. The permeability of the cell envelope is altered, the drugs interfere with the RNA and protein synthesis, the influence on the lipid metabolism and on the intracytoplasmic membrane systems seem to be late stages of drug effects.

Clotrimazole

This was the first drug of the group which has been investigated for its effects on the ultrastructure of *Candida albicans. In vitro* the integrity of the nuclear membrane is quickly lost, the cytoplasmic membrane is damaged.[134, 136, 327, 329] *Candida* cells in clotrimazole-treated mice reveal a progressive disintegration of the cytoplasmic structures with rapidly increasing vacuolization. The number of ribosomes decreases. The cytoplasmic membrane becomes more distinct and invaginated, and later disintegrates. The area between the cytoplasmic membrane and the cell wall contains an increasing number of lomasomes. The final stage shows electron-dense deposits in the cell wall (Figures 6.17, 6.18). The damaged *Candida* cells are rapidly removed from the infected tissue by phagocytosing cells. [311, 312, 316] Corresponding observations are made with human vaginal infections by yeasts,[312, 313, 315, 317] in skin infections caused by keratinophilic dermatophytes,[312, 315] as well as in *Aspergillus fumigatus*-infected and clotrimazole-treated mice.[314]

Figure 6.17. Effect of clotrimazole on *Trichophyton mentagrophytes* (W.-H. Voigt). After 96 hours at a concentration of 0·01 mg *in vitro*, the main changes are irregular undulating cytoplasmic membrane with increased number of lomasomes (lo), vacuoles (v), and swollen mitochondria (m), and a decreased number of ribosomes. cw: cell wall. (× 60,000)

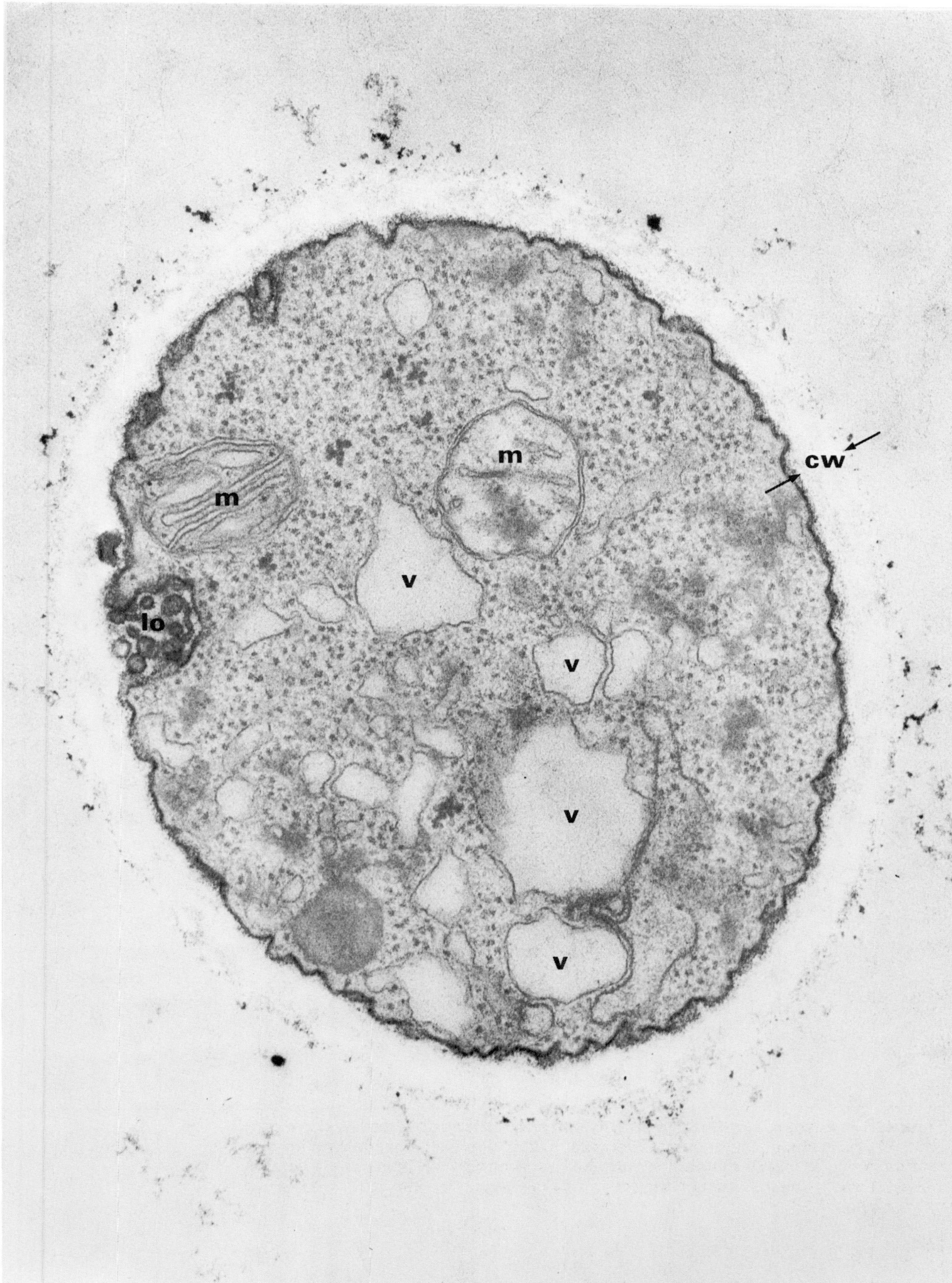
m
m
cw
v
lo
v
v
v

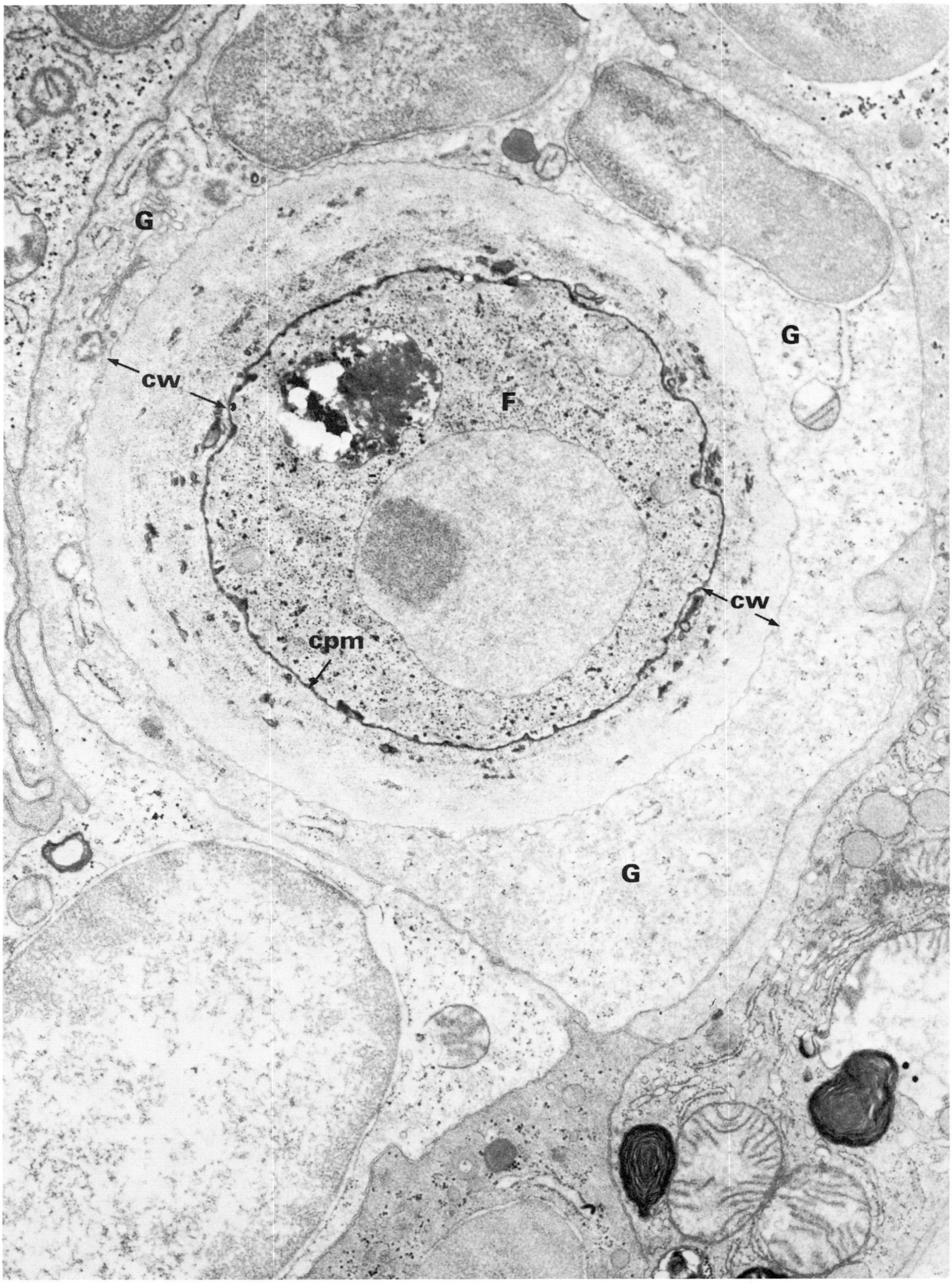
G
G
cw
F
cw
cpm
G

Miconazole

The effect of miconazole has been studied with *Candida albicans.*[63, 160] The changes observed are similar to those caused by other imidazoles. *Candida albicans* cells in SEM assays present rough surface structures and multiple disorientated buds and bud scars, and they form small clusters of interconnected cells. Vesicular structures on the cell surface represent cytoplasmic remnants derived from broken cells which can be confirmed by parallel TEM.[64] The morphological changes in *Candida albicans* cells treated with miconazole or clotrimazole, in high dose levels, differ markedly with respect to the number of cells involved, the intensity of lesions, and their subcellular location.[65] Additional cytochemical studies concerning some oxidative and perioxidative enzymes lead to an alternative hypothesis of action mechanisms of miconazole, proposing hydrogen peroxide intoxication as the cause of fungus death (Figure 6.19(a)–(c)).[28, 65]

Econazole

Studies on the effect of econazole emphasize the differences in toxicity directed against keratinophilic fungi on the one hand[103, 242] and human pathogenic yeasts on the other hand.[243] The ultrastructural manifestations of the econazole effect resemble those caused by other imidazole compounds. Special interest is devoted to the impairment of the lipid metabolism,[242, 243, 245] to cell wall changes,[247] and to the drug influence on the internal architecture of cytomembranes investigated by freeze-etching methods (Figure 6.20(a),(b)).[246] *Candida* cell wall fissures and the formation of cell debris after econazole treatment have been proved by SEM.[9] The ultrastructural changes in *Pityrosporum orbiculare*, the causative agent of pityriasis versicolor, under econazole treatment have been investigated *in vitro* as well as in specimens of infected human skin.[263]

6.4.4. 5-Fluorocytosine

This compound is actively transported into the fungus cell by a permease, is deaminated to 5-fluorouracil which is incorporated into the RNA where it replaces the uracil, thus interfering with the protein synthesis. A second drug effect is the inhibition of thymidylatsynthetase which leads to a reduction of the DNA synthesis.

In *Cryptococcus neoformans* 5-FC causes a higher electron density of the cytoplasm; the ribosome-bearing membranes around the mitochondria disappear; the differentiated nuclear structures are replaced by a uniform fine granular

Figure 6.18. Effect of clotrimazole on *Candida albicans* (W.-H. Voigt).
After four oral applications of 100 mg clotrimazole per kg to a mouse experimentally infected with *Candida albicans* (F) main changes in the *Candida* cell engulfed by granulocytes (G) are irregular undulating cytoplasmic membrane (cpm), extremely enlarged cell wall with depositions of contrasted material, and electron-dense substances in the cytoplasm. cw: cell wall. (× 24,000)

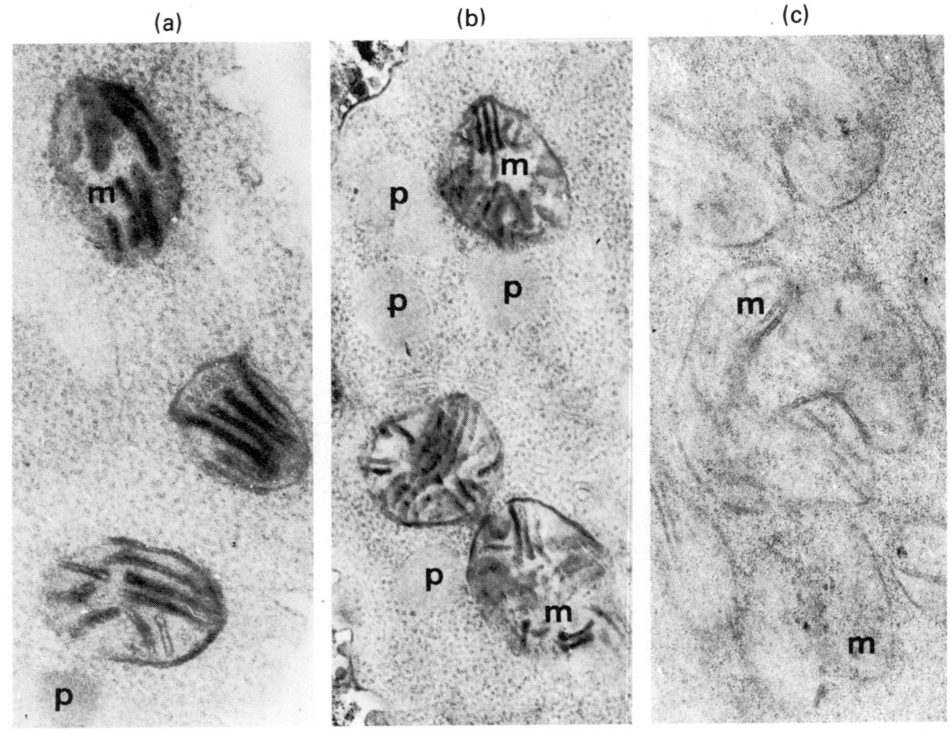

Figure 6.19. Effect of miconazole on the activity of cytochrome-c-peroxidase in *Candida albicans in vitro* (M. Borgers).

(a) Control cell. The cristae as well as the limiting membrane of the mitochondria (m) are strongly reactive. Note the absence of reaction product in the peroxisome (p). (× 55,350)

(b) Exposure to 10^{-6} M miconazole for 24 hours. The activity is still localized on the mitochondrial christae (m), but it is already slightly reduced as compared with that in the previous picture. p: peroxisome. (× 35,600)

(c) Exposure to 10^{-5} M miconazole for 24 hours. The mitochondria (m) are devoid of reaction product. (× 47,500)

nucleoplasm.[166] In *Candida albicans* and *Saccharomyces cerevisiae* the cells are enlarged and the cell wall becomes thinner under drug influence. The changes in nuclear structures resemble those caused by DNA synthesis inhibitors. The thinner cell wall of 5–FC-treated *Candida* cells may reflect an impaired thymidylate synthesis (Figure 6.21).[13] The metabolic breakdown finally results in the formation of cytoplasmic lipid drops.[214, 311] *Candida* strains sensitive to 5–FC possess a cell wall layer of high electron density which disappears after treatment with trypsin and seems to lead to a reversible resistance to 5–FC.[72, 189, 190] The investigation of specimens from two patients with

Figure 6.20. Effect of econazole on *Candida albicans in vitro* (H.-J. Preusser).
Influence on the cytoplasmic membrane, demonstrated with freeze-etching technique.

(a) P face of the *Candida* plasmalemma showing a surface which is tightly covered with globular protein particles. Numerous membrane invaginations appear on the convex P face. (× 46,000)

(b) P face after 25 hour treatment with 10 ppm econazole nitrate. The globular proteins on the P face are decreased in number; the remaining particles are arranged in groups partly forming a hexagonal pattern. (× 60,000)

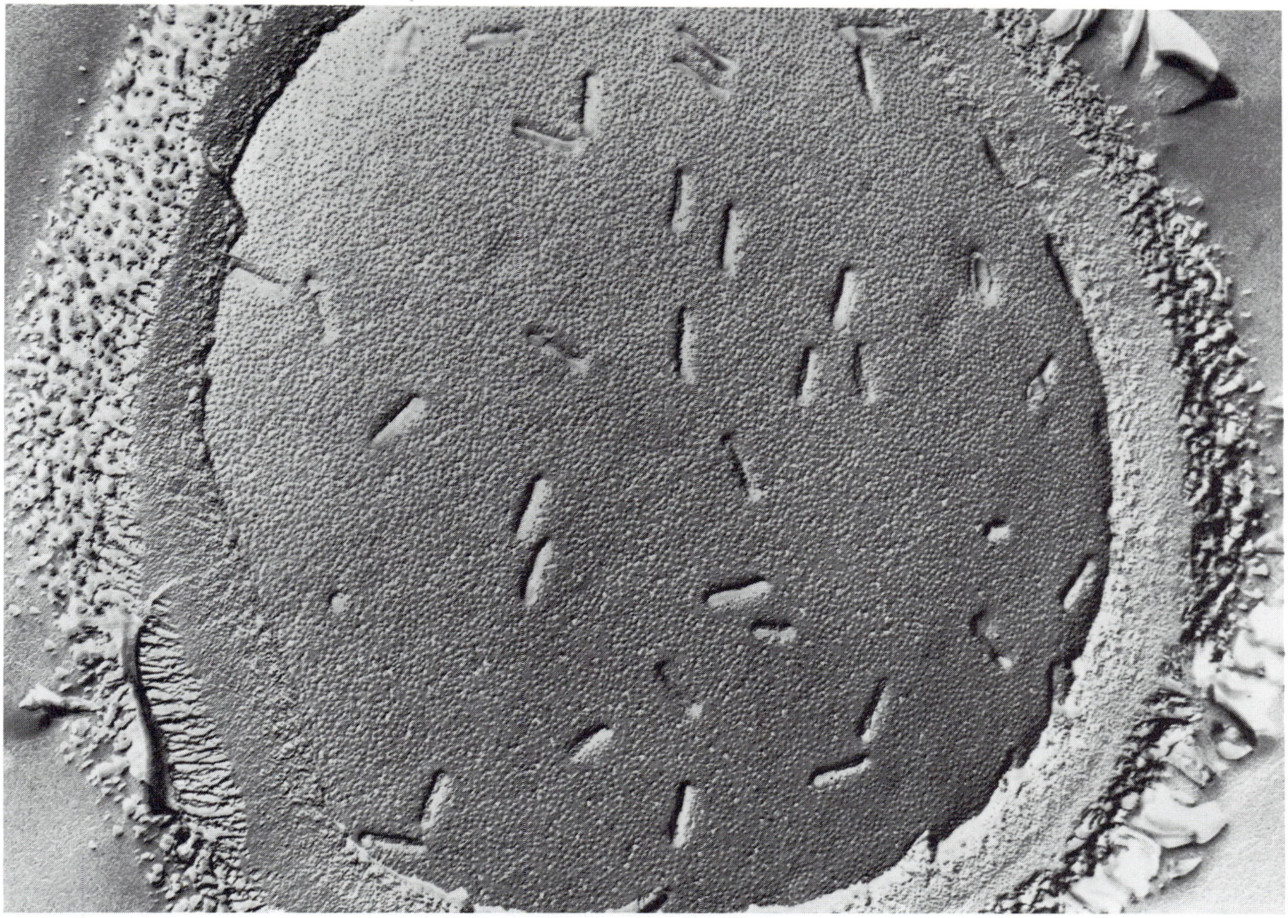
(a)

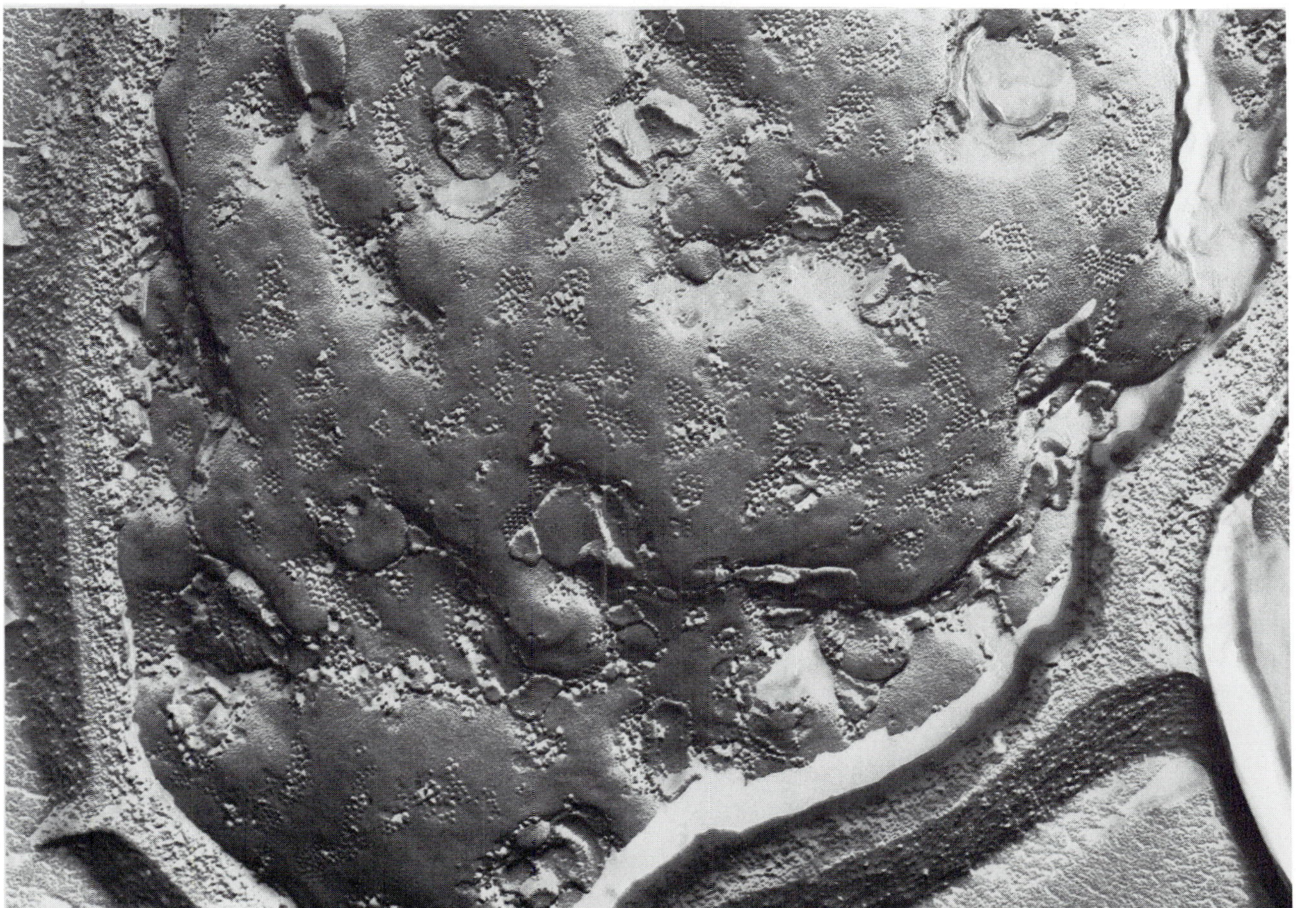
(b)

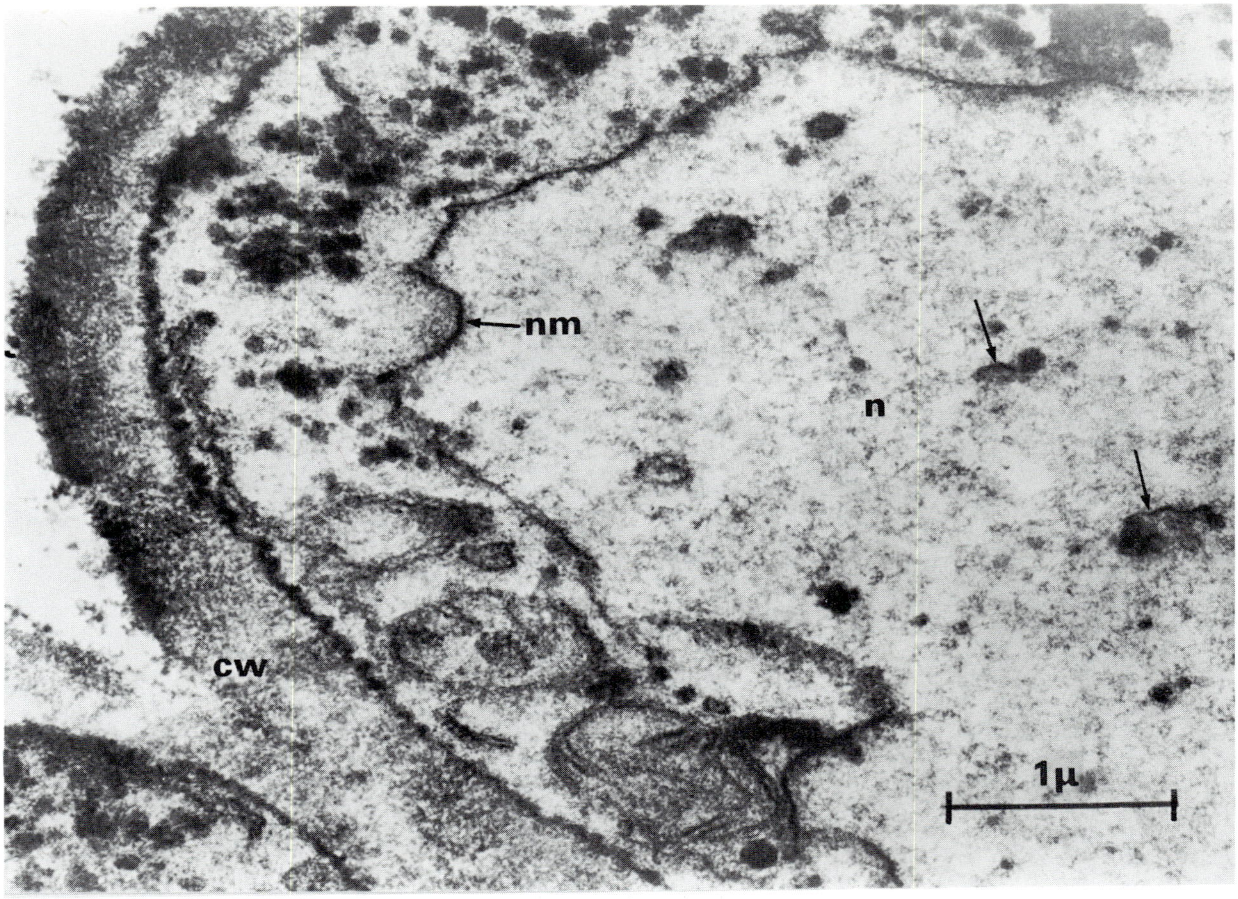

Figure 6.21. Effect of 5-fluorocytosine on *Candida albicans in vitro* (T. Arai).
After 12 hours exposure *in vitro* to a concentration of 10 µg 5-fluorocytosine per ml, the cell wall (cw) becomes very thin and the nucleus (n) is enlarged and shows a fine filamentous structure (arrows). The features of the changes indicate the inhibition of the DNA metabolism by the drug. nm: nuclear membrane. (× 30,000)

5–FC-treated chromomycosis (one caused by *Phialophora dermatitidis*, the other by *Phialophora verrucosa*) proves the rapid destruction of the fungus showing shrunk cytoplasma, swollen mitochondria, and a plasmalemma taken off from the cell wall.[305]

6.4.5. Other compounds

The effect of pyrrolnitrin on the dermatophyte *Microsporum audouinii* has been examined by TEM and SEM.[45] The cytology of *Candida albicans*, as influenced by benzalkonium chloride and filipin,[87] sodium caprylate[1, 318] and thiobenzoate,[88] has been explored in early studies. Observation of the effect of 8-methoxypsoralen, combined with ultraviolet irradiation to *Pityrosporum orbiculare in vitro*, displays early damage to mitochondrial structures and a later formation of irregular cytoplasmic membrane systems.[262] Other drugs have been used in a series of studies due to their well-known action mechanism for the proof of certain metabolic pathways in some fungi.[226, 228, 229]

References to Part Six

1. Adams, J. N., Painter, B. G., and Payne, W. J., "Effect of sodium caprylate on *Candida albicans.* I. Influence of concentration on ultrastructure," *J. Bact.*, **86**, 748, 1963.
2. Adler, J. P., "Screening for viruses in human pathogenic fungi,"*12th International Congress of Microbiology, Munich, 1972.*
3. Agar. H. D. and Douglas, H. C., "Studies of budding and cell wall structure of yeast," *J. Bact.*, **70**, 427, 1955.
4. Ainsworth, G. C. and Sussman, A. S., *The Fungi. An Advanced Treatise*, vol. 1, *The Fungal Cell*, Academic Press, New York, 1965.
5. Akisada, T., Tokunaga, J., and Kobayashi, M., "Studies on the structure of the cell of yeast like fungi by electron microscopy. 1. On the fine structure of chlamydospore of *Candida albicans* and *Candida stellatoidea*," *Jap. J. Bact.*, **21**, 1, 1966.
6. Albouy, J., "Morphological changes in fungal ultrastructure caused by fungal viruses," *12th International Congress of Microbiology, Munich, 1978.*
7. Al-Doory, Y., "The ultrastructure of *Cryptococcus neoformans*," *Sabouraudia*, **9**, 113, 1971.
8. ——— and Baker, C. A., "Comparative observations of ultrastructure of five species of *Candida*," *Mycopath. Mycol. appl.*, **44**, 355, 1971.
9. Ånsehn, S. and Winblad, B., "Surface morphology of *Candida albicans* after treatment with antimycotic drugs," *Mykosen*, suppl. 1, 322, 1978.
10. ———, Schönebeck, J., and Winblad, B., "Human and experimental renal candidosis. A light and scanning electron microscopic study," *6th International Congress ISHAM, Tokyo, 1975.*
11. Arai, T. and Mikami, Y., "Cellular and humoral defense mechanisma of *Candida albicans* infection," in K. Iwata (ed.), *Yeasts and Yeast-like Microorganisms in Medical Science, Proceedings 2nd International Spec. Symposium on Yeasts*, University of Tokyo Press, Tokyo, 1976, pp. 191–200.
12. ———, ———, and Yokoyama, K., "Phagocytosis of *Candida albicans* by rabbit alveolar macrophages and guinea pig neutrophils," *Sabouraudia*, **15**, 171, 1977.
13. ———, ———, ———, Kawata, T., and Masuda, K., "Morphological changes in yeasts as a result of the action of 5-fluorocytosine," *Antimicrob. Agents Chemother.*, **12**, 255, 1977.
14. Arnold, W., Garrison, R. G., and Boyd, K. S., "Periplasmic structure in *Saccharomyxes rouxii* (Boutroux), an osmophil," *Microbiology*, **28**, 1047, 1974.

15. Aronson, J. M., "The cell wall," in G. C. Ainsworth and A. S. Sussman (eds.), *The Fungi*, vol. 1, Academic Press, New York, 1965.
16. Avram, A. and Grupper, C., "Aspects tridimensionels des champignons dans les tissues et en culture," *6th International Congress ISHAM, Tokyo, 1975.*
17. Bader, G., *Die viszeralen Mykosen*, Fischer, Jena, 1965.
18. Baker, R. D., "The pathologic anatomy of mycoses. Human infection with fungi, actinomycetes and algae," in E. Uehlinger (ed.), *Handbuch der speziellen pathologischen Anatomie und Histologie*, Band III, *Atmungswege und Lungen*, Teil 5, Springer, Berlin, New York, 1971.
19. Bartholomew, J. W. and Mittwer, T., "Demonstration of yeast bud scars with the electron microscope," *J. Bact.*, **65**, 272, 1953.
20. Bastide, M., Bastide, J.-M., Morin, O., Miègeville, M., Hadibi, E., and Vermeil, C., "Etude morphologique des protoplastes de diverses espèces appartenaut au genre *Cryptococcus*," *Bull. Soc. fr., Mycol. Méd.*, **6**, 53, 1977.
21. Baxter, M. and Mann, P. R., "Electron microscopic studies of the invasion of human hair *in vitro* by three keratinophilic fungi, *Sabouraudia*, **7**, 33, 1969.
22. Beckett, A., Heath, I. B., and McLaughlin, D. J., *An Atlas of Fungal Ultrastructure*, Longman, London, 1974.
23. Belcher, R. W., Carney, J. F., and Monahan, F. G., "An electron microscopic study of phagocytosis of *Candida albicans* by polymorphonuclear leucocytes," *Lab. Invest.*, **29**, 620, 1973.
24. Berge, T. and Kaplan, W., "Systemic candidiasis with asteroid body formation," *Sabouraudia*, **5**, 310, 1966.
25. Biguet, J., *Lungenaspergillose*, Film, Lille, 1971.
26. Blank, H., Taplin, D., and Roth, F. J., Jr., "Electron microscopic observation of the effects of griseofulvin on dermatophytes," *Arch. Derm.*, **81**, 667, 1960.
27. Borgers, M. and De Nollin, S., "The preservation of subcellular organelles of *Candida albicans* with conventional fixatives," *J. Cell Biol.*, **62**, 574, 1974.
28. ——— and ———, "Ultrahistochemistry of the fungal cell," *12th International Congress of Microbiology, Munich, 1978.*
29. Bossanyi, G. and Olah, G. M., "Activité phophatasique acide en relation avec la pénétration intracellulaire du parasite *Puccina graminis* Var. *tritici* dans la cellule du blé," *Mycopath. Mycol. appl.*, **54**, 161, 1974.
30. Bracker, C. E., "Ultrastructure of fungi," *A. Rev. Phytopathol.*, **5**, 343, 1967.
31. ———, "Structure and transformation of chitin synthetase particles (chitosomes) during microfibril synthesis *in vitro*," *Proc. natn. Acad. Sci. U.S.A.*, **73**, 4570, 1976.
32. Breathnach, A. S., Gross, M., and Martin, B., "Freeze-fracture replication of cultured *Pityrosporum orbiculare*," *Sabouraudia*, **14**, 105, 1976.
33. Breslau, A. M., Hensley, T. J., and Erickson, J. O., "Electron microscopy of cultured spherules of *Coccidioides immitis*," *J. biophys. biochem. Cytol.*, **9**, 627, 1961.

34. Campbell, C. K., Electron microscopy of aspergillosis in fowl chicks," *Sabouraudia*, **8**, 133, 1970.
35. ———, Fine structure and physiology of conidial germination in *Aspergillus fumigatus*," *Trans. Br. mycol. Soc.*, **57**, 393, 1971.
36. Carbonell, L. M., "Ultraestructura y gemación comparativa del *Paracoccidioides brasiliensis* y del *Blastomyces dermatitidis*," *Acta cient. venez.*, **3**, 277, 1967.
37. ———, "Cell wall changes during the budding process of *Paracoccidioides brasiliensis* and *Blastomyces dermatitidis*," *J. Bact.*, **94**, 213, 1967.
38. ———, "Ultrastructure of human pathogenic fungi and their mycoses," in E. Uehlinger (ed.), *Handbuch der speziellen pathologischen, Band III, Anatomie und Histologie*, Ţeil 5; R. D. Baker (ed.), *The Pathologic Anatomy of Mycoses*, Springer, Berlin, New York, 1971.
39. ——— and Angulo, O. A., "Ultraestructura del *Histoplasma capsulatum* en histoplasmomas," *Revta. Lat.-am. Anat. Patol.*, **5**, 26, 1961.
40. ——— and Berliner, M., "Regeneration of protoplasts of *Histoplasma capsulatum.* An electron microscopy study," *Proceedings 5th International Congress ISHAM, Paris, 1971*, p. 15.
41. ——— and Pollak, L., "Ultraestructura del *Paracoccidiodes brasiliensis* en cultivas," *Revta. Lat.-am. Anat. patol.*, **5**, 26, 1961.
42. ——— and ———, "'Myelin figures' in yeast culture of *Paracoccidioides brasiliensis*," *J. Bact.*, **83**, 1356, 1962.
43. ——— and ———, "Ultraestructura del *Paracoccidioides brasiliensis* en cultivos de la fase levaduriforme," *Mycopathologia*, **19**, 184, 1963.
44. ——— and Rodriguez, J., "Transformation of mycelial and yeast forms of *Paracoccidioides brasiliensis* in cultures and in experimental inoculations," *J. Bact.*, **90**, 504, 1956.
45. Carlone, N. A. and Scannerini, S., "Scanning and transmission electron microscopy evidence of the cytological effect of Pyrrolnitrin on *Microsporum audouinii in vitro*," *Mycopath. Mycol. appl.*, **53**, 111, 1974.
46. Cassone, A., "Chemical dissection and ultrastructure of the cell wall, as related to morphogenesis in *Candida albicans*," *2nd International Mycology Congress, Tampa, 1977.*
47. ———, Simonetti, N., and Strippoli, V., "Ultrastructural changes in the wall during germ-tube formation from blastospores of *Candida albicans*," *J. gen. Microbiol.*, **77**, 417, 1973.
48. ———, ———, and ———, "Wall structure and bud formation in *Cryptococcus neoformans*," *Arch. Microbiol.*, **95**, 205, 1974.
49. ———, ———, and ———, "Capsule and wall structure in *Cryptococcus neoformans* grown *in vivo*," *J. Submicroscop. Cytol.*, **6**, 113, 1974.
50. ———, ———, and ———, "Production, ultrastructure, and germination of *Candida albicans* chlamydospores, Spores VI," *Am. Soc. Microbiol.*, **172**, 1975.
51. Cawson, R. A. and Rajasingham, K. C., "Comparative ultrastructure of *Candida* hyphae grown *in vitro* and in human lesions," *Int. Ass. Dent. Res., Brit. Div.*, 1972.

52. ——— and ———, "Ultrastructural features of the invasive phase of *Candida albicans*," *Br. J. Derm.*, **87**, 435, 1972.
53. ——— and ———, "Variation of intracytoplasmic membrane structures during invasive activity by hyphae of *Candida albicans*," *Int. Ass. Dent. Res., Brit. Div.*, 1973.
54. Chong, K. C., Adam, B. A., and Soo-Hoo, T. S., "Morphology of *Piedra hortai*," *Sabouraudia*, **9**, 157, 1975.
55. Cole, G. T., "Conidiogenesis in pathogenic hyphomycetes. *I. Sporothrix, Exophiala, Geotrichum* and *Microsporum*," *Sabouraudia*, **14**, 81, 1976.
56. ———, "Blastic and thallic conidial development in the fungi imperfecti," *2nd International Mycology Congress, Tampa, 1977.*
57. ———, "Conidial ontogeny of some dematious fungi pathogenic to man," *2nd International Mycology Congress, Tampa, 1977.*
58. Collins, D. N. and Edwards, M. R., "Filamentous forms of *Blastomyces dermatitidis* in mouse lung," *Sabouraudia*, **7**, 237, 1970.
59. ———, Oppenheim, I. A., and Edwards, M. R., "*Cryptococcosis* associated with systemic lupus erythematosus. Light and electron microscopy observations on a morphologic variant," *Archs. Path.*, **91**, 78, 1971.
60. Cook, W. L., Felton, F. G., Muchmore, H. G., and Rhoades, E. R., "Cell wall differences of patient and soil isolates of *Cryptococcus neoformans*," *Sabouraudia*, **7**, 257, 1969.
61. De Brito, T., Furtado, J. S., Castro, R. M., and Manini, M., "Intraepithelial parasitism as an infection mechanism in human paracoccidioidomycosis (South American blastomycosis)," *Virchows. Arch. path. Anat. Physiol.*, **361**, 129, 1973.
62. Delektorsky, W. W., Scheklakow, N. D., and Schadyew, C. H., "Untersuchung der Ultrastruktur von *M. canis* im Raster-(REM) und Transmissionsmikroskop (TEM)," *Castellania*, **4**, 119, 1976.
63. De Nollin, S. and Borgers, M., "The ultrastructure of *Candida albicans* after *in vitro* treatment with Miconazole," *Sabouraudia*, **12**, 341, 1974.
64. ——— and ———, "Scanning electron microscopy of *Candida albicans* after *in vitro* treatment with Miconazole," *Antimicrob. Agents Chemother.*, **7**, 704, 1975.
65. ——— and ———, "An ultrastructural and cytochemical study of *Candida albicans* after *in vitro* treatment with imidazoles," *Mykosen*, **19**, 317, 1976.
66. ———, Thone, F., and Borgers, M., "Enzyme cytochemistry of *Candida albicans*," *J. Histochem. Cytochem.*, **23**, 758, 1975.
67. De Vries, G. A., Kemp, R. F. O., and Speller, D. C. E., "Endocarditis caused by *Coprinus delicatulus* (= *C. cinereus*)," *Proceedings 5th International Congress ISHAM, Paris, 1971*, pp. 185–186.
68. Djaczenko, W. and Cassone, A., "Visualization of new ultrastructural components in the cell wall of *Candida albicans* with fixatives containing TAPO," *J. Cell Biol*, **52**, 186, 1972.
69. Dorn, M. and Roehnert, K., "Scanning electron microscopy of *Pityrosporum furfur*," *Mykosen*, suppl. 1, 141, 1978.
70. ——— and Wolf, H. H., "Onychomykose: Licht- und Rasterelektronenmikroskopie, *Mykosen*," **20**, 147, 1977.

71. Drexler, H., Melchinger, H. and Müller, J., "Asteroid bodies an Fadenpilzen in Organen einer Leukämie-Patientin," *Schwerpunktmedizin* **2**, 42, 1979.
72. Drouhet, E. and Montplaisir, S., "Phénotypes de résistance des *Candida* à la 5-fluorocytosine et relation avec l'ultrastructure de la paroi," *Bull. Soc. fr. Mycol. Méd.*, **2**, 115, 1975.
73. Dumon, H. and Quilici, M., "Activité candidacide des polynucléaires neutrophiles humains: Étude ultrastructurale et effet du traitement préalable des levures," *Bull. Soc. fr. Mycol. Méd.*, **2**, 185, 1975.
74. Edwards, G. A. and Edwards, M. R., "The intracellular membrane of *Blastomyces dermatitidis*," *Am. J. Bot.*, **47**, 622, 1960.
75. ——, ——, and Hazen, E. L., "Electron microscopic study of *Histoplasmia* in mouse spleen," *J. Bact.*, **77**, 429, 1959.
76. Edwards, M. R., "Internal and external fine structure of the yeast *Cryptococcus neoformans*," *6th International Congress on Electron Microscopy, Kyoto, 1966.*
77. ——, Gordon, M. A., Lapa, E. W., and Ghiorse, W. C., "Micromorphology of *Cryptococcus neoformans*," *J. Bact.*, **94**, 766, 1967.
78. ——, Hazen, E. L., and Edwards, G. A., "The fine structure of the yeast-like cells of *Histoplasma* in cultures," *J. gen. Microbiol.*, **20**, 496, 1959.
79. Elkhouly, A. E., "Studies of nystatin action on *Candida albicans* using scanning electron microscope," *Mykosen*, **21**, 318, 1978.
80. Erickson, J. O. and Breslau, A. M., "Electron microscopy of *Coccidioides immitis*," *Fed. Proc.*, **19**, 243, 1960.
81. Evans, E. G. V. and Hose, H., "Degradation of native keratin by dermatophytes. A scanning electron microscopy study," *2nd International Mycology Congress, Tampa, 1977.*
82. Fukui, S., Kawamoto, S., Yasuhara, S., and Tanaka, A., "Microbody of methanol-grown yeasts. Localization of catalase and flavin-dependant alcohol oxidase in the isolated microbody," *Eur. J. Biochem.*, **59**, 561, 1975.
83. —— Tanaka, A., Kawamoto, S., Yasuhara, S., Teranishi, Y., and Osumi, M., "Ultrastructure of methanol-utilizing yeast cells: Appearance of microbodies in relation to high catalase activity," *J. Bact.*, **123**, 318, 1975.
84. Furtado, J. S. and De Brito, T., "Ultrastructure of *Paracoccidioides brasiliensis* and *Paracoccidioides loboi* in human tissue," *WHO Report: Paracoccidioidomycosis*, Washington, 1972, pp. 29–30.
85. ——, ——, and Freymuller, E., "The fine structure and reproduction of *Paracoccidioides brasiliensis* in human tissue," *Sabouraudia*, **5**, 226, 1967.
86. ——, —— and ——, "Structure and reproduction of *Paracoccidioides loboi*," *Mycologia*, **59**, 286, 1967.
87. Gale, G. R., "Cytology of *Candida albicans* as influenced by drugs acting on the cytoplasmic membrane," *J. Bact.*, **86**, 151, 1963.
88. —— and McLain, H. H., "Effect of thiobenzoate on cytology of *Candida albicans*," *J. gen. Microbiol.*, **36**, 297, 1964.

89. Garrison, R. G. and Boyd, K. S., "Ultrastructural studies of induced morphogenesis by *Aspergillus parasiticus*," *Sabouraudia*, **12**, 179, 1974.
90. —— and ——, "Observations of a naturally-occurring lattice array of ribosomes in ultrathin sections of *Ajellomyces dermatitidis*," *Mycopath. Mycol. appl.*, **54**, 335, 1974.
91. —— and ——, "Aspects of the dimorphism of *Histoplasma farciminosum* (later corrected: *Paracoccidioides brasiliensis*): A light and electron microscopic study, *Sabouraudia*, **13**, 174, 1975.
92. ——, ——, and Mariat, F., "Ultrastructural studies of the mycelium-to-yeast transformation of *Sporothrix schenckii*," *J. Bact.*, **124**, 959, 1975.
93. —— and Lane, J. W., "Yeastlike to mycelial phase transformation of *Histoplasma capsulatum* as observed by scanning electron microscopy," *Mycopathol. Mycol. appl.*, **43**, 185, 1971.
94. —— and ——, "Scanning-beam electron microscopy of the conidia of the brown and albino filamentous varieties of *Histoplasma capsulatum*," *Mycopath. Mycol. app.*, **49**, 185, 1973.
95. —— and ——, "The fine structure of the microconidium of *Blastomyces dermatitidis*," *Mycopath. Mycol. appl.*, **52**, 93, 1974.
96. ——, ——, and Field, M. F., "Ultrastructural changes during the yeast-like to mycelial phase conversion of *Blastomyces dermatitidis* and *Histoplasma capsulatum*," *J. Bact.*, **101**, 628, 1970.
97. ——, ——, and Johnson, D. R., "Electron microscopy of the transitional conversion cell of *Histoplasma capsulatum*," *Mycopath. Mycol. appl.*, **44**, 121, 1971.
98. ——, —— and ——, "Ultrastructural studies on the cleistothecium of *Ajellomyces dermatitidis*," *Sabouraudia*, **11**, 131, 1973.
99. ——, Mariat, F., Boyd, K. S., and Tally, J. F., "Ultrastructural and electron cytochemical studies of *Entomophthora coronata*," *Ann. Microbiol., Institut Pasteur*, (*Paris*), **126B**, 149, 1975.
100. Gibson, R. K. and Peberdy, J. F., "Fine structure of protoplasts of *Aspergillus nidulans*," *J. gen. Microbiol.*, **72**, 529, 1972.
101. Glick, A. D. and Kwon-Chung, K. J., "Ultrastructural comparison of coils and ascospores of *Emmensiella capsulata* and *Ajellomyces dermatitidis*," *Mycologia*, **65**, 216, 1973.
102. Gross, U. and Gross, W., "Zur Wirkung des Nystatins," *Arch. Derm. Forsch.*, **242**, 389, 1972.
103. Grupper, C. and Avram, A., "Un nouvel antifongique à large spectre: l'éconazole. Action *in vitro* et *in vivo*," *6th International Congress ISHAM, Tokyo, 1975.*
104. Gull, K. and Trinci, A. P. J., "Septal ultrastructure in *Basidiobolus ranarum*," *Sabouraudia*, **13**, 49, 1975.
105. Gutierrez, F., Fu, Y. S., and Lurie, H. I., "Cryptococcosis histologically resembling Histoplasmosis. A light and electron microscopical study," *Archs Path.*, **100**, 347, 1975.
106. Hasegawa, T., "Submicroscopic localization of succinic dehydrogenase in *Candida albicans*," *Jap. J. Derm.*, **B75**, 165, 1965.

107. ———, "Demonstration of acid polysaccharides in *Trichophyton rubrum*," *J. Histochem. Cytochem.*, 235, 1972.
108. ———, "Freeze etching observations of *Candida albicans*," *J. Derm.*, **2**, 159, 1975.
109. ———, "*Trichophyton rubrum* and *Trichophyton mentagrophytes* studied by freeze-etching," *Sabouraudia*, **13**, 241, 1975.
110. ———, Nakai, E., and Rajan, V.S., "Freeze-etching observations of *Trichophyton tonsurans*," *6th International Congress ISHAM, Tokyo, 1975.*
111. ——— and Rajan, V. S., "Freeze-etching observations of *Microsporum gypseum*," *Nishinihon J. Derm.*, **38**, 263, 1976.
112. ———, So, Y., Nakai, E., and Rajan, V. S., "Freeze-etching observations of *Trichophyton mentagrophytes*," *Nishinihon J. Derm.*, **37**, 780, 1975.
113. ———, Takeuchi, M., Nakai, E., and Rajan, V. S., "Freeze-etching observations of *Fonsecaea pedrosoi*," *J. Derm.*, **4**, 1, 1977.
114. Hawker, L. E., "Fine structure of fungi as revealed by electron microscopy," *Biol. Rev.*, **40**, 52, 1965.
115. Hejtmanek, M., "Raster-Elektronenmikroskopie des Adiaspiromykoseerregers," *Mykosen* **19**, 77, 1976.
116. ———, "Conversion of adiasporic and mycelial stages of *Emmonsia crescens* examined in the scanning electron microscope," *Folia Parasit. (Praha)*, **23**, 165, 1976.
117. ———, "Examination of human pathogenic fungi in the scanning electron microscope," *Acta Universitatis Palackianae Olomucensis* (*Sb. Praci přir. Fak. palack, Univ. Olomouci*), **76**, 21, 1976.
118. ———, "Scanning electron microscopy of experimental adiaspiromycosis," *Mycopathologia*, **58**, 91, 1976.
119. ——— and Bartek, J., "Mutants of *Emmonsia crescens* – Their pathogenicity and size of adiaspores *in vivo*," *Folia Microbiol.*, **21**, 297, 1976.
120. ——— and ———, "Zur Ultrastruktur von *Nannizzia incurvata*," *Mykosen*, **11**, 285, 1968.
121. ———, Necas, O., and Havelkova, M., "The ultrastructure of the spherules of *Emmonsia crescens* studied by the technique of freeze-etching," *Mycopath. mycol. appl.*, **54**, 79, 1974.
122. Hill, T. W., "Ultrastructure of Ascosporogenesis in *Nannizzia gypsea*," *J. Bact.*, **122**, 743, 1975.
123. Hirai, M., Shimizu, S., Teranishi, Y., Tanaka, A., and Fukui, S., "Effects of hydrocarbons on the morphology of *Candida tropicalis* pK 233," *Agric. Biol. Chem. Jap.*, **36**, 2335, 1972.
124. Hirano, T., Lindegren, C. C., and Bang, Y. N., "Electron microscopy of virus-infected yeast cells," *J. Bact.*, **83**, 1363, 1962.
125. Hoch, H. C., "Cytological approaches to hydrolytic enzyme localizations in fungal-host interactions, *2nd International Mycology Congress, Tampa, 1977.*
126. Houwink, A. L. and Kreger, D. R., "Observations on the cell wall of yeast. An electron microscope and X-ray diffraction study," *Antonie van Leeuwenhoek*, **19**, 1, 1953.

127. Ikeda, H., "An electron microscopic study of *Trichophyton rubrum*," *Jap. J. Derm.*, **74**, 269, 1964.
128. Ito. Y., Nozawa, Y., Suzuki, H., and Setoguti, T., "Surface structure of dermatophytes as seen by the scanning electron microscope," *Sabouraudia*, **7**, 270, 1969.
129. ———, Setoguti, T., Nozawa, Y., and Sakurai, S., "An electron microscopic observation of *Trichophyton violaceum*," *J. invest. Derm.*, **48**, 124, 1967.
130. ———, ———, ———, ———, and Kitajima, Y., "Comparative studies on ultrastructure of the pigmented and albino hyphae of *Microsporum cookei*," *Sabouraudia*, **8**, 60, 1970.
131. ———, ———, ———, ———, and Yoneya, T., "Further observations on the fine structure of *Trichophyton violaceum*: non-pigmented, mutant mycelia," *Sabouraudia*, **7**, 138, 1969.
132. Iwata, K. and Hirata, T., "Studies on the fine structures of the cell of pathogenic fungi by electron microscope. I. On the fine structure of *Candida albicans*," *Jap. J. Bact.*, **18**, 393, 1963.
133. ——— and ———, "On the structures of mitochondria and intracytoplasmic membrane system of *Candida albicans*," *Jap. J. Bact.*, **21**, 533, 1966.
134. ———, Kanda, Y., Yamaguchi, H., and Osumi, M., "Electron microscopic studies on the mechanism of action of Clotrimazole on *Candida albicans*," *Sabouraudia*, **11**, 205, 1973.
135. ———, Uchida, K., Hiratani, T., Tokunaga, J., Tokunaga, M., and Edanaga, M., "Observation by S.E. microscopy on the behavior of *Candida albicans* cells in the kidney of mice following intravenous challenge," *6th International Congress ISHAM, Tokyo, 1975.*
136. ———, Yamaguchi, H., Kanda, Y., and Osumi, M., "Mechanism of action of Clotrimazole on *Candida albicans*: An electron microscopic study," in K. Iwata (ed.), *Yeasts and Yeast-like Microorganisms in Medical science, Proceedings 2nd International Spec. Symposium on Yeasts*, University of Tokyo Press, Tokyo, 1976, pp. 335–338.
137. Joshi, K. R., Gavin, J. B., and Wheeler, E. E., "A scanning electron microscopic study of the morphogenesis of *Candida albicans in vitro*," *Sabouraudia*, **11**, 263, 1973.
138. Jouvert, S., Miègeville, M., Morin, O., Hadibi, E., Vermeil, C., and Bastide, M., "Morphological study of heterobasidiomycetous yeasts protoplasts by scanning electron microscopy," *2nd International Mycology Congress, Tampa, 1977.*
139. Kanetsuna, F. and Carbonell, L. M., "Glycogen in yeast form of *Paracoccidioides brasiliensis, Nature, Lond.*, **208**, 686, 1965.
140. ——— and ———, "Enzymes in glycolysis and the citric acid cycle in the yeast and mycelial forms of *Paracoccidioides brasiliensis*," *J. Bact.*, **92**, 1315, 1966.
141. Kawakami, N., "Electron microscopy of ascomycetes. Investigation of fine structures and classification of sporogenous yeasts," *Mem. Fac. Eng., Hiroshima Univ.*, **1**, 207, 1960.

142. Keddie, F. M., "Electron microscopy of *Malassezia furfur* in *Tinea versicolor*," *Sabouraudia*, **5**, 134, 1966.
143. ——, "Further investigations of the electron microscopic examination of human hair naturally infected by *Microsporum gypseum*," *Sabouraudia*, **6**, 1, 1967.
144. —— and Barajas, L., "Three-dimensional reconstruction of *Pityrosporum* yeast cells based on serial section electron microscopy," *J. Ultrastruct. Res.*, **29**, 260, 1969.
145. —— and ——, "Quantitative ultrastructural variations between *Pityrosporum ovale* and *P. orbiculare* based on serial section electron microscopy," *J. Derm.*, **11**, 40, 1972.
146. Koch-Svoboda, Y. and Rademacher, K. H., "Morphological and biochemical investigations on the cell wall of yeasts by scanning electron microscopy," in E. K. Novak (ed.), *Proceedings 5th International Spec. Symposium on Yeasts*, part I, Budapest, 1977, p. 41.
147. Kodousek, R., Hejtmanek, M., Havelkova, M., and Strachova, Z., "A survey of the ultrastructural organization of spherules of fungus *Emmonsia crescens* Emmons et Jellison 1960," *Acta Universitatis Palacckianae Olomucensis, Fac. Med. (Sb. Praci přir. Fak. plack. Univ. Olomouci)*, **63**, 87, 1972.
148. Kopp, F., "Electronmicroscopy of yeasts," in A. Prescott (ed.), *Methods in Cell Biology*, vol. IX, *Yeast Cells*, Academic Press, New York, 1975.
149. Kreger, D. R. and Kopecká, M., "On the nature and formation of the fibrillar nets produced by protoplasts of *Saccharomyces cerevisiae* in liquid media: an electronmicroscopic, X-ray diffraction and chemical study," *J. gen. Microbiol.*, **92**, 207, 1975.
150. Kuntert, J. and Hejtmanek, M., "Raster-Elektronenmikroskopie der menschlichen Haare, befallen durch den Dermatophyten *Microsporum gypseum* (Bodin) Guiart und Grigorakis *in vitro*," *Mykosen*, **19**, 329, 1976.
151. Kuttin, E. S., Müller, J., Takamiya, H., Melchinger-Müller, H., and Jaeger, R., "Electronmicroscopic studies of *Candida albicans* infected kidneys in non-immunized and immunized hosts," *14. Tagg. Deutschsprach. Mykolog. Ges., Tübingen, 1978*; *Mykosen* (in press).
152 ——, ——, ——, ——, and ——, "Electronmicroscopic studies of peritoneal lavages after intraperitoneal application of *Candida albicans* to non-immunized and immunized hosts," *Mykosen* (in press).
153. Kwon-Chung, K. J. and Popkin, T. J., "Ultrastructure of septal complex in *Filobasidiella neoformans* (*Cryptococcus neoformans*)," *J. Bact.*, **126**, 524, 1976.
154. Laden, E. L. and Erickson, J. O., "Electron microscope study of *Epidermophyton floccosum*," *J. invest. Derm.*, **31**, 55, 1958.
155. Lampen, J. O., "Amphothericin B and other polyene antifungal antibiotics," *Am. J. clin. Path.*, **52**, 138, 1969.
156. Lane, J. W. and Garrison, R. G., "Electron microscopy of the yeast to mycelial phase conversion of *Sporotrichum schenckii*," *Can. J. Microbiol.*, **16**, 747, 1970.

157. —— and ——, "Electron microscopy of selfparasitism by *Histoplasma capsulatum* and *Blastomyces dermatitidis*," *Mycopath. Mycol. appl.*, **40**, 271, 1970.
158. ——, ——, and Field, M. F., "Ultrastructural studies on the yeast like and mycelial phase of *Sporotrichum schenckii*," *J. Bact.*, **100**, 1010, 1969.
159. ——, ——, and Johnson, D. R., "Drug-induced alterations in the ultrastructural organization of *Histoplasma capsulatum* and *Blastomyces dermatitidis*," *Mycopath. Mycol. appl.*, **48**, 289, 1972.
160. Lauwerys-De Nollin, S. and Borgers, M., "Ultrastructur van Gistcellen en de Veranderingen na *in vitro*-Behandlung met Miconazole," *Vlaams Diergeneeskundig Tijdschrift*, **3**, 77, 1975.
161. Lindegren, C. C., Bang, Y. N., and Hirano, T., "Progress report on the zymophage," *Trans. N. Y. Acad. Sci. Ser. II*, **24**, 540, 1962.
162. —— and Lindegren, G., "Modifications of the yeast cell produced by different substances," paper without reference.
163. Lurie, H. I., "Histopathology of sporotrichosis. Notes on the nature of the asteroid body," *Archs Path.*, **75**, 421, 1963.
164. ——, Shadomy, H. J., and Still, W. J. S., "An electron microscopic study of *Cryptococcus neoformans* (Coward strain)," *Sabouraudia*, **9**, 15, 1971.
165. —— and Still, W. J. S., "The 'capsule' of *Sporotrichum schenckii* and the evolution of the asteroid body. A light and electron microscopic study," *Sabouraudia*, **7**, 64, 1969.
166. Mahvi, T. A., Heere, L. J., Spicer, S. S., and Annable, M., "Biochemical and ultrastructural changes in *Cryptococcus neoformans* in the presence of 5-fluorocytosin," in E. Drouhet (ed.), *Proceedings 5th International Congress ISHAM*, Paris, 1971, pp. 283–284.
167. Malfati, M. G. and Zapata, R., "El dimorfismo de algunos hongos patógenos observados a travéz del microscopio electrónico," *Prensa méd. argent.*, **41**, 3869, 1954.
168. Malinsky, J., Hejtmanek, M., and Kodousek, R., "Electron microscopy of experimental adiaspiromycosis," *Acta Universitatis Palackianae Olomucensis, Fac. med.* (*Sb. Praci přir. Fak. palack. Univ. Olomouci*), **63**, 99, 1972.
169. Marquardt, H., "Der Feinbau von Hefezellen im Elektronenmikroskop. I. *Rhodotorula rubra*," *Z. Naturforschg.*, **17b**, 42, 1962.
170. Marquardt, H., "Der Feinbau von Hefezellen im Elektronenmikroskop. II. *Saccharomyces cerevisiae*-Stämme," *Z. Naturforschg.*, **17b**, 689, 1962.
171. Matsuzaki, O., "Observation of fungi in the cerumen of otomycosis," *6th International Congress ISHAM, Tokyo, 1975.*
172. Meinhof, W., "Untersuchungen zur Struktur und Funktion hautpathogener Pilze unter besonderer Berücksichtigung histochemischer und elektronenmikroskopischer Befunde," Habilitationsschrift Universität Marburg 1966.
173. ——, "Untersuchungen zur Ultrastruktur von *Keratinomyces ajelloi*," *Arch. klin. exp. Derm.*, **226**, 33, 1966.
174. ——, "Untersuchungen zur Ultrastruktur von *Microsporum gypseum*. I.

Aufbau der Zellwand, der Septen und der cytoplasmatischen Membran," *Arch. klin. exp. Derm.*, **228**, 111, 1967.

175. ——, "Untersuchungen zur Ultrastruktur von *Microsporum gypseum.* II. Aufbau und Vermehrung der Mitochondrien," *Arch. klin. exp. Derm.*, **228**, 122, 1967.
176. ——, "Untersuchungen zur Ultrastruktur von *Trichophyton verrucosum* und *Trichophyton schönleinii*," *Arch. klin. exp. Derm.*, **228**, 265, 1967.
177. ——, "Elektronenmikroskopische und enzymhistochemische Untersuchungen zur Zytologie pathogener Pilze," *Fortschr. Med.*, **85**, 597, 1967.
178. ——, "Zur funktionellen Organisation der Dermatophyten-Zelle," *Mykosen*, **10**, 355, 1967.
179. —— and Vogell, W., "Ultrastructure of dermatophytes," *13. Int. Kongr. Derm. München, 1967.*
180. Melchinger-Müller, H., Müller, J., Takamiya, H., and Nold, B., "*Candida* asteroid bodies in klinischen Untersuchungsmaterialien: Immunelektronenmikroskopische Untersuchungen an Vaginalmaterial von Kolpitis-Patientinnen," *Mykosen*, **23**, 161, 1980.
181. Miègeville, M. and Morin, O., "Observations de différentes souches de levures et de leurs protoplastes en microscopie électronique à balayage," *C.r. hebd. Séanc. Acad. Sci., Paris*, **283D**, 417, 1976.
182. ——, ——, Hadibi, E., and Bastide, M., "Apport de la microscopie électronique à balayage pour l'étude morphologique des levures du genre *Cryptococcus* et de leurs protoplastes," *Bull. Soc. fr. Mycol. Méd.*, **6**.
183. Miyake, S., Iwamoto, Y., Nagao, M., Sugimura, T., and Ohsumi, M., "Phenotype of a temperature-sensitive, respiration-deficient (cyt) mutant of yeast," *J. Bact.*, **109**, 409, 1972.
184. Mizuno, N. and Montes, L. F., "Oxidative enzyme activity in *Candida albicans*," *Sabouraudia*, **5**, 46, 1966.
185. Montes, L. F., "Fungi and fungal infections," in A. Zelickson (ed.), *Ultrastructure of Normal and Abnormal Skin*, Lea and Febiger, Philadelphia, 1967, pp. 347–364.
186. ——, "Systemic abnormalities and the intracellular site of infections of the *stratum corneum*," *JAMA*, **213**, 1469, 1970.
187. ——, Patrick, T. A., Martin, S. A., and Smith, M. S., "Ultrastructure of blastospores of *Candida albicans* after permanganate fixation," *J. invest. Derm.*, **45**, 227, 1965.
188. —— and Wilborn, W. H., "Ultrastructural features of host-parasite relationship in oral candiasis," *J. Bact.*, **96**, 1349, 1968.
189. Montplaisir, S., Drouhet, E., and Nabarra, B., "Sensibilité et résistance des *Candida* à la 6-fluorocytosine. Relation avec l'ultrastructure de la paroi," *C. r. Séanc. Acad. Sci., Paris*, **280D**, 2211, 1975.
190. ——, Nabarra, B., and Drouhet, E., "Susceptibility and resistance of *Candida* to 5-fluorocytosine in relation to the cell wall ultrastructure," *Antimicrob. Agents Chemother.*, **9**, 1028, 1976.
191. Moor, H. and K. Mühlethaler, "Fine structure in frozen-etched yeast cells," *J. Cell Biol.*, **17**, 609, 1963.
192. Moore, R. T., "The ultrastructure of fungal cells," in G. C. Ainsworth

and A. S. Sussman (eds.), *The Fungi*, vol. 1, Academic Press, New York, London, 1965, pp. 95–118.
193. —— and McAlear, J. H., "Fine structure of mycota. 5. Lomasomes previously uncharacterized hyphal structures," *Mycologia*, **53**, 194, 1961.
194. —— and ——, "Fine structure of mycota. Reconstruction from skipped serial sections of the nuclear envelope and its continuity with the plasma membrane," *Expl. Cell Res.*, **24**, 588, 1961.
195. —— and ——, "Fine structure of mycota. 7. Observation on septa of *Ascomycetes* and *Basidiomycetes*," *Am. J. Bot.*, **49**, 86, 1962.
196. —— and ——, "Fine structure of mycota. 9. Fungal mitochondria," *J. Ultrastruct. Res.*, **8**, 144, 1963.
197. Mori, K. and Kuida, K., "Virus-like particles in several mushrooms," *2nd International Mycology Congress, Tampa, 1977.*
198. Müller, J., "Immunobiological aspects of *Candida* mycoses – a review of electron microscopic studies," *Mykosen*, suppl. 1, 289, 1978.
199. ——, Douchet, C., and Sakuma, S., "Etudes au microscope électronique des reins de souris infectées par *Candida albicans*," in E. Drouhet (ed.), *Proceedings 5th International Congress ISHAM*, Paris, 1971, p. 8.
200. ——, ——, and ——, "Candidose expérimentale de la souris. Etude au microscope électronique des relations hôtes-parasites au niveau du tissu rénal," *Bull. Soc. fr. Mycol. Méd.*, **1**, 75, 1972.
201. ——, ——, and ——, "Elektronenmikroskopische Untersuchungen zur Wirt-Parasit-Auseinandersetzung bei der experimentallen Candida-Infektion der Maus: Singuläre Candida-Zellen im Nierengewebe," *Derm. Mschr.*, **159**, 327, 1973.
202. ——, ——, and ——, "Electronmicroscopic studies on the host-parasite relationship of the experimental *Candida albicans* infection of the mouse: The Candida granuloma in the kidney," in K. Iwata (ed.), *Yeast and Yeast-like Microorganisms in Medical science, Proceedings 2nd International Spec. Symposium on Yeasts*, University of Tokyo Press, Tokyo, 1976, pp. 178–183.
203. ——, ——, and ——, "A propos de l'ultrastructure des organes infectés par *Candida albicans*: Cellules de *Candida* et lysosomes," *Bull. Soc. fr. Mycol. Méd.*, **2**, 103, 1973.
204. ——, Kuttin, E. S., Takamiya, H., Melchinger-Müller, H., and Jaeger, R., "Immunoelectronmicroscopic studies of *Candida albicans* infected tissues in non-immunized and immunized hosts," *Sabouraudia* (in press).
205. ——, Sattmann, G., Takamiya, H., and Melchinger-Müller, H., "Immunbiologische Asperkte bei kommensaler Sproßpilzebsiedlung des Menschen," "Dissertation Med. Fak. Freiburg," 1980.
206. ——, Takamiya, H., and Jaeger, R., "Elektronenmikroskopische Darstellung von Immunreaktionen an Candida-Zellen: IV. Nachweis von Candida-zellgebundenen Antikörpern in Urinproben von Nephritis-Patienten," *12. Tagg. Deutschsprach. Mykolog. Ges., Wien, 1975.*
207. ——, ——, and ——, "Elektronenmikroskopische Darstellung von Immunreaktionen an *Candida*-Zellen; Asteroid Bodies bei *Candida albicans* im Urin von Nephritis-Patienten," *Sabouraudia*, **15**, 87, 1977.

208. ——, ——, Nold, B., and Jaeger, R., "Electronmicroscopic proof of immunoreactions on Candida cells: III. Evidence of anti-Candida antibodies *in vivo* by the indirect immunoferritin test," in K. Iwata (ed.), *Recent Advances in Medical and Veterinary Mycology*, University of Tokyo Press, Toykyo, 1976, pp. 59–65.
209. ——, ——, —— and ——, "Elektronenmikroskopische Darstellung von Immunreaktionen an *Candida*-Zellen: V. Zur Freisetzung von Candida-Zellwand-Antigenen im Zuge der Wirt-Parasit-Auseinandersetzung," *13. Tagg. Deutschsprach. Mykolog. Ges., Hannover, 1976.*
210. ——, ——, Vogt, A., and ——, "Elektronenmikroskopische Darstellung von Immunreaktionen an Candida-Zellen: I. *In Vitro*-Inkubation von *Candida albicans* mit Anti-Candida-Humanserum," *Mykosen*, **19**, 295, 1976.
211. ——, Tokunaga, J., and Takamiya, H., "Electronmicroscopic proof of immunoreactions on Candida cells: II. Analysis of *Candida albicans* cell wall antigens by the direct immunoferritin test," *6th International Congress ISHAM, Tokyo, 1975.*
212. Nickerson, W. J. and Falcone, G., "Identification of protein disulfide reductase as a cellular division enzyme in yeasts," *Science*, **124**, 722, 1956.
213. Nozawa, Y., Kitajima, Y., and Ito, Y., "Isolation, ultrastructure and chemical composition of the outer-most layer of the *Epidermophyton floccosum* cell wall," *6th International Congress ISHAM, Tokyo, 1975.*
214. Normark, F. and Schönebeck, J., "*In vitro* studies of 5-fluorocytosine resistance in *Candida albicans* and *Torulopsis glabrata*," *Antimicrob. Agents Chemother.*, **2**, 114, 1972.
215. O'Hern, E. M. and Henry, B. S., "A cytological study of *Coccidioides immitis* by electron microscopy," *J. Bact.*, **72**, 632, 1956.
216. Okuda, K., "Immunophagocytosis of *Candida albicans* by neutrophils," *Jap. J. med. Mycol.*, **15**, 59, 1974.
217. Olah, G. M. and Reisinger, O., "Étude ultrastructural et cytochimique de l'appareil sporifère chez *Phialophora richardsiae*," *Can. J. Botany*, **52**, 2473, 1974.
218. Osumi, M. and Fukui, S., "Relation between the function and the ultrastructure of profusely appearing microbodies in hydrocarbon-utilizable yeasts," *30th Annual Proceedings of Electron Microscopy Society of America, Los Angeles, 1972.*
219. —— and ——, "Surface structure of yeast cells growing on *n*-alkane," *6th International Congress ISHAM, Tokyo, 1975.*
220. ——, ——, and Nagatani, T., "Ultrastructure of hydrocarbon-utilizable yeast cells," *Proceedings 8th International Congress of Electron Microscopy*, vol. II, Canberra, 1974, pp. 578–579.
221. ——, Fukuzumi, F., Teranishi, Y., Tanaka, A., and Fukui, S., "Development of microbodies in *Candida tropicalis* during incubation in *n*-alkane medium," *Arch. Microbiol.*, **103**, 1, 1975.
222. ——, ——, and Yamada, N., "Scanning electron microscopy of hydrocarbon-growing *Candida* yeasts," *Jap. Women's Univ. J.*, **21**, 169, 1974.

223. ———, ———, ———, Nagatani, Teranishi, Y., Tanaka, A., and Fukui, S., "Surface structure of some *Candida* yeast cells grown on *n*-alkanes," *J. Ferment. Technol. Osaka*, **53**, 244, 1975.
224. ———, Imazumi, F., Imai, M. Sato, H., and Yamaguchi, H., "Isolation and characterization of microbodies from *Candida tropicalis* PK 233 cells grown on normal alkanes," *J. gen. appl. Microbiol.*, **21**, 375, 1975.
225. ———, ———, and Sato, H., "A new technique for the fixation of microbodies isolated from *Candida* yeast cells grown on *n*-alkane," *Jap. Women's Univ. J.*, **22**, 175, 1975.
226. ——— and Kitsutani, S., "Aspects of mitochondriogenesis in yeast. II. The effect of chloramphenicol on *Rhodotorula mucilaginosa*," *J. Electron Microsc.*, **20**, 23, 1971.
227. ———, Miwa, N., Teranishi, Y., Tanaka, A., and Fukui, S., "Ultrastructure of *Candida* yeasts grown on *n*-alkanes. Appearance of microbodies and its relationship to high catalase activity," *Arch. Microbiol.*, **99**, 181, 1974.
228. ——— and Mura, M., "Mitochondriogenesis in yeast cells in the presence of Bleomycin during respiratory adaption," in K. Iwata (ed.), *Yeasts and Yeast-like Microorganisms in Medial Science, Proceedings 2nd International Spec. Symposium on Yeasts*, University of Tokyo Press, Tokyo, 1976, pp. 46–52.
229. ———, ———, and Aoki, M., "Aspects of mitochondriogenesis in yeast. III. Effects of Bleomycin on *Saccharomyces cerevisiae* during respiratory adaption," *Jap. Women's Univ. J.*, **20**, 77, 1973.
230. Ota, F., Fukui, K., and Yoshida, N., "Requirements for regeneration of yeast protoplasts and their ultrastructure during reversion," in K. Iwata (ed.), *Yeasts and Yeast-like Microorganisms in Medical Science, Proceedings 2nd International Spec. Symposium on Yeasts*, University of Tokyo Press, Tokyo, 1976, pp. 39–45.
231. Partridge, B. M. and Drewe, J. A., "*Candida* protoplasts and their ultrastructure," *Sabouraudia*, **12**, 166, 1974.
232. Pore, R. S., D'Amato, R. F., and Ajello, L., "*Fissuricella* gen. nov.: A new taxon for *Prototheca filamenta*," *Sabouraudia*, **15**, 69, 1977.
233. Poulain, D., "Contribution à l'étude ultrastructurale de l'invasion des téguments par les dermatophytes. Travail préliminaire concernant la structure de *Trichophyton mentagrophytes* et les modifications épidermiques qu'il provoque chez le cobaye," Diplome d'études approfondies, Université de Lille, 1972.
234. ———, "Contribution à l'étude des relations hôte-parasite au cours de la dermatophytie expérimental du cobaye par *Trichophyton mentagrophytes*. Etude de l'immunité à médiation cellulaire *in vitro* par tests immunologiques et *in vivo* en microscopie électronique. Etude cytochimique ultrastructurale de la paroi du dermatophyte," Thèse de troisième cycle, Université de Lille, 1975.
235. ———, Andrieu, S., and Biguet, J., "Ultrastructure de la paroi de *Candida albicans*. Note préliminaire," *Bull. Soc. fr. Mycol. Méd.*, **4**, 107, 1975.

236. —— and Biguet, J., "Etude en microscopie électronique de l'action kératinolytique de *Trichophyton mentagrophytes* dans le poil de cobaye infecté expérimentalement," *Sabouraudia*, **12**, 1, 1974.
237. ——, ——, Deblock, S., and Vernes, A., "La paroi mycélienne de *Trichophyton mentagrophytes*. Etude ultrastructurale de ses couches constitutives aprés application d'une méthode de mise en évidence des polysaccharides," *Sabouraudia*, **13**, 244, 1975.
238. ——, ——, and Vernes, A., "Aspects ultrastructuraux des relations hôte-parasite au niveau des téguments parasités par *Trichophyton mentagrophytes*," *Ann. Microbiol., Institut Pasteur (Paris)*, **127B**, 473, 1976.
239. ——, Desseaux, G., and Vernes, A., "Variation de la structure antigénique pariétale des blastospores de *Candida albicans*. II. Appréciation de la richesse de souches clonées en antigènes P, influence de milieu de culture," *Bull. Soc. fr. Mycol. Méd.*, **7**, 37, 1977.
240. ——, Fruit, J., and Vernes, A., "Variations de la structure antigénique pariétale de *Candida albicans*. I. Mise en évidence au niveau des blastospore d'un antigène P dépendant de leur origine," To be published in *Sabouraudia*, **17**, 1979.
241. ——, Tronchin, G., Dubremetz, J. F., and Biguet, J., "Ultrastructure of the cell wall of *Candida albicans* blastospores. Study of its constitutive layers by the use of a cytochemical technique revealing polysaccharides," *Ann. Microbiol., Institut Pasteur (Paris)*, **129A**, 141, 1978.
242. Preusser, H.-J., "Die Wirkung von Econazol auf die Feinstruktur der Zellen von *Trichophyton rubrum*," *Mykosen*, **18**, 453, 1975.
243. ——, "Effects of *in vitro* treatment with Econazole on the ultrastructure of *Candida albicans*," *Mykosen*, **19**, 304, 1976.
244. ——, "The influence of preparation of the ultrastructure of fungi," *Mykosen,* suppl. 1, 43, 1978.
245. ——, "Gefrierätz-Untersuchungen der Cytoplasmamembran von *Candida albicans* nach Einwirkung von Econazol," *14. Tagg. Deutschsprach. Mykolog. Ges., Tübingen, 1978. Sabouraudia*, **17**, 389, 1979.
246. —— and Rostek, H., "Effects of the imidazole derivative econazole on the cytomembranes of fungi," *2nd International Mycology Congress, Tampa, 1977.*
247. —— and ——, "An unusual method of reproduction of *Candida albicans* and *Trichophyton rubrum* electron micrographs," *Castellania*, **5**, 87, 1977.
248. Pugh, D. and Cawson, R. A., "The cytochemical localization of phospholipase A and lysophospholipase in *Candida albicans*," *Sabouraudia*, **13**, 110, 1975.
249. —— and ——, "The cytochemical localization of phospholipase in *Candida albicans* infecting the chick chorioallantoic membrane," *Sabouraudia*, **15**, 29, 1977.
250. —— and ——, "The effect of nystatin on the morphology of *Candida albicans*," *Sabouraudia*, **16**, 307, 1978.
251. Rajasingham, K. C., "Ultrastructural studies of *Candida albicans* with

special reference to oral lesions of candidosis and the problem of fixation techniques," M. Phil. Thesis, University of London, 1974.

252. ——— and Cawson, R. A., "Intracellular glycogen of *Candida albicans* and response of intact organisms to amylase," *Int. Ass. Dent. Res., Brit. Div.*, 1973.

253. ——— and ———, "Ultrastructural localization of glycogen in *Candida albicans*, Session," *Br. Soc. Mycopathol.*, 1975.

254. ——— and ———, "Cell wall ultrastructure in *Candida albicans*, Session," *Br. Soc. Mycopathol.*, 1976.

255. Rambourg, A., "An improved silver methenamine technique for the detection of periodic acid-reactive complex carbohydrates with the electron microscope," *J. Histochem. Cytochem.*, **15**, 409, 1967.

256. Reisinger, O. and Olah, G. M., "Etude ultrastructurale et cytochimique de la conidiogénèse chez *Beauveria bassiana* (Bals.) Vuill.," *Can. J. Microbiol.*, **20**, 1387, 1974.

257. Reiss, F. and Leonard, L., "Electron microscopic studies of ultrasonic irradiated *Trichophyton mentagrophytes*," *Dermatologica*, **117**, 401, 1958.

258. Ribi, E., Hoyer, B. H., and Goode, G., "The fine structure of the cell wall of *Histoplasma capsulatum* as revealed by electron microscopy and X-ray diffraction studies," *Fed. Proc.*, **15**, 105, 1956.

259. ——— and Salvin, S. B., "Antigens from the yeast phase of *Histoplasma capsulatum*. I. Morphology of the cell as revealed by the electron microscope," *Expl. Cell Res.*, **10**, 394, 1956.

260. San-Blas, G. and Carbonell, L. M., "Chemical and ultrastructural studies on the cell walls of the yeast-like and mycelial forms of *Histoplasma farciminosum*," *J. Bact.*, **119**, 602, 1974.

261. Scheklakow, N. D., Delektorski, W. W., and Babajan, K. R., "Einige Besonderheiten der Ultrastruktur des *Epidermophyton floccosum* und seine Lagerung in den Geweben der Kranken mit Epidermophytia inguinalis," *Castellania*, **5**, 131, 147, 1977.

262. Scherwitz, C., "Wachstumsverhalten und Ultrastruktur von *Pityrosporum orbiculare* nach Einwirkung von 8-Methoxypsoralen und UV–A (365 nm)-Licht," *14. Tagg. Deutschsprach. Mykolog. Ges., Tübingen, 1978, Mykosen* (in press).

263. ——— and Martin, R., "*In vitro* and *in vivo* ultrastructural changes in *Pityrosporum orbiculare* after treatment with Econazole," *Mykosen*, suppl. 1, 328, 1978.

264. ———, ———, and Ueberberg, H., "Ultrastructural investigation of the formation of *Candida albicans* germ tubes and septa," *Sabouraudia*, **16**, 115, 1978.

265. Schnell, J. D. and Voigt, W.-H., "Das Verhalten von Sproßpilzen am nicht verhornenden Plattenepithel," *Arch. Gynaek.*, **217**, 377, 1974.

266. Schönebeck, J., Winblad, B., and Ånsehn, S., "Renal candidosis complicating caeco-cystoplasty. Case report with scanning electron microscopic studies," *Scand. J. Urol. Nephrol.*, **6**, 129, 1972.

267. Shahar, A., Kletter, Y., and Aronson, M., "Cinematographic and scanning electron microscopy of leucocytic rings formed *in vitro* around *Cryp-*

tococcus neoformans," in *Contributions in Microbiology* and *Immunology*, vol. 3, Karger, Basel, 1977, pp. 60–65.

268. Shetsiruli, L. T. and Apridonidze, K. G., "Ultrastructural aspects of mechanisms of appearance of generalized forms of mycoses caused by *Trichophyton rubrum*," *6th International Congress ISHAM, Tokyo, 1975*.
269. —— and ——, "Studies of the mycotic process caused by *Trichophyton rubrum* (*in vitro*) with transmission and scanning electron microscopy," *Castellania*, **4**, 177, 1976.
270. Shimizu, R., "Studies on the germ tube-like filaments formation of *Candida albicans*. 1. Experiment on the germ tube-like filaments formation *in vitro*," *J. Kyushu Dent. Soc.*, **23**, 507, 1969.
271. ——, "Studies on the germ tube-like filaments formation of *Candida albicans*. 2. A study of germ tube-like filaments formation and its virulence *in vivo*," *J. Kyushu Dent. Soc.*, **23**, 529, 1969.
272. Simonetti, N., Strippoli, V., and Cassone, A., "Yeast-mycelial conversion induced by *n*-actyl-*d*-glucosamin in *Candida albicans*," *Nature, Lond.*, **250**, 344, 1974.
273. Stoetzner, H. and Kemmer, C., "The morphology of *Candida albicans* in human candidosis. A light and electron microscopic study," *Mykosen*, **18**, 511, 1975.
274. Stoian, M. and Grigoriu, D., "Some ultrastructural aspects of the pathogenic fungi with reference to genus *Microsporum*," in E. Drouhet (ed.), *Proceedings 5th International Congress ISHAM, Paris, 1971*, pp. 11–12.
275. Sun, S. H. and Huppert, M., "A cytological study of morphogenesis in *Coccidioides immitis*," *Sabouraudia*, **14**, 185, 1976.
276. ——, Sekhon, S. S., and Huppert, M., "Electron microscopic studies of saprophytic and parasitic forms of *Coccidioides immitis*," *2nd International Mycology Congress, Tampa, 1977*.
277. Takeo, K., "Ultrastructure of chlamydospore of *Candida albicans* as observed by freeze-etching," *Proceedings 8th International Congress on Electron Microscopy*, vol. II, Canberra, 1974, pp. 580–581.
278. ——, "Ultrastructure of polymorphic *Mucor* as observed by means of freeze-etching. II. Vegetative yeast form grown under anaerobic conditions," *Arch. Microbiol.*, **99**, 91, 1974.
279. ——, "Ultrastructure of polymorphic *Mucor* as observed by means of freeze-etching. III. Dormant Sporangiospore," *Arch. Microbiol.*, **99**, 99, 1974.
280. ——, Hsu, H., and Uesaka, I., "Fine structure of *Aspergillus fumigatus* as observed by freeze-etching," *6th International Congress ISHAM, Tokyo, 1975*.
281. —— and Nishiura, M., "Ultrastructure of polymorphic *Mucor* as observed by means of freeze-etching. I. Vegetative growth of mycelium and arthrospore formation in submerged and aerated cultures," *Arch. Microbiol.*, **98**, 175, 1974.
282. ——, Uesaka, I., and Nishiura, M., "The wall particle of the genus *Cryptococcus*: Large size and characteristic distribution," *J. gen. Microbiol.*, **84**, 223, 1974.

283. ———, ———, Uehira, K., and Nishiura, M., "Fine structure of *Cryptococcus neoformans* grown *in vitro* as observed by freeze etching," *J. Bact.*, **113**, 1442, 1973.
284. ———, ———, ———, and ———, "Fine structure of *Cryptococcus neoformans* grown *in vivo* as observed by freeze-etching," *J. Bact.*, **113**, 1449, 1973.
285. Thyagarajan, T. R., Conti, S. F., and Naylor, H. B., "Electron microscopy of *Rhodotorula glutinis*," *J. Bact.*, **83**, 381, 1962.
286. Tokunaga, J., Fujita, T., Hattori, A., and Müller, J., "Scanning electron microscopic observation of immunoreactions on the cell surface: Analysis of *Candida albicans* cell wall antigens by the immunoferritin method," *Proceedings 9th Annual SEM Symposium*, Chicago 1976, pp. 301–310.
287. ———, Harada, K., and Matsumoto, H., "The preparation of fungus cell for scanning electron microscopy," *Jap. J. med. Mycol.*, **11**, 49, 1970.
288. ——— and Tokunaga, M., "SEM observation of spiny colonies of *Candida albicans* at low temperature," in K. Iwata (ed.), *Yeasts and Yeast-like Microorganisms in Medical Science, Proceedings 2nd International Spec. Symposium on Yeasts*, University of Tokyo Press, Tokyo, 1976, pp. 30–38.
289. ——— and ———, "Cryo-scanning microscopy of conidiospores formations in *Aspergillus niger*," *Jeol News*, **11e**, 51, 1973.
290. ———, ———, and Egashira, T., "Electron microscopical studies on growing fungal cells: II. Cell wall formation in budding and germinating blastospores of genus *Candida*," *Jap. J. Bact.*, **24**, 676, 1969.
291. ———, ———, and ———, "Electron microscopical studies on growing fungal cells: III. Germination ability of chlamydospores of *Candida stellatoidea* and *Candida albicans*, and their fine structure during germination," *Jap. J. Bact.*, **24**, 683, 1969.
292. ———, ———, and ———, "Ultrastructural changes during germination process of fungus cells," *Jap. J. med. Mycol.*, **11**, 83, 1970.
293. ———, ———, ———, and Harada, K., "Electron microscopical studies on growing fungal cells. I. Nuclear division and accompanying behavior of mitochondria in the genus *Candida*," *Jap. J. Bact.*, **24**, 671, 1969.
294. ———, ———, Müller, J., Takamiya, H. and Vogt, A. "Ultrastructural host-fungus relationships: fungi pathogenic for warm-blooded hosts," *12th International Congress of Microbiology, Munich, 1978.*
295. ———, ———, Yotsumoto, H., and Hataba, Y., "Cryo-scanning electron microscopy of conidiospore formation in *Aspergillus niger*," *Jap. J. Med. Mycol.*, **14**, 39, 1973.
296. Tokunaga, M. and Tokunaga, J., "Fine structure of chlamydospores in the genus *Candida*," in K. Iwata (ed.), *Yeasts and Yeast-like Microorganisms in Medical Science, Proceedings 2nd International Spec. Symposium on Yeasts*, University of Tokyo Press, Tokyo, 1976, pp. 24–29.
297. ———, ———, and Harada, K., "Scanning and transmission electron microscopy of sterigma and conidiospore formation in *Aspergillus* group," *J. Electron Microsc.*, **22**, 27, 1973.

298. Tomomatsu, S., "A study on griseofulvin. I. Comparison of electron microscopic observations of effects on griseofulvin with that of fungicidal drugs," *Bull. pharm. Res. Inst.*, **26**, 11, 1960.
299. Tsukahara, T., "Cytological structures of *Cryptococcus neoformans*," *Jap. J. Microbiol.*, **7**, 53, 1963.
300. ———, "Electron microscopy of swelling and germinating conidiospores of *Aspergillus niger*," *Sabouraudia*, **6**, 185, 1967.
301. ———, "Electron microscopy of conidiospore formation in *Aspergillus niger*," *Sabouraudia*, **8**, 93, 1970.
302. ———, Sato, A., and Okada, R., "Electron microscopic studies on the cytological structure of *Trichophyton mentagrophytes*," *Jap. J. Microbiol.*, **8**, 83, 1964.
303. Tubaki, K., "Annellation in fungi and the arsenic reaction," in K. Iwata (ed.), *Recent Advances in Medical and Veterinary Mycology*, University of Tokyo Press, Tokyo, 1976, pp. 277–282.
304. Uetsuka, A., Sato, S., Okazaki, N., Ohno, Y., Itoh, M., and Yoshimura, K., "Morphological changes of yeast and yeastlike fungi in tissue culture," *6th International Congress ISHAM, Tokyo, 1975.*
305. Urabe, H., Suenaga, M., Yasuda, M., and Koda, H., "Treatment of cutaneous candidiasis and chromomycosis with 5-fluorocytosine, in K. Iwata (ed.), *Recent Advances in Medical and Veterinary Mycology*, University of Tokyo Press, Tokyo, 1976, pp. 97–106.
306. Uyeda, K., Sotomatsu, S., and Washidzu, R., "Black grain mycetoma. Report of a case and electron microscopic observation of *Madurella mycetomi*," *6th International Congress ISHAM, Tokyo, 1975.*
307. Vanbreuseghem, R., "Ultrastructure of *Rhinosporidium seeberi* and epidemiology of Rhinosporidiosis," *6th International Congress ISHAM, Tokyo, 1975.*
308. Van Rijn, H. J. M., Linnemans, W. A. M., and Boer, P., "Localization of acid phosphatase in protoplasts from *Saccharomyces cerevisiae*," *J. Bact.*, **123**, 1144, 1975.
309. Venezia, R. A. and Lachapelle, R. C., "The use of ferritin conjugated antibodies in the study of cell wall components of *Candida albicans*," *Can. J. Microbiol.*, **19**, 1445, 1973.
310. Visset, M. F. and Vermeil, C., "Contribution à la connaissance des fungi kératinophiles de l'ouest de la France. V. Etude au microscope électronique à balayage," *Mycopath. Mycol. appl.*, **49**, 89, 1973.
311. Voigt, W.-H., "Zur Wirkung von Antimykotika auf die Ultrastruktur von Pilzzellen," in J. Hartung and D. Lubach (eds.), *Mykosen. Systematik, Klinik, Therapie*, Thieme, Stuttgart, 1975, pp. 115–120.
312. ———, "Clotrimazole (Canesten) – Electron microscope studies of the action on the ultrastructure of pathogenic fungi," in A. G. Bayer (ed.), *Broad-Spectrum Antimycotic Canesten, Experience and Outlook*, Leverkusen, 1975, pp. 12–19.
313. ———, "Demonstration der Mykose-Therapie: Elektronenmikroskopische Studie zur Canesten-Wirkung auf Pilze am Vaginalepithel," *Sexualmedizin*, **4**, 702, 1975.

314. ———, "Die Wirkung von Clotrimazol (Canesten) auf die Ultrastruktur von Schimmelpilzen (*Aspergillus fumigatus*) im infizierten Tier," *Mykosen*, **19**, 345, 1976.
315. ———, "Elektronenmikroskopischer Nachweis der antimykotischen Wirkung von Canesten auf die Ultrastruktur der Pilzzelle," *Münch. med. Wschr.*, **118**, suppl. 1, 16, 1976.
316. ——— and Plempel, M., "Elektronenmikroskopische Untersuchungen an humanpathogenen Pilzen. 1. Mitteilung: Ultrastrukturelle Veränderungen von *Candida albicans*-Zellen durch Clotrimazol im Tierexperiment," *Arzneim.-Forsch. (Drug. Res.)*, **24**, 508, 1974.
317. ——— and Schnell, J. D., "Elektronenmikroskopische Untersuchungen an humanpathogenen Pilzen. 2. Mitteilung: Ultrastrukturelle Veräderungen von *Candida albicans*-Zellen am menschlichen Vaginalepithel unter der Behandlung mit Clotrimazol," *Arzneimittel-Forsch. (Drug Res.)*, **24**, 508, 1974.
318. Watt, L. S., Adams, J. N., and Payne, W. J., "Cytological and physiological effects of sodium caprylate on *Candida albicans*," *Antibiot. Chemother.*, **12**, 173, 1962.
319. Werner, H. J., Catsulis, C., Jolly, H. W., and Carpenter, C. L., "Electron microscope observations of the fine structure of *Microsporum gypseum*," *J. invest. Derm.*, **48**, 481, 1967.
320. ———, Jolly, H. W., and Lee, J. H., "Electron microscopic observations of *Epidermophyton floccosum*," *J. invest. Derm.*, **43**, 139, 1964.
321. ———, ———, and Spurlock, B. O., "Electron microscope observations of the fine structure of *Microsporum canis*," *J. invest. Derm.*, **46**, 130, 1966.
322. Wilborn, H. W. and Montes, L. F., "Ultrastructure of host-parasite relationship in human and experimental candidiasis," in E. Drouhet (ed.), *Proceedings 5th International Congress ISHAM*, Paris, 1971, pp. 9–10.
323. Winner, H., "Electron microscopy of *Candida albicans in vitro* and *in vivo*," *3rd International Congress ISHAM, Edinburgh, 1964.*
324. ———, "Recent advances in systemic candidosis," in *Contributions in Microbiology and Immunology*, vol. 4, Karger, Basel, 1977, pp. 64–76.
325. ——— and Drewe, J. A., "The cell wall of *Candida albicans*," in E. Drouhet (ed.), *Proceedings 5th International Congress ISHAM, Paris, 1971*, p. 7.
326. ——— and Hurley, R., *Candida albicans*, Churchill, London, 1964.
327. Yamaguchi, H. and Iwata, K., "Effect of various surfactants on anti-*Candida* action of Clotrimazole with special reference to fungicidal synergy with dioctyl sodium sulfosuccinate," *6th International Congress ISHAM, Tokyo, 1975.*
328. ———, Kanda, Y., and Iwata, K., "Macromolecular structure and morplogy of native glycogen particles isolated from *Candida albicans*," *J. Bact.*, **120**, 441, 1974.
329. ———, ———, ———, and Osumi, M., "Electronmicroscopical studies on the fungitoxocity of bisphenyl-(2-chlorphenyl)-1-imidazolylmethan (Bay b 5097): Preferential effect on the nuclear envelope in *Candida albicans* cells," *J. Electron Microsc.*, **22**, 167, 1973.
330. Yoshii, Z., Tokunaga, J., and Tawara, J., *Atlas of Scanning Electron Microscopy in Microbiology*, Williams and Wilkins, Baltimore, 1976.

Index

Acanthamoeba culbertsoni, 283, 288
Acholeplasma, 69, 77
Acholeplasma axanthum, 77
Acholeplasma granularum, 77
Acholeplasma laidlawii, 77, 224, 226
Acholeplasma modicum, 77
Acholeplasmataceae, 69
Acid phosphatase, 304
Actinomyces, 112
Actinomyces dentocariosus, 132
Actinomyces israelii, 112, 115, 120
Actinomyces naeslundii, 112, 115, 118, 120
Actinomyces odontolyticus, 112, 114, 120
Actinomyces viscosus, 112, 115, 116, 118, 120
Actinomycetes, 112–20
Actinoplanes, 112
Activated lymphocyte, 353
Acute necrotizing ulcerative gingivitis, 220, 222
Adeleina, 308
Adeno-satellite virus, 31
Adenoviridae, 33
Adenovirus, 3, 4, 32, 33
Adiaspiromycosis, 337
Aeromonas, 186
African Swine fever, 36
Aggregata, 308
Ajellomyces, 337
Ajellomyces dermatitidis, 337
Alkaline phosphatase, 340
Allantoic fluid, 14
Alphavirus, 38
Alysiella filiformis, 200, 202
Amastigote, 292
Amoeba, 283–88
  cell coat, 284
  cell membrane, 284
  external appearance and life cycle, 284
  mode of division of, 286
  ultrastructure, 284
Amoeba histolytica, 281
Amoeba proteus, 283
Amphotericin B, 362
Antibody attachment, 56
Antigen-antibody (-complement) complexes, 353, 356
Antimycotic drugs, 335
Appareil élaborateur singulier, 306
Arboviruses, 38
Arizona, 186
Arthrobacter, 112
Arthroderma, 336
Artifacts, 282
Aspergillus, 343, 358, 361
Aspergillus fumigatus, 341, 345, 364
Aspergillus nidulans, 341
Aspergillus niger, 341
Aspergillus oryzae, 341
Aspergillus parsiticus, 341
Asteroid bodies, 358, 361
ATPase, 76
Aviadenovirus, 33
Avian infectious bronchitis, 34

Babes-Ernst bodies, 130
Bacilli, 105–10
Bacillus anthracis, 105
Bacillus anthracis Sterne, 106, 108
Bacillus fastidiosus, 93, 107, 108, 110
Bacillus macerans, 110
Bacillus macroides, 105, 106
Bacillus polymyxa, 109, 110
Bacillus subtilis, 93
Bacteria, 91
  comparison of procaryotic and eucaryotic ultrastructure, 101–3
Bacteriophages, 49
  morphological classes, 50
  problems of structure analysis, 49
Bacteroides, 162–64
Bacteroides fragilis, 162
Bacteroides insolitus, 162
Bacteroides melaninogenicus, 162
Bacteroides melaninogenicus subspecies *asaccharolyticus*, 162, 164, 167
Bacteroides oralis, 162
Bang's disease, 154
Bartonella bacilliformis, 244
Bartonellaceae, 244
Base plate carries, 54
Basidiobolus, 342
Beneckea, 186
Benzalkonium chloride, 370
Bifidobacterium, 103, 112, 124–28
Bifidobacterium bifidum, 124, 128
Biochemical processing, 282
Blastocrithidia, 294
Blastodinium, 291
Blastomyces dermatitidis, 362
Blastomycosis, 347
Blepharoplast, 299
Blue tongue virus, 31
Bodonia, 291
Bordet-Gengot bacillus, 154
Bordetella, 154–58
Bordetella bronchiseptica, 158
Bordetella pertussis, 154
Borrelia, 222, 224
Borrelia anserina, 218
Borrelia hermsii, 218
Borrelia recurrentis, 218, 220
Brain tissue, 18
Branhamella, 164

Branhamella catarrhalis, 173
Brucella, 154
Brucella abortus, 154, 156
Brucella canis, 154
Brucella melitensis, 154, 156
Brucella neotomae, 154
Brucella ovis, 154
Brucella pertussis, 158
Brucella suis, 154

Calicivirus, 29
Campylobacter fetus, 214
Candida albicans, 338–42, 345–52, 355, 357, 358, 362, 364, 367, 368, 370
Capnocytophaga, 184–86, 188
Capsids, 50
Capsomeres, 4
Capsule phage, 61
Carinamoeba, 315
Caseous lymphadenitis, 130
Caulobacter, 60
Central nervous system, 39
Cercopithecus, 82
Chemical fixation, 281
Chemotaxis, 347, 350
Chlamydia psittaci, 248, 252, 254, 256, 258
Chlamydia trachomatis, 248
Choanomastigote, 292, 296
Chorioallantoic membrane (CAM), 14–15
Chromatoid bodies, 288
Chromomycosis, 337, 347
Citrobacter, 186
Clostrida, 105–10
Clostridium, 105
Clostridium botulinum, 105, 110
Clostridium botulinum Holland 17B, 108
Clostridium botulinum Type E, 106, 109, 110
Clostridium botulinum Type 33A, 109
Clostridium histolyticum, 105, 110
Clostridium novyi, 105
Clostridium perfringens, 105, 110, 111
Clostridium septicum, 105
Clostridium sporogenes, 105, 106, 110
Clostridium tetani, 105–7
Clotrimazole, 364, 367
Coccidia, 308, 309
Coccidioides immitis, 337
Coccidiomorpha, 313
Coccodinium, 291
Coliform-enteric bacilli, 186
Common wart virus, 30, 31, 39
Conoid, 311
Contractile vacuole, 288
Coronaviridae, 38
Coronavirus, 14, 38
Corynebacterium, 112, 128–32
Corynebacterium diphtheriae, 129, 130, 132
Corynebacterium gravis, 130
Corynebacterium intermedius, 130
Corynebacterium minutissimum, 132
Corynebacterium mitis, 130
Corynebacterium ovis, 130
Coxiella, 232
Coxiella (Rickettsia) burnetti, 236, 240, 244, 248, 250
Coxsackie virus, 26
Cristispira, 218
Crithidia, 294, 296
Critical point drying (CPD), 101
Cryptococcus neoformans, 101, 337, 345, 351
Cubic viruses, 25–36
Cyclops, 291
Cyst structures, 289–90
Cytochrome-c-peroxidase, 368
Cytomegalovirus, 13–15, 34
Cytopathic effect (CPE), 13
Cytoplasmic structures, 304–6

Dane particles (HBcAG), 15, 33
Deletion loops, 55
Dermacentor andersoni, 238, 242
Dermatomycoses, 342
Dinoflagellida, 291
Diplococcus, 152
Diplococcus pneumoniae, 152
Diplomonadida, 292
Distemper viruses, 21
DNA, 49, 50, 52, 55, 60, 61, 75, 76, 300–3, 368, 370
DNA viruses
  cubic, with lipoprotein envelope, 31–36
  cubic, without lipoprotein component, 31
  nonsymmetrical, 39
Dodecahedron, 3
Doderlein's bacillus, 135
Drug action on fungi, 362–70

Ebola virus, 25
ECHO virus, 26
Econazole, 367, 368
EDIM virus, 31
Edwardsiella, 186
Eimeria, 308
Eimeriina, 308–15
  cell wall, 310–11
  general fine structural characteristics, 310–12
Emmonsia crescens, 337
Emonsiella capsulata, 337
Endocytosis, 297
Endocytotic inhibitor, 307
Endocytotic marker, 307
Endomycoses, 343–47
Entamoeba histolytica, 283–88
Entamoeba invadens, 283
Enterobacter, 186
Enterobacteriaceae, 186
Enterovirus, 26
Entomophthora, 341
Enzyme localization, 76–77, 97
Epidermal lesions, 17, 39
Epidermophyton floccosum, 336, 343
Epimastigote, 292, 294
Epstein-Barr (EB) virus, 34
Erwinia, 186
Erythrocytic stages, 316
Escherichia, 145
Escherichia coli, 57, 59, 97, 124, 186, 190
Esterases, 95
Ether-sensitivity, 19
Eucaryotic strains, 261–65
Eucaryotic structure, 101–3
Euglena, 299
Exoerythrocytic stages, 319

Feces, 18–19
Feline respiratory virus, 27

Filipin, 362, 370
Filopodium-like processes, 299
Fissuricella, 342
Flagella, 19
Flagellar phages, 60
Flagellar pocket, 299
Flavivirus, 38
5-Fluorocytosine, 367–70
Fonsecaea pedrosoi, 337
Francisella tularensis, 198
Freeze-etching, 52, 71, 93, 95, 99, 101, 138, 252, 256, 281, 335–37, 341
Friedlander's bacillus, 190
Fungi, 335
  chronic disease, 353
  destruction in immunized host, 352, 355
  differences between pathogenic and nonpathogenic, 346
  drug action on, 362–70
  evidence of effects on host, 342–47
  evidence of host effects on, 347–58
  keratinophilic, 336
  morphological investigations, 335
  nonkeratinophilic, 343
  opportunistic pathogenic, 338–42
  pathogenic role of enzyme patterns, 346
  phagocytosis of, 347–52
  tissue areas infected by, 343
  ultrastructure, 335–42
  see also Mycoses
Fusobacterium fusiforme, 180
Fusobacterium necrophorum, 180
Fusobacterium nucleatum, 180
Fusospirochaetal infections, 220

Gaffkya tetragena, 140
Gametocyte, 319
Gastroenteritis, 40
Gemella haemolysans, 173
Gene products, 53, 56
Giardia, 292
Glycogen bodies, 287
Gram-negative procaryotes, 154–254
Gram-positive bacterium, 124
Gram-positive procaryotes, 105–10
Granhamella, 244
Gregarinia, 308
Griseofulvin, 362
Gumboro disease virus, 18, 31

Haemamoeba, 315
Haemophilus, 154–58
Haemophilus aegyptius, 154
Haemophilus ducreyi, 154, 158, 160
Haemophilus influenzae, 154, 158, 160, 161
Haemophilus parainfluenzae, 154
Haemophilus suis, 154
Haemophilus vaginalis, 154, 158, 160
Haemosporina, 308, 315
Hartmanella castellani, 285, 286
Helical symmetry, 19
Helical virus, 24
Hepatitis A virus, 11, 12, 28, 30
Hepatitis B antigen, 15–16, 33, 36
Hepatitis B virus, 36
Hepatitis virus, 18
Herpes particles, 18
Herpes simplex types I and II (herpesvirus hominis I and II), 34
Herpes virion, 35
Herpesvirus, 4, 15, 33–36
Herpetomonas, 294
Herpetoviridae, 34–36
Heterakis gallinae, 291
Heteroduplex phage DNA molecules, 55–57
Hexamita, 292
Histoid nodule, 126
Histomonas meleagridis, 291
Histoplasma, 345
Histoplasma capsulatum, 336, 342, 362
Histoplasma farciminosum, 337
Histoplasmosis, 347
Hoarella, 308
Hodgkin's disease, 122
Homogenization, 15, 17, 19
Huffia, 315
Hydrogenosomes, 292
Hydrolytic enzymes, 361

Icosahedron, 3
Icterovirus, 30
Imidazole antimycotics, 364
Immune electron microscopy, 10–12, 19, 40, 335
Infectious respiratory disease of cats, 29
Infective stages, 312
Influenza, 14
Influenza virus, 21–23
Iridoviridae, 36
Iridovirus, 36
Isospora, 308, 311
Isospora gondii, 313

Jaundice, 218

Kellenberger-Ryter-Séchaud osmium technique, 135
Keratinomyces, 336
Keratinophilic dermatophytes, 336
Kinetoplast, 301
Kinetoplastida, 291, 297
Klebs-Loffler bacillus, 128
Klebsiella, 186
Klebsiella aerogenes, 194, 196
Klebsiella pneumoniae, 190, 194

Lactobacillus, 135
Lactobacillus acidophilus, 135, 136
Lactobacillus bifidus, 124
Lambda, 50
Lambdoid phages, 57
Lankesterella, 308
Laverania, 315
L-cells, 252, 254
Lead citrate, 95
Legionella pneumophilia, 260
Legionnaires bacterium, 252–54
Leishmania, 292–308
  cell membrane, 296–99
  external appearance, 296
  terminology, 292–96
  ultrastructure, 296–306
Leishmania ctenocephali, 309
Leishmania donovani, 302
Leishmania mexicana, 294, 301
Lepra cells, 224

Leptomonas ctenocephali, 298
Leptospira, 222, 224
Leptospira icterohemorrhagiae, 220
Leptospira interrogans, 218, 220
Leptospira pomona, 218
Leptospirosis, 220
Leptotrichia buccalis, 180, 184
Leukemia, 122
Lipid vacuoles, 304
Lipopolysaccharidemycolic acid, 124
Lipoprotein membrane, 19
Listeria, 112
Listeria monocytogenes, 112, 113
Liver cell, 103
Louping ill, 38
Lymphocytotoxic effects, 352
Lymphosarcoma, 122
Lysophospholipase, 345
Lyssavirus, 24

Macrogametes, 316
Malarial parasites, 315
Mammary tumour virus, 18
Marburg virus, 24, 26
Marek's disease virus, 34
Mastadenovirus, 33
Mastigote, 292
Measles, 21
Membrane adhesion, 60
Meningopneumonitis, 252, 258
Merozoites, 313, 316, 317
Mesosomes, 138
8-Methoxypsoralen, 370
Miconazole, 367, 368
Micrococcus mucilaginosus, 138, 140
Micrococcus radiodurans, 138, 140
Microgametes, 314–16
Microgametocyte, 319
Microgamonts, 314
Micromastigote, 292, 294
Micromonospora, 112
Micronemes, 312, 320
Micropore, 312
Microsporum audouinii, 370
Microsporum canis, 362
Microsporum cookei, 336
Microsporum gypseum, 343
Microtubules, 311
Milkers' nodules, 39
Minute virus of mice, 31
Mitochondria, 286, 301
Mollicutes, 69
Molluscum contagiosum, 39
Moraxella, 164
Moraxella lacunata, 173
Moraxella nonliquefaciens, 173, 180
Morbillivirus, 21
Mosquito stages, 320
Mucor, 341, 342
Mumps virus, 19
Mycobacteria, 122–24
  acid-fast, 123, 124, 126
  atypical, 122
Mycobacterium, 112
Mycobacterium, sp. strain NQ, 124
Mycobacterium bovis, 122, 180
Mycobacterium leprae, 122, 124, 289
Mycobacterium lepraemurium, 122
Mycobacterium phlei, 122, 124
Mycobacterium smegmatis, 122, 124
Mycobacterium tuberculosis, 122, 123
Mycoplasma, 50, 69, 77, 101, 226–32
  basic structural components, 73–77
  cell ultrastructure, 73–81
  cellular morphology, 71
  electron microscopy, 70–71
  'fried-egg' appearance, 69
  morphology and structure, 72
  nuclear area, 75–76
  phages, 81
  reproduction, 71
  rho (ρ)-forms, 79–81
  specialized structures, 78–81
  terminal structures, 78
  visualization in tissues and cell cultures, 81–82
Mycoplasma arthriditis, 230
Mycoplasma buccale, 69
Mycoplasma dispar, 73
Mycoplasma faucium, 69
Mycoplasma fermentans, 69, 226
Mycoplasma gallisepticum, 73, 75, 76, 78, 230
Mycoplasma hominis, 69, 81, 226, 228, 230, 233
Mycoplasma hyopneumoniae, 73
Mycoplasma laidlawii, 226, 230
Mycoplasma meleagridis, 73, 76, 228, 233
Mycocplasma mycoides, 69, 74, 76, 230
Mycoplasma mycoides subsp. *capri*, 74
Mycoplasma mycoides subsp. *mycoides*, 73
Mycoplasma neurolyticum, 228, 230
Mycoplasma orale, 69, 72, 224, 228, 230, 235
Mycoplasma pneumoniae, 69, 71–73, 78, 79, 82, 226–28, 230
Mycoplasma pulmonis, 73, 82, 230
Mycoplasma salivarium, 69, 72, 73, 230
Mycoplasmataceae, 69
Mycoses:
  acute, primary infection, 358
  causative agents, 336–37
  chronic disease, 361
  evidence for pathogenetic concepts in, 358–61
  ultrastructural features of host-fungus relationships in, 342–61
  see also Fungi
Mycotic diseases, 335
Mycotic lesions, 335
Mycotic tissue foci, 335

Naegleria, 283, 288–90
  cysts, 289–90
  flagellate form, 290
  pathogenic microorganisms, 289
Naegleria fowleri, 283, 285, 286, 288, 290, 291
Naegleria gruberi, 289, 290
Naegleria jadini, 290
Nannizzia gypsea, 336
Negative staining, 5–6, 93, 94
  addition of specific antibody, 11
  advantages of, 7
  egg-grown material, 14
  equipment for, 8
  general methods, 7–9
  'on the grid' method, 7, 10
  progress in, 40
  results obtained from, 39–40
  techniques of, 6–12
  tissue culture specimens, 13–14

Negative staining—*continued*
  types of specimen examined by, 12–19
  see also Immuno-electron microscopy
Neisseria, 164–73
Neisseria caviae, 173
Neisseria gonorrhoeae, 167, 168, 172, 173
Neisseria haemolysans, 173
Neisseria meningitidis, 167, 172, 173
Neisseria ovis, 173
Neisseriaceae, 180
Newcastle disease virus, 19
Nicotinamide adenine dinucleotide (NADH)
    oxidase, 77
Nocardia, 112, 120–22
Nocardia asteroides, 120–22
Nocardia dentocariosus, 132
Nocardia salivae, 132
Nucleic acid, 6, 19, 26, 27, 31, 33, 75, 94
Nucleosidases, 95
Nutrient uptake and digestion, 307–8
Nystatin, 364

Oceanospirillum atlanticum, 212, 214
Octahedron, 3
Oocysts, 316, 320
Ookinetes, 320
Opisthomastigote, 292, 293
Orbivirus, 31
Orf particles, 36, 39
Ornithodorus, 220
Orthomyxoviridae, 21–23
Orthopoxvirus, 39
Osmium tetroxide, 94
ØX174, 50, 57

Paired organelles, 311
Papilloma viruses, 31
Papovaviridae, 31
Papovavirus, 18
Paracoccidioides brasiliensis, 337
Paracocci-dioidomycisis, 347
Parainfluenza, 14
Parainfluenza I, 18
Parainfluenza virus, 5, 19, 21
Paramyxoviridae, 19, 21
Paranuclear body, 312
Parapoxvirus, 39
Parvoviridae, 31
Parvovirus, 30, 31
Pasteurella pestis, 198
Pediculous humanus, 220
Pelomyxa palustris, 286
Peptococcus, 143
Peptostreptococcus, 143
Periplast, 296
Phagocytosis of fungi, 347–52
Phialophora, 337
Phialophora dermatitidis, 370
Phialophora verrucosa, 370
Phlebotomus, 244
Phosphatases, 95
Phosphate buffered saline (PBS), 14, 15
Phosphates, 97
Phosphotungstic acid, 6–7, 16, 28
Photobacterium, 184
Picornaviridae, 26–30
Piedra hortai, 343
Pili, 60
Pityriasis versicolor, 341, 367
Pityrosporum, 343
Pityrosporum orbiculare, 341, 367, 370
Plasma membrane, 73–74
Plasmodium, 315, 316
Plasmodium berghei, 317, 319
Plasmodium cathemerium, 319
Plasmodium coatneyi, 317
Plasmodium cynomolgi, 319
Plasmodium falciparum, 281, 316
Plasmodium gallinaceum, 319
Plasmodium gonderi, 317
Plasmodium lophurae, 317
Plasmodium malariae, 316
Plasmodium ovale, 316, 317
Plasmodium vinckei, 317
Plasmodium vivax, 316, 317
Pleuropneumonia organisms (PPLO), 226
Pneumococcus, 145
Pneumovirus, 21
Podoravirus, 28
Polar bodies, 130
Polar ring, 311
Poliomyelitis particles, 27
Poliovirus, 26
Polyene antimycotics, 362
Polyoma, 32
Polyomavirus, 32
Polyribosomes, 75
Porphyridium cruentum, 101, 105
Positive staining, 93, 94
Potassium permanganate, 94
Potassium tellurite, 76
Poxviridae, 39
Primary amoebic meningoencephalitis, 288
Proactinomycetes, 112
Procaryotic strains, 261–65
Procaryotic structure, 101–3
Progressive multifocal leukoencephalopathy
    (PML), 18, 32, 39
Promastigote, 292, 293, 296
Pronase, 73
Propionibacterium, 112
Protaminobacter ruber, 93
Proteus, 186
Proteus mirabilis, 190, 192, 194
Proteus morganni, 194
Proteus vulgaris, 190, 192, 194
Protococcidia, 308
Protozoa, 101, 281
  classification of, 282–83
  morphology of, 281
Protozoology, 281
Providencia, 186
Pseudomonas, 214–18
Pseudomonas aeruginosa, 214, 216, 218
Pseudomonas cepacia, 214, 216
Pseudomonas mallei, 216
Pseudomonas multivoras, 214
Pseudomonas pseudomallei, 216
Pyrrolnitrin, 370
Pythonella, 308

Q-fever, 236, 240

Rabies virus, 24, 25
Refractile body, 312

Relapsing fever, 219, 220
Reoviridae, 31
Reovirus, 29
Respiratory syncytial virus, 21
Reticuloendothelial system, 122
Retortamonadida, 292
Rhabdoviridae, 24–25
Rhinovirus, 27–28
Rhizomastigida, 291
Rhizopodea, 283
ρ-fiber, 80, 81
Rhoptries, 311, 320
Ribonucleic acid, 19
Ribonucleoprotein (RNP), 5, 21
Ribosomes, 74–76, 103, 135, 288
Rickettsia, 232, 236
Rickettsia burnetii, 240
Rickettsia canada, 238, 240, 242, 244
Rickettsia prowazeki, 236, 238
Rickettsia quintana, 232, 247
Rickettsia quintana Fuller, 247
Rickettsia rickettsii, 238
Rickettsia sennetsu, 238, 242
Rickettsia tsutsugamuschi, 238, 240
Rinderpest, 21
RNA, 31, 50, 60, 75, 342
RNA viruses, 5, 30
  cubic, 26–31
  nonsymmetrical, 38
Rochalimaea, 232
Rochalimae quintana, 240
Rosette forms, 308
Rotavirus, 29, 31, 40
Rothia, 132
Rothia dentocariosa, 132, 134
Rubella virus, 11, 35, 38
Rubivirus, 38
Ruthenium red, 95

Saccharomyces cerevisiae, 368
Salmonella, 145
Salmonella typhimurium, 190, 192, 194
Sanramoeba, 315
Sarcina maxima, 140, 141
Sarcina ventriculi, 140, 141
Scabs, vesicle fluid, 16–17
Schizonts, 316, 318
Selenomonas, 200
Sendai virus, 18
Serratia, 186
Serratia marcescens, 198–200
Serum, 15–16
Sex pili, 60
Sexual forms, 319
Sexual reproductive cells, 314
Shadowcasting, 91, 93, 94
Shigella, 145, 186
Simonsiella crassa, 200
Simonsiella filiformis, 200
Simonsiella muelleri, 200
Simonsiella Schmid, 200
Sindbis virus, 35, 38
Slime-appendages, 200
Sodium caprylate, 370
Solid tissue specimens, 17–18
Sphaerophorus necrophorus, 180
Spherules, 73
Spirillum serpens, 212, 214
Spirochaeta, 224
Spirochaeta aurantia, 218
Spirochaeta stenostrepta, 218, 220, 222
Spirochaetales, 218, 222
Spirochetes, 218–26
Spiroplasma, 69, 101
Spiroplasma citri, 70, 71, 232, 236
Spiroplasmataceae, 69
Sporoblasts, 320
Sporothrix schenkii, 337, 342, 345, 358
Sporozoa, 308
Sporozoites, 312, 316, 320
Spotted fever, 236
Staphylococcus, 145
Staphylococcus aureus, 135, 136, 138
Staphylococcus epidermidis, 135, 136, 138
Stools, 18–19
Streptococcus, 143–52
Streptococcus faecalis, 143, 144
Streptococcus mutans, 143, 145, 150
Streptococcus pneumoniae, 152, 190
Streptococcus pyogenes, 143, 146
Streptococcus salivarius, 143, 145, 148
Streptomyces, 112
Streptosporangium, 112
Strigomonas, 296
Strigomonas oncopelti, 304, 306
Subacute sclerosing panencephalitis (SSPE), 18, 39
Subpellicular microtubules, 296
Substitution loops, 55
Surface lipopolysaccharide, 61
Surface replication, 93
SV40, 32
Syphilis, 218

T2, 52, 55, 60
T3, 57
T4, 50, 52–56, 59
T7, 57
Tail fibers, 55
Tail pins, 55
'Technikosomes', 138
Telosporea, 308
T-even phages, 56, 57, 59
Thermoactinomyces, 112
Thiobenzoate, 370
Tinea versicolor, 343
Tipula iridescent virus, 36
Tissue contrast, 95
Togaviridae, 38
Torulopsis glabrata, 342
Toxonemes, 312
Toxoplasma, 301
Trench fever, 236
Trench mouth, 222
Treponema, 222
Treponema buccale, 218
Treponema calligyrum, 218
Treponema denticola, 218
Treponema genitalis, 222
Treponema microdentium, 218
Treponema pallidum, 218, 220, 224
Treponema pertenue, 218
Treponema refringens, 218
Treponema vincentii, 222
Trichomonadida, 292
Trichomonas gallinae, 101, 103
Trichomonas vaginalis, 99, 292

Trichophyton, 336, 337
Trichophyton ajelloi, 342
Trichophyton mentagrophytes, 336, 342–44, 362, 364
Trichophyton rubrum, 342, 343, 362
Trophozoite, 316, 317
Trypanosematina, 291
Trypanosoma, 292–308
cell membrane, 296–99
external appearance, 296
nucleus, 300–1
terminology, 292–96
ultrastructure, 296–306
Trypanosoma brucei, 295, 298, 299, 301, 302, 306
Trypanosoma congolense, 298, 301, 306
Trypanosoma cruzi, 281, 301, 302, 304
Trypanosoma equinum, 301, 303
Trypanosoma equiperdum, 303
Trypanosoma gambiense, 298, 301
Trypanosoma mega, 305
Trypanosoma noctuae, 301
Trypanosoma olimorphon, 301
Trypanosoma rhodesiense, 281, 301
Trypomastigote, 292, 295, 297, 303
Typhus, 236

Ulcerative lymphangitis, 130
Ultrahistochemistry, 335
Ultrastructural autoradiography, 335
Uranyl acetate, 95
Ureaplasma, 69, 226
Ureaplasma urealyticum, 69, 72–77, 228
Urine, 15

Vaccinia virus, 36
Vacuolar apparatus, 287, 303
Varicella-zoster virus, 34
Variola virus, 16
Veillonella, 164, 180
Veillonella alcalescens, 182
Vesicle fluid scabs, 16–17
Vesicular exanthema of swine, 29
Vesicular stomatitis virus (VSV), 24
Vesiculovirus, 24
Vi antigen, 61
Vibrio, 186, 200–14
Vibrio sp 4716, 206
Vibrio cholerae, 200, 203, 204, 206, 212
Vibrio comma, 200
Vibrio eltor, 204, 210
Vibrio fetus, 212
Vibrio parahaemolyticus, 200, 203, 204, 206, 208
Vietnamese time bomb, 216
Vincent's angina, 220, 222
Vinckeia, 315
Viral entry, 57–61
Viral hepatitis A, 40
Viral morphogenesis, 53–55
Virion, 50–55
Virus morphology, 5, 19–40
Virus particles, 22
Viruses, 3
essential nature of, 3
helical, 5
icosahedral shell, 4
investigation techniques, 3
spherical, 3
symmetry, 3
Volutin granules, 130

Weil's disease, 218
Whipple's disease, 103

X-ray diffraction, 3, 26, 50

Yaws, 218
Yellow fever, 38
Yersinia, 103, 145, 186
Yersinia enterocolitica, 198, 199
Yersinia pestis, 198, 199
Yersinia pseudotuberculosis, 198
Yersinia tularensis, 198

Zoomastigophorea, 290–92
Zygomycetes, 341, 343
Zygotes, 316